Musculoskeletal Regeneration Medicine

Paul Lee
Editor

Musculoskeletal Regeneration Medicine

Regenerative Medicine Across AI, Engineering, and Clinical Orthopaedics

Springer

Editor
Paul Lee
Honorary Professor, College of Health and Science
University of Lincoln
Lincoln, UK

ISBN 978-3-031-77455-3 ISBN 978-3-031-77456-0 (eBook)
https://doi.org/10.1007/978-3-031-77456-0

This Springer imprint is published by the registered company Springer Nature Switzerland AG
The registered company address is: Gewerbestrasse 11, 6330 Cham, Switzerland

To my mentor Professor James Richardson, for your enduring contributions to Cartilage Regeneration.

Kar Hao Teoh, in memory of my good friend and my Co-architect in Regenerative Medicine.

Contents

MSK Regeneration Principle, Engineering, Biology and Medicine

Paul Lee and Tanvi Verma

Abstract This chapter explores the interdisciplinary principles of musculoskeletal (MSK) regeneration, integrating engineering, biology, and physics to address orthopaedic conditions. It highlights the limitations of traditional mechanistic approaches, emphasising the need for systematic thinking to achieve long-term regenerative success. Current medical interventions, including stem cell therapies, platelet-rich plasma (PRP), and growth factors, are examined for their role in enhancing tissue repair. The chapter delves into the importance of early interventions and proactive care in preventing degeneration. A detailed discussion on the role of physics, chemistry, and biology in tissue regeneration reveals how mechanical stress, biochemical modulation, and cellular pathways interact to drive healing. Additionally, advancements in personalised healthcare solutions, such as artificial intelligence (AI) and digital twin technologies, are explored, demonstrating how they optimise treatment strategies through real-time analysis and patient-specific data. Finally, the chapter discusses emerging innovations in biomaterials, 3D printing, and AI-driven solutions that are poised to revolutionise MSK regenerative medicine. The importance of interdisciplinary collaboration is emphasised as the key to advancing future therapies, ensuring sustainable and effective regeneration for improved patient outcomes and longevity.

Keywords Systematic regeneration · Proactive medicine · Personalised healthcare · Physiological time · Biochemical modulation

P. Lee (✉)
London Cartilage Clinic, London, UK

MSK Doctors, MSK House, Sleaford, UK

College of Health and Science, University of Lincoln, Lincoln, UK

MSK Doctors, ICRS Center of Excellence for Cartilage Regeneration & Joint Preservation, Grantham, UK
e-mail: leep@mskdoctors.com

T. Verma
College of Health and Science, University of Lincoln, Lincoln, UK
e-mail: tanvi@mskdoctors.com

P. Lee (ed.), *Musculoskeletal Regeneration Medicine*,
https://doi.org/10.1007/978-3-031-77456-0_1

1 Introduction

1.1 Overview of Regenerative Medicine

Regenerative medicine is a transformative field focused on repairing, replacing, or regenerating damaged tissues by harnessing the body's natural healing processes. This approach holds immense potential for musculoskeletal (MSK) health, which encompasses bones, muscles, tendons, and cartilage—tissues that often have limited regenerative capacity. Regenerative techniques such as stem cell therapy, growth factor delivery, and tissue engineering aim to restore the normal function of MSK structures, which are frequently compromised by injury, degeneration, or disease (Pittenger et al. 1999).

The Role of Regeneration in MSK Health
The musculoskeletal system is fundamental to human mobility, and conditions such as osteoarthritis, tendon injuries, and cartilage degeneration often result in chronic pain and disability. Traditional treatments like joint replacements and pain management therapies focus primarily on alleviating symptoms but fail to address underlying tissue damage. Regenerative medicine offers a more comprehensive approach by directly targeting the damaged tissues to stimulate repair and regeneration. For example, mesenchymal stem cells (MSCs) can differentiate into bone, cartilage, and muscle cells, making them valuable for regenerating these tissues. Moreover, the application of engineered scaffolds mimicking the extracellular matrix (ECM) can enhance cellular repair and regeneration in tendons and ligaments.

Regeneration as a Proactive Approach
Current approaches to musculoskeletal conditions are often reactive, focusing on treating symptoms after significant damage has occurred. However, regenerative medicine promotes a proactive strategy that intervenes early in the disease process, aiming to slow or reverse degeneration before it necessitates invasive treatments like surgery. Early intervention therapies, such as platelet-rich plasma (PRP) or stem cell injections, have shown promise in preventing the progression of cartilage and bone degeneration, potentially delaying or even avoiding the need for joint replacements. By emphasising early diagnosis and regenerative intervention, this proactive model ensures better long-term outcomes for patients (Clarke et al. 2018).

Vision and Mission of Regeneration

- **Vision**: The overarching vision of regenerative medicine is to transform health by integrating science and engineering principles to redefine how we treat tissue damage and aging. By focusing on regeneration, rather than replacement alone, this vision seeks to improve both the quality and longevity of life for individuals suffering from musculoskeletal conditions.
- **Mission**: The mission is to pioneer regenerative therapies that combine biology, chemistry, physics, and engineering to optimise human health. The approach not only aims to develop cutting-edge treatments but also to create personalised,

preventive strategies that help patients maintain musculoskeletal health and function long into the future.

1.2 Integration of Science and Engineering

The Regeneration philosophy emphasises the need for a proactive approach to medicine. This approach involves early intervention and personalised care to prevent irreversible damage to MSK tissues, rather than relying solely on traditional reactive methods that treat symptoms after significant degeneration has occurred. This interdisciplinary methodology aligns with the growing recognition that the body's repair mechanisms can be augmented through the precise application of scientific and engineering principles (Clarke et al. 2018).

The Six Steps Approach

Central to the regeneration philosophy is the six-step model, it provides a comprehensive approach to enhancing tissue repair and regeneration, combining biological, chemical, and physical interventions. Each element of the science and engineering equation plays a role in maximising the efficacy of treatment and is based on scientific principles that guide the body's healing processes. The steps are as follows:

1. **Minimised Disturbance**
   Minimise the unnecessary physical or surgical disruptions to injured or degenerating tissues. This principle is based on the idea that excessive mechanical or biological interference can hinder the body's natural repair mechanisms. By reducing the external disturbances to the tissue environment, regenerative therapies can work more effectively in promoting healing with fewer complications. For instance, less invasive procedures can prevent scarring or additional tissue damage, which could impede the natural regenerative processes.
2. **Modify Chemistry**
   The biological microenvironment surrounding damaged tissues plays a key role in the success of regenerative treatments. Modifying the local chemistry by introducing bioactive agents such as growth factors, cytokines, or stem cells can significantly enhance the body's ability to repair and regenerate tissues. These chemical modifications aim to activate and guide the natural regenerative processes by promoting cellular proliferation, migration, and differentiation within the damaged area. This step-controlled manipulation of biochemical pathways to create an optimal environment for tissue recovery.
3. **Maximise Biological Benefits**
   Biological enhancement is a cornerstone of regenerative therapies. The focus here is on utilising the body's inherent healing potential by boosting the effects of biological processes that promote regeneration. This can be achieved through the administration of stem cells, the application of growth factors, or other methods that encourage tissue repair at a cellular level. Stem cells, in particular, are

critical in this stage due to their ability to differentiate into various tissue types, such as bone, cartilage, or muscle, depending on the treatment area. The goal is to harness and amplify the body's natural biological responses for more efficient and effective regeneration.

4. **Determine the Physics**

   The physical environment of the tissues is just as important as the chemical and biological factors. Tissue regeneration is greatly influenced by mechanical forces such as pressure, tension, and load-bearing. Understanding how these forces affect the healing process is essential for optimising recovery. For instance, applying controlled mechanical stimuli can promote bone growth or cartilage repair by stimulating the cells' mechano-transduction pathways. This step involves determining the ideal physical conditions under which regenerative therapies should be applied to maximise their effectiveness.

5. **Digital Body Banking (Digital Twin)**

   A Digital Twin is a virtual model of the patient's body that can be used to simulate various treatment scenarios, predict outcomes, and personalise regenerative strategies. This technology allows clinicians to monitor the progress of regeneration in real-time, adjust therapies as needed, and ensure that treatments are tailored to the patient's unique physiological conditions. By integrating AI and machine learning into regenerative medicine, Digital Twins offer a forward-thinking approach to precision healthcare.

6. **Decisions for Personalised Healthcare**

   The final step in the regeneration equation is about making informed, data-driven decisions that personalise healthcare for each patient. Through the integration of biology, chemistry, physics, and digital technology, regenerative medicine can be customised to meet the specific needs of an individual's condition and recovery trajectory. Personalised healthcare ensures that treatments are not only effective but also optimised for the patient's unique biological makeup, lifestyle, and health history (Lee et al. 2022). This personalised approach improves patient outcomes by reducing the risk of complications and maximising the regenerative potential of the therapies used.

1.3 The Science Behind Regeneration: Physics, Chemistry, and Biology

Physics in Musculoskeletal Regeneration

Physics plays an essential role in musculoskeletal (MSK) regeneration, particularly through the impact of mechanical forces on tissue repair. Understanding how mechanical stress, energy transfer, and physical forces affect tissue healing and regeneration is a critical component of regenerative medicine. Mechanical loading, including compression, tension, and shear stress, plays a significant role in tissue development and repair. In articular cartilage, for instance, compressive forces stimulate the synthesis of extracellular matrix (ECM) components such as collagen and proteoglycans, which help maintain the biomechanical integrity of the tissue. This process, known as mechanotransduction, converts mechanical forces into biochemical signals, influencing gene expression and tissue remodeling. Mechanical loading, whether through daily activities or controlled rehabilitation exercises, can stimulate cellular processes that promote tissue repair. For example, cartilage responds to compressive forces by enhancing chondrocyte activity and extracellular matrix (ECM) production, which are essential for tissue maintenance and regeneration.

Medical devices, such as mechanical stimulators and electro-magnetic technologies, are being developed to deliver these forces in a controlled manner, enabling more effective MSK regeneration. Recently, electromagnetic fields have shown promise in MSK regeneration. These fields have been used to stimulate osteoblast activity, leading to improved bone healing, particularly in cases where fractures have failed to unite with traditional treatments. The application of electromagnetic fields has also been explored for soft tissue regeneration, adding another dimension to physical interventions in regenerative therapies.

Chemistry's Role in Regenerative Medicine

The chemical environment is fundamental to effective tissue regeneration. Modifying this environment can significantly enhance therapeutic outcomes. Controlled release systems for therapeutic agents, such as bioactive growth factors or anti-inflammatory compounds, are essential for maintaining an optimal healing environment. These systems ensure the gradual release of these agents, preventing premature degradation while enhancing tissue regeneration.

In MSK regenerative therapies, the introduction of bone morphogenetic proteins (BMPs) is widely used to stimulate bone formation, while transforming growth factor-beta (TGF-β) is helpful for cartilage repair. Both are used to create a biochemical environment conducive to the healing and growth of damaged tissues.

Recently, Polymeric scaffolds, infused with bioactive molecules, have shown promise in promoting cellular differentiation and tissue growth. This chemical modulation strategy is vital in ensuring that cells receive the necessary cues for tissue repair, making it a cornerstone of modern regenerative medicine.

Biological Processes in Regeneration

Biological processes such as hormonal regulation, sleep, rest, sex, and psychological well-being (happiness) play an underexplored yet crucial role in augmenting tissue repair. These processes interact with traditional cellular mechanisms to enhance the body's regenerative capacity, thus contributing to more effective outcomes in MSK regenerative therapies.

Growth hormone (GH), primarily released during deep sleep phases, is paramount for stimulating collagen production, for the healing of tendons, muscles, and cartilage. Testosterone promotes muscle hypertrophy, aiding in bone density, and supporting the repair of micro-damages in tissues. Conversely, elevated cortisol levels—often linked to chronic stress—can undermine tissue repair by promoting a catabolic state. Therefore, factors contributing to hormonal imbalance disrupts the body's inflammatory response, slowing the healing process.

Sleep, rest, another biological factor, is not merely restorative for cognitive and physical energy; it is also a key period for anabolic processes that facilitate tissue repair. Deep sleep, or slow-wave sleep, enhances the secretion of anabolic hormones such as GH, directly aiding in tissue regeneration. Moreover, sleep plays an indirect role by reducing pro-inflammatory cytokines, which can otherwise delay tissue repair. Inadequate sleep or rest, conversely, elevates these inflammatory markers, prolonging recovery times and increasing susceptibility to chronic MSK conditions like osteoarthritis.

Rest intervals, particularly in rehabilitation settings, are essential for allowing tissues to undergo the necessary repair processes without overstressing the affected regions. Structured rest following controlled mechanical loading or post-surgical interventions ensures that the body can utilise the regenerative stimuli optimally, whether from stem cell therapies or bioengineered scaffolds.

The psychological state specifically positive emotional states, including happiness, reduce systemic levels of cortisol and other stress-related hormones, creating a favourable environment for tissue regeneration. Happiness and psychological well-being are linked to increased levels of endorphins and oxytocin, both of which lower inflammation and aid in pain relief, indirectly promoting better tissue healing. Moreover, chronic stress and depression have been shown to elevate inflammatory markers such as interleukin-6 (IL-6), which can exacerbate conditions like tendinopathy or osteoarthritis, hampering the efficacy of regenerative treatments.

Sexual activity also influences regenerative biology through the release of oxytocin, known to reduce stress and improve mood, which in turn helps to create a more conducive environment for healing. Sex contributes to overall hormonal balance, reducing inflammation and positively impacting recovery timelines in MSK injuries.

Timing in Regeneration: Chronological Versus Physiological Time

Timing is an important factor in the success of regenerative therapies, particularly the distinction between chronological time (the age of the patient) and physiological time (the biological age of the tissues). Younger individuals tend to have a higher regenerative capacity, allowing for faster and more complete tissue repair. As individuals age, however, their regenerative potential diminishes. This discrepancy between chronological and physiological time is crucial for determining the appropriate timing of interventions.

Early interventions—before significant degeneration occurs—are often more effective, particularly in treating conditions like osteoarthritis. Proactive treatments that initiate healing before major tissue loss can prevent the need for invasive procedures and enhance long-term MSK health by significantly slowing disease progression.

Effective regeneration strategies must be tailored to both the patient's physiological state and the nature of the injury. By optimising the timing of interventions, clinicians can significantly improve the outcomes of regenerative therapies, ensuring faster recovery and better tissue function.

1.4 Principles of Engineering in Musculoskeletal (MSK) Regeneration

Tissue Engineering for MSK Repair

Tissue engineering plays a vital role in musculoskeletal (MSK) repair by combining biomaterials, scaffold design, and advanced technologies like 3D printing. The goal is to restore or regenerate damaged tissues, such as bone, cartilage, and muscle, by providing structural and biochemical support. Scaffolds act as temporary matrices, guiding cells to grow and proliferate, ultimately forming new tissues.

Scaffold-based treatments have evolved to utilise both cell-free scaffolds and cell-seeded scaffolds. Cell-free scaffolds aim to recruit the body's endogenous cells to the defect site, while cell-seeded scaffolds involve pre-loading stem cells to enhance regeneration. For example, mesenchymal stem cells (MSCs) are commonly used in these applications due to their ability to differentiate into various cell types, including osteoblasts, chondrocytes, and myocytes (Kon et al. 2021). Biomaterials like collagen, hyaluronic acid, and synthetic polymers such as polylactic acid (PLA) and polyglycolic acid (PGA) are commonly used in these scaffolds due to their biocompatibility and biodegradability. These materials degrade over time, allowing the newly formed tissue to take over the structure as the scaffold dissolves (Kon et al. 2021; Zhang et al. 2020).

Mechanical Engineering and Biomechanics

Mechanical forces are critical to the success of MSK tissue regeneration. Mechanical loading influences tissue healing through the process of mechanotransduction, where cells convert physical stimuli into biochemical responses. This is especially relevant for tissues like cartilage and bone, which rely on mechanical stress to maintain their integrity and function.

Biomechanics also plays an essential role in optimising the structure and function of 3D-printed scaffolds. Scaffolds must be designed with the correct pore size, mechanical strength, and elasticity to mimic the properties of native tissues. In bone regeneration, for example, scaffolds with specific mechanical properties promote osteointegration and new bone formation, ensuring that the material can withstand physiological loads while the tissue heals (Kon et al. 2021).

Electromagnetic technologies have also been investigated for their potential to accelerate muscle, bone, and joint regeneration. Electromagnetic stimulation has been shown to enhance osteoblast activity, promoting bone healing, particularly in non-union fractures, where traditional methods often fail (Zhang et al. 2020).

Digital Twins and AI in Regenerative Medicine

The integration of artificial intelligence (AI) and digital body banking into MSK regenerative medicine has revolutionised personalised healthcare. A digital twin is a virtual model of a patient's body, created by using real-time data and AI algorithms to simulate various treatment outcomes. These real-time simulations allow clinicians to predict how different interventions—such as scaffold implantation or stem cell therapies—will perform in an individual patient (Kon et al. 2021).

Through machine learning models, AI can predict disease progression, optimise rehabilitation protocols, and even prevent MSK injuries before they occur. One of the most promising applications is the ability to assess patient movement patterns, allowing for more precise tracking of joint function and musculoskeletal health. This approach not only enhances the accuracy of treatment plans but also reduces the risk of complications and improves long-term outcomes (Zhang et al. 2020).

1.5 Mechanistic Versus Systematic Thinking in Regenerative Medicine

Challenges of Traditional Mechanistic Approaches

Traditional approaches to regenerative medicine, especially in treating musculoskeletal (MSK) conditions, often follow a mechanistic or reductionist framework. This method breaks down biological systems into isolated components and treats individual symptoms or damages separately. For example, conventional treatments for osteoarthritis, such as microfracture surgery or cartilage debridement, are designed to address localised tissue damage by mechanically inducing a healing response (Fan et al. 2016). However, these methods frequently result in incomplete tissue regeneration and fail to restore full function because they overlook the complex interplay of cells, tissues, and mechanical forces involved in complete regeneration (Bakhtiary et al. 2021).

A primary limitation of mechanistic approaches is their inability to address the full complexity of human biology. Musculoskeletal tissues, such as cartilage, bone, and muscle, operate in highly integrated environments, where cellular interactions, mechanical forces, and biochemical signals continuously influence one another. Treating a single component in isolation often leads to short-term results and incomplete healing, as it fails to account for the broader systemic factors that drive regeneration. In particular, treatments based on singular therapies, like cell injections without considering the broader structural and mechanical context, may not provide the regenerative cues required for long-lasting results (Trumbull et al. 2016).

Additionally, these approaches are predominantly reactive—implemented after significant damage has already occurred. This reactive nature can delay effective intervention and miss the opportunity for early, proactive treatments that could prevent the progression of degeneration. For instance, by the time visible symptoms of osteoarthritis appear, significant cartilage and joint damage may already be irreversible, limiting the effectiveness of mechanistic treatments (Kim et al. 2014).

Systematic Thinking for True Regeneration

In contrast, systematic thinking embraces an interdisciplinary and holistic approach to regeneration, considering the body as a complex system where biological, physical, and chemical processes are deeply interconnected. This method emphasises the integration of engineering, biology, and physics to create regenerative strategies that address not just the damaged tissue, but also the mechanical and biochemical environment that supports healing (Ingber 2003).

For instance, scaffold-based therapies used in conjunction with stem cells and growth factors provide structural support while stimulating cellular activity to promote tissue regeneration (Jones et al. 2023). These scaffolds mimic the extracellular matrix (ECM), providing a platform for cell attachment and growth while delivering bioactive molecules that promote differentiation and tissue repair. By combining materials engineering with biological principles, such as using biodegradable polymers for scaffolds that integrate naturally into the body, the systematic approach provides a more sustainable and effective regenerative solution (Narayanan et al. 2016).

Furthermore, mechanotransduction—the process by which cells convert mechanical stimuli into biochemical signals—highlights the importance of combining physics with biological regeneration. By optimising mechanical forces, such as load-bearing exercises or using engineered scaffolds to apply controlled mechanical stress, the healing process can be significantly enhanced (Zhang and Habibović 2022). This interplay between physical forces and biological processes is key to achieving long-term regenerative success, as it promotes both the growth of new tissues and the restoration of their functional properties.

Proactive and Personalised Approaches

One of the key aspects of systematic thinking in regenerative medicine is the shift from reactive to proactive care. Early intervention can help prevent severe damage to MSK tissues and maintain joint functionality. Systematic approaches advocate for the use of advanced diagnostics and personalised treatment plans that identify early signs of tissue degeneration and initiate regenerative therapies before irreversible damage occurs (Lyu et al. 2023; Su et al. 2018).

By employing techniques like digital twins and artificial intelligence (AI), clinicians can simulate treatment outcomes and optimise personalised healthcare strategies. AI-driven tools allow for precise control over the delivery of regenerative agents and scaffold design, ensuring that interventions are tailored to the patient's unique physiological conditions. The combination of digital tools and systematic regenerative strategies enhances the likelihood of successful outcomes and long-term recovery (Trumbull et al. 2016; Zhang and Habibović 2022).

Ultimately, systematic thinking recognises that regeneration is a multi-faceted process, requiring the coordination of biological signals, mechanical forces, and material properties. By addressing these factors together, regenerative therapies can be more effective, sustainable, and tailored to the individual needs of patients, offering greater potential for long-term recovery and tissue health.

1.6 Medical Applications of MSK Regeneration

Current Medical Interventions in Regenerative Medicine

The advancement of regenerative medicine has revolutionised treatments for musculoskeletal (MSK) conditions, particularly in orthopaedic care. Modern interventions include the use of stem cell therapies, platelet-rich plasma (PRP), and growth factors, which aim to repair and regenerate damaged tissues rather than simply managing symptoms.

- **Stem/ medicinal Cell Therapies**: Mesenchymal stem cells /medicinal signalling cells (MSCs), derived from sources like bone marrow, adipose tissue, and even tendon-derived cells, have become central in treating conditions like osteoarthritis, tendon injuries, and cartilage damage (Fan et al. 2016; Ni et al. 2021). MSCs promote tissue regeneration by differentiating into various cell types, including osteoblasts and chondrocytes, while secreting bioactive molecules that modulate inflammation and support tissue repair (Lyu et al. 2023). These therapies have been shown to reduce pain, improve joint function, and, in some cases, delay the need for more invasive interventions such as joint replacement (Zhang et al. 2020).
- **Platelet-Rich Plasma (PRP)**: PRP therapy utilises a concentrated mixture of platelets derived from the patient's blood. Platelets release growth factors such as platelet-derived growth factor (PDGF) and transforming growth factor-beta (TGF-β), which enhance tissue repair and promote cellular activity (Kim et al. 2014). PRP is often used in conjunction with other therapies like stem cells or scaffolds, helping to accelerate healing in tendons, ligaments, and cartilage injuries. The strength of PRP is dependent on the quality and the quantity of the growth factors and not the total count of the platelets.
- **Growth Factors**: The administration of exogenous growth factors like insulin-like growth factor-1 (IGF-1) (Qi et al. 2018) or bone morphogenetic proteins (BMPs) can further stimulate cellular responses and improve tissue repair. These growth factors are often delivered via scaffolds or hydrogels to enhance the local regenerative environment, promoting long-term healing in bones, muscles, and cartilage (Gan et al. 2021).

Combining Biology and Engineering for MSK Restoration

One of the critical innovations in MSK regenerative medicine is the integration of biological and engineering principles. Scaffolds, designed to mimic the extracellular matrix (ECM), provide structural support to cells, facilitating cell attachment, migration, and differentiation. When seeded with stem cells or combined with growth factors, these scaffolds enhance tissue regeneration, providing a more comprehensive treatment for MSK conditions (Narayanan et al. 2016; Wu et al. 2021).

For example, bioengineered scaffolds used in cartilage repair can be customised through 3D printing technologies to match the patient's anatomical structure, allowing for precise and personalised treatments. The combination of biodegradable polymers like polycaprolactone (PCL) with bioactive agents creates an environment conducive to tissue regeneration, while simultaneously ensuring that the scaffold will degrade as the tissue heals (Petretta et al. 2021).

Proactive Regenerative Medicine

In recent years, regenerative medicine has shifted towards a proactive approach aimed at maintaining MSK health before irreversible damage occurs. The focus has moved from treating advanced degenerative conditions to early interventions that help prevent or slow tissue degradation. This shift is particularly important in managing chronic diseases like osteoarthritis and tendonitis, where early application of regenerative therapies can preserve tissue function and delay the progression of degeneration (Su et al. 2018).

- **Early Interventions**: Applying stem cell therapies, PRP, or growth factors in the early stages of joint deterioration has been shown to promote cartilage repair and slow the onset of osteoarthritis symptoms. Proactive regenerative strategies focus on tissue preservation, reducing inflammation and enhancing the body's natural healing mechanisms (Bakhtiary et al. 2021).
- **Maintaining Vitality**: In addition to addressing tissue damage, regenerative medicine emphasises the maintenance of long-term vitality. This includes regular monitoring of joint health, optimising physical activity, and using personalised regenerative treatments that are tailored to an individual's specific musculoskeletal condition. By intervening early and maintaining joint function, patients can delay or avoid the need for invasive surgeries like joint replacements (Trumbull et al. 2016).

1.7 The Future of MSK Regenerative Therapies

Emerging Innovations in Regenerative Medicine

The future of musculoskeletal (MSK) regenerative therapies is shaped by groundbreaking advancements in biomaterials, stem cell technologies, and artificial intelligence (AI)-driven solutions. These innovations aim to enhance the precision, efficacy, and scalability of treatments for MSK disorders. The continuous development of bioactive scaffolds, such as biodegradable polymers and hydrogels, provides structural support to the tissues while delivering growth factors or stem cells to accelerate the healing process. For instance, bioactive materials are being explored for their ability to stimulate endogenous repair mechanisms in bone and cartilage injuries.

Stem cell therapies remain a cornerstone of future MSK regenerative treatments. Ongoing research is optimising the use of mesenchymal stem cells (MSCs), derived from various tissues such as bone marrow, adipose tissue, and synovium, for targeted tissue regeneration. These stem cells are not only capable of differentiating into multiple lineages, but also have potent immunomodulatory and anti-inflammatory effects, enhancing tissue recovery and minimising the progression of degenerative conditions.

Moreover, advancements in 3D bioprinting are unlocking new possibilities in creating personalised implants for MSK repair. This technology allows for the fabrication of custom-fit scaffolds that can be tailored to the patient's specific anatomy, ensuring better integration with the host tissue and optimising the healing process.

2 Conclusion

The future of MSK regenerative medicine depends heavily on interdisciplinary collaboration across various fields such as engineering, biology, chemistry, and clinical medicine. This collaborative effort ensures that new technologies and biological insights are translated into practical clinical applications. For example, the integration of biomechanics and regenerative biology has led to the development of more sophisticated rehabilitation protocols that better support tissue healing post-surgery.

Additionally, AI-driven tools are being increasingly applied to regenerative medicine to predict treatment outcomes, model tissue regeneration, and personalise patient care. AI-based systems are also crucial in designing patient-specific interventions, enabling real-time adjustments based on individual biological responses.

As these technological innovations converge with biological research, the potential for long-term success in regenerating MSK tissues grows exponentially. The interdisciplinary approach is key to overcoming the challenges posed by complex musculoskeletal conditions, paving the way for more effective and sustainable therapies.

References

Bakhtiary N, Liu C, Ghorbani F. Bioactive inks development for osteochondral tissue engineering: a mini-review. Gels. 2021;7(4):274. https://doi.org/10.3390/gels7040274.

Clarke G, Harley P, Hubber EL, Manea T, Manuelli L, Read E, Watt FM. Bench to bedside: Current advances in regenerative medicine. Curr Opin Cell Biol. 2018;55:59–66. https://doi.org/10.1016/j.ceb.2018.05.006. Epub 2018 Jul 11. PMID: 30007127.

Fan J, Wang D, Liu H, Fan H, Yang F. Stem cells in musculoskeletal regeneration: from benchtop to bedside. Stem Cells Int. 2016;2016(1) https://doi.org/10.1155/2016/8432314.

Gan Q, Choy K, Foo C, Leong P, Cheong S. Incorporating insulin growth factor-1 into regenerative and personalised medicine for musculoskeletal disorders: a systematic review. J Tissue Eng Regen Med. 2021;15(5):419–41. https://doi.org/10.1002/term.3192.

Ingber D. Mechanobiology and diseases of mechanotransduction. Ann Med. 2003;35(8):564–577. https://doi.org/10.1080/07853890310016333.

Jones C, Penney B, Theodossiou S. Engineering cell–ecm–material interactions for musculoskeletal regeneration. Bioengineering. 2023;10(4):453. https://doi.org/10.3390/bioengineering10040453.

Kim D, Choi Y, Kim S, Lee J, Kim S, Kim Y, et al. In vivo effects of adipose-derived stem cells in inducing neuronal regeneration in Sprague-dawley rats undergoing nerve defect bridged with polycaprolactone nanotubes. J Korean Med Sci. 2014;29(Suppl 3):S183. https://doi.org/10.3346/jkms.2014.29.s3.s183.

Kon E, Altomare D, Dorotei A, Di Matteo B, Marcacci M. Scaffolds for Cartilage repair. In: Krych AJ, Biant LC, Gomoll AH, Espregueira-Mendes J, Gobbi A, Nakamura N, editors. Cartilage injury of the knee. Springer; 2021. https://doi.org/10.1007/978-3-030-78051-7_21.

Lyu F, Peng S, Li J. Editorial: Progress on musculoskeletal disorders and stem cell therapies. Front Bioeng Biotechnol. 2023;11 https://doi.org/10.3389/fbioe.2023.1153525.

Narayanan G, Vernekar V, Kuyinu E, Laurencin C. Poly (lactic acid)-based biomaterials for orthopaedic regenerative engineering. Adv Drug Deliv Rev. 2016;107:247–76. https://doi.org/10.1016/j.addr.2016.04.015.

Ni M, Sun W, Li Y, Ding L, Lin W, Peng H, et al. Sox11 modified tendon-derived stem cells promote the repair of osteonecrosis of femoral head. Cell Transplant. 2021;30:096368972110538. https://doi.org/10.1177/09636897211053870.

Petretta M, Gambardella A, Desando G, Cavallo C, Bartolotti I, Shelyakova T, et al. Multifunctional 3D-printed magnetic polycaprolactone/hydroxyapatite scaffolds for bone tissue engineering. Polymers. 2021;13(21):3825. https://doi.org/10.3390/polym13213825.

Pittenger MF, Mackay AM, Beck SC, Jaiswal RK, Douglas R, Mosca JD, Moorman MA, Simonetti DW, Craig S, Marshak DR. Multilineage potential of adult human mesenchymal stem cells. Science. 1999;284(5411):143–7. https://doi.org/10.1126/science.284.5411.143. PMID: 10102814.

Qi Z, Xia P, Zheng S, Fu C, Chang Y, et al. Combined treatment with electrical stimulation and insulin-like growth factor-1 promotes bone regeneration in vitro. PLoS One. 2018;13(5):e0197006. https://doi.org/10.1371/journal.pone.0197006.

Su P, Tian Y, Yang C, Ma X, Wang H, Pei J, et al. Mesenchymal stem cell migration during bone formation and bone diseases therapy. Int J Mol Sci. 2018;19(8):2343. https://doi.org/10.3390/ijms19082343.

Trumbull A, Subramanian G, Yildirim-Ayan E. Mechanoresponsive musculoskeletal tissue differentiation of adipose-derived stem cells. Biomed Eng Online. 2016;15(1) https://doi.org/10.1186/s12938-016-0150-9.

Wu R, Li H, Yang Y, Zheng Q, Li S, Chen Y. Bioactive silk fibroin-based hybrid biomaterials for musculoskeletal engineering: recent progress and perspectives. ACS Appl Bio Mater. 2021;4(9):6630–46. https://doi.org/10.1021/acsabm.1c00654.

Zhang Y, Habibović P. Delivering mechanical stimulation to cells: state of the art in materials and devices design. Adv Mater. 2022;34(32) https://doi.org/10.1002/adma.202110267.

Zhang J, et al. Human adipose-derived mesenchymal progenitor cells plus microfracture and hyaluronic acid for cartilage repair: a phase IIa trial. Regen Med. 2020;15 https://doi.org/10.2217/rme-2019-0068.

Basic Science in Cartilage Regeneration

Dzhian Abazovic, Oliver Dulić, and Jovana Vuković

Abstract Cartilage regeneration is a promising field within orthopedic and tissue engineering disciplines, aiming to address the challenges posed by cartilage injuries and degenerative diseases. This abstract explores recent advancements in basic science that have significantly contributed to our understanding of cartilage biology and the development of regenerative strategies. Key areas of focus include the cellular and molecular mechanisms involved in cartilage formation, maintenance, and repair, as well as the role of stem cells, growth factors, and biomaterials in promoting cartilage regeneration. Understanding the intricate interplay between chondrocytes, mesenchymal stem cells, and signaling pathways such as TGF-β and BMP has provided crucial insights for designing innovative approaches to stimulate cartilage repair.

Moreover, advancements in tissue engineering techniques, including scaffold-based and cell-based therapies, have shown promising results in preclinical and clinical studies. This abstract highlights the importance of bridging basic science research with translational efforts to develop effective strategies for cartilage repair and regeneration, ultimately aiming to improve clinical outcomes and enhance the quality of life for patients suffering from cartilage-related conditions.

Keywords Cartilage · Regeneration · Scaffold · Tissue engineering · Matrix · Cells · Growth factors

D. Abazovic (✉)
School of Medicine, University of Belgrade, Belgrade, Serbia

ABA Medica, Ulcinj, Montenegro; https://invivox.com/en/profile/8731

O. Dulić · J. Vuković
School of Medicine, University of Belgrade, Belgrade, Serbia

Department for Orthopedic Surgery and Traumatology, Clinical Center of Vojvodina, Novi Sad, Serbia

Medical Faculty, University of Novi Sad, Novi Sad, Serbia

P. Lee (ed.), *Musculoskeletal Regeneration Medicine*,
https://doi.org/10.1007/978-3-031-77456-0_2

1 Introduction

1.1 Structure of Cartilage

Cartilage, ro cartilaginous tissue, embodies a specialized form of connective tissue renowned for its intricate extracellular matrix (ECM) dynamics. Integral to skeletal formation alongside bone tissue, cartilage is notably avascular and aneural, relying on passive diffusion for nutrient and oxygen exchange from its perichondrium or, within articulating surfaces, the synovial fluid and bone vasculature.

Positioned between dense connective and osseous tissue, its biomechanical properties render it both resilient and pliable (Lačković et al. 2019).

Central to its structural integrity are its cellular constituents and ECM intricacies. Chondrogenic cells, chondroblasts, and chondrocytes orchestrate within lacunae, forming isogenic clusters. Surrounding these clusters, the ECM meticulously organizes into a pericellular capsule, while territorial ECM delineates the boundaries. The amalgamation of isogenic clusters and their ECM domains form the fundamental chondron structure, while inter-chondron interactions shape the interterritorial matrix.

Comprising collagenous and elastic fibers embedded within a ground substance rich in proteoglycans and glycoproteins, the ECM confers cartilage with a balance of strength and flexibility (Lačković et al. 2019).

Typified by their spatial distribution, cellular arrangement, and ECM composition, three principal cartilage types emerge—hyaline, elastic, and fibrous, each tailored to specific anatomical niches and physiological demands, serving as pivotal components in structural and functional contexts within the body (Lačković et al. 2019).

Clinical Importance

Osteoarthritis, often affecting the hip and knee joints, is a common consequence of the degenerative changes that occur in cartilage with aging. These changes primarily involve chondrocytes and the extracellular matrix, leading to the fragmentation of hyaline cartilage within the joint (Lačković et al. 2019). Displacement of the peripheral portion of the intervertebral disc, predominantly composed of fibrous cartilage and typically occurring in the posterior region, can result in nerve compression, potentially causing neurological symptoms (Lačković et al. 2019).

In cases of cartilage damage or complete loss, ankylosis may occur, characterized by the fusion of jointed bones, ultimately resulting in the complete immobility of the affected joint (Lačković et al. 2019).

Hyaline Cartilage

Hyaline cartilage, serving as a foundational element in the intricate process of embryonic skeletogenesis, assumes a critical role in providing structural support and facilitating skeletal development. However, as the developmental journey progresses from the embryonic phase to postnatal stages, hyaline cartilage undergoes a remarkable transformation known as endochondral ossification. This transformative process entails the gradual replacement of cartilaginous tissue with bone, ultimately resulting in the formation of the mature skeletal framework (Lačković et al. 2019).

In the continuum of physiological development, hyaline cartilage persists into adulthood primarily in the form of articular cartilage. This specialized tissue blankets the articulating surfaces of joints, providing crucial cushioning and lubrication essential for smooth and pain-free movement.

Moreover, it extends its presence to anatomical structures beyond the realm of joints, encompassing regions such as the rib ends and various components of the respiratory system, including the nose, larynx, trachea, and bronchi (Lačković et al. 2019).

Characterized by its distinct visual appearance—a semi-translucent quality imbued with a delicate milky-blue hue—hyaline cartilage is enveloped by a protective sheath known as the perichondrium.

Within this fibrous connective tissue layer resides a reservoir of chondrogenic progenitor cells, which play an indispensable role in the ongoing maintenance and repair of cartilaginous tissue throughout life (Lačković et al. 2019).

Remarkably, while the majority of hyaline cartilage is encased by the perichondrium, articular cartilage presents a notable exception, lacking this outer protective layer. This unique architectural feature underscores the specialized nature of articular cartilage and its distinct functional role within the broader landscape of the body's connective tissue matrix (Lačković et al. 2019).

In essence, hyaline cartilage embodies a remarkable journey—from its foundational role in embryonic skeletogenesis to its pivotal function as articular cartilage in adulthood—signifying its enduring importance in maintaining skeletal integrity and facilitating seamless movement throughout the lifespan (Lačković et al. 2019).

Cells of Hyaline Cartilage and ECM

Chondrogenic cells derive from mesenchymal stem cells and exhibit characteristics akin to fibroblasts. They possess the dual capability of proliferation and differentiation into chondroblasts (Lačković et al. 2019).

Chondroblasts serve as the primary architects of the extracellular matrix (ECM), responsible for its synthesis and secretion. Morphologically, they present as oval or elongated cells with short projections. Within their cytoplasm, a complex network of organelles including rough endoplasmic reticulum (RER), Golgi apparatus, and secretory vesicles facilitates the production of ECM components such as type II collagen and ground substance. Upon encapsulation by ECM, chondroblasts transition into chondrocytes (Lačković et al. 2019).

Chondrocytes, the matured form of chondroblasts, typically display an oval morphology and can reach sizes of up to 30 micrometers. Their cytoplasmic contents vary depending on activity level, with active chondrocytes exhibiting organelles involved in ECM synthesis while inactive ones accumulate storage materials like glycogen and lipid droplets (Lačković et al. 2019).

Residing within lacunae of the ECM, chondrocytes proliferate through mitotic division, forming multicellular aggregates known as isogenous clusters. These clusters, in conjunction with the surrounding ECM, constitute the chondron (Lačković et al. 2019).

The ECM of hyaline cartilage, considerably more abundant than cellular constituents, encompasses distinct structural elements such as the pericellular capsule, territorial matrix, and interterritorial matrix, primarily composed of collagen fibers and ground substance (Lačković et al. 2019).

Comprising predominantly type I collagen (90–95%), collagen fibers intricately interlace within a diffuse network. The ground substance, containing proteoglycans, glycoproteins (including chondronectin and chondrocalcin), and a substantial aqueous fraction, plays a crucial role in maintaining cartilage integrity and function (Lačković et al. 2019).

In hyaline cartilage, glycosaminoglycans display distinct regional qualities. Close to the lacunae within isogenous groups, an abundance of acidic glycosaminoglycans, particularly chondroitin sulfates, contributes to heightened basophilia, defining this area as the territory (Lačković et al. 2019).

The glycoproteins present in hyaline cartilage, such as chondronectin and chondrocalcin, play vital roles. Chondronectin acts as an adhesive glycoprotein, facilitating connections between integrins on the cell membranes of chondrocytes and chondroblasts with collagen fibers, chondroitin sulfate, and hyaluronic acid. Meanwhile, chondrocalcin is involved in calcium ion binding (Lačković et al. 2019).

Water, comprising a significant portion of fresh cartilage weight, primarily binds between glycosaminoglycan side chains, acting as a biomechanical spring. Additionally, free water exists between proteoglycan molecules, allowing metabolite diffusion between perichondrium blood vessels and chondrocytes (Lačković et al. 2019).

The compression of articular cartilage matrix due to locomotor activity compresses glycosaminoglycan side chains, releasing bound water from the matrix, reducing its volume and increasing free water volume. Aging cartilage experiences reduced water and glycosaminoglycan concentrations in the matrix, leading to decreased elasticity and increased susceptibility to fractures (Lačković et al. 2019).

Elastic Cartilage

Elastic cartilage is the least abundant type of cartilage in the human body, found in structures such as the outer ear, auditory tube, epiglottis, and certain parts of the larynx (cartilago cuneiformis, cartilago corniculata). It is composed of cells and extracellular matrix (ECM) and is surrounded by perichondrium. Histologically, it closely resembles hyaline cartilage, with some notable differences:

- Chondrocytes exhibit a larger and more spherical morphology.
- Isogenous groups comprise a reduced number of cells, typically ranging from 2 to 4.
- The extracellular matrix (ECM) shows decreased ground substance content but is rich in elastic fibers in addition to predominantly type II collagen fibers. These elastic fibers, distinctly visible under light microscopy, significantly contribute to the heightened elasticity of elastic cartilage.
- In contrast to hyaline cartilage, elastic cartilage does not undergo ossification (Lačković et al. 2019).

Fibrous Cartilage

Fibrous cartilage is found mainly where tendons and ligaments attach to bones. It's also in some joints like the temporomandibular and sternoclavicular joints, as well as in structures like the intervertebral discs, pubic symphysis, and knee menisci. Here's how it differs from hyaline cartilage:

- Absence of a perichondrium.
- Chondrocytes within fibrous cartilage form isogenous groups aligned in rows, arranged in parallel arrays with a longitudinal orientation.
- The extracellular matrix of fibrous cartilage closely resembles dense (fibrous) connective tissue, with minimal ground substance and abundant wide parallel bundles of collagen fibers, predominantly consisting of type I collagen.
- When observed under a microscope, the territorial matrix surrounding cells appears basophilic, while the interterritorial matrix demonstrates an acidophilic staining pattern (Lačković et al. 2019).

1.2 The Incapacity of Articular Cartilage for Regeneration

Articular cartilage repair poses significant challenges due to its unique characteristics and limited regenerative capacity. Unlike many other tissues, articular cartilage exhibits a low cell to matrix ratio, minimal mitotic activity, and slow turnover rate. Moreover, aging further diminishes its reparative potential by reducing the number of marrow mesenchymal stem cells and chondrocyte activity.

Consequently, spontaneous repair in mature cartilage typically results in the formation of fibrous tissue or fibrocartilage, both of which are unable to withstand the high compressive loads experienced at articulating surfaces (Buckwalter and Mankin 1997).

There are two main types of cartilage defects: chondral, which are superficial and do not penetrate the subchondral bone, and osteochondral, which are full-thickness defects that extend into the underlying bone. While superficial defects are less likely to cause disability, osteochondral defects offer a potential pathway for spontaneous repair as they expose the defect to a blood supply, facilitating the formation of a fibrin clot and triggering a response from mesenchymal stem cells in the marrow (Chu et al. 1999).

Repair mechanisms can be intrinsic, involving mesenchymal stem cells, or extrinsic, involving tissue from the articular surface. However, the presence of proteoglycans in cartilage may inhibit mesenchymal stem cell adhesion, hindering the repair process. Current research endeavors focus on innovative approaches such as biodegradable scaffolds, which provide a synthetic matrix for chondrocytes while recruiting host cells into the defect. Additionally, other methods like grafting, transplants, microfracture, electrical stimulation, and laser therapy are being explored (Athanasiou et al. 1996; Hunziker and Rosenberg 1996).

Despite these efforts, the outcomes of articular cartilage repair procedures remain inconsistent. Poor organization and integration between host and new cartilage often characterize the results, with the quality of neocartilage ranging from suboptimal to comparable to native cartilage in terms of morphological, biochemical, and biomechanical properties. Thus, while advancements in research offer promising avenues for improving cartilage repair, the quest for effective and reliable methods continues (Brittberg et al. 1994a; Sams and Nixon 1995).

1.3 Tissue Engineering

Tissue engineering stands as a rapidly advancing domain, serving as a transformative approach where cells proliferate and arrange their extracellular matrix within a three-dimensional lattice, culminating in clinically viable tissue mirroring the histochemical, biochemical, and biomechanical attributes of native tissue. This technology eliminates the necessity for tissue transplantation by reconstructing mammalian tissues structurally and functionally. It offers promising avenues for treating damaged and diseased articular cartilage.

A pivotal focus lies in producing *ex vivo* grown articular cartilage that mirrors the characteristics of native cartilage. Typically, three core components form the foundation of tissue engineering: a matrix scaffold, cells, and signaling molecules like growth factors or genes. This endeavor has spurred vigorous research activities, utilizing bioreactors as pivotal tools in this pursuit. Researchers employ rotating bioreactors, to subject cell-seeded polyglycolic acid scaffolds to diverse stimuli. These rotating bioreactors provide dynamic conditions, facilitating nutrient and gas exchange while promoting mechanical stimulation. Additionally, some employed a method involving the immobilization of cell- laden polyglycolic acid scaffolds within polycarbonate capsules to direct and control fluid-induced shear within the lattice. Analysis following weeks of culture in these bioreactors has revealed organized extracellular matrix components akin to articular cartilage, underscoring the importance of mechanical stimulus in promoting cartilage matrix synthesis (Dunkelman et al. 1995; Freed et al. 1998; Freed et al. 1993).

Bioreactors offer advantages such as ensuring homogeneous nutrient distribution, vital to surmount the diffusion barrier posed by synovial fluid and cartilage layers. Hindered diffusion, which could impede cellular uptake of essential nutrients and gases, is effectively mitigated by bioreactors.

Researchers assert that tissue engineering seeks to emulate embryonic or fetal cell/tissue differentiation processes. However, the microenvironment within mature tissue presents obstacles to such recapitulation. Although embryonic, fetal, and adult organisms share common signaling substances, the target cells and their responsiveness evolve throughout ontogenesis. Consequently, fundamental adjustments are necessary to realize successful tissue engineering in adult organisms (Athanasiou et al. 1998; Caplan et al. 1997; Hollinger et al. 2000; Jackson and Simon 1999; Reddi 2000, 2001).

How Does it Work?

The prevailing direction in tissue engineering is shifting towards the development of constructs comprising not just matrices and cells but also signaling molecules. These molecules are aimed at guiding differentiation towards desired outcomes. This trend arises from the realization that durable, high-quality repair tissue resembling hyaline cartilage can only be consistently formed by streamlining the healing response. This involves implanting a more uniform cell population and inhibiting the spontaneous repair reaction, which involves uncontrolled activities from various cell types and signaling substances originating from the subchondral bone-tissue spaces.

Two potential tissue engineering approaches can be pursued.

In the first approach, repair tissue is entirely engineered *in vitro*, with the fully differentiated construct then implanted into the defect void. This method offers better control over cell metabolism and differentiation *in vitro*, provided appropriate bioreactor systems, growth factors, and delivery systems are available. However, challenges arise concerning tissue integration, mechanical fixation, and replicating ideal mechanical loading conditions. Moreover, concerns regarding biocompatibility and immunological reactions often emerge, along with difficulties associated with implanting engineered implants into joint surfaces.

The second, more commonly adopted approach focuses on engineering the basic building block: a matrix scaffold containing a homogeneous cell population and signaling molecules entrapped within a suitable delivery system. This approach aims for *in vivo* differentiation and remodeling of repair tissue under physiological mechanical loading conditions. Repair tissue formed in situ is more likely to adhere to and integrate with native cartilage, naturally adapting to joint contours. However, controlling long-term cell activity and preventing contamination from spontaneous healing responses pose challenges.

Both approaches have primarily been tested *in vitro* and using experimental. The choice between them depends on the type of defect being treated and the surrounding microenvironment's biology.

Successful translation to clinical application hinges on resolving challenges such as immunological rejection, mechanical fixation, and ensuring tissue integration and adequate physical properties.

Additionally, advancements in understanding the roles of growth factors in chondrogenesis and exploring gene therapy offer promising avenues for improving articular cartilage repair. Animals (Mow et al. 1991; Ochi et al. 2001; Schreiber et al. 1999a; b; Schreiber and Ratcliffe 2000; Weisser et al. 2001).

Some reviews have comprehensively covered the effectiveness of different signaling substances, particularly growth factors, in inducing chondrogenesis. Primarily assessed *in vitro* and to a limited extent *in vivo*, these substances have been evaluated through diverse combinations of matrices, cell types, and growth factors in various experiments. Examples include collagen matrices containing bone-marrow stromal cells and TGF-alpha 1, agarose matrices containing chondrocytes and FGF-2, fibrin matrices containing chondrocytes and IGF-1, and polylacetate matrices containing perichondrial cells and TGF- alpha 1. Successful testing *in vivo* has been accomplished using horse and mature miniature pig models. These studies highlight the disparity between *in vivo* and *in vitro* conditions, with primarily only growth factors of the TGF-alpha superfamily showing effectiveness *in vivo*, despite various types supporting chondrogenic activities under *in vitro* conditions (Andrades et al. 1999; Angele et al. 1999; Arnoczky et al. 1988; Fortier et al. 1999; Fortier et al. 1997; Glowacki 2000; Hunziker 2001; Mosca et al. 2000; Reddi 1998a, b; Suh and Matthew 2000; van den Berg et al. 2001).

Some authors have extensively reviewed gene therapy and gene transfer technologies, which offer potent new possibilities for managing articular cartilage repair responses. *In vitro* experiments have successfully transfected genes carrying chondrogenic information into various cell types, including chondrogenic precursor (bone-marrow stromal) cells, chondrocytes, perichondrial cells, and periosteal cells. *In vivo*, there was a successful attempt to induce a repair response using transfected mesenchymal cells or chondrocytes. The transfected cells in this study were shown to produce ample amounts of BMP- 2, effectively stimulating the formation of cartilage tissue without the risk of unintended vector dissemination or triggering immunological responses (Mosca et al. 2000; Banfi et al. 2000; Evans and Robbins 1999; Evans et al. 2000; Gelse et al. 2001; Goomer et al. 2001; Kang et al. 2000; Prockop 1997).

The subsequent text will delve further into the second approach, given its more widespread usage.

2 Scaffolds

Scaffolds made of natural biomaterials have attracted significant attention for the repair of osteochondral defects due to their beneficial features. These biomaterials, such as collagens, hyaluronan, alginate, periosteum, and coral, exhibit suitable adhesion properties, increased biocompatibility, and reduced toxicity during scaffold degradation. They have shown promise in both *in vivo* and *in vitro* experiments, particularly when used as three-dimensional scaffolds to transport cells into articular cartilage defects (Kreder et al. 1994; Shahgaldi 1998; Solchaga et al. 1999; van Susante et al. 1995).

For their efficacy in facilitating or promoting articular cartilage repair, many different types of matrix have been tested *in vitro*, as well as in experimental animals and in human patients. These matrices can be broadly categorized according to their chemical nature into:

1. Protein-based polymers (fibrin, collagen, gelatine)
2. Carbohydrate-based ones (polylactic acid, polyglycolic acid, hyaluronan, agarose, alginate, chitosan)
3. Artificial materials (dacron, teflon, carbon fibers, polyesterurethane, polybutyric acid, polyethylmethacrylate, hydroxyapatite)
4. Within/between classes (crosslinkage, chemical modifications, geometrical modifications, matrix combinations)

The ideal matrix should possess several key characteristics to effectively support tissue repair and regeneration. Firstly, it should be porous to allow the migration of cells, either loaded onto the matrix or native ones from the surrounding tissue, through its interstices. Additionally, the surface properties of the matrix should promote cell adhesion, which can be achieved by incorporating cell adhesion-promoting peptides like Arg-Gly-Asp into the basic matrix material (Suh and Matthew 2000; Ryd et al. 1997).

Moreover, the matrix must exhibit biocompatibility, meaning it should induce minimal adverse immunological reactions in the tissue where it's implanted. In some cases, the matrix may need to facilitate the proliferation of one cell population while inhibiting another. It should also be biodegradable, capable of being replaced by physiological extracellular components over time. However, it's essential to consider that degradation products from some matrices can have cytotoxic or nephrotoxic effects.

Certain physical characteristics, such as deformability, elasticity, and volume stability, are also important considerations for the matrix. While it may not be realistic to expect a single material to meet all these requirements, combining different materials and making chemical modifications can help approximate the ideal matrix.

The value of a matrix lies in its ability to enhance or promote a spontaneous repair response within critical-sized defects, where natural healing alone may be insufficient. When a porous matrix is implanted into such a defect, it absorbs blood and becomes permeated by it, allowing for the expansion of the hematoma and facilitating cell migration into the matrix. This process supports and enhances spontaneous healing responses, ultimately aiding in tissue repair and regeneration.

In the field of tissue engineering for articular cartilage, the challenge lies in understanding chondrocyte attachment to scaffolds such as polylactides-polyglycolides without the aid of serum. Serum, abundant in fibronectin and vitronectin, plays a crucial role in facilitating cell adhesion. However, concerns regarding health risks associated with human and animal-derived proteins have prompted a reevaluation of serum's role in these applications.

Despite the potential of agents like polylysine to induce cell adhesion through charge-mediated interactions, they may not fully support essential cellular attributes such as well-organized spreading, actin cytoskeleton formation, or the development of focal adhesions. Consequently, the pursuit of serum elimination may lead to the exploration of synthetic peptides engineered to enhance cell attachment on scaffold surfaces (LeBaron and Athanasiou 2000). These synthetic peptides, tailored to interact with cell surface adhesion molecules like integrins and proteoglycans, are currently in development and are poised to revolutionize cartilage tissue engineering (LeBaron and Athanasiou 2000; Dee and Andersen 1998; Pierschbacher and Ruoslahti 1984).

Articular cartilage, intricately fashioned to withstand biomechanical strains, is constantly subjected to complex stress dynamics. Embedded within this matrix are chondrocytes, the cellular constituents, which are continually exposed to substantial mechanical forces, potentially influencing the synthesis and organization of cartilage. Alongside an array of cell types, these chondrocytes forge direct connections with their surroundings through focal adhesions. These adhesive structures, discrete regions of the cell's plasma membrane binding to extracellular elements, effectively anchor the cell within its microenvironment. Beyond their role in providing structural support and linkage between cells and the extracellular matrix, focal adhesions serve as conduits for signal transduction, transmitting external stimuli. Despite the intricate nature of mechanotransductional pathways, it is evident that mechanical cues wield profound influence over the synthesis and sculpting of articular cartilage.

In the realm of tissue engineering studies targeting articular cartilage, externally applied mechanical forces emerge as pivotal. Notably, within polyglycolic acid scaffolds seeded with bovine chondrocytes, subjected to either turbulent flow or rotational settings, there is a conspicuous enhancement in the deposition of a mechanically superior matrix, contrasting with static conditions. This contrast is underscored by the findings of several earlier studies demonstrating a reduction in proteoglycan synthesis under static conditions (Gray et al. 1988; Hunziker 2002; Jones and Klamfeldt 1982; Sah et al. 1989).

Moreover, mechanical stimuli exert a discernible impact on proteoglycan synthesis within native cartilage, where increased load magnitude or prolonged loading durations correlate with diminished proteoglycan synthesis in bovine articular cartilage disks. Consequently, the metabolic activity of chondrocytes appears significantly modulated by varying mechanical parameters, including load duration and magnitude. Interestingly, cellular metabolic activity demonstrates distinct responses to diverse loading regimens. Specifically, cyclic loading over short durations impedes cell division while concurrently fostering the synthesis of sulfated glycosaminoglycans and other vital components (Palmoski and Brandt 1984; van Kampen et al. 1985; Veldhuijzen et al. 1987).

A particular study highlighted the significant impact of *in vivo* applied loading on cartilage synthesis. It involved implanting small titanium chambers, one equipped with a mobile piston for applying external compressive loads, into the tibiae of rats. These chambers were allowed to host infiltrated host mesenchymal cells for a duration of 3 weeks. Subsequently, for a period of 7 weeks, twice daily, 20 cycles of load stimuli were administered to the mesenchymal tissue within the load chamber. Upon histological examination, both chambers were found to contain bone tissue; however, in the load chamber, cartilage formation was observed adjacent to the piston, while the control chamber, devoid of loading, exhibited bone tissue exclusively. This suggests that cyclic loading induced differentiation of the mesenchymal tissue into cartilage. Therefore, the application of mechanical stimuli, whether through hydrostatic pressure, fluid flow-induced shear, or interaction with solid objects, generally facilitates the expression of cartilage molecules and the synthesis of cartilaginous material. Consequently, when designing bioreactors aimed at fostering cartilage genesis *in vitro*, careful consideration should be given to the method of conveying stimuli (Tagil and Aspenberg 1999).

2.1 The Subsequent Text Will Delve into Selected Natural Materials Utilized for Scaffolds. Protein-Based Polymers

Collagen

Collagen matrices, also referred to as collagen sponges, have typically been utilized not only as scaffolds but also as carriers for various cell types. These include allogeneic chondrocytes or mesenchymal stem cells in rabbits, embryonic chondrocytes in chickens, or a combination of cells and growth factors. Although research has demonstrated their ability to enhance the natural healing response in osteochondral defects in rabbits, their clinical use in human patients remains limited, except in cosmetic facial surgery (Frenkel et al. 1997; Im et al. 2001; Kawamura et al. 1998; Perka et al. 2000a; Speer et al. 1979; Toolan et al. 1996).

It's noteworthy that in all these studies, the defects were full-thickness osteochondral, suggesting that observed repairs likely resulted from a combination of the intervention and the natural healing response from blood-borne cells and signaling substances from the subchondral bone space. Studies examining the fate of matrix-borne cells have indicated that they constitute only a small portion of the final repair tissue population, with blood and bone marrow-derived cells being predominant (Grande et al. 1989).

Despite promising outcomes, biocompatibility concerns are often overlooked, with adverse reactions sometimes underestimated. Our laboratory's systematic analysis has revealed significant biocompatibility issues. However, collagen is a natural bodily constituent, providing a natural adhesion surface for cells and carrying biological information for their activity. Additionally, collagen degradation products are non-toxic. Therefore, efforts to enhance its biocompatibility, potentially through minor structural modifications, are warranted given the numerous available possibilities.

Gelatin

Gelatin, essentially denatured collagen, has not seen widespread use as scaffolding structures to enhance spontaneous repair in articular cartilage defects. Even when utilized as carriers, such as for mesenchymal stem cells, its investigation has been limited. Yet, it has garnered interest as a substrate for fostering chondrocyte proliferation, both within controlled laboratory environments and within living organisms. When exposed to vascularized tissue compartments like bone and bone marrow spaces, gelatin is expected to provoke significant inflammatory responses and foreign body giant cell reactions, especially if commercial products like Gelfoam® are employed. It's important to note that Gelfoam® is primarily designed for use as a hemostatic surgical sponge rather than as a matrix for carrying cells (Pesakova et al. 1990; Ponticiello et al. 2000).

Fibrin

Fibrinogen and its polymerized form, fibrin, are natural components of the intravascular space, i.e., the blood, representing the major extracellular component of this 'fluid' connective tissue. Fibrinogen exists physiologically in a liquid state within the bloodstream, polymerizing only in instances of vascular lesioning or other pathological situations to form a three-dimensional solid matrix (fibrin) preventing blood loss into the extravascular space. While fibrin lacks mechanical function, it plays an important inductive role by facilitating and promoting tissue healing activities within the extravascular space. Fibrin is proinflammatory and induces its own degradation and substitution by cellular components of the extravascular tissue spaces. Its degradation products are physiological and non-toxic (Brandstedt et al. 1980; van Hinsbergh et al. 2001).

Fibrin has been extensively used not only as a naked scaffolding material but also as a carrier for cells and growth factors, available on both an autologous and allogeneic basis. It regulates spontaneous repair activities within full-thickness articular cartilage defects, effectively utilized in surgical strategies without active biologics.

Spontaneous healing responses are limited to defects falling within a critical size range.

Exogenous fibrin clots, like collagen matrices, promote the spreading of endogenous blood over a larger defect volume than would otherwise be occupied by the spontaneously developing hematoma, maximizing natural repair activities. Incorporating chondrocytes into exogenous fibrin clots, both *in vitro* and *in vivo*, enhances the healing response. Growth factors further enhance the healing response, demonstrated in equine and canine models for articular cartilage repair and meniscus repair in canines. Immunological reactions to exogenous fibrin have been observed in some animal studies (Arnoczky et al. 1988; Fortier et al. 1997, 1999; Haisch et al. 2000; Kawabe and Yoshinao 1991; Nixon et al. 1999).

Some researchers have used fibrin glues to enhance articular cartilage repair. However, due to their exceedingly high (unphysiological) concentrations and protein densities, they impede rather than facilitate cell invasion, not supporting a healing response (Brittberg et al. 1997; Henning et al. 1990; Kaplonyi et al. 1984, 1988; Schlag and Redl 1988; van Susante et al. 1999).

Despite extensive experimental research with fibrin and some clinical experience in veterinary medicine, its widespread use in human patients, except for meniscal repair, has been limited. Clinicians seem satisfied with results elicited by the spontaneous healing response, considering the improvement from exogenous fibrin in animal experiments insufficiently dramatic or convincing to warrant a more elaborate approach in humans (Henning et al. 1990).

2.2 *Carbohydrate Based Ones*

Agarose

Agarose, also known as sepharose, is a polysaccharide derived from certain Asian seaweeds, containing L- and D-galactose residues. Its significance lies in serving as a matrix for culturing chondrocytes, allowing them to express their typical gene activities and secrete key extracellular components. Since then, it has been utilized in numerous foundational *in-vitro* studies exploring chondrocyte autoregulatory signaling mechanisms, differentiation from mesenchymal stem cells, and extracellular matrix biomechanics and calcification (Aulthouse et al. 1992; Buschmann et al. 1992; Hunter et al. 1993; Inoue et al. 1997; Mauck et al. 2000; Skantze et al. 1985; Thompson et al. 1985).

Agarose's inertness has made it valuable for studying chondrocyte physiological activities *in vitro*. However, its poor biodegradability, lacking appropriate enzymatic degradation systems in mammalian tissues, diminishes its attractiveness as a matrix candidate for *in vivo* cartilage repair studies. Some research showed that agarose, when implanted alone within rabbit knee-joint osteochondral defects, impedes spontaneous repair processes, likely due to extensive foreign body giant cell reactions within the bony compartment. This inhibitory response was partly alleviated by incorporating allogeneic chondrocytes into the agarose matrix (Rahfoth et al. 1998; Verbruggen et al. 1990).

Due to the lack of extensive experimental animal data regarding agarose's utilization in articular cartilage repair and its limited potential stemming from inadequate biodegradability, it's understandable that this matrix hasn't been employed in human patients.

Alginate

Alginate, a gelatinous carbohydrate derived from brown algae through isolation and purification, undergoes polymerization in the presence of calcium ions. The resulting meshwork's stiffness is influenced by its guluronic acid content. It serves as a matrix favoring chondrogenesis and maintaining chondrocytic phenotypicity in three-dimensional culturing systems. Alginate has also been instrumental in fundamental studies, particularly concerning the temporal formation of cartilage matrix, as it readily depolymerizes (in the absence of calcium ions) and promptly repolymerizes (Wong et al. 2001) (Atala et al. 1993).

Injectable forms of alginate, when seeded with chondrocytes, are used in treating vesicoureteral reflux. Within three-dimensional alginate cultures, dedifferentiated chondrocytes can redifferentiate with appropriate nutrient supply. Similarly, bone marrow-derived mesenchymal stem cells can differentiate into chondrocytes when seeded within this matrix under suitable nutritional and stimulatory conditions, both *in vitro* and *in vivo*. Moreover, adult human chondrocytes cultured within alginate synthesize cartilage matrix components resembling those produced in the native tissue environment (Atala et al. 1994; Bonaventure et al. 1994; Diduch et al. 2000; Hauselmann et al. 1992, 1996; Fragonas et al. 2000).

Despite promising indications *in vitro*, *in vivo* applications of alginate matrices have yielded disappointing results. When implanted alone, alginate inhibits spontaneous repair responses. Similarly, alginate-based cartilage tissue engineered *in vitro* induces severe foreign body giant cell reactions and immunological responses, besides undergoing degradation when implanted within full-thickness defects in experimental animals. Consequently, alginate matrices have not been utilized in human patients for articular cartilage repair (Diduch et al. 2000; Fragonas et al. 2000; Paige et al. 1996; Perka et al. 2000b).

Hyaluronan

Hyaluronan, a physiological component of the articular cartilage matrix, forms macromolecules of tremendous length and molecular weight, which are perfectly biocompatible and biodegradable. In theory, unmodified hyaluronan would be an ideal matrix to support articular cartilage repair. However, to achieve the physicochemical properties and structural organization necessary for this purpose, it is usually cross-linked by esterification or other means, compromising its biocompatibility (Barbucci et al. 2000; Goa and Benfield 1994).

Matrices composed of hyaluronan alone have not been applied alone to enhance spontaneous repair responses, but they have been frequently used as carriers for chondrocytes or bone-marrow-derived mesenchymal stem cells. Hyaluronan matrices loaded with such cells have been shown to elicit the deposition of a cartilage-like extracellular matrix *in vitro*, and there is some evidence of this occurring also *in vivo*. Such constructs have likewise been shown to support cell differentiation processes. However, the hyaluronan degradation products yielded may lead to chondrolysis under certain conditions (Solchaga et al. 1999; Brun et al. 1999; Knudson et al. 2000; Kujawa et al. 1986; Maleski and Knudson 1996a; Maleski and Knudson 1996b; Radice et al. 2000; Solchaga et al. 2000).

In an unmodified form, hyaluronan has proved to be an unsuitable matrix. Only modified (cross- linked) versions have found a use in articular cartilage repair, and these are poorly biocompatible and chondrolytic. Hence, until further chemical modifications can be elicited to produce a more functionalized derivative, this type of matrix is unlikely to be used in human articular cartilage repair (Bulpitt and Aeschlimann 1999).

Chisotan

Chitosan, also known as polyglusam, is a copolymer of glucosamine and N-acetylglucosamine, capable of being cross-linked with polyanionic chondroitin sulfate to form a hydrogel. *In-vitro* studies have presented evidence of its potential as a matrix to facilitate articular cartilage repair, as it efficiently supports chondrogenic activities and promotes the *in-vitro* expression of cartilage extracellular matrix proteins by human chondrocytes. Moreover, it can serve as a carrier for growth factors. Despite not yet being employed as a matrix in animal or human articular cartilage repair studies, its sole clinical application thus far has been in preventing post-surgical adhesions. However, with growing awareness among investigators, it is likely to be embraced more enthusiastically in the future due to its excellent biodegradability (Suh and Matthew 2000; Costain et al. 1997; Lahiji et al. 2000; Mattioli-Belmonte et al. 1999; Sechriest et al. 2000; Senkoylu et al. 2001).

2.3 *Artifical Materials*

Dacron and Teflon

Meshes composed of dacron (polyethylene terephthalates) or teflon (polytetrafluoroethylene), akin to those made of carbon fibers, have been utilized in experimental animal studies to enhance and spatially expand the spontaneous repair reaction within full-thickness articular cartilage defects. In experiments involving rabbits, these matrices have resulted in a significant increase in the volume of the lesion void filled with repair tissue. However, the composition of this tissue varied greatly, ranging from highly vascularized fibrous connective tissue to fibrocartilage. Additionally, these materials have been successfully used to replace meniscal tissue in rabbits (Berman and Bronstein 1990; Hanff et al. 1990; Messner and Gillquist 1993a; Messner and Gillquist 1993b).

In human patients, there is a study that discusses the application of polyurethane-coated teflon meshes, leading to pain relief and improved joint functionality falling within the typical range (70–80%). However, a higher incidence of joint stiffness was observed in this patient population (Defrere and Franckart 1992).

When utilized as suturing materials, artificial ligaments, or membranes for abdominal surgery, dacron and teflon are known to provoke severe foreign body giant cell reactions and inflammatory responses. Such reactions would also be anticipated in full-thickness osteochondral defects. This could be a significant factor contributing to the limited pursuit of testing these matrices by scientists. Unless novel and more functionalized derivatives are developed in the future, these materials are unlikely to garner much attention.

Several other experimental matrices are currently in development or have been available for some time, but they cannot be recommended for veterinary or human use until certain serious issues are addressed.

2.4 *Within/Between Classes*

Matrix Combinations and Chemical Modification

The potential for chemical modification of the described basic matrices is virtually limitless. This includes combinations of multiple matrix types, the formation of copolymers, and cross-linking with various substrates to alter three-dimensional organization and geometry. One commonly employed approach involves cross-linking matrices, such as those based on collagen. Additionally, the cross-linking or functionalization of hyaluronan-based matrices has been proposed to enhance their effectiveness in articular cartilage repair. Similarly, promising outcomes have been achieved through the formation of copolymers between different chemical species, such as collagen and alginate or polyurethane and polylactic acid. Another effective strategy is the functionalization of a matrix by incorporating agents to promote cell adhesion or growth factors, turning it into a delivery system for improved performance in articular cartilage repair. Moreover, geometric modifications, such as producing sponge or foam-like matrices, have been shown to enhance the differentiation potential of chondroprogenitor cells and the metabolic activity of mature chondrocytes, in that order (Solchaga et al. 2000; Bulpitt and Aeschlimann 1999; Ahsan et al. 1999; Chvapil 1977; de Chalain et al. 1999; Lohmann et al. 2000a; Lohmann et al. 2000b; Riesle et al. 1998; Veth et al. 1986; Woo Lee et al. 2002).

3 The Use of Cells

3.1 Cells in Suspension

For quite some time, the concept of utilizing transplanted cells for articular cartilage repair has been present. The aim is to furnish the defect site with a homogeneous population of chondrocytes capable of generating a robust cartilage matrix.

Chondrocytes, along with chondroblasts, were initially utilized for transplantation, always embedded within a matrix. The transplantation of autologous chondrocytes suspended at low densities in culture medium was suggested as an empirical approach. However, there is no experimental evidence to support this approach, as dedifferentiated chondrocytes have not shown redifferentiation in two- dimensional culturing systems or aqueous suspension, except under high-density micromass conditions- only within a matrix. The use of dedifferentiated chondrocytes is necessary for *in-vitro* expansion to obtain the high numbers needed to fill large defect volumes (Grande et al. 1989; Benya and Shaffer 1982; Bradham et al. 1995; Brittberg et al. 1994b; Grande and Pitman 1988; Grande et al. 1987; Helbing 1982; Itay et al. 1987; Johnstone et al. 1998a; Katsube et al. 2000; Pitman et al. 1989; Solursh 1991).

Challenges arise from the application of this methodology in animal models, leading to the formation of fibrocartilage-like repair tissue. However, close examination reveals problematic aspects, including the dominance of the periosteal flap tissue within the defect, obscuring the true origin of the repair tissue. Studies have shown that transplanted autologous chondrocytes contribute minimally to the repair response, if at all. The detachment of periosteal flaps within days post-surgery suggests that neither this tissue nor the transplanted cells significantly contribute to the repair response (Breinan et al. 1997).

Another complication is the possibility of a triple chondrogenic system, with repair responses potentially originating from transplanted cells, the cambial layer of the periosteal flap, or spontaneous healing from the subchondral bone-tissue spaces. The exact source remains unclear, but spontaneous responses mediated by blood clot formation appear most likely (Gilbert 1998; Newman 1998).

Despite these uncertainties, autologous chondrocyte transplantation has been introduced into clinical practice, yielding reportedly good to excellent results in terms of pain relief and improved joint functionality in many cases. However, no prospective, double-blinded clinical trials have compared this methodology with alternative, established treatment protocols (Brittberg et al. 1994b).

3.2 Cells in a Matrix

Various pools of chondroprogenitor cells exist within adult mammalian organisms, capable of differentiating into chondrocytes and forming cartilage tissue for transplantation. These include cells within the cambial layers of the perichondrium and periosteum, adult cartilage (chondrocytes), bone marrow stroma (mesenchymal stem cells), and the synovial membrane. Fetal and embryonic precursor cells, such as fetal stem cells and chondroblasts, are also viable options. Over the past decades, numerous empirical approaches have been developed to assess the efficacy of these cell pools in inducing repair responses within articular cartilage defects. These approaches involve transplanting cells within a matrix- carrier system and testing them in full-thickness articular cartilage defects, superimposing this method upon the spontaneous healing response from subchondral bone-tissue spaces. Additionally, it's crucial to acknowledge that blood seeping into the defect void carries various cell types and signaling molecules, likely playing a role in triggering the chondrogenic switch.

3.3 Specific Cells

Chondrocytes and Chondroblasts

Within a collagenous matrix, chondrocytes sourced from both autogenic and allogeneic origins, along with alternative matrices like fibrin, carbon fibers, and agarose, have undergone testing. Some studies have explored the use of fetal or embryonic chondrocytes or chondroblasts. Across these investigations, researchers have observed an enhancement in the healing process, resulting in predominantly fibrocartilaginous repair tissue. The introduction of transplanted cells has been noted to augment the spontaneous repair response, albeit with lingering concerns regarding the long-term stability and immunoreactivity of the resultant tissue. Despite these findings, such cell-matrix constructs have yet to be applied clinically in human patients (Frenkel et al. 1997; Kawamura et al. 1998; Perka et al. 2000a; Kawabe and Yoshinao 1991; Rahfoth et al. 1998; Benya and Shaffer 1982; Helbing 1982; Chu et al. 1995a; Hemmen et al. 1991; Robinson et al. 1989; Sams et al. 1995; Silas 1998; Wakitani et al. 1989).

Since chondrocytes from surrounding cartilage do not migrate into the defect area, both primary and passaged cells have been seeded onto scaffolds for cartilage repair. Direct use of chondrocytes from donated cartilage is limited due to overharvesting issues. Serial passage of chondrocytes as monolayers provides a means to acquire a large number of cells that generally retain the chondrocyte phenotype.

These cells can then be easily seeded onto three-dimensional material to achieve a homogeneous distribution of chondrocytes and cartilage extracellular matrix material throughout the scaffold dimension (Dunkelman et al. 1995).

Perichondrial/Periosteal Cells

Over the years, scientists have recognized the ability of certain cells found in the perichondrium and periosteum to develop into cartilage-forming cells. These cells, located in the cambial layer of these tissues, have been collected from both the individual's own body (autogenic) and from donors (allogeneic) for use in tissue engineering projects. After rigorous testing in lab settings to confirm their ability to produce cartilage, these cells have been incorporated into various materials, including those made of polylactic acid. Notably, cells from the individual's own perichondrium tend to perform better than those from donors (Chu et al. 1995a; Chu et al. 1995b; Chu et al. 1997; De Bari et al. 2001a; Dounchis et al. 2000a; Dounchis et al. 2000b; Dounchis et al. 1997).

When these cell-matrix combinations are implanted into animals with cartilage defects for research purposes, they significantly improve the healing process of the cartilage. However, it appears that these perichondrial and periosteal cells do not survive for long periods in the repaired tissue, and their numbers decrease over time. While the body can recruit some replacement cells from bone marrow and other bone tissues, this doesn't lead to stable long-term repair (Ostrander et al. 2001).

Despite the potential benefits of using these precursor cells in human patients due to their ability to grow and differentiate into cartilage cells regardless of age, they have not been widely used in clinical practice. This may be due to inconsistent results observed in animal studies, the instability of repair tissue over time, and challenges related to using cells from donors. Additionally, the complexity of the procedure, which involves multiple surgeries and complex cell isolation processes, as well as the lack of a simple, commercially available solution, may also hinder their clinical application (De Bari et al. 2001a; Ostrander et al. 2001).

Bone Marrow Stroma

The bone-marrow stroma harbors multipotential precursor cells. They garnered significant interest with their multipotent lineage capabilities. These cells hold immense value for tissue engineering due to their pluripotency. However, their numbers decline and their potential to proliferate and differentiate deteriorate with age in humans (Ostrander et al. 2001; Cancedda et al. 2000; Caplan 1990, 1991; Johnstone and Yoo 1999; Pittenger et al. 1999, 2000). Researchers have explored the application of mesenchymal stem cells from bone-marrow stroma to induce repair in articular cartilage across various animal models, often in conjunction with different matrices like collagen-type-I gels or modified hyaluronan. While *in-vitro* studies have tracked the differentiation of these stem cells and assessed the impact of growth factors, *in-vivo* experiments have largely relied on natural signaling substances. Some studies, like that by Lennon et al., have incorporated fibroblasts into their systems to bolster tissue differentiation (Solchaga et al. 1999; Im et al. 2001; Johnstone et al. 1998a; Pittenger et al. 2000; Quarto et al. 1995; Dennis et al. 1999a;

Gimble et al. 1994; Kadiyala et al. 1997; Lennon et al. 2000; Mackay et al. 1998; Martin et al. 1998; Rogers et al. 1995; Wakitani et al. 1994).

The outcomes achieved with bone-marrow stromal cells have mirrored those seen with other cell- matrix systems, with the resulting repair tissue predominantly fibrocartilaginous in nature, albeit with varying quality and durability. One advantage of using such cells is their relative immunological tolerance, making allogeneic cells a feasible commercial option. This would eliminate the need for bone- marrow tissue biopsies, potentially avoiding secondary pathologies and logistical challenges associated with transport and sterility issues.

While this type of cell-matrix construct has been applied in humans using autologous bone- marrow stromal cells in one instance, its broader clinical application awaits further exploration and refinement (Quarto et al. 2001).

Synovial Cells

Articular cartilage chondrocytes and synovial cells are closely related, originating from a common precursor pool, as revealed by studies on the developmental trajectory of synovial joints. This functional relationship persists from fetal development through adulthood. Interestingly, synovial cells have displayed significant chondrogenic potential under various pathological conditions, including within chondromatous tumors and chondroid osteophytes within the synovial space (Apte and Athanasou 1992; Bandara et al. 1991; Bertoni et al. 1991; Elford et al. 1992; Iwata et al. 1993; Miyamoto et al. 2000; O'Connell 2000; Pacifici et al. 2000; Sim et al. 1977; Szendroi and Deodhar 2000; Van den Berg et al. 1993).

Additionally, research has shown that chondroprogenitor cells derived from synovial tissue maintain their proliferative capacity and ability to differentiate into chondrocytes regardless of age (Van den Berg et al. 1993; De Bari et al. 2001b).

While chondroprogenitor cells from synovial tissue have not been directly isolated for use in articular cartilage repair studies, experiments on adult rabbits, miniature pigs, and goats have indirectly demonstrated the recruitment of synovial cells for repairing small partial-thickness defects. This recruitment occurs when an appropriate matrix containing growth factors from the TGF-alpha superfamily is provided to stimulate their proliferation, migration, and differentiation (Nishimura et al. 1999).

For larger partial-thickness defects, adjustments to the methodology would be necessary due to the prolonged healing time required. In the case of full-thickness defects, precautions must be taken to prevent osseous upgrowth into the cartilaginous compartment, as observed in previous studies (Hunziker 1999, 2002; Hunziker et al. 2001a; Hunziker and Driesang 2001).

Conclusive findings are notably absent in the majority of studies exploring different cell types combined with various matrices. This deficiency often arises from inadequate study designs that overlook crucial controls. Leaving defects untreated is insufficient for meaningful conclusions.

Progenitor Cells

Experiments exploring the use of mesenchymal-derived progenitor cells for repairing osteochondral defects are yielding encouraging results. These cells, sourced from humans, mice, and rabbits, can be cultured and expanded *in vitro* while retaining their capacity to differentiate and generate cartilage and bone material. Understanding the precise molecular mechanisms governing differentiation pathways becomes increasingly significant in light of this phenomenon (Gray et al. 1988; Hunziker et al. 2001b; Dennis et al. 1999b; Dulbecco et al. 1959).

In a study conducted by Grande et al., mesenchymal stem cells were cultured within polyglycolic acid polymer matrices and then implanted into full-thickness defects in the articular cartilage of rabbits. Histological examination after 12 weeks revealed promising outcomes. Defects containing polyglycolic acid polymer matrices seeded with mesenchymal stem cells exhibited tissue closely resembling normal cartilage. The repair tissue's surface layer displayed cartilaginous material comparable in thickness to normal articular cartilage, with the defect area also containing subchondral bone. These findings suggest successful differentiation of mesenchymal stem cells into osteoblasts and chondrocytes within the matrix, effectively reconstructing the three-dimensional structure of osteochondral tissue (Johnstone et al. 1998b).

As progenitor cells continue to be isolated and characterized, their potential as an increasingly valuable resource for engineering articular cartilage should be fully realized.

Among the growing prospects in orthopedic research is the utilization of cartilage from regions with inherent regenerative capabilities, such as the mandibular condylar cartilage. This specific type of cartilage houses a layer of prechondroblastic mesenchymal stem cells, potentially facilitating chondrogenesis and the development of hyaline cartilage (Grande et al. 1995).

Another avenue of exploration involves embryonic stem cells, which exhibit pluripotency, capable of generating diverse cell types under the influence of specific differentiation factors. Namba et al.'s study, for instance, showcased the remarkable reparative potential of young cells by inducing superficial defects in fetal cartilage, which healed without fibrous scar tissue formation within 28 days. However, this reparative ability diminishes as progenitors differentiate and age (Brittberg 1999).

In another study by Martin et al., chick embryo bone marrow stromal cells, when seeded onto polyglycolic scaffolds supplemented with fibroblast growth factor-2, successfully differentiated into epiphyseal cartilage. The increasing number of reports emphasizing progenitor cells as a potential cell source for orthopedic research underscores the promising trajectory of further exploration (Martin et al. 1998).

Nevertheless, several considerations loom large, including the age of the cells, the regulation of their differentiation pathways, and ethical concerns surrounding the sourcing and collection methods of embryonic cells. These pressing issues continue to be at the forefront as researchers endeavor to deepen our understanding of these cells and their practical application in orthopedic research.

3.4 Seeding Density

In the approach to tissue engineering of articular cartilage, the necessity of cells has become increasingly apparent, particularly considering its remarkably low cellular density. These cells can either be applied onto scaffolds for subsequent implantation into cartilage defects or integrated into matrices to produce cartilage constructs *ex vivo*. An essential aspect of this process is determining the initial cell density required per scaffold or defect volume. However, despite its significance, systematic studies investigating the effects of initial cell density on cartilage formation and identifying the optimal cell numbers have been lacking. Notably, both the quantity and integrity of the construct increase as the initial cell density rises (Namba et al. 1998).

3.5 Media Composition

In the quest to generate fully functional articular cartilage *ex vivo*, the use of appropriate culture media is equally important as cell density. The growth medium for chondrocyte cultures should contain essential components such as inorganic salts, amino acids, vitamins, and a buffering agent like bicarbonate buffer or HEPES. Additionally, D-glucose and sodium pyruvate serve as energy sources, while glutamine acts as both an energy and carbon source. Various supplements beneficial for chondrocyte culture are also incorporated.

Serum, typically constituting 10% of the growth medium volume, is included to provide growth and attachment factors. To support cartilage synthesis, supplements like ascorbic acid, additional proline, and a blend of nonessential amino acids are included. Ascorbic acid, a water-soluble vitamin (vitamin C), aids in the folding and stabilization of newly synthesized collagen chains, as well as the formation of intra- and intermolecular collagen crosslinks. The regular addition of 50 microg/mL fresh ascorbic acid every 72 h to cell-seeded scaffolds is essential for effective cartilage synthesis.

Dulbecco's Modified Eagle Medium (DMEM) is a suitable growth medium for chondrocyte cultures and cartilage synthesis. A recommended cartilage-specific media formulation may comprise DMEM supplemented with 10% fetal bovine serum, 2 mM L-glutamine, nonessential amino acids, 50 pg/mL proline, and 1 mM sodium pyruvate (Dunkelman et al. 1995; Puelacher et al. 1994).

4 Growth and Differentiation Factors

Influencing cellular functions within articular cartilage, bioactive molecules such as fibroblast growth factors (FGF) and bone morphogenetic proteins (BMP) play vital roles. They impact processes like chondrogenesis and cell metabolism. However, excessive doses of TGF-beta, for instance, can be harmful.

A study conducted by Athanasiou et al. demonstrated the potential of utilizing biodegradable scaffolds containing human recombinant TGF-beta1 to enhance the quality of repair cartilage in goats with 7-mm defects. In another investigation, the application of recombinant human BMP-2 with a collagen carrier showed promising effects on the healing of full-thickness osteochondral defects in rabbits. This approach notably accelerated the formation of new subchondral bone and improved cartilage histological appearance (Iwata et al. 1993; Athanasiou et al. 1997; Sellers et al. 1997).

Advancing our understanding of the effects of growth factors and cytokines allows for combining these factors within scaffolds to modulate various cell functions and enhance articular cartilage synthesis.

Biodegradable scaffolds engineered to release bioactive agents at controlled doses and temporal increments are expected to play a crucial role in advancing cartilage repair techniques in the near future.

5 Future Possibilities

The eficient use of scaffolds in articular cartilage repair hinges on careful consideration of biomechanical factors. Scaffold materials, including polylactic acid and polyglycolic acid, along with their copolymers, play a significant role in determining the scaffold's functionality across different environments. By regulating mechanical properties such as porosity, the rate of scaffold degradation can be controlled through hydrolysis. Biomechanical design, incorporating both elastic and viscoelastic properties, is crucial for scaffold biocompatibility and compressive stiffness modulation, impacting mechanotransduction and cell apoptosis. Viscoelastic properties also influence fluid flow dynamics within the scaffold, affecting mechanotransduction and cell apoptosis induced by hydrostatic pressure and shear forces. Scaffold permeability is equally important, regulating fluid flow and impacting pH levels, degradation rates, and mechanotransduction processes (Athanasiou et al. 1998; Therin et al. 1992; Vert et al. 1994).

Creating successful tissue-engineered articular cartilage relies heavily on scaffolds, which play a pivotal role. An ideal scaffold design encompasses various factors such as biocompatibility, bioabsorbability, and specific material properties like porosity, degradation rate, and mass loss. Surface modifications can enhance scaffold effectiveness by promoting cell adhesion through chemically reactive moieties. Engineered mechanical characteristics including stiffness, compressibility, and permeability are crucial for achieving successful tissue engineering outcomes. Matching scaffold properties with those of native tissue enhances compatibility with the host physiology, while ensuring permeability and porosity fosters cellular mechanotransduction, essential for efficient tissue repair (LeBaron and Athanasiou 2000).

6 Conclusion

Cartilage regeneration remains one of the most complex and pressing challenges in orthopaedics and tissue engineering. This chapter has explored the multifactorial strategies underpinning current research, from the fundamental biology of cartilage cells and extracellular matrix dynamics to innovative therapeutic interventions including biomaterials, scaffolds, bioreactors, and progenitor cell applications. Despite promising advances in both basic science and translational approaches, consistent and durable clinical outcomes remain elusive.

References

Ahsan T, Lottman LM, Harwood F, Amiel D, Sah RL. Integrative cartilage repair: inhibition by betaaminopropionitrile. J Orthopaed Res. 1999;17:850–7.

Andrades J, Han B, Becerra J, Sorgente N, Hall F, Nimni M. A recombinant human TGF-beta1 fusion protein with collagen-binding domain promotes migration, growth, and differentiation of bone-marrow mesenchymal cells. Exp Cell Res. 1999;250:485–98.

Angele P, Kujat R, Nerlich M, Yoo J, Goldberg V, Johnstone B. Engineering of osteochondral tissue with bone-marrow mesenchymal progenitor cells in a derivatized hyaluronan-gelatin composite sponge. Tissue Eng. 1999;5:545–54.

Apte S, Athanasou N. An immunohistological study of cartilage and synovium in primary synovial chondromatosis. J Pathol. 1992;166:277–81.

Arnoczky SP, Warren RF, Spivak JM. Meniscal repair using an exogenous fibrin clot. An experimental study in dogs. J Bone Joint Surg Am. 1988;70:1209–17.

Atala A, Cima L, Kim W, Paige K, Vacanti J, Retik A, et al. Injectable alginate seeded with chondrocytes as a potential treatment for vesicoureteral reflux. J Urol. 1993;150:745–7.

Atala A, Kim W, Paige K, Vacanti C, Retik A. Endoscopic treatment of vesicoureteral reflux with a chondrocyte-alginate suspension. J Urol. 1994;152:641–3. discussion 4, 3; discussion 644.

Athanasiou KA, Niederauer GG, Agrawal CM. Sterilization, toxicity, biocompatibility, and clinical applications of polylactic acid/polyglycolic acid copolymers. Biomaterials. 1996;17:93–102.

Athanasiou KA, Korvick D, Schenck RC. Biodegradable implants for the treatment of osteochondral defects in a goat model. Tissue Eng. 1997;3:363–74.

Athanasiou KA, Schmitz JP, Agrawal CM. The effects of porosity on in vitro degradation of PLA-PGA implants used in repair of articular cartilage. Tissue Eng. 1998;4:5343–63.

Aulthouse A, Carubelli C, Dow T, Ziegelmayer C, Beck M. Influence of retinol on human chondrocytes in agarose culture. Anat Rec. 1992;232:52–9.

Bandara G, Georgescu H, Lin C, Evans C. Synovial activation of chondrocytes: evidence for complex cytokine interactions. Agents Actions. 1991;34:285–8.

Banfi A, Muraglia A, Dozin B, Mastrogiacomo M, Cancedda R, Quarto R. Proliferation kinetics and differentiation potential of ex vivo expanded human bone-marrow stromal cells: implications for their use in cell therapy. Exp Hematol. 2000;28:707–15.

Barbucci R, Magnani A, Rappuoli R, Lamponi S, Consumi M. Immobilisation of sulphated hyaluronan for improved biocompatibility. J Inorg Biochem. 2000;79:119–25.

Benya PD, Shaffer JD. Dedifferentiated chondrocytes reexpress the differentiated collagen phenotype when cultured in agarose gels. Cell. 1982;30:215–24.

Berman D, Bronstein S. Osteophytic reaction to a polytetrafluoroethylene temporomandibular joint implant. Report of a case. Oral Surg Oral Med Oral Pathol. 1990;69:20–3.

Bertoni F, Unni K, Beabout J, Sim F. Chondrosarcomas of the synovium. Cancer. 1991;67:155–62.

Bonaventure J, Kadhom N, Cohen-Solal L, Ng K, Bourguignon J, Lasselin C, et al. Reexpression of cartilage-specific genes by dedifferentiated human articular chondrocytes cultured in alginate beads. Exp Cell Res. 1994;212:97–104.

Bradham DM, Passaniti A, Horton WE. Mesenchymal cell chondrogenesis is stimulated by basement membrane matrix and inhibited by age-associated factors. Matrix Biol. 1995;14:561–71.

Brandstedt S, Rank F, Olson P. Wound healing and formation of granulation tissue in normal and defibrinogenated rabbits. An experimental model and histological study. Eur Surg Res. 1980;12:12–21.

Breinan H, Minas T, Hsu H, Nehrer S, Sledge C, Spector M. Effect of cultured autologous chondrocytes on repair of chondral defects in a canine model. J Bone Joint Surg. 1997;79A:1439–51.

Brittberg M. Autologous chondrocyte transplantation. Clin Orthop Relat Res. 1999;367:S147–S55.

Brittberg M, Lindahl A, Nilsson A. Treatment of deep cartilage defects in the knee with autologous chondrocyte transplantation. N Engl J Med. 1994a;331(2):389–894.

Brittberg M, Lindahl A, Nilsson A, Ohlsson C, Isaksson O, Peterson L. Treatment of deep cartilage defects in the knee with autologous chondrocyte transplantation. N Engl J Med. 1994b;331:879–95.

Brittberg M, Sjogren-Jansson E, Lindahl A, Peterson L. Influence of fibrin sealant (Tisseel) on osteochondral defect repair in the rabbit knee. Biomaterials. 1997;18:235–42.

Brun P, Abatangelo G, Radice M, Zacchi V, Guidolin D, Gordini DD, et al. Chondrocyte aggregation and reorganization into three-dimensional scaffolds. J Biomed Mater Res. 1999;46:337–46.

Buckwalter JA, Mankin HJ. Articular cartilage. Part I: tissue design and chondrocyte matrix interactions. J Bone Joint Surg. 1997;79A:600–11.

Bulpitt P, Aeschlimann D. New strategy for chemical modification of hyaluronic acid: preparation of functionalized derivatives and their use in the formation of novel biocompatible hydrogels. J Biomed Mater Res. 1999;47:152–69.

Buschmann M, Gluzband Y, Grodzinsky A, Kimura J, Hunziker E. Chondrocytes in agarose culture synthesize a mechanically functional extracellular matrix. J Orthop Res. 1992;10:745–58.

Cancedda R, Castagnola P, Cancedda F, Dozin B, Quarto R. Developmental control of chondrogenesis and osteogenesis. Int J Dev Biol. 2000;44:707–14.

Caplan A. Stem cell delivery vehicle. Biomaterials. 1990;11:44–6.

Caplan A. Mesenchymal stem cells. J Orthop Res. 1991;9:641–50.

Caplan AI, Elyaderani M, Mochizuki Y, et al. Principles of cartilage repair and regeneration. Clin Orthop. 1997;342:256–69.

Chu C, Monosov A, Amiel D. In situ assessment of cell viability within biodegradable polylactic acid polymer matrices. Biomaterials. 1995a;16:1381–4.

Chu CR, Coutts RD, Yoshioka M, Harwood FL, Monosov AZ, Amiel D. Articular cartilage repair using allogeneic perichondrocyte-seeded biodegradable porous polylactic acid (PLA): a tissue-engineering study. J Biomed Mater Res. 1995b;29:1147–54.

Chu CR, Dounchis JS, Yoshioka M, Sah RL, Coutts RD, Amiel D. Osteochondral repair using perichondrial cells: a 1-year study in rabbits. Clin Orthop. 1997;340:220–9.

Chu CR, Convery FR, Akeson WH, et al. Articular cartilage transplantation. Clinical results in the knee. Clin Orthop. 1999;360:159–68.

Chvapil M. Collagen sponge: theory and practice of medical applications. J Biomed Mater Res. 1977;11:721–41.
Costain D, Kennedy R, Ciona C, McAlister V, Lee T. Prevention of postsurgical adhesions with N,Ocarboxymethyl chitosan: examination of the most efficacious preparation and the effect of N, Ocarboxymethyl chitosan on postsurgical healing. Surgery. 1997;121:314–9.
De Bari C, Dell Accio F, Luyten FP. Human periosteum-derived cells maintain phenotypic stability and chondrogenic potential throughout expansion regardless of donor age. Arthritis Rheum. 2001a;44:85–95.
De Bari C, Dell Accio F, Tylzanowski P, Luyten FP. Multipotent mesenchymal stem cells from adult human synovial membrane. Arthritis Rheum. 2001b;44:1928–42.
de Chalain T, Phillips J, Hinek A. Bioengineering of elastic cartilage with aggregated porcine and human auricular chondrocytes and hydrogels containing alginate, collagen, and kappa-elastin. J Biomed Mater Res. 1999;44:280–8.
Dee KC, Andersen TT. Bizios R design and function of novel osteoblast-adhesive peptides for chemical modification of biomaterials. J Biomed Mater Res. 1998;40371-377:371–7.
Defrere J, Franckart A. Teflon/polyurethane arthroplasty of the knee: the first 2 years preliminary clinical experience in a new concept of artificial resurfacing of full thickness cartilage lesions of the knee. Acta Chir Belg. 1992;92:217–27.
Dennis JE, Merriam A, Awadallah A, Yoo JU, Johnstone B, Caplan AI. A quadripotential mesenchymal progenitor cell isolated from the marrow of an adult mouse. J Bone Miner Res. 1999a;14:700–9.
Dennis JE, Merriam A, Awadallah A, et al. A quadripotential mesenchymal progenitor cell isolated from the marrow of an adult mouse. J Bone Miner Res. 1999b;14:700–9.
Diduch D, Jordan L, Mierisch C, Balian G. Marrow stromal cells embedded in alginate for repair of osteochondral defects. Arthroscopy. 2000;16:571–7.
Dounchis J, Goomer R, Harwood F, Khatod M, Coutts R, Amiel D. Chondrogenic phenotype of perichondrium-derived chondroprogenitor cells is influenced by transforming growth factor-beta. J Orthop Res. 1997;15:803–7.
Dounchis J, Bae W, Chen A, Sah R, Coutts R, Amiel D. Cartilage repair with autogenic perichondrium cell and polylactic acid grafts. Clin Orthop. 2000a;377:248–64.
Dounchis J, Coutts R, Amiel D. Cartilage repair with autogenic perichondrium cell/polylactic acid grafts: a two-year study in rabbits. J Orthop Res. 2000b;18:512–5.
Dulbecco R, Freeman G, Plaque formation by the polyoma virus. Virology. 1959;8:396–7.
Dunkelman NS, Zimber MP, Le Baron RG, et al. Cartilage production by rabbit articular chondrocytes on polyglycolic acid scaffolds in a closed bioreactor system. Biotechnol Bioeng. 1995;46:299–305.
Elford PR, Graeber M, Ohtsu H, Aeberhard M, Legendre B, Wishart WL, et al. Induction of swelling, synovial hyperplasia and cartilage proteoglycan loss upon intra-articular injection of transforming growth factor-beta-2 in the rabbit. Cytokine. 1992;4:232–8.
Evans CH, Robbins PD. Genetically augmented tissue engineering of the musculoskeletal system. Clin Orthop Related Res. 1999;307:S410–S8.
Evans CH, Ghivizzani SC, Smith P, Shuler FD, Mi Z, Robbins PD. Using gene therapy to protect and restore cartilage. Clin Orthop Relat Res. 2000;00:S214–S9.
Fortier L, Nixon A, Mohammed H, Lust G. Altered biological activity of equine chondrocytes cultured in a three-dimensional fibrin matrix and supplemented with transforming growth factor beta-1. Am J Vet Res. 1997;58:66–70.
Fortier L, Lust G, Mohammed H, Nixon A. Coordinate upregulation of cartilage matrix synthesis in fibrin cultures supplemented with exogenous insulin-like growth factor-I. J Orthop Res. 1999;17:467–74.
Fragonas E, Valente M, Pozzi-Mucelli M, Toffanin R, Rizzo R, Silvestri F, et al. Articular cartilage repair in rabbits by using suspensions of allogenic chondrocytes in alginate. Biomaterials. 2000;21:795–801.

Freed LE, Marquis JC, Nohria A, et al. Neocartilage formation in vitro and in vivo using cells cultured on synthetic biodegradable polymers. J Biomed Mater Res. 1993;27:11–23.
Freed LE, Hollander AP, Martin I, et al. Chondrogenesis in a cell-polymer-bioreactor system. Exp Cell Res. 1998;240:58–65.
Frenkel S, Toolan B, Menche D, Pitman M, Pachence J. Chondrocyte transplantation using a collagen bilayer matrix for cartilage repair. J Bone Joint Surg. 1997;79B:831–6.
Gelse K, Jiang QJ, Aigner T, Ritter T, Wagner K, Poschl E, et al. Fibroblast-mediated delivery of growth factor complementary DNA into mouse joints induces chondrogenesis but avoids the disadvantages of direct viral gene transfer. Arthritis Rheum. 2001;44:1943–53.
Gilbert J. Current treatment options for the restoration of articular cartilage. Am J Knee Surg. 1998;11:42–6.
Gimble JM, Wanker F, Wang CS, Bass H, Wu X, Kelly K, et al. Regulation of bone-marrow stromal cell differentiation by cytokines whose receptors share the gp130 protein. J Cell Biochem. 1994;54:122–33.
Glowacki J. In vitro engineering of cartilage. J Rehabil Res Dev. 2000;37:171–7.
Goa KL, Benfield P. Hyaluronic acid—a review of its pharmacology and use as a surgical aid in ophthalmology, and its therapeutic potential in joint disease and wound healing. Drugs. 1994;47:536–66.
Goomer R, Deftos L, Terkeltaub R, Maris T, Lee M, Harwood F, et al. High-efficiency non-viral transfection of primary chondrocytes and perichondrial cells for ex-vivo gene therapy to repair articular cartilage defects. Osteoarthr Cart. 2001;9:248–56.
Grande DA, Pitman MI. The use of adhesives in chondrocyte transplantation surgery. Preliminary studies. Bull Hosp Jt Dis Orthop Inst. 1988;48:140–8.
Grande DA, Singh IJ, Pugh J. Healing of experimentally produced lesions in articular cartilage following chondrocyte transplantation. Anat Rec. 1987;218:142–8.
Grande DA, Pitman MI, Peterson L, Menche D, Klein M. The repair of experimentally produced defects in rabbit articular cartilage by autologous chondrocyte transplantation. J Orthop Res. 1989;7:208–18.
Grande DA, Southerland SS, Manji R, et al. Repair of articular cartilage defects using mesenchymal stem cells. Tissue Eng. 1995;1:345–53.
Gray ML, Pizzanelli AM, Grodzinsky AJ, et al. Mechanical and physiochemical determinants of the chondrocyte biosynthetic response. J Orthop Res. 1988;6:777–92.
Haisch A, Loch A, David J, Pruss A, Hansen R, Sittinger M. Preparation of a pure autologous biodegradable fibrin matrix for tissue engineering. Med Biol Eng Comput. 2000;38:686–9.
Hanff G, Sollerman C, Abrahamsson S, Lundborg G. Repair of osteochondral defects in the rabbit knee with Gore-Tex (expanded polytetrafluoroethylene). An experimental study. Scand J Plast Reconstr Surg Hand Surg. 1990;24:217–23.
Hauselmann H, Aydelotte M, Schumacher B, Kuettner K, Gitelis S, Thonar E. Synthesis and turnover of proteoglycans by human and bovine adult articular chondrocytes cultured in alginate beads. Matrix. 1992;12:116–29.
Hauselmann H, Masuda K, Hunziker E, Neidhart M, Mok S, Michel B, et al. Adult human chondrocytes cultured in alginate form a matrix similar to native human articular cartilage. Am J Phys. 1996;271:C742–52.
Helbing G. Transplantation of isolated chondrocytes in articular cartilage defects. Regeneration of adult hyaline cartilage with fetalchondrocytes. Fortschr Med. 1982;100:83–7.
Hemmen B, Archer CW, Bentley G. Repair of articular cartilage defects by carbon fibre plugs loaded with chondrocytes. In: Caterson B, Jared KJ, editors. Transactions of the 37th annual meeting, Orthopaedic Research Society. Park Ridge: Orthopaedic Research Society; 1991. p. 278.
Henning CE, Lynch MA, Yearout KM, Vequist SW, Stallbaumer RJ, Decker KA. Arthroscopic meniscal repair using an exogenous fibrin clot. Clin Orthop. 1990;252:64–72.
Hollinger JO, Winn S, Bonadio J. Options for tissue engineering to address challenges of the aging skeleton. Tissue Eng. 2000;6:341–50.

Hunter G, Holmyard D, Pritzker K. Calcification of chick vertebral chondrocytes grown in agarose gels: a biochemical and ultrastructural study. J Cell Sci. 1993;104:611–8.
Hunziker EB. Biologic repair of articular cartilage— defect models in experimental animals and matrix requirements. Clin Orthop Relat Res. 1999;367:S135–S46.
Hunziker EB. Growth-factor-induced healing of partial-thickness defects in adult articular cartilage. Osteoarthr Cart. 2001;9:22–32.
Hunziker EB. Clinically applicable animal models in tissue engineering. Lippincott; 2002.
Hunziker EB, Driesang IMK. Functional barrier principle of growth-factor-based articular cartilage repair. Clin Orthop Rel Res. 2001;391:S182–9.
Hunziker EB, Rosenberg LC. Repair of partial-thickness defects in articular cartilage: cell recruitment from the synovial membrane. J Bone Joint Surg. 1996;78:78A721–33.
Hunziker EB, Driesang IM, Morris EA. Chondrogenesis in cartilage repair is induced by members of the transforming growth factor-beta superfamily. Clin Orthop. 2001a;391:S171–81.
Hunziker EB, Driesang IM, Saager C. Structural barrier principle for growth factor-based articular cartilage repair. Clin Orthop. 2001b;391:S182–9.
Im G, Kim D, Shin J, Hyun C, Cho W. Repair of cartilage defect in the rabbit with cultured mesenchymal stem cells from bone-marrow. J Bone Joint Surg. 2001;83B:289–94.
Inoue H, Kondo J, Koike T, Shukunami C, Hiraki Y. Identification of an autocrine chondrocyte colonystimulating factor: chondromodulin-I stimulates the colony formation of growth plate chondrocytes in agarose culture. Biochem Biophys Res Commun. 1997;241:395–400.
Itay S, Abramovici A, Nevo Z. Use of cultured embryonal chick epiphyseal chondrocytes as grafts for defects in chick articular cartilage. Clin Orthop. 1987;220:284–303.
Iwata H, Ono S, Sato K, et al. Bone morphogenetic protein-induced muscle- and synovium-derived cartilage differentiation in vitro. Clin Orthop. 1993;296:295–300.
Jackson DW, Simon TM. Tissue engineering principles in orthopaedic surgery. Clin Orthop. 1999;367:S31–45.
Johnstone B, Yoo J. Autologous mesenchymal progenitor cells in articular cartilage repair. Clin Orthop. 1999;367:S156–62.
Johnstone B, Hering T, Caplan A, Goldberg V, Yoo J. In vitro chondrogenesis of bone-marrow-derived mesenchymal progenitor cells. Exp Cell Res. 1998a;238:265–72.
Johnstone B, Hering TM, Caplan AI, et al. In vitro chondrogenesis of bone marrowderived mesenchymal progenitor cells. Exp Cell Res. 1998b;238:265–72.
Jones IL, Klamfeldt A. Sandstrom T the effect of continuous mechanical pressure upon the turnover of articular cartilage proteoglycans in vitro. Clin Orthop. 1982;165:283–9.
Kadiyala S, Young R, Thiede M, Bruder S. Culture expanded canine mesenchymal stem cells possess osteochondrogenic potential in-vivo and in-vitro. Cell Transplant. 1997;6:125–34.
Kang R, Ghivizzani SC, Muzzonigro TS, Herndon JH, Robbins PD, Evans CH. Orthopaedic applications of gene therapy—from concept to clinic. Clin Orthop Relat Res. 2000;375:324–37.
Kaplonyi G, Zimmermann I, Farkas T, Viola T. Repair of cartilage injuries and osteochondral fractures with fibrin glue. Orv Hetil. 1984;125(2237–40):43.
Kaplonyi G, Zimmerman I, Frenyo A, Farkas T, Nemes G. The use of fibrin adhesive in the repair of chondral and osteochondral injuries. Injury. 1988;19:267–72.
Katsube K, Ochi M, Uchio Y, Maniwa S, Matsusaki M, Tobita M, et al. Repair of articular cartilage defects with cultured chondrocytes in Atelocollagen gel. Comparison with cultured chondrocytes in suspension. Arch Orthop Trauma Surg. 2000;120:121–7.
Kawabe N, Yoshinao M. The repair of full-thickness articular cartilage defects. Immune responses to reparative tissue formed by allogeneic growth plate chondrocyte implants. Clin Orthop. 1991;268:279–93.
Kawamura S, Wakitani S, Kimura T, Maeda A, Caplan AI, Shino K, et al. Articular cartilage repair: rabbit experiments with a collagen gel-biomatrix and chondrocytes cultured in it. Acta Orthop Scand. 1998;69:56–62.

Knudson W, Casey B, Nishida Y, Eger W, Kuettner K, Knudson C. Hyaluronan oligosaccharides perturb cartilage matrix homeostasis and induce chondrocytic chondrolysis. Arthritis Rheum. 2000;43:1165–74.

Kreder HJ, Moran M, Keeley FW, et al. Biologic resurfacing of a major joint defect with cryopreserved allogeneic periosteum under the influence of continuous passive motion in a rabbit model. Clin Orthop. 1994;300:288–96.

Kujawa MJ, Carrino DA, Caplan AI. Substrate bonded hyaluronic acid exhibits a size-dependent stimulation of chondrogenic differentiation of stage 24 limb mesenchymal cells in culture. Dev Biol. 1986;114:519–28.

Lačković V, Nikolić I, Todorović V, et al. General and oral histology and embryology. 2019;6:102–5.

Lahiji A, Sohrabi A, Hungerford D, Frondoza C. Chitosan supports the expression of extracellular matrix proteins in human osteoblasts and chondrocytes. J Biomed Mater Res. 2000;51:586–95.

LeBaron RG, Athanasiou KA. Extracellular matrix cell adhesion peptides: functional applications in orthopaedic materials. Tissue Eng. 2000;6(2):85–103.

Lennon D, Haynesworth S, Arm D, Baber M, Caplan A. Dilution of human mesenchymal stem cells with dermal fibroblasts and the effects on in-vitro and in-vivo osteochondrogenesis. Dev Dyn. 2000;219:50–62.

Lohmann CH, Schwartz Z, Niederauer GG, Carnes DL, Dean DD, Boyan BB. Pretreatment with platelet derived growth factor-BB modulates the ability of costochondral resting zone chondrocytes incorporated into PLA/PGA scaffolds to form new cartilage in-vivo. Biomaterials. 2000a;21:49–61.

Lohmann C, Schwartz Z, Niederauer G, Boyan B. Degree of differentiation of chondrocytes and their pretreatment with platelet-derived-growth factor. Regulating induction of cartilage formation in resorbable tissue carriers in-vivo. Orthopade. 2000b;29:120–8.

Mackay A, Beck S, Murphy J, Barry F, Chichester C, Pittenger M. Chondrogenic differentiation of cultured human mesenchymal stem cells from marrow. Tissue Eng. 1998;4:415–28.

Maleski M, Knudson C. Hyaluronan-mediated aggregation of limb bud mesenchyme and mesenchymal condensation during chondrogenesis. Exp Cell Res. 1996a;225:55–66.

Maleski M, Knudson C. Matrix accumulation and retention in embryonic cartilage and in-vitro chondrogenesis. Connect Tissue Res. 1996b;34:75–86.

Martin I, Padera RF, Vunjak-Novakovic G, et al. In vitro differentiation of chick embryo bone marrow stromal cells into cartilaginous and bone-like tissues. J Orthop Res. 1998;16:181–9.

Mattioli-Belmonte M, Gigante A, Muzzarelli RAA, Politano R, De Benedittis A, Specchia N, et al. N,N-dicarboxymethyl chitosan as delivery agent for bone morphogenetic protein in the repair of articular cartilage. Med Biol Eng Comput. 1999;37:130–4.

Mauck RL, Soltz MA, Wang CCB, Wong DD, Chao PHG, Valhmu WB, et al. Functional tissue engineering of articular cartilage through dynamic loading of chondrocyte-seeded agarose gels. J Biomech Eng. 2000;122:252–60.

Messner K, Gillquist J. Prosthetic replacement of the rabbit medial meniscus. J Biomed Mater Res. 1993a;27:1165–73.

Messner K, Gillquist J. Synthetic implants for the repair of osteochondral defects of the medial femoral condyle: a biomechanical and histological evaluation in the rabbit knee. Biomaterials. 1993b;14:513–21.

Miyamoto H, Sakashita H, Wilson D, Goss A. Synovial chondromatosis of the temporomandibular joint. Br J Oral Maxillofac Surg. 2000;38:205–8.

Mosca J, Hendricks J, Buyaner D, Davis-Sproul J, Chuang L, Majumdar M, et al. Mesenchymal stem cells as vehicles for gene delivery. Clin Orthop. 2000;379:S71–90.

Mow V, Ratcliffe A, Rosenwasser M, Buckwalter J. Experimental studies on repair of large osteochondral defects at a high weight bearing area of the knee joint: a tissue engineering study. J Biomech Eng. 1991;113:198–207.

Namba RS, Meuli M, Sullivan KM, et al. Spontaneous repair of superficial defects in articular cartilage in a fetal lamb model. J Bone Joint Surg. 1998;80A:4–10.

Newman A. Articular cartilage repair. Am J Sports Med. 1998;26:309–24.
Nishimura K, Solchaga L, Caplan A, Yoo J, Goldberg V, Johnstone B. Chondroprogenitor cells of synovial tissue. Arthritis Rheum. 1999;42:2631–7.
Nixon A, Fortier L, Williams J, Mohammed H. Enhanced repair of extensive articular defects by insulin-like growth factor-I-laden fibrin composites. J Orthop Res. 1999;17:475–87.
O'Connell J. Pathology of the synovium. Am J Clin Pathol. 2000;114:773–84.
Ochi M, Uchio Y, Tobita M, Kuriwaka M. Current concepts in tissue engineering technique for repair of cartilage defect. Artif Organs. 2001;25:172–9.
Ostrander R, Goomer R, Tontz W, Khatod M, Harwood F, Maris T, et al. Donor cell fate in tissue engineering for articular cartilage repair. Clin Orthop. 2001;389:228–37.
Pacifici M, Koyama E, Iwamoto M, Gentili C. Development of articular cartilage: what do we know about it and how may it occur? Connect Tissue Res. 2000;41:175–84.
Paige KT, Cima LG, Yaremchuk MJ, Schloo BL, Vacanti JP, Vacanti CA. De novo cartilage generation using calcium alginate-chondrocyte constructs. Plast Reconstr Surg. 1996;97:168–78.
Palmoski MJ, Brandt KD. Effects of static and cyclic compressive loading on articular cartilage plugs in vitro. Arthritis Rheum. 1984;27:675–81.
Perka C, Schultz O, Lindenhayn K, Spitzer RS, Muschik M, Sittinger M, et al. Joint cartilage repair with transplantation of embryonic chondrocytes embedded in collagen-fibrin matrices. Clin Exp Rheumatol. 2000a;18:13–22.
Perka C, Schultz O, Spitzer R, Lindenhayn K. The influence of transforming growth factor beta1 on mesenchymal cell repair of full-thickness cartilage defects. J Biomed Mater Res. 2000b;52:543–52.
Pesakova V, Stol M, Adam M. Comparison of the influence of gelatine and collagen substrates on growth of chondrocytes. Folia Biol (Praha). 1990;36:264–70.
Pierschbacher MD, Ruoslahti E. Cell attachment activity of fibronectin can be duplicated by small synthetic fragments of the molecule. Nature. 1984;309:30–3.
Pitman MI, Menche D, Song EK, Ben Yishay A, Gilbert D, Grande DA. The use of adhesives in chondrocyte transplantation surgery: in-vivo studies. Bull Hosp Jt Dis Orthop Inst. 1989;49:213–20.
Pittenger M, Mackay A, Beck S, Jaiswal R, Douglas R, Mosca J, et al. Multilineage potential of adult human mesenchymal stem cells. Science. 1999;284:143–7.
Pittenger M, Mosca J, McIntosh K. Human mesenchymal stem cells: progenitor cells for cartilage, bone, fat and stroma. Curr Top Microbiol Immunol. 2000;251:3–11.
Ponticiello MS, Schinagl RM, Kadiyala S, Barry FP. Gelatin-based resorbable sponge as a carrier matrix for human mesenchymal stem cells in cartilage regeneration therapy. J Biomed Mater Res. 2000;52:246–55.
Prockop D. Marrow stromal cells as stem cells for nonhematopoietic tissues. Science. 1997;276:71–4.
Puelacher WC, Kim SW, Vacanti JP, et al. Tissue-engineered growth of-cartilage: the effect of varying the concentration of chondrocytes seeded onto synthetic polymer matrices. Int J Oral Maxillofac Surg. 1994;23:49–53.
Quarto R, Thomas D, Liang C. Bone progenitor cell deficits and the age-associated decline in bone repair capacity. Calcif Tissue Int. 1995;56:123–9.
Quarto R, Mastrogiacomo M, Cancedda R, Kutepov S, Mukhachev V, Lavroukov A, et al. Repair of large bone defects with the use of autologous bone marrow stromal cells. N Engl J Med. 2001;344:385–6.
Radice M, Brun P, Cortivo R, Scapinelli R, Battaliard C, Abatangelo G. Hyaluronan-based biopolymers as delivery vehicles for bone-marrow-derived mesenchymal progenitors. J Biomed Mater Res. 2000;50:101–9.
Rahfoth B, Weisser J, Sternkopf F, Aigner T, vonderMark K, Brauer R. Transplantation of allograft chondrocytes embedded in agarose gel into cartilage defects of rabbits. Osteoarthr Cart. 1998;6:50–65.

Reddi A. Cartilage-derived morphogenetic proteins and cartilage morphogenesis. Microsc Res Tech. 1998a;43:131–6.
Reddi A. Role of morphogenetic proteins in skeletal tissue engineering and regeneration. Nat Biotechnol. 1998b;16:247–52.
Reddi AH. Morphogenetic messages are in the extracellular matrix: biotechnology from bench to bedside. Biochem Soc Trans. 2000;28:345–9.
Reddi AH. Bone morphogenetic proteins: from basic science to clinical applications. J Bone Joint Surg. 2001;83A:S1–6.
Riesle J, Hollander A, Langer R, Freed L, Vunjak NG. Collagen in tissue-engineered cartilage: types, structure, and crosslinks. J Cell Biochem. 1998;71:313–27.
Robinson D, Halperin N, Nevo Z. Fate of allogeneic embryonal chick chondrocytes implanted orthotopically, as determined by the host's age. Mech Ageing Dev. 1989;50:71–80.
Rogers J, Young H, Adkison L, Lucas P, Black A. Differentiation factors induce expression of muscle, fat, cartilage, and bone in a clone of mouse pluripotent mesenchymal stem cells. Am Surg. 1995;61:231–6.
Ryd L, Karrholm J, Ahlvin P. Knee scoring systems in gonarthrosis. Evaluation of interobserver variability and the envelope of bias. Score assessment group. Acta Orthop Scand. 1997;68:41–5.
Sah RL, Kim YJ, Doong JY, et al. Biosynthetic response of cartilage explants to dynamic compression. J Orthop Res. 1989;7:619–36.
Sams AE, Nixon AJ. Chondrocyte-laden collagen scaffolds for resurfacing extensive articular cartilage defects. Osteoarthr Cartil. 1995;3:47–59.
Sams A, Minor R, Wootton J, Mohammed H, Nixon A. Local and remote matrix responses to chondrocyteladen collagen scaffold implantation in extensive articular cartilage defects. Osteoarthr Cart. 1995;3:61–70.
Schlag G, Redl H. Fibrin sealant in orthopedic surgery. Clin Orthop. 1988;227:269–85.
Schreiber R, Ratcliffe A. Tissue engineering of cartilage. Methods Mol Biol. 2000;139:301–9.
Schreiber R, Dunkelman N, Naughton G, Ratcliffe A. A method for tissue engineering of cartilage by cell seeding on bioresorbable scaffolds. Ann N Y Acad Sci. 1999a;875:398–404.
Schreiber R, Ilten-Kirby B, Dunkelman N, Symons K, Rekettye L, Willoughby J, et al. Repair of osteochondral defects with allogeneic tissue engineered cartilage implants. Clin Orthop. 1999b;367:S382–95.
Sechriest V, Miao Y, Niyibizi C, Westerhausen-Larson A, Matthew H, Evans C, et al. GAG- augmented polysaccharide hydrogel: a novel biocompatible and biodegradable material to support chondrogenesis. J Biomed Mater Res. 2000;49:534–41.
Sellers RS, Peluso D, Morris EA. The effect of recombinant human bone morphogenetic protein-2 (rhBMP-2) on the healing of full-thickness defects of articular cartilage. J Bone Joint Surg Am. 1997;79:1452–63.
Senkoylu A, Simsek A, Sahin FI, Menevse S, Ozogul C, Denkbas EB, et al. Interaction of cultured chondrocytes with chitosan scaffold. J Bioact Compat Polym. 2001;16:136–44.
Shahgaldi BF. Coral graft restoration of osteochondral defects. Biomaterials. 1998;19:205–13.
Silas S. Chondrocyte transplantation using a collagen bilayer matrix for cartilage repair. J Bone Joint Surg Br. 1998;80:743–4.
Sim F, Dahlin D, Ivins J. Extra-articular synovial chondromatosis. J Bone Joint Surg Am. 1977;59:492–5.
Skantze K, Brinckerhoff C, Collier J. Use of agarose culture to measure the effect of transforming growth factor beta and epidermal growth factor on rabbit articular chondrocytes. Cancer Res. 1985;45:4416–21.
Solchaga LA, Dennis JE, Goldberg VM, et al. Hyaluronic acid-based polymers as cell carriers for tissue-engineered repair of bone and cartilage. J Orthop Res. 1999;17:205–13.
Solchaga L, Yoo J, Lundberg M, Dennis J, Huibregtse B, Goldberg V, et al. Hyaluronan-based polymers in the treatment of osteochondral defects. J Orthop Res. 2000;18:773–80.
Solursh M. Formation of cartilage tissue in-vitro. J Cell Biochem. 1991;45:258–60.

Speer D, Chvapil M, Volz R, Holmes M. Enhancement of healing in osteochondral defects by collagen sponge implants. Clin Orthop. 1979;144:326–35.
Suh J, Matthew H. Application of chitosan-based polysaccharide biomaterials in cartilage tissue engineering: a review. Biomaterials. 2000;21:2589–98.
Szendroi M, Deodhar A. Synovial neoformations and tumours. Baillieres Best Pract Res Clin Rheumatol. 2000;14:363–83.
Tagil M, Aspenberg P. Cartilage induction by controlled mechanical stimulation in vivo. J Orthop Res. 1999;17:200–4.
Therin M, Christel P, Li S, et al. In vivo degradation of massive poly (alpha-hydroxy acids): validation of in vitro findings. Biomaterials. 1992;13:594–600.
Thompson A, Piez K, Seyedin S. Chondrogenesis in agarose gel culture. A model for chondrogenic induction, proliferation and differentiation. Exp Cell Res. 1985;157:483–94.
Toolan B, Frenkel S, Pachence J, Yalowitz L, Alexander H. Effects of growth-factor-enhanced culture on a chondrocyte-collagen implant for cartilage repair. J Biomed Mater Res. 1996;31:273–80.
Van den Berg WB, Van Osch GJVM, Van der Kraan PM, Van Beuningen HM. Cartilage destruction and osteophytes in instability-induced murine osteoarthritis—role of Tgf beta in osteophyte formation. Agents Actions. 1993;40:215–9.
van den Berg W, van der Kraan P, Scharstuhl A, van Beuningen H. Growth factors and cartilage repair. Clin Orthop. 2001;391:S244–50.
van Hinsbergh V, Collen A, Koolwijk P. Role of fibrin matrix in angiogenesis. Ann N Y Acad Sci. 2001;936:426–37.
van Kampen GP, Veldhuijzen JP, Kuijer R, et al. Cartilage response to mechanical force in high-density chondrocyte cultures. Arthritis Rheum. 1985;28:419–24.
van Susante JL, Buma P, van Osch GJ, et al. Culture of chondrocytes in alginate and collagen carrier gels. Acta Orthop Scand. 1995:66549–56.
van Susante J, Buma P, Schuman L, Homminga G, van den Berg WB, Veth R. Resurfacing potential of heterologous chondrocytes suspended in fibrin glue in large full-thickness defects of femoral articular cartilage: an experimental study in the goat. Biomaterials. 1999;20:1167–75.
Veldhuijzen JP, Huisman AH, Vermeiden JP, et al. The growth of cartilage cells in vitro and the effect of intermittent compressive force. A histological evaluation. Connect Tissue Res. 1987;16:187–96.
Verbruggen G, Veys E, Wieme N, Malfait A, Gijselbrecht L, Nimmegeers J, et al. The synthesis and immobilisation of cartilage-specific proteoglycan by human chondrocytes in different concentrations of agarose. Clin Exp Rheumatol. 1990;8:371–8.
Vert M, Mauduit J, Li S. Biodegradation of PLA/GA polymers: increasing complexity. Biomaterials. 1994;15:1209–13.
Veth R, Jansen H, Leenslag J, Pennings A, Hartel R, Nielsen H. Experimental meniscal lesions reconstructed with a carbon fiber-polyurethanepoly(L-lactide) graft. Clin Orthop. 1986;202:286–93.
Wakitani S, Kimura T, Hirooka A, Ochi T, Yoneda M, Yasui N, et al. Repair of rabbit articular surfaces with allograft chondrocytes embedded in collagen gel. J Bone Joint Surg Br. 1989;71:74–80.
Wakitani S, Goto T, Pineda S, Young R, Mansour J, Caplan A, et al. Mesenchymal cell-based repair of large, full-thickness defects of articular cartilage. J Bone Joint Surg. 1994;76A:579–92.
Weisser J, Rahfoth B, Timmermann A, Aigner T, Brauer R, von dMK. Role of growth factors in rabbit articular cartilage repair by chondrocytes in agarose [in process citation]. Osteoarthr Cart. 2001;9(Suppl A):S48–54.
Wong M, Siegrist M, Wang X, Hunziker E. Development of mechanically stable alginate/chondrocyte constructs: effects of guluronic acid content and matrix synthesis. J Orthop Res. 2001;19:493–9.
Woo Lee J, Ning Qi W, Scully SP. The involvement of â1 integrin in the modulation by collagen of chondrocyte-response to transforming growth factor- â1. J Orthop Res. 2002;20:66–75.

Biomechanics in MSK Regeneration

Sedeek Mosaid

Abstract This chapter discusses the role of biomechanics in regenerating bone and musculoskeletal tissues, emphasising the incorporation of mechanical characteristics into tissue-engineered scaffolds. The unique structure of bone is described, highlighting its viscoelastic and other properties crucial for tissue regeneration and scaffold development. Developments in scaffold materials, including natural, synthetic and composite varieties, and the significance of mechanotransduction in tissue integration and engineering are discussed. In addition, the biomechanical aspects of cartilage, tendon, and muscle are discussed, reflecting their importance in musculoskeletal repair Emerging technologies that accurately replicate the mechanical characteristics are also discussed, with a focus on their potential to enhance outcomes in regenerative medicine.

Keywords Bone regeneration · Musculoskeletal biomechanics · Scaffold · Mechanotransduction · Stem cells

1 The Role of Biomechanics in Bone Regeneration

Tissue engineering involves the repair and replacement of tissue and organs through the use of transplanted cells, scaffolds, DNA, proteins and protein constituents (Amini et al. 2012). Biomechanics is relevant in musculoskeletal regeneration, particularly bone regeneration. It considers the impact of mechanical forces, motion and loading on tissue healing. These are fundamental to clinical and experimental therapeutic approaches to obtain optimal regeneration and tissue restoration (Al-Nbaheen et al. 2012).

S. Mosaid (✉)
Department of Trauma & Orthopaedics, Lincoln County Hospital, United Lincolnshire Teaching Hospitals NHS Trust, Lincoln, UK

P. Lee (ed.), *Musculoskeletal Regeneration Medicine*,
https://doi.org/10.1007/978-3-031-77456-0_3

Developing bone tissue replacements and scaffolds that replicate natural bone requires first defining and understanding the mechanical behaviours and properties of bone. Bone is a highly dynamic, structured collagen-hydroxyapatite composite that provides flexibility and rigidity. The Bone contains Haversian canals which are integral to the structure and function of osteons, while the core exhibits a lattice-like microstructure of trabeculae (Alvarez and Nakajima 2009; Schwarcz et al. 2017).

From the macro- to the nanoscale, this hierarchical structure is the basis of the remarkable adaptability of bone, enabling it to withstand various mechanical loads, whether compression, tension or shear forces. The recapitulation of such properties in the tissue-engineered constructs is essential to enable integration with native bone and support physiological loads (Fraldi et al. 2022; Lopes et al. 2018).

This anisotropy refers to the directional dependence of its mechanical and structural properties, meaning that the tissue exhibits different characteristics when measured along different axes. In turn, this depends on the organisation of its constituents into fibrous and lamellar structural units formed by collagen fibres distributed in sheets or layers of mineralised crystallised matter (Fraldi et al. 2022; Lopes et al. 2018).

Anisotropy is crucial in the design of tissue-engineered scaffolds because it mimics the mechanical properties of native bone and is vital in inducing cell differentiation and tissue regeneration (Fraldi et al. 2022; Lopes et al. 2018).

Advances in 3D printing and fabrication have enabled the creation of anisotropic scaffolds at the microstructural level, significantly improving the mechanical behaviour of the engineered bone and outcomes in tissue engineering applications (Winarso et al. 2022; Bobyn et al. 2009).

Bone behaves as a viscoelastic material, absorbing energy and shocks. This behaviour depends on the combination of the properties of the organic matrix (mainly collagen, as well as other proteins) and the mineral compartment (hydroxyapatite crystals). When deformed by an external force, collagen fibres stretch, giving bone flexibility, while hydroxyapatite crystals resist deformation, increasing bone stiffness (Frassica et al. 1997; Butler et al. 2000). The primary challenge is to replicate these features in synthetic or bioengineered materials to create bone replacements that accurately mimic natural bone behaviour and merge successfully with host tissue (Butler et al. 2000; Niu et al. 2023).

The complex interactions of bone remodelling, Wolff's law and mechanotransduction are fundamental to bone biomechanics, offering insights into therapeutic bone repair and regeneration strategies (Hadjidakis and Androulakis 2006).

Bone remodelling depends on the coordinated activity of osteoclasts and osteoblasts in three main phases: activation, resorption and formation. The activation step begins with the recruitment of osteoclast precursors to the bone surface. This is followed by the resorption phase, during which osteoclasts degrade the mineralised bone matrix. Finally, in the formation step, osteoblasts deposit new bone matrix, which is subsequently mineralised (Hadjidakis and Androulakis 2006).

Bone remodelling adapts the bone shape to mechanical stress and repairs microdamage, strengthening the bone against fracture. Adaptation to mechanical stress is a hallmark of bone health and a basis for maintaining bone mass and function over a lifetime. Since the skeleton acts as a valuable calcium repository, bone remodelling is an important part of calcium homeostasis because it reacts to hormonal fluctuations and systemic calcium requirements (Hadjidakis and Androulakis 2006).

Wolff's law states that bone adapts to the loads under which it is placed, meaning that bone density and structure change in response to mechanical stress. In orthopaedic surgery, the positioning of prosthetic implants must consider the distribution of mechanical forces to avoid stress shielding (Chen et al. 2009). Characterised by a reduction in bone density because of inadequate mechanical stress, stress shielding often arises from excessive load bearing rather than load sharing by implants (Chen et al. 2009).

Bone engineering uses Wolff's law to create more complex and sophisticated materials to repair and regenerate bones. One factor in bone tissue engineering is the design of scaffolds to mimic the native mechanical environment for bone development at the repair site. The scaffolds must be designed with the exact mechanical properties for initial support. Simultaneously, scaffolds require precise mechanical properties to provide temporary support while facilitating gradual load transfer to newly formed bone tissue (Fraldi et al. 2022; Chen et al. 2009).

Mechanical loading accelerates the differentiation of stem cells into osteoblasts (bone-generating cells), which is relevant for bone regeneration because mechanical stimulation facilitates the repair of fractures and bone defects. Mesenchymal stem cells (MSCs) are multipotent stromal cells and can differentiate and develop into osteoblasts. The connection between osteoblasts and mechanical stimuli occurs through multiple mechanosensitive pathways (Freitas et al. 2019). Two key pathways involve integrins and mechanosensitive ion channels, such as Piezo1 and transient receptor potential vanilloid 4 (TRPV4). Integrins are essential for the cell response to the extracellular matrix (ECM) because of their ability to transduce mechanical signals. Osteogenesis activation is underpinned by the RhoA and mitogen-activated protein kinase (MAPK) pathways. Piezo1 channels are also essential for osteogenic differentiation of MSCs under mechanical loading conditions, these channels act as mechanosensors that convert mechanical stimuli into biochemical signals, which promote osteogenesis (O'Conor et al. 2016; Hynes 2002; Sun et al. 2019).

The mechanical environment affects the signalling pathways controlling osteoblast activity, which directly affects regenerative therapy. For instance, the application of fluid shear stress to osteoblast-like cells activates numerous pathways, including transforming growth factor beta (TGF-β)/Smad, NF-κB and bone morphogenetic protein signalling, which are essential for bone production and remodelling. These findings highlight the significance of the relationship between mechanical stimuli and biological responses, as well as the critical role of Wolff's law in informing regenerative treatments (You et al. 2000; Tang et al. 2009).

Mechanotransduction, the mechanism by which cells sense and respond to mechanical stimuli and convert them to biochemical signals, underlies several aspects of osseous development, remodelling and repair (Kahla et al. 2020). Therefore, a comprehensive understanding of mechanotransduction processes is essential for developing successful techniques to build viable bone structures in tissue engineering .

Osteocytes, the principal cells responsible for detecting mechanical alterations in bone, are buried within this matrix and are central to sensing mechanical stimuli. They interact with osteoblasts and osteoclasts to regulate bone formation and resorption, maintaining bone quality and the ability to adapt to mechanical loads (Han et al. 2018; Duncan and Turner 1995).

Osteocytes are entirely embedded in the bone matrix during skeletal maturation. Their unique stellar or spider-like shape allows them to form an extensive interconnected network of long-branched cellular processes within a fluid-filled canalicular system. This network is essential for their mechanosensory function (Kahla et al. 2020).

Integrin receptors of osteocytes are an essential link between the ECM and the cytoskeleton and are fundamental in bone mechanotransduction. Mechanical loads initiate a cascade of events, including the generation of fluid flow within the lacuna-canalicular system. This activates mechanosensitive ion channels and G-protein-coupled receptors to release signalling molecules, including prostaglandins, nitric oxide and adenosine 5′-triphosphate. In turn, these signals influence osteoblast and osteoclast activities and the complex balance between bone formation and resorption (Hynes 2002; Elder and Athanasiou 2009).

2 Biochemical Reactions to Mechanical Stimuli

In response to mechanical stimulation, osteocytes secrete the glycoprotein sclerostin to suppress osteoblast differentiation by antagonising the Wnt signalling pathway, modulating the formation of bone. Mechanical loading decreases sclerostin and stimulates bone formation. Conversely, unloading increases sclerostin expression, causing bone resorption. Activating these pathways in response to mechanical stimuli exemplifies the critical role of the MAPK and PI3K/Akt pathways in cell proliferation and differentiation. The bone remodelling process involves a complex network of signalling pathways, including the RANKL pathway, which is crucial for osteoclast differentiation and activity. The interplay between these pathways ensures the coordinated regulation of bone resorption and formation. By modulating the Wnt pathway, sclerostin indirectly affects other signaling pathways involved in bone remodelling (Kahla et al. 2020; Guntur and Rosen 2012).

3 Fundamentals of Orthopaedic Tissue Engineering

In tissue engineering, scaffolds, signals and cells are assembled individually or in combination to create various tissue types.

3.1 Scaffolds

Tissue-engineered scaffolds are essential to regenerative medicine and bone tissue development. These biocompatible materials act as temporary scaffolds to facilitate cell adhesion, proliferation and differentiation and guide the formation of new tissue and its integration with host tissue (Tatara and Mikos 2016).

3.2 Types of Tissue-Engineered Scaffolds

Natural Polymer Scaffolds

With their biocompatibility and similarity to the ECM, natural polymers such as collagen, gelatin, chitosan and alginate hold much promise for tissue engineering. Collagen scaffolds are a promising research direction for bone repair due to their ability to enhance cell attachment and differentiation. However, natural polymers usually have inferior properties to synthetic polymers, which could restrict their use in load-bearing applications (Winarso et al. 2022; Bhattarai et al. 2006).

Synthetic Polymer Scaffolds

Synthetic polymers such as polylactic acid, polyglycolic acid and polycaprolactone enable greater control of mechanical properties and degradation rates. These scaffolds can be engineered to achieve precise biomechanical properties that best match the needs of the bone tissue, providing significant support during the early stages of bone regeneration. Moreover, their uniformity and adaptability make them highly appealing for clinical use (Diez-Pascual 2017; Ma et al. 2012).

Composite Scaffolds

Composite scaffolds comprising natural and synthetic polymers result in an intimate synergy that optimises the performance of each component. For instance, the mechanical strength of gelatin is improved by combining it with chitosan, which enhances cell adhesion and biological activities (Adel et al. 2022; Asadpour et al. 2019). The degradation rate can also be further slowed by combination with silk fibroin to achieve mechanical robustness, providing potential applicability for long-term bone tissue engineering. More precise reproduction of the natural microenvironment of bone might improve the regenerative performance (Adel et al. 2022; Asadpour et al. 2019).

Ceramic-Derived Scaffolds

Ceramics, including hydroxyapatite, tricalcium phosphate and bioactive glass, have mechanical strength and osteoconductivity, which together encourage bone cell attachment and proliferation, enhancing the development of bone-like tissue. However, their limitations include brittleness and slow degradation, requiring the integration of other materials. Consequently, the strength and stiffness of the inorganic component are balanced by incorporating an organic polymer such as polylactic acid to form a hybrid material with the toughness of the polymer (Gao et al. 2014; Losquadro et al. 2009).

4 Biomechanical Factors in Scaffold Design

4.1 Mechanical Strength and Elastic Modulus

The mechanical strength and elastic modulus of scaffolds define the ability of the scaffolds to transmit physiological loads without deformation or failure. During the early stages of healing, the mechanical properties of scaffolds should mimic those of native bone to provide adequate support and enable appropriate load transmission. Differences in the elastic moduli of scaffolds and host bones could cause stress shielding, involving a reduction in the stress on native bone, which results in bone resorption or stress risers where the probability of fracture at the native bone/scaffold interface increases (Winarso et al. 2022; Bružauskaitė et al. 2015).

4.2 Porosity and Interconnectivity

Porosity is a critical factor influencing cell mobility, the transport of nutrients and the elimination of waste in scaffolds. Interconnectivity involves the structural connections between pores; it is essential for vascularisation and the ingrowth of new tissue, as it enables blood vessels to penetrate the scaffold (Adel et al. 2022; Karageorgiou and Kaplan 2005). These two properties must be carefully balanced because increased porosity may decrease the mechanical strength and weaken the ability of the scaffold to bear loads. This requires careful consideration and a balance between maintaining mechanical integrity and ensuring biological functionality (Karageorgiou and Kaplan 2005; Abbasi et al. 2020).

4.3 Rate of Degradation

The scaffold's degradation rate must correspond to the pace of new tissue creation (Winarso et al. 2022). If a scaffold degrades rapidly, it may fail to provide adequate support during the early stages of bone healing. Conversely, if the degradation is too slow, the scaffold may interfere with new tissue integration and remodelling (Zhang et al. 2014). The adaptability of biodegradable synthetic polymers enables the modulation of the degradation rate, making them appropriate for diverse bone tissue engineering uses (Zhang et al. 2014).

4.4 Surface Characteristics

Roughness, chemistry, topography and other surface features are essential factors in inducing cell adhesion, proliferation and differentiation, facilitating efficient bone regeneration. Osteointegration is dramatically improved when these scaffolds are coated with hydroxyapatite because their bioactive surfaces enable the bone adhesion and proliferation of cells (Gao et al. 2014; Abbasi et al. 2020).

4.5 Impact of Cyclic Loading on Scaffolds and Bone Tissue Engineering

Cyclic loading, similar to the natural mechanical forces experienced by bone, is required for initiating cellular responses that lead to tissue growth and remodelling. Cyclic loading induces ECM and mineralised tissue formation and is required for properly oriented self-organisation of the regenerated tissue so that the de novo tissue aligns with the native tissue and structurally and functionally incorporates into the host (Roshan-Ghias et al. 2010; Rubin and Lanyon 1984).

Bone cells, including osteoblasts, osteoclasts and osteocytes, react to cyclic loading via several mechanotransduction processes. Osteoblasts, the cells responsible for bone formation, increase their activity in response to cyclic loading, promoting bone matrix deposition and mineralisation. Osteocytes, the principal mechanosensors in bone, transmit mechanical signals to osteoblasts and osteoclasts, coordinating bone remodelling and preserving bone mass and integrity (Hadjidakis and Androulakis 2006; Kahla et al. 2020).

4.6 Impact of Shear Stress on Scaffolds and Bone Tissue Engineering

Shear stress generated within the pores of scaffolds by fluid flow is crucial in cellular behaviour. It provides nutrition and waste removal for optimal cell growth and differentiation. An effective scaffold design must use shear stress to optimise bone regeneration (Zhao et al. 2015). Shear stress affects the organisation and function of osteoblasts, causing their proliferation and differentiation and thereby enhancing bone regeneration (Zhao et al. 2015).

4.7 Impact of Compression and Tension on Scaffolds and Bone Tissue Engineering

Compressive and tensile stresses are also crucial in maintaining bone mass and integrity. They stimulate the mechanosensitive osteocytes within the bone to modulate remodelling through the signalling pathways regulating osteoblast and osteoclast function. Designing scaffolds to accommodate these stresses will enhance natural bone healing processes, this promotes tissue remodelling, rather than simply supporting final tissue form and function (Amini et al. 2012; Kahla et al. 2020). Compression enhances osteoblast differentiation and thus new bone formation, whereas tension helps determine the proper alignment and orientation of the collagen fibres in the bone matrix. Therefore, biomaterial scaffold design should enable the generation and transmission of these forces to afford optimal bone repair (Amini et al. 2012; Chen et al. 2009).

4.8 Cellular Structures

Cell incorporation is essential for tissue development and integration in orthopaedic tissue engineering. There are different ways to recruit cells to implanted scaffolds include releasing chemokines, these molecules help draw cells to the scaffold, aiding in integration. Attaching cell ligands to the scaffolds is another way for incorporating cells, these ligands help cells adhesion. Osteoconduction and osteoinduction are also techniques for recruiting cells, as scaffolds can be made to enhance bone growth. MSCs are ideal for tissue engineering because they can differentiate into many adult cells based on their microenvironment. However, adipose-derived stem cells are increasingly popular because of their easy availability, low harvesting cost and ease of growth. Amniotic fluid-derived MSCs have produced promising bone and cartilage regeneration results in small-animal models. MSCs from different sources can have different differentiation potentials, which can have important implications for the quality of tissue repair (Tatara and Mikos 2016; Kwon et al. 2018; Zuk et al. 2001).

4.9 Signals

In tissue engineering, signals are the environmental cues that influence tissue regeneration. Signals in scaffolds are classified as biological, chemical, mechanical or electrical. Platelet-rich plasma and its variants are common biological signals in tissue engineering protocols. Chemical signals can accelerate the healing process by modulating the inflammatory pathway or acting directly on cells, such as in bisphosphonate therapy, which targets osteoclasts. Mechanical cues have been traditionally used to stimulate bone formation, such as in distraction osteogenesis. Passive mechanical signals derived from scaffolds, including substrate stiffness and roughness, can direct the differentiation of stem cells into distinct lineages (Tatara and Mikos 2016; Laurencin et al. 2014). While electrical cues are important in directing the formation of functional skeletal muscle tissue, they have not been explored to the same extent as other cellular cues for engineering orthopaedic tissue (Qazi et al. 2015).

5 Biomechanics of Tissue Engineering of Articular Cartilage

Articular cartilage is an avascular connective tissue that covers the ends of bones in diarthrodial (mobile) joints and is devoid of blood vessels, nerves and lymphatic vessels, which hinders its natural healing capabilities. Articular cartilage comprises water, collagen and proteoglycans, and the only cell type present in this tissue is the chondrocyte (Fox et al. 2009).

The ECM retains the architecture and biomechanical properties of the cartilage. Type II collagen offers an interconnected network to provide tensile strength and, ultimately, the mechanical architecture of the tissue. Proteoglycans, highly glycosylated proteins, confer cartilage resistance to compression by binding and retaining water (Fox et al. 2009; Poole et al. 2001; Kim et al. 2010).

Chondrocytes are metabolically active cells with a homeostatic role in the development, maintenance and repair of the ECM. Chondrocytes are derived from MSCs and represent approximately 2% of the total articular cartilage volume. The shape, number and size of the chondrocytes vary according to their anatomical location in the articular cartilage (Fox et al. 2009). Each chondrocyte creates a niche and orchestrates the turnover of ECM in its immediate vicinity. The niche also encapsulates the chondrocyte in its matrix and prevents its migration to other areas of cartilage (Fox et al. 2009).

Mechanotransduction is a process that converts mechanical inputs into biochemical responses. Chondrocytes are the primary cells in cartilage that sense and respond to mechanical forces.This complex process can occur via multiple pathways, including ion channels, integrins and the cytoskeleton, influencing cellular responses to mechanical pressures and altering gene expression and protein synthesis in response to mechanical loads (Fox et al. 2009; Wang et al. 2023).

Chondrocytes interact with the ECM via integrins, which link the ECM to the cytoskeleton and initiate intracellular signalling pathways that regulate chondrocyte proliferation, differentiation and matrix synthesis. Primary cilia are essential mechanosensors in chondrocytes, detecting mechanical stresses in the cartilage microenvironment. These primary cilia modulate chondrocyte proliferation and differentiation, which helps maintain cartilage function. TRPV4 is expressed on the primary cilia of chondrocytes. TRPV4 is a calcium-permeable ion channel that is highly expressed in articular chondrocytes and involved in sensing osmotic pressure changes within the cartilage matrix. The primary cilia are engaged in crucial signalling pathways, such as the Hedgehog pathway, which regulates chondrocyte function and homeostasis of the cartilage matrix, modulated by mechanical strain. As primary cilia can sense mechanical loading, such responses can change cilia structure and function, changing chondrocyte activity and matrix production. This helps explain the role of primary cilia as mechanosensors (Dieterle et al. 2021; O'Conor et al. 2016).

5.1 Biomechanical Environment

Articular cartilage is in a highly dynamic biomechanical environment involving complex loading forces including compression, shear and tension, which are crucial for preserving cartilage health and functionality (Fox et al. 2009; Bader et al. 2011). Compression is the principal load-bearing motion in cartilage. It enables the movement of water and solutes throughout the ECM and is essential for nutrition distribution and waste elimination. Shear forces arise from the joint's sliding motions. Tension occurs because of the collagen network's resistance to deformation.

5.2 Scaffolds for Cartilage Tissue Engineering

Scaffolds provide structural support and facilitate cell growth and tissue formation. An ideal scaffold material requires the following characteristics to ensure proper regeneration. The first requirement is controlled biodegradation to nontoxic degradation products. Second, the scaffold must contain a porous structure for the diffusion of nutrients and metabolites from the tissue. Third, it must support the survival, proliferation, differentiation and production of ECM by the cells. Fourth, the scaffold must be anchored to the cartilage defect site and integrated into the surrounding tissue. Finally, it must provide mechanical support to the newly regenerated tissue (Karageorgiou & Kaplan 2005; Steele et al. 2022).

5.3 *Types of Scaffolds*

Natural Matrix Materials

These materials are derived from natural sources such as cartilage matrix, periosteum, collagen, hyaluronic acid and cellulose; they are found as water-absorbent hydrogels and water-laden porous matrices. These matrices can all be readily designed for injectability and integration into the cartilage because their high water content makes them compatible with the natural environment of cartilage repair cells (Chen et al. 2020).

The polysaccharide hyaluronic acid (HA) comprises alternating units of D-glucuronic acid and N-acetylglucosamine, which are naturally found in cartilage and synovial fluid. HA's ability to attract water molecules helps to maintain the viscoelastic properties of cartilage, which are vital for its resilience and ability to withstand mechanical stress. HA controls the synthesis of cartilage, healing of cartilage injuries and apoptosis (Dovedytis et al. 2020). Adding HA to tissue engineering scaffolds dramatically improves the structural and functional characteristics of regenerated cartilage. Its strong water-binding capacity and high viscoelasticity help mimic the structures of the native ECM and are essential in providing an adequate microenvironment for chondrogenesis and helping maintain the chondrocyte phenotype (Dovedytis et al. 2020; Chen et al. 2017). Bioinks are materials adapted to 3D bioprinting, which produces artificial tissues and organs by printing inks with biomaterials on a substrate layer by layer. Bioinks are central to 3D bioprinting approaches in regenerative medicine. HA is promising for bioinks due to its unique biological and physicochemical properties. The rheological characteristic of HA, such as shear-thinning behaviour, is critical for bioink application. Shear-thinning behaviour allows the bioink to flow easily under the pressure of extrusion but maintains its shape once printed, which is essential for creating stable and precise 3D structures (Chen et al. 2023).

Hybrid hydrogels are composite materials that combine different types of polymers to enhance their properties. Double-network hydrogels are a type of advanced hydrogel structure that consists of two interpenetrating polymer network. Hybrid hydrogels containing HA combined double-network hydrogels, have shown improved mechanical properties and applicability for the transplantation of stem cells for cartilage repair (Wang et al. 2022).

Chitosan is a naturally occurring biopolymer and one of the most promising materials in cartilage tissue engineering. It is structurally similar to glycosaminoglycans (GAGs), one of the main components of cartilage ECM, and promotes cartilage resistance against mechanical challenges such as shear and compression forces. Therefore, chitosan is a suitable candidate for cartilage tissue engineering scaffolds. Chitosan-based hydrogels have gained attention. For example, catechol-modified chitosan hydrogels can enhance the proliferation and chondrogenic differentiation ability of bone marrow-derived stem cells (BMSCs) (Man et al. 2016; Chen et al. 2024).

Synthetic Materials

These are the most adaptable materials and include carbon fibre, hydroxyapatite, polylactic acid, polyglycolic acid and sodium alginate gel. The porosity and degradability of these materials are customisable to enable appropriate proliferation and growth of chondrocytes (Chen et al. 2024).

Composite Scaffolds

These are created by combining the advantages of natural and synthetic materials.

3D DCECM scaffolds are three-dimensional structures made from decellularized cartilage extracellular matrix (DCECM). These scaffolds are designed to mimic the natural environment of cartilage tissue. 3D DCECM scaffolds modified by graphene oxide improved cell adhesion, proliferation and differentiation, successfully bridging cartilage defects with hyaline cartilage after implantation (Gong et al. 2020).

Bioactive Scaffolds

Bioactive scaffolds actively participate in the healing process by releasing molecules that promote cell growth and differentiation. For instance, bioactive magnesium-containing glass nanosphere composite scaffolds stimulate cartilage and subchondral bone regeneration (Chen et al. 2024).

6 Cells

The cell source is critical in cartilage functional tissue engineering (CFTE). The ideal cartilage cell source is easily isolated and expanded and produces ECM proteins characteristic of the same type of cartilage, such as aggrecan and type II collagen. Chondrocytes and stem cells are the most studied cell sources for their potential in cartilage tissue engineering (Kock et al. 2011).

6.1 Chondrocytes

Articular chondrocytes, the main cellular component of cartilage, may be chosen for CFTE since they synthesise cartilage-specific components such as type II collagen and proteoglycans. However, problems obtaining cells from small cartilage pieces and linear differentiation during ex vitro expansion have limited their use (Kock et al. 2011).

Nasal chondrocytes have emerged as an alternative to articular chondrocytes. These cells are easier to obtain. Harvesting and expanding them is less invasive, and they have shown promising results in cartilage production. Nasal chondrocytes retain their cartilage formation capacity during the expansion period and can produce type II collagen and GAG-rich ECM (Mumme et al. 2016).

Autologous chondrocyte implantation (ACI) is a technology designed to repair cartilage defects by implanting chondrocytes at the target sites. Its first-generation technology used periosteum sealing; however, this was limited by its thinness, vulnerability to rupture and hypertrophy (Sayed et al. 2010). The periosteum was later replaced by a collagen membrane for sealing and fixing chondrocytes, leading to the emergence of the second generation of ACI technology. While ACI has shown satisfactory long-term clinical outcomes, some drawbacks remain. For example, chondrocytes may leak from the graft area and uneven cell distributions in the graft area can result in poorly regenerated cartilage surfaces. The clinical application of autologous chondrocyte transplantation still faces the challenges of high cost, technical complexity and joint incisions (Marlovits et al. 2005; Brittberg 1999).

The third generation of ACI addresses the limitation of chondrocyte delivery. Chondrocytes are precombined with a carrier and implanted into the cartilage defect area. This enables the previously fixed cultured cells to be implanted into the target area, effectively increasing the survival rate of chondrocytes after surgery. The limitation of this technique is that the source of chondrocytes is limited, necessitating two surgical procedures (Chen et al. 2024; Bright and Hambly 2014).

6.2 Multipotent Stem Cells

MSCs are multipotent stem cells with the potential for self-renewal and differentiation into various cell lineages, including cartilage. They also regulate local metabolism and repair, making them one of the most promising cell sources for cartilage regeneration (Kwon et al. 2018).

6.3 Bone Marrow-Derived Stem Cells

BMSCs are considered the current gold standard of cartilage repair because of their excellent chondrogenic potential. They are the most thoroughly investigated cell type for CFTE.

However, BMSCs face challenges such as poor cell adhesion and retention, phenotypic alterations, and lower engraftment rates after implantation. These issues limit their effectiveness, particularly in treating early to middle-stage mild injuries. The retention time of BMSCs in vivo is relatively short, necessitating multiple injections to achieve a better therapeutic effect (Guo et al. 2023).

6.4 Adipose-Derived MSCs

These possess many characteristics of BMSCs with lower chondrogenesis capacity; however, they are more accessible and easier to collect, making them an appealing alternative (Guo et al. 2023).

6.5 Umbilical Cord-Derived MSCs

These cells have increased proliferative capacity and can differentiate into chondrocytes, making them a suitable option for cartilage regeneration (Baksh et al. 2007).

6.6 Mechanical Stimuli in Cartilage Tissue Engineering

Recreating the mechanical environment in tissue engineering is crucial to creating engineered cartilage construct that is as close as possible to natural tissue.

Effects of Compression

Compression is a critical mechanical force influencing cartilage. Dynamic compression is physiologically relevant and improves the mechanical properties of synthetic cartilage by stimulating matrix deposition by cells and aligning collagen fibres along the direction of the applied load, strengthening and reinforcing the tissue. Dynamic compression is similar to what occurs in cartilage in vivo. Low strain and precise compression loads can assist with ECM growth and protection against cell injury. They preserve the integrity of cartilage tissue by promoting healthy cellular metabolism and ECM generation. When subjected to static compression, the volume of cartilage tissue is reduced due to the compaction of the collagen matrix, which influences molecular transport and biosynthesis patterns and diminishes chondrocyte activity (Xie et al. 2019; Connizzo et al. 2023).

Shear Stress

Shear stress, induced by fluid dynamics or sliding motions, significantly affects chondrocyte metabolism and ECM formation in cartilage tissue engineering. Shear stress is a crucial element in cartilage tissue engineering that enhances the synthesis of GAGs and collagen. Shear stress also affects the orientation and morphology of chondrocytes, leading to a more organised tissue architecture (Connizzo et al. 2023).

Tensile Strain

Tensile strain elongates cartilage tissue and stimulates the creation of collagen and other vital components. Furthermore, tensile strain promotes chondrocyte proliferation and differentiation, which is crucial for enhancing tissue development and maturation. This mechanical force helps organise the tissue structure and improves its mechanical properties (Connizzo et al. 2023; Sanchez-Adams et al. 2014).

7 Biomechanics of Tissue Engineering of Tendons

Tendons are fibrous connective tissues that carry the forces generated by the muscle to the bone, mediating body movement and stability. Tenocytes, a specialised fibroblast cell type, are the principal cells of tendons. They secrete ECM and are responsible for maturing collagen fibrils; these cells are typically arranged in longitudinal rows near collagen fibrils (Franchi et al. 2007; Asahara et al. 2017). Collagen is the main component of the tendon ECM. Its fibres are densely packed in a network of fibrous connective tissue that is crucial to tendon function and mechanical characteristics (Franchi et al. 2007; Asahara et al. 2017).

The ultimate tensile strength is the highest stress a tendon can withstand before rupture. High tensile strength is necessary for tendons to withstand the forces muscles produce during movement. This strength mainly originates from the arrangement and packing of collagen fibres. The hierarchical organization of tendons, which comprises collagen microfibrils, fibrils and fibres, contributes to their biomechanical characteristics. Having these components parallel increases the mechanical rigidity and stiffness of the tendon. Type I collagen is the most abundant form in tendons. It increases tensile strength due to its triple-helix structure and high degree of crosslinking (Asahara et al. 2017; Wang et al. 2011; Lavagnino et al. 2015).

Stiffness is a fundamental biomechanical property that contributes to the tendon's capacity to resist deformation when exposed to a load. It is represented by the slope of the stress-strain curve in the linear region. Tendons must be sufficiently stiff to transmit forces with an efficient range of motion. Factors affecting a tendon's stiffness include the alignment of collagen fibres and crosslinking of collagen and other ECM components, including elastin, proteoglycans and other proteins (Asahara et al. 2017; Wang et al. 2011).

The mechanical properties of tendons include viscoelasticity, the ability of tendons to exhibit elastic and viscous behaviour under mechanical stress; mechanical behaviour is dependent on the rate of mechanical strain. This property allows tendons to be more deformable at low strain rates and less deformable at high strain rates. This is determined by stress relaxation, creep and hysteresis of the tendon. Stress relaxation occurs when the stress gradually decreases over time while the tendon is held under constant strain. Creep is the gradual increase in strain of a tendon exposed to constant stress. Hysteresis is the energy loss in the tendon during repeated loading and unloading, represented by the difference between the loading and unloading curves on the stress-strain graph. These viscoelastic behaviours of tendons have clinical implications: they contribute to the functional biological properties that reduce fatigue failure and are vital to dissipate energy and prevent damage during repetitive movements. For instance, the peak stress in tissue substance of tendons is reduced during each cycle because of stress relaxation, which is observed during walking and jogging (Wang et al. 2011; Wang 2005; Jung et al. 2009).

Fatigue resistance is the ability to withstand repeated loading and unloading cycles without losing significant mechanical properties. Tendons experience many loading and unloading cycles during normal daily activities; therefore, fatigue resistance is important (Wang 2005; Jung et al. 2009).

The stress-strain curve is a standard measure of the mechanical properties of tendons. This curve has three key sections: toe, linear and failure. The toe region is the first part of the tendon's response when it is stretched. This includes the initial stretching of the collagen fibril crimp pattern observed at strains reaching 2%, after which the crimps are fully extended. The linear area follows; here, the fibres respond linearly to strain. If the strain is below 4%, the tendon returns to its original length when the external force is removed. The slope of this linear region corresponds to the Young's modulus of the tendon. Young's modulus is a measure of the stiffness of a material. It is defined as the ratio of stress to strain. Microscopic tearing of tendon fibres typically occurs when the tendon is stretched beyond 4%. Macroscopic failure often occurs beyond 8–10% strain; further stretching causes tendon rupture (Wang 2005; Woo 1986).

7.1 Tendon Healing Process

Tendon healing is a complex process comprising three phases. The inflammation phase, which occurs immediately after the tendon injury and lasts a few days, is characterised by the recruitment of inflammatory cells to the injury site, which clear debris and express cytokines and growth factors to initiate the healing process. In the proliferation phase, a few days after the injury and lasting several weeks, fibroblasts migrate to the injured site to synthesise collagen and other ECM components, including proteoglycans. The final stage is the remodelling phase, during which the collagen fibres mature and become arranged in the proper direction (Wang 2005; Woo et al. 1999). Remodelling is a crucial step in reestablishing the tendon's mechanical properties. Furthermore, the matrix becomes denser, collagen turnover, water content, and the ratio of collagen Types I to III begins to approach normal levels. An integrated sequence of biochemical and biomechanical signals is critical for remodeling. These signals regulate the expression of structural and enzymatic proteins, including degradation enzymes such as collagenase (Wang 2005; Woo et al. 1999).

Biomechanics is crucial in the tendon healing process, mainly through mechanical loading during therapy. Applying controlled loading forces during therapy can stimulate tenocytes, increasing their production and promotes the alignment of collagen. Tenocytes can upregulate collagen synthesis when subjected to tensional forces. While controlled loading is beneficial, inappropriate or excessive loading can cause re-injury or conditions characterised by persistent pain and dysfunction (Benjamin et al. 2008; Durgam and Stewart 2017).

Mechanical stress induces the expression of many genes vital for tendon repair, including those encoding the type I and III collagens, decorin, tenascin-C and matrix metalloproteinases. Mechanical loads modulate the expression of key growth factors required for tendon regeneration and angiogenesis, including TGF-β, insulin-like growth factor 1 and vascular endothelial growth factor (Lavagnino et al. 2015; Wang 2005; Citeroni et al. 2020).

Considering the mechanical properties of native tendons is essential for designing tissue engineering scaffolds for tendons. This will enable the creation of constructs that can withstand physiological stresses and induce collagen fibre alignment. Natural polymers (e.g. collagen and silk) or synthetic polymers (e.g. polylactic acid) and polycaprolactone are widely used as tendon scaffolds in tissue engineering (Bianchi et al. 2021; Bagnaninchi et al. 2006). Scaffolds with similar mechanical properties such as natural tendons enable cell differentiation and the formation of ECM. Mechanical loading of scaffolds during the culture period improves the mechanical characteristics and efficacy of implanted tendons. Such mechanical cues are required to develop scaffolds exhibiting anisotropic and viscoelastic properties resembling those of native tendons (Pien et al. 2023).

Due to the complex hierarchical structure of tendons, which is critical to their mechanical performance, the scaffolds must be aligned at the micro- and nanoscales to guide the same scale organisation of structure in the regenerated tissue after implantation. Although different types of fully aligned topography currently exist in scaffolds. However, current scaffolds for tendon regeneration are more concerned with aligned topography, which does not entirely reproduce the actual microstructure of tendons. Therefore, to mimic native tendon ECM characteristics, the fiber and the ECM must be identical. Both the fiber scale and corrugated superstructure should not be overlooked (Huang et al. 2023).

The greatest challenge is to design and fabricate a suitable 3D scaffold for tendon tissue engineering that can simultaneously stimulate tendon tissue regeneration and exhibit suitable mechanical properties to function as a biomimetic scaffold. The emerging methods of 3D bioprinting, electrospinning and wet-spinning provide greater control over the scaffold's hierarchy. The advantage of 3D bioprinting is that different natural and synthetic materials can be used to generate structures that regenerate the extracellular matrix and offer structural integrity. It also allows for the direct binding of cells and growth factors essential for the tendon. However, these advantages remain limited by process automation and structural constraints. Finally, the scaffolds must degrade at an appropriate speed based on the tissue regeneration rate (Ruiz-Alonso et al. 2021).

MSCs are commonly used because they can differentiate into tenocytes. MSCs are harvested from bone marrow, adipose tissue and other tissues, providing a rich source of cells to regenerate new tissues (Citeroni et al. 2020; Pien et al. 2023).

7.2 Differences Between Ligaments and Tendons

Ligaments are dense connective tissues that connect bones to stabilise joints. In contrast to the less elastic tendons, ligaments are more elastic, enabling them to be elongated under load and then return to their original shape, this increases joint mobility and assists in impact absorption. Being more elastic, ligaments are more susceptible to damage. Ligaments are mainly exposed to tensile forces while providing joint stability. However, tendons withstand tensile and compressive forces while transmitting muscle-generated loads to bones (Viidik 1990).

Stress-strain curves for ligaments and tendons are distinct, reflecting their anatomical differences and functions. Tendons typically function at the lower range of the stress-strain curve to minimise the risk of failure; however, ligaments function at the higher range of the curve. Ligaments have a longer toe region because their collagen fibres are more crimped and disorganised, enabling increased elongation. In the linear region, ligaments have a shallower slope, reflecting their ability to stretch. Concerning the failure point, ligaments can tolerate higher strains before they fail compared to tendons (Yang et al. 2013; Biewener 2008).

7.3 Skeletal Muscle

Human skeletal muscle comprises long, cylindrical muscle fibres surrounded by connective tissues. Each muscle fibre is composed of myofibrils divided into sarcomeres, the primary contractile units of muscle. The principal role of skeletal muscle relies on the coordinated contraction and relaxation of sarcomeres, facilitated by the sliding-filament mechanism (Lieber 2002).

Mechanical forces are fundamental to muscle growth and rehabilitation. Exercise imposes mechanical stresses on muscles, supporting muscle growth and promoting muscular hypertrophy and strength. However, a lack of mechanical stimuli, as in muscle disuse, results in muscle shrinkage and reduced strength (Bodine 2013).

Following a muscle injury, the body undergoes a complex healing process involving inflammation, removal of damaged tissue and the regeneration of new muscle fibres. Satellite cells—the principal muscle stem cells located between the basal lamina and the sarcolemma of muscle fibres—are crucial to this healing process. Following damage, satellite cells activate, proliferate and differentiate into myoblasts, ultimately fusing to create new muscle fibres or repair old ones (Baoge et al. 2012).

Muscle tissue scaffolds used in engineering must account for the mechanical properties of natural skeletal muscle. The scaffolds should be strong enough to support cell proliferation and resist daily physiological loading, while remaining sufficiently flexible to accommodate muscle shortening and lengthening. Scaffolds should have a suitable microarchitecture to support the future development of muscle tissue. The scaffold should be porous to enable cell movement, nutrient distribution and waste removal. The alignment of scaffold fibres can affect the orientation of muscle cells, promoting the formation of aligned muscle fibres similar to those observed in natural muscle tissue. This alignment is essential for the optimal performance of designed muscle tissue since it improves its contraction capability (McCullen et al. 2009).

While recent advances in 3D printing have improved scaffold architecture, existing synthetic polymers are often too rigid or have poor surface and degradation profiles, and natural polymers used as hydrogels lack the mechanical stability required for muscle contraction. Gokyer et al. (2021) presented a novel approach to 3D printing biodegradable polyurethaneurea elastomers that possess high elasticity and controlled biodegradability, which can significantly enhance the development of effective skeletal muscle tissue engineering scaffolds. This advance allows for better integration and replacement of damaged muscle tissue (McCullen et al. 2009; Gokyer et al. 2021).

Mechanical stimulation is crucial to the growth and functionality of muscle tissues. Applying cyclic loading or compressive stresses to cell-seeded scaffolds via bioreactors can enhance muscle fibre alignment, myofibril density and contractile characteristics. The direct activation of mechanotransduction pathways by mechanical stress activates muscle-specific genes and proteins essential for tissue development. Exposing muscle to mechanical loads in vitro reproduces the effects of exercise in vivo by stimulating muscle growth and increased strength (Rindom and Vissing 2016).

8 Conclusion

Biomechanics is essential for the effective regeneration of bone and musculoskeletal tissues. Recognising the mechanical properties of the tissue and its reaction to mechanical stresses is essential for the construction of efficient tissue-engineered scaffolds. Advances in scaffold technology have markedly improved the replication of natural musculoskeletal tissue properties. Combining mechanical stimulation and mechanotransduction pathways is essential for enhancing cell differentiation, tissue integration and, ultimately, the efficacy of bone repair and regeneration. Future research will enhance these methodologies, resulting in better-functioning biomimetic musculoskeletal tissues.

References

Abbasi N, et al. Porous scaffolds for bone regeneration. J Sci Adv Mater Dev. 2020;5(1):1–9. https://doi.org/10.1016/j.jsamd.2020.01.007.

Adel IM, ElMeligy MF, Elkasabgy NA. Conventional and recent trends of scaffolds fabrication: a superior mode for tissue engineering. Pharmaceutics. 2022;14(2):306. https://doi.org/10.3390/pharmaceutics14020306.
Al-Nbaheen M, Vishnubalaji R, Ali D, Bouslimi A, Al-Jassir F, Megges M, Prigione A, Adjaye J, Kassem M, Aldahmash A. Human stromal (mesenchymal) stem cells from bone marrow, adipose tissue and skin exhibit differences in molecular phenotype and differentiation potential. Stem Cell Rev Rep. 2012;9(1):32–43. Available at:. https://doi.org/10.1007/s12015-012-9365-8.
Alvarez K, Nakajima H. Metallic scaffolds for bone regeneration. Materials. 2009;2(3):790–832.
Amini RA, Laurencin TC, Nukavarapu PS. Bone tissue engineering: recent advances and challenges. Begell House. 2012;40(5):363–408. Available at: https://doi.org/10.1615/critrevbiomedeng.v40.i5.10.
Asadpour S, et al. Natural biomacromolecule based composite scaffolds from silk fibroin, gelatin and chitosan toward tissue engineering applications. Int J Biol Macromol. 2019;154:1285–94. https://doi.org/10.1016/j.ijbiomac.2019.11.003.
Asahara H, Inui M, Lotz MK. Tendons and ligaments: connecting developmental biology to musculoskeletal disease pathogenesis. J Bone Miner Res. 2017;32(9):1773–82. https://doi.org/10.1002/jbmr.3199.
Bader DL, Salter DM, Chowdhury TT. Biomechanical influence of cartilage homeostasis in health and disease. Arthritis. 2011;2011:1–16. https://doi.org/10.1155/2011/979032.
Bagnaninchi PO, et al. Tissue engineering for tendon repair * COMMENTARY. Br J Sports Med. 2006;41(8):e10. https://doi.org/10.1136/bjsm.2006.030643.
Baksh D, Yao R, Tuan RS. Comparison of proliferative and multilineage differentiation potential of human mesenchymal stem cells derived from umbilical cord and bone marrow. Stem Cells. 2007;25(6):1384–92. https://doi.org/10.1634/stemcells.2006-0709.
Baoge L, et al. Treatment of skeletal muscle injury: a review. ISRN Orthop. 2012;2012:1–7. https://doi.org/10.5402/2012/689012.
Benjamin M, Kaiser E, Milz S. Structure-function relationships in tendons: a review. J Anat. 2008;212(3):211–28. https://doi.org/10.1111/j.1469-7580.2008.00864.x.
Bhattarai N, et al. Alginate-based Nanofibrous Scaffolds: structural, mechanical, and biological properties. Adv Mater. 2006;18(11):1463–7. Available at: https://doi.org/10.1002/adma.200502537.
Bianchi E, et al. Innovative strategies in tendon tissue engineering. Pharmaceutics. 2021;13(1):89. https://doi.org/10.3390/pharmaceutics13010089.
Biewener AA. Tendons and ligaments: structure, mechanical behavior and biological function. Springer eBooks; 2008. p. 269–84. https://doi.org/10.1007/978-0-387-73906-9_10.
Bobyn JD, McKenzie K, Karabasz D, Krygier JJ, Tanzer M. Locally delivered bisphosphonate for enhancement of bone formation and implant fixation. J Bone Joint Surg. 2009;91(Supplement 6):23–31. Available at:. https://doi.org/10.2106/jbjs.i.00518.
Bodine SC. Disuse-induced muscle wasting. Int J Biochem Cell Biol. 2013;45(10):2200–8.
Bright P, Hambly K. A systematic review of reporting of rehabilitation in articular-cartilage-repair studies of third-generation autologous chondrocyte implantation in the knee. J Sport Rehabil. 2014;23(3):182–91. https://doi.org/10.1123/jsr.2013-0045.
Brittberg M. Autologous chondrocyte transplantation. Clin Orthop Relat Res. 1999;367:S147–55. https://doi.org/10.1097/00003086-199910001-00016.
Bružauskaitė I, et al. Scaffolds and cells for tissue regeneration: different scaffold pore sizes—different cell effects. Cytotechnology. 2015;68(3):355–69. https://doi.org/10.1007/s10616-015-9895-4.
Butler DL, Goldstein SA, Guilak F. Functional tissue engineering: the role of biomechanics. J Biomech Eng. 2000;122(6):570–5. Available at:. https://doi.org/10.1115/1.1318906.
Chen J-H, et al. Boning up on Wolff's Law: mechanical regulation of the cells that make and maintain bone. J Biomech. 2009;43(1):108–18. https://doi.org/10.1016/j.jbiomech.2009.09.016.
Chen Q, et al. Recent advances in polysaccharides for osteoarthritis therapy. Eur J Med Chem. 2017;139:926–35. https://doi.org/10.1016/j.ejmech.2017.08.048.

Chen L, et al. Growth factor and its polymer scaffold-based delivery system for cartilage tissue engineering. Int J Nanomedicine. 2020;15:6097–111. https://doi.org/10.2147/ijn.s249829.
Chen H, et al. 3D printed scaffolds based on hyaluronic acid bioinks for tissue engineering: a review. Biomater Res. 2023;27(1) https://doi.org/10.1186/s40824-023-00460-0.
Chen M, et al. Advancements in tissue engineering for articular cartilage regeneration. Heliyon. 2024;10(3):e25400. https://doi.org/10.1016/j.heliyon.2024.e25400.
Citeroni MR, et al. In vitro innovation of tendon tissue engineering strategies. Int J Mol Sci. 2020;21(18):6726. https://doi.org/10.3390/ijms21186726.
Connizzo BK, Han L, Sah RL. Electromechanobiology of cartilage and osteoarthritis. In: Advances in experimental medicine and biology; 2023. https://doi.org/10.1007/978-3-031-25588-5.
Dieterle MP, et al. Integrins, cadherins and channels in cartilage mechanotransduction: perspectives for future regeneration strategies. Expert Rev Mol Med. 2021;23 https://doi.org/10.1017/erm.2021.16.
Diez-Pascual A. Tissue engineering bionanocomposites based on poly(propylene fumarate). Polymers. 2017;9(12):260. https://doi.org/10.3390/polym9070260.
Dovedytis M, Liu ZJ, Bartlett S. Hyaluronic acid and its biomedical applications: a review. Eng Regen. 2020;1:102–13. https://doi.org/10.1016/j.engreg.2020.10.001.
Duncan RL, Turner CH. Mechanotransduction and the functional response of bone to mechanical strain. Calcif Tissue Int. 1995;57(5):344–58.
Durgam S, Stewart M. Cellular and molecular factors influencing tendon repair. Tissue Eng Part B Rev. 2017;23(4):307–17. https://doi.org/10.1089/ten.teb.2016.0445.
Elder SH, Athanasiou KA. Synergistic and additive effects of hydrostatic pressure and growth factors on tissue formation. PLoS One. 2009;4(4):e6791.
Fox AJS, Bedi A, Rodeo SA. The basic science of articular cartilage: structure, composition, and function. Sports Health Multidiscipl Appro. 2009;1(6):461–8. https://doi.org/10.1177/1941738109350438.
Fraldi M, et al. Toughening and mechanosensing in bone: a perfectly balanced mechanism based on competing stresses. Cornell University; 2022. Available at: https://doi.org/10.48550/arxiv.2206.14493.
Franchi M, et al. Collagen structure of tendon relates to function. Sci World J. 2007;7:404–20. https://doi.org/10.1100/tsw.2007.92.
Frassica JF, et al. Skeletal system: biomechanical concepts and relationships to normal and abnormal conditions. Elsevier BV. 1997;27(4):321–7. Available at: https://doi.org/10.1016/s0001-2998(97)80004-9
Freitas J, et al. Genetically engineered-MSC therapies for non-unions, delayed unions and critical-size bone defects. Int J Mol Sci. 2019;20(14):3430. https://doi.org/10.3390/ijms20143430.
Gao C, et al. Current progress in bioactive ceramic scaffolds for bone repair and regeneration. Int J Mol Sci. 2014;15(3):4714–32. https://doi.org/10.3390/ijms15034714.
Gokyer S, et al. 3D printed biodegradable polyurethaneurea elastomer recapitulates skeletal muscle structure and function. ACS Biomater Sci Eng. 2021;7(11):5189–205. https://doi.org/10.1021/acsbiomaterials.1c00703.
Gong M, et al. Graphene oxide–modified 3D acellular cartilage extracellular matrix scaffold for cartilage regeneration. Mater Sci Eng C. 2020;119:111603. https://doi.org/10.1016/j.msec.2020.111603.
Guntur RA, Rosen JC. Bone as an Endocrine organ. Elsevier BV. 2012;18(5):758–62. Available at: https://doi.org/10.4158/ep12141.ra
Guo X, et al. Regeneration of articular cartilage defects: therapeutic strategies and perspectives. J Tissue Eng. 2023;14:204173142311647. https://doi.org/10.1177/20417314231164765.
Hadjidakis D, Androulakis I. Bone Remodeling. Wiley. 2006;1092(1):385–96. Available at: https://doi.org/10.1196/annals.1365.035.
Han Y, You X, Xing W, et al. Paracrine and endocrine actions of bone—the functions of secretory proteins from osteoblasts, osteocytes, and osteoclasts. Bone Res. 2018;6:16. https://doi.org/10.1038/s41413-018-0019-6.

Huang L, et al. Biomimetic scaffolds for tendon tissue regeneration. Biomimetics. 2023;8(2):246. https://doi.org/10.3390/biomimetics8020246.

Hynes RO. Integrins. Cell. 2002;110(6):673–87. https://doi.org/10.1016/s0092-8674(02)00971-6.

Jung H-J, Fisher MB, Woo SL. Role of biomechanics in the understanding of normal, injured, and healing ligaments and tendons. BMC Sports Sci Med Rehabil. 2009;1(1) https://doi.org/10.1186/1758-2555-1-9.

Kahla RB, et al. Cell interaction and mechanobiological modeling of bone remodeling process. IntechOpen eBooks; 2020. https://doi.org/10.5772/intechopen.95045.

Karageorgiou V, Kaplan D. Porosity of 3D biomaterial scaffolds and osteogenesis. Biomaterials. 2005;26(27):5474–91. https://doi.org/10.1016/j.biomaterials.2005.02.002.

Kim D-D, Kim D-H, Son Y-J. Three-dimensional porous scaffold of hyaluronic acid for cartilage tissue engineering. In: Zilberman M, editor. Active implants and scaffolds for tissue regeneration. Studies in mechanobiology, tissue engineering and biomaterials. Berlin: Springer; 2010. p. 329–49. Available at:. https://doi.org/10.1007/8415_2010_51.

Kock L, Van Donkelaar CC, Ito K. Tissue engineering of functional articular cartilage: the current status. Cell Tissue Res. 2011;347(3):613–27. https://doi.org/10.1007/s00441-011-1243-1.

Kwon SG, et al. Recent advances in stem cell therapeutics and tissue engineering strategies. Biomater Res. 2018;22(1) https://doi.org/10.1186/s40824-018-0148-4.

Laurencin CT, et al. Delivery of small molecules for bone regenerative engineering: preclinical studies and potential clinical applications. Drug Discov Today. 2014;19(6):794–800. https://doi.org/10.1016/j.drudis.2014.01.012.

Lavagnino M, et al. Tendon mechanobiology: current knowledge and future research opportunities. J Orthop Res. 2015;33(6):813–22. https://doi.org/10.1002/jor.22871.

Lieber RL. Skeletal muscle structure, function, and plasticity: the physiological basis of rehabilitation. 2nd ed. Baltimore: Lippincott Williams & Wilkins; 2002.

Lopes D, et al. Bone physiology as inspiration for tissue regenerative therapies. Biomaterials. 2018;185:240–75. https://doi.org/10.1016/j.biomaterials.2018.09.028.

Losquadro WD, et al. Polylactide-co-glycolide Fiber–Reinforced calcium phosphate bone cement. Arch Facial Plast Surg. 2009;11(2) https://doi.org/10.1001/archfacial.2008.512.

Ma H, et al. Preparation and cytocompatibility of polylactic acid/hydroxyapatite/graphene oxide nanocomposite fibrous membrane. Springer Nature. 2012;57(23):3051–8. Available at: https://doi.org/10.1007/s11434-012-5336-3.

Man Z, et al. Transplantation of allogenic chondrocytes with chitosan hydrogel-demineralized bone matrix hybrid scaffold to repair rabbit cartilage injury. Biomaterials. 2016;108:157–67. https://doi.org/10.1016/j.biomaterials.2016.09.002.

Marlovits S, et al. Cartilage repair: generations of autologous chondrocyte transplantation. Eur J Radiol. 2005;57(1):24–31. https://doi.org/10.1016/j.ejrad.2005.08.009.

McCullen DS, et al. Nanofibrous composites for tissue engineering applications. Wiley. 2009;1(4):369–90. Available at: https://doi.org/10.1002/wnan.39

Mumme M, et al. Nasal chondrocyte-based engineered autologous cartilage tissue for repair of articular cartilage defects: an observational first-in-human trial. Lancet. 2016;388(10055):1985–94. https://doi.org/10.1016/s0140-6736(16)31658-0.

Niu Y, Du T, Liu Y. Biomechanical characteristics and analysis approaches of bone and bone substitute materials. J Funct Biomater. 2023;14(4):212. https://doi.org/10.3390/jfb14040212.

O'Conor CJ, et al. Cartilage-specific knockout of the mechanosensory ion channel TRPV4 decreases age-related osteoarthritis. Sci Rep. 2016;6(1) https://doi.org/10.1038/srep29053.

Pien N, et al. Tissue engineering of skeletal muscle, tendons and nerves: a review of manufacturing strategies to meet structural and functional requirements. Appl Mater Today. 2023;31:101737. https://doi.org/10.1016/j.apmt.2023.101737.

Poole AR, Kojima T, Yasuda T, Mwale F, Kobayashi M, Laverty S. Composition and structure of articular cartilage: a template for tissue repair. Clin Orthop Relat Res. 2001;391:S26–33.

Qazi TH, et al. Biomaterials based strategies for skeletal muscle tissue engineering: existing technologies and future trends. Biomaterials. 2015;53:502–21. https://doi.org/10.1016/j.biomaterials.2015.02.110.

Rindom E, Vissing K. Mechanosensitive molecular networks involved in transducing resistance exercise-signals into muscle protein accretion. Front Media. 2016;7. Available at: https://doi.org/10.3389/fphys.2016.00547.

Roshan-Ghias A, et al. In vivo cyclic loading as a potent stimulatory signal for bone formation inside tissue engineering scaffold. Eur Cells Mater. 2010;19:41–9. https://doi.org/10.22203/ecm.v019a05.

Rubin CT, Lanyon LE. Regulation of bone formation by applied dynamic loads. J Bone Joint Surg. 1984;66(3):397–402. https://doi.org/10.2106/00004623-198466030-00012.

Ruiz-Alonso S, et al. Tendon tissue engineering: cells, growth factors, scaffolds and production techniques. J Control Release. 2021;333:448–86. https://doi.org/10.1016/j.jconrel.2021.03.040.

Sanchez-Adams J, et al. The Mechanobiology of articular cartilage: bearing the burden of osteoarthritis. Curr Rheumatol Rep. 2014;16(10) https://doi.org/10.1007/s11926-014-0451-6.

Sayed KE, et al. Heterotopic autologous chondrocyte transplantation—a realistic approach to support articular cartilage repair? Tissue Eng Part B Rev. 2010;16(6):603–16. https://doi.org/10.1089/ten.teb.2010.0167.

Schwarcz HP, Abueidda D, Jasiuk I. The ultrastructure of bone and its relevance to mechanical properties. Front Phys. 2017;5 https://doi.org/10.3389/fphy.2017.00039.

Steele JAM, et al. In vitro and in vivo investigation of a zonal microstructured scaffold for osteochondral defect repair. Biomaterials. 2022;286:121548. https://doi.org/10.1016/j.biomaterials.2022.121548.

Sun W, et al. The mechanosensitive Piezo1 channel is required for bone formation. elife. 2019;8 https://doi.org/10.7554/elife.47454.

Tang Y, et al. TGF-β1–induced migration of bone mesenchymal stem cells couples bone resorption with formation. Nat Med. 2009;15(7):757–65. https://doi.org/10.1038/nm.1979.

Tatara AM, Mikos AG. Tissue engineering in orthopaedics. J Bone Joint Surg. 2016;98(13):1132–9. https://doi.org/10.2106/jbjs.16.00299.

Viidik A. Structure and function of normal and healing tendons and ligaments. Springer eBooks; 1990. p. 3–38. https://doi.org/10.1007/978-1-4612-3448-7_1.

Wang JHC. Mechanobiology of tendon. J Biomech. 2005;39(9):1563–82. https://doi.org/10.1016/j.jbiomech.2005.05.011.

Wang JH, Guo Q, Li B. Tendon biomechanics and mechanobiology—a minireview of basic concepts and recent advancements. J Hand Ther. 2011;25(2):133–41. https://doi.org/10.1016/j.jht.2011.07.004.

Wang M, et al. Designing functional hyaluronic acid-based hydrogels for cartilage tissue engineering. Mater Today Bio. 2022;17:100495. https://doi.org/10.1016/j.mtbio.2022.100495.

Wang N, et al. Mechanotransduction pathways in articular chondrocytes and the emerging role of estrogen receptor-α. Bone Res. 2023;11(1) https://doi.org/10.1038/s41413-023-00248-x.

Winarso R, et al. Application of fused deposition modeling (FDM) on bone scaffold manufacturing process: a review. Elsevier BV. 2022;8(11):e11701. Available at: https://doi.org/10.1016/j.heliyon.2022.e11701.

Woo SLY. Biomechanics of tendons and ligaments. Springer eBooks; 1986. p. 180–95. https://doi.org/10.1007/978-1-4612-4866-8_14.

Woo SL, et al. Tissue engineering of ligament and tendon healing. Clin Orthop Relat Res. 1999;367:S312–23. https://doi.org/10.1097/00003086-199910001-00030.

Xie Y, et al. Proper mechanical stimulation improve the chondrogenic differentiation of mesenchymal stem cells: improve the viscoelasticity and chondrogenic phenotype. Biomed Pharmacother. 2019;115:108935. https://doi.org/10.1016/j.biopha.2019.10893.

Yang G, Rothrauff BB, Tuan RS. Tendon and ligament regeneration and repair: clinical relevance and developmental paradigm. Birth Defects Res. 2013;99(3):203–22. https://doi.org/10.1002/bdrc.21041.

You J, et al. Substrate deformation levels associated with routine physical activity are less stimulatory to bone cells relative to Loading-Induced oscillatory fluid flow. J Biomech Eng. 2000;122(4):387–93. https://doi.org/10.1115/1.1287161.
Zhang H, Zhou L, Zhang W. Control of scaffold degradation in tissue engineering: a review. Tissue Eng Part B Rev. 2014;20(5):492–502. https://doi.org/10.1089/ten.teb.2013.0452.
Zhao F, Vaughan TJ, McNamara LM. Quantification of fluid shear stress in bone tissue engineering scaffolds with spherical and cubical pore architectures. Biomech Model Mechanobiol. 2015;15(3):561–77. https://doi.org/10.1007/s10237-015-0710-0.
Zuk PA, et al. Multilineage cells from human adipose tissue: implications for cell-based therapies. Tissue Eng. 2001;7(2):211–28. https://doi.org/10.1089/107632701300062859.

Sleep Disorders and Regenerative Medicine

Sara McNeillis

Abstract Sleep is a part of a circadian rhythm with an approximate occurrence of once in 24 hours. It is timed to the light/dark cycle with sleep occurring in the period of darkness for humans and during the period of lightness in rats and mice. There are two processes controlling the sleep period. Process C which is governed by Clock genes and Process S which is governed by rising levels of adenosine. These processes also influence hormone release and gene expression in cells to promote proteolysis or proteogenesis. The cellular processes in musculoskeletal tissues promote fibroblast, chondrocyte, osteoblast and osteoclast proliferation and differentiation. In addition, cellular processes also control the release of cytokines and the stimulation of the unfolded protein process. Sleep loss has been shown to have adverse effects on these processes. Hypoxia stimulates the release of vascular derived endothelial growth factor which should have beneficial effects on wound healing but can be countered by pro-inflammatory cytokines. Manipulating and improving sleep and hypoxia may lead to stable and well-timed cellular process to encourage healing and improve stem cell survival.

Keywords Circadian · Hormones · Sleep loss · Regenerative process · Stem cells

1 Introduction

1.1 The Connection Between Sleep and Regenerative Medicine

In, early depiction of art, sleep was a state in the hands of gods. In, Greek mythology, the god of Hypnos was assigned as the provider and protector of sleep. Scientific description in 1834 stated sleep to be a state between wakefulness and death, with death being the complete suspension of the waking state. Sleep is described as a

S. McNeillis (✉)
University College London Hospitals, London, UK

P. Lee (ed.), *Musculoskeletal Regeneration Medicine*,
https://doi.org/10.1007/978-3-031-77456-0_4

reversible state which occurs every 24 hours for a period where consciousness is suspended, and postural muscles are relaxed. In, the average adult human's sleep is termed monophasic because the sleep period occurs once in 24 hours. Timing natural sleep coincides with the 24-hour earth cycle and occurs during the period of darkness. There are two processes which assist sleep to occur: Process C and Process S.

Process C is termed the circadian process (Fig. 1). It was found that the suprachiasmatic nucleus was the location of the circadian pacemaker (Fuller et al. 1981; Drucker-Colin et al. 1984). This process is controlled by Clock genes (Vitaterna 1994). Clock (circadian locomotor out-put cycles kaput) and BMAL1 (aryl hydrocarbon receptor nuclear translocator-like protein 1 (ARNTL), or brain and muscle ARNT-like 1) are located in the nucleus and transcript the proteins *clock* and *bmal* which form a dimer. These cross into the cytoplasm and induce the genes PER and CRY. Proteins *per* and *cry* inhibit the further production of proteins *clock* and *bmal1*, thus forming a negative feedback loop which takes approximately 24 hrs (Bae et al. 1998).

Legendre et al. in 1910 described an experiment in which dogs were infused with serum from dogs who were sleep deprived. The study dogs fell asleep (Legendre

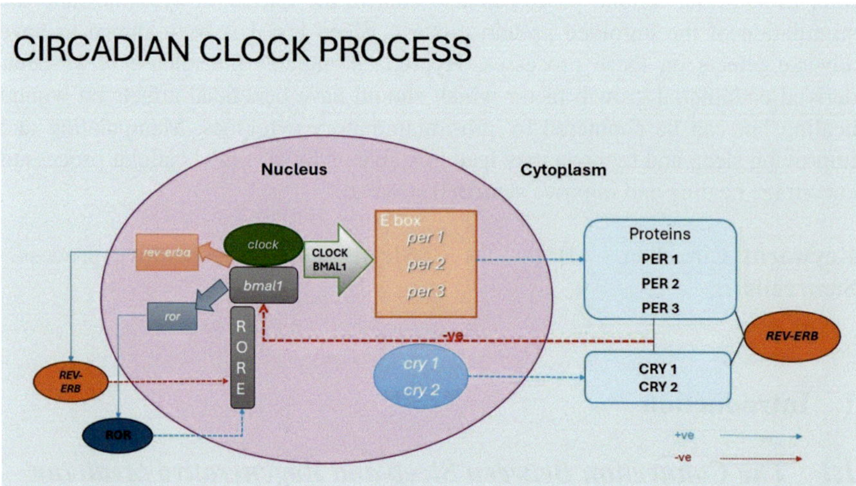

Fig. 1 Circadian clock process. The *Clock* and *Bmal1* genes transcript their respective proteins CLOCK and BMAL1 which form a dimer. This interacts at E-boxes on *per* and *cry* genes to increase transcription of the PER and CRY proteins which are released into the cytoplasm. They undergo phosphorylation in the cytoplasm to form heterodimer with other proteins and feedback to inhibit further transcription of CLOCK AND BMAL1. Thus, Per1/Cry1 inhibit their own transcription by inhibiting *Clock bmal1* and Per2/Cry2 inhibits its own transcription. This feedback process resulting in cyclical increase and decrease in levels occurs over 24 hrs. Meanwhile a second feedback loop which acts on bmal1 is the rev-erb α/β and ROR α/β orphan nuclear-receptor genes which form the REVERB and ROR proteins. These then bind to the (Retinoic acid-related orphan receptor responses element) RORE binding site within the promoter of bmal1 and either initiate or inhibit bmal1 transcription

and Piéron 1910). Further sleep deprivation experiments were conducted and the two process model for sleep was described with the second process being called Process S (Borbély and Neuhaus 1979). Process S substrate is now known to be Adenosine (Porkka-Heiskanen et al. 1997). Adenosine is a substrate of cell metabolism which builds up during the wakefulness due to the utilisation of ATP it is found within the cells and on the surface of the cells (Zimmermann 2000). Adenosine is maintained across the intra and extracellular spaces in equilibrium by nucleoside transporters (Gu et al. 1995) (Fig. 2). Adenosine concentrations accumulate progressively during wakefulness and reduce during sleep by acting on the basal forebrain to increase slow wave sleep. Extracellular adenosine is released when astrocytes covert ATP to adenosine which was thought to act on A1R receptors to induce sleep. Adenosine is a potent inhibitor of neural transmission (Porkka-Heiskanen et al. 1997; Schmitt et al. 2012).

Sleep is not a continuous state. Sleep progresses from wakefulness to deep sleep and then cycles into REM sleep. Overall, the average human adult will have sleep cycles each of approximately 90 mins with a greater proportion of deep sleep at the start of the night to a greater propensity for REM sleep towards the end of the night (McCarley). A description of sleep stages also known as sleep architecture, shows a healthy young adult would spend approximately 15–20% of the total sleep time in slow wave sleep, 20–25% in REM sleep, 50–60% in light sleep and a small proportion in wakefulness. Sleep onset is usually within 15 mins (Feinberg et al. 1967; Dumermuth et al. 1972).

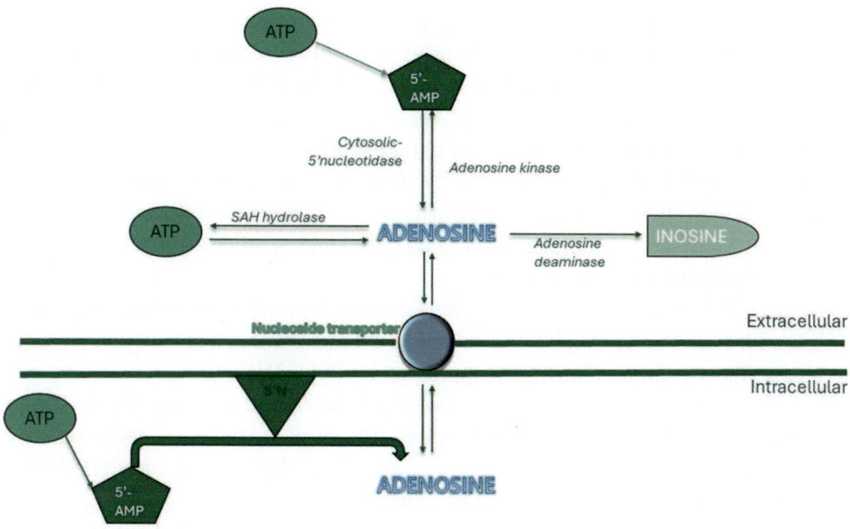

Fig. 2 Adenosine pathway. Adenosine is a byproduct of cyclical adenosine monophosphate (cAMP) and levels are maintained by the nucleoside transporter. Adenosine is also a substrate for the synthesis of adenosine triphosphate

In 1856 the benefits of a good night's sleep were described by attributing cognitive dysfunction due to prolonged wakefulness and restoration of this by recovery sleep. Eight hours sleep was advised by the author. Below, we describe the benefits of sleep on more cellular functions of regeneration.

2 Natural Sleep and Regenerative Processes

2.1 Understanding How Sleep Affects Tissue Regeneration

The circadian rhythm is the 24-hour biological cycle that is derived from the clock genes. The body rhythm is controlled by the suprachiasmatic nucleus which operates as a master clock sending signals to the hypothalamus. Therefore, the circadian rhythm also influences the release of many hormones.

One of the key anabolic hormones, growth hormone, is released in slow wave sleep and is tightly controlled to this stage by cellular mechanisms. Growth hormone is released from the pituitary gland which receives signals from the hypothalamus, namely the growth hormone releasing hormone. The positive actions of growth hormone include an increase in muscle cell protein synthesis and lipolysis which enable the use of free fatty acids. It stimulates the production of insulin like growth factor and enables the increase in intracellular glucose and glucose-6-phosphate. Overall growth hormone increased the muscle mass. Insulin like growth factor 1 increases protein synthesis and reduces proteolysis (Boyar and Kapen 1977).

Catabolic hormones such as cortisol and thyroid stimulating hormone are also influenced by sleep. Over the course of the day thyroid stimulating hormone increases to a peak in early evening. During sleep this is reduced. Sleep deprivation studies in shift workers showed continual increase in TSH during the night whilst they were awake (Goichot et al. 1998). Although TSH increases the metabolic rate, it does this by increasing glycolysis and via prostaglandins to increase heart rate and temperature. Cortisol is less influenced by sleep, however there is a burst of cortisol prior to waking (Hurwitz et al. 2004). Cortisol secretion is under the influence of the sympathetic nervous system via the adrenal gland and is released in stress. Cortisol increases extracellular glucose levels.

Melatonin is a hormone which is synthesised in the pineal gland from tryptophan (Fig. 3). The secretion of melatonin occurs in darkness and peaks between midnight and 3 am. Its secretion is however suppressed by light. Retinal photoreceptive ganglionic cells transmit light signals to the suprachiasmatic nucleus via the retino-hypothalamic tract and onto the pineal gland. Pineal gland cells are photosensitive, thus the light signal from the retina suppresses the enzyme arylalkylamine N-acetyltransferase which acetylates serotonin.

Adrenergic output also influences the enzyme. The pineal gland is not the only site where melatonin is secreted, extra pineal sources have been found in the body systems (Vasey et al. 2021). In the body melatonin acts on MT1 and MT 2 receptors.

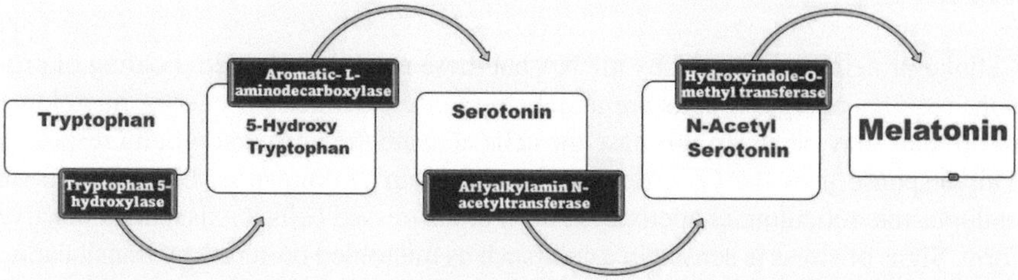

Fig. 3 Melatonin synthesis from tryptophan. Light signals from phototic cells cause suppression of the Arylalkylamine N-acetly transferase enzyme which is a critical step in the synthesis of melatonin

It has been implicated to have an influence in fertility processes in males and females and especially of small mammals. Melatonin improves the sleep state, it also appears to have effects on other body systems. Experiments on melatonin deficiency have shown not only altered sleep state but also cellular integrity and fertility to be dysfunctional leading to poor healing and carcinogenic potential. Melatonin in musculoskeletal regeneration has a role to play. It is a powerful antioxidant and can reduce free radical related damage to cellular processes. Adequate but not high doses of melatonin have been seen to be beneficial in studies investigating the potential recovery from peripheral nerve injuries (Behram Kandemir and Sarikcioglu 2015). Melatonin has also been shown to improve and enhance survival of adipose derived mesenchymal stem cells when placed in an area of ischaemia (Lee et al. 2017). In humans Melatonin was used in combination with strontium Vitamin D3 and Vitamin K2 to improve osteopenia revealed an increase in osteoblastogenesis in stem cell cultures and reduced osteoclastogenesis, whilst the combination improved overall mood and sleep quality in the subjects (Maria et al. 2017).

2.2 *Role of Sleep in Healing and Cellular Regeneration*

Most humans are active during the day and asleep at night. The number of hours of sleep decreases from newborn to adulthood and a little further into old age (Tune 1968). The function of sleep was unknown until recently although it was postulated by Smith et al. that sleep indeed appeared to enhance protein synthesis in the rat brain (Smith 1985). Restoration of brain energy metabolism was hypothesised to occur by EEG synchronisation in slow wave sleep (Benington and Heller 1995) with glycogen stores being repleted during sleep. Unfortunately, many animal studies showed that this was not proven in sleep deprivation studies.

2.3 Unfolded Protein Process

Cellular proteins are formed by mRNA but these need to be folded. Folding of proteins requires energy. If cells are under stress or challenged for energy, misfolding of proteins may occur. In this case the cells activate the unfolded protein response. This response uses the Glucose Regulated Protein 78 known as BiP which is an endoplasmic reticulum chaperone. In normal unstressed cells it exists in its inactive form. Signs of stress is sensed in a cell (such as misfolded proteins) by translocating proteins PERK (protein kinase RNA like endoplasmic reticulum kinase), IRE1 (inositol requiring enzyme 1) and ATF6 (Activating transcription factor 6). These then ensure BiP is activated and attaches to the endoplasmic reticulum to unfold and degrade the misfolded protein. The aim is also to assist the endoplasmic reticulum to increase its folding capacity. If this cannot be achieved the BiP induces cell apoptosis or elicits an immune response by expressing at the cell surface, where immunoglobulins can recognise the damaged cell. In sleep BiP levels are at baseline (Naidoo et al. 2007; Vasey et al. 2021). Thus, the unfolded protein process must function effectively to maintain the integrity of the cell (Fig. 4).

2.4 Gene Expression in Sleep

Studies elucidating gene expression between wakefulness and sleep reveal interesting results. Genes relating to biosynthesis of macromolecular proteins and biosynthetic pathways were upregulated in sleep. Of note genes for haem and porphyrin synthesis were upregulated in the cerebral cortex. In addition, genes encoding cellular and mitochondrial ribosomes were also upregulated as were genes encoding cellular and mitochondrial molecular transport. Thus, genes encoding protein synthesis appear to be upregulated in sleep.

2.5 Immune Function in Sleep

Cytokines have a role in sleep (Fig. 5). Inflammatory cytokines are released in various conditions and promote sleep for example tumour necrosis factor α (TNF α) induces NREM sleep. Interleukin 1 (IL-1) also induces NREM sleep (Moldofsky 1995). Sleep onset however results in a reduction of catecholamines and IL-2 (Irwin et al. 1999).

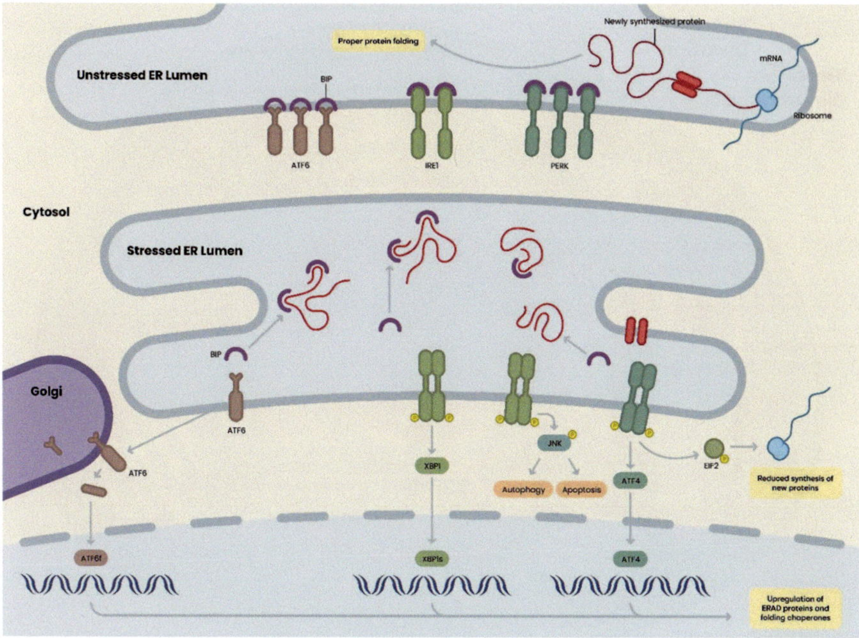

Fig. 4 Unfolded protein process is stimulated when there is a misfolding of proteins. BiP which is normally inactivated and attached to AT6/IRE1 and PERK will disassociate and attach to the misfolded protein to alert the cellular mechanisms to recognise and remove these

3 Sleep Loss and Impact on Regeneration

3.1 Effects of Sleep Deprivation on Regeneration

Sleep loss was known to be detrimental to mind and body functions as early as 1850s. It was only in the late twentieth century that studies on sleep deprivation in rats and the associated effects on regenerative processes was conducted. Here we summarise these studies.

Sleep deprivation in rats was shown to be fatal. Rats subjected to 3 weeks of sleep deprivation succumbed to lethal septicaemia. Sleep deprivation in human studies up to 11 days did not show any fatalities, however. It is possible that oxidative stress is increased if sleep deprived, and this may lead to irreparable cellular damage and translocation of bacteria (Everson and Toth 2000; Cirelli et al. 2004; Naidoo et al. 2005).

An early study in 1997 with 72 hours sleep deprivation did not however show any differences on subcutaneously created wounds between rats who were sleep deprived and those who were not sleep deprived. Again, REM sleep deprivation studies also did not show a difference in wound healing. The addition of nicotine and sleep deprivation study in 2004 did show that there was an decrease in

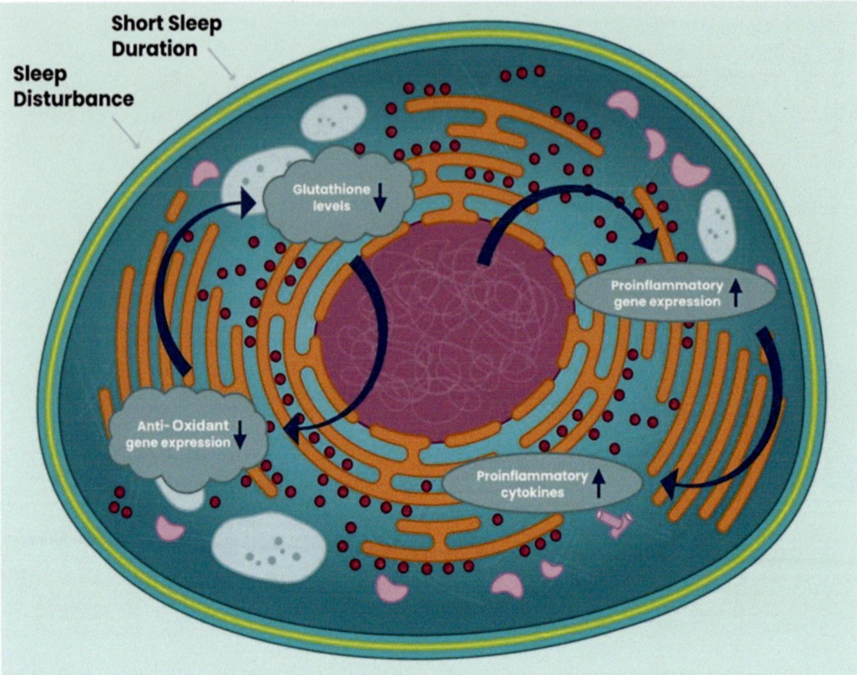

Fig. 5 Sleep loss leads to a cascade of proinflammatory gene expression and formation of proinflammatory cytokines. It also reduces the expression of anti-inflammatory gene expression and glutathione synthesis

fibroblasts and capillaries in this group (Landis and Whitney 1997; Mostaghimi et al. 2005; Gümüştekín et al. 2004). Sleep deprivation in combination with morphine was also noted to have a detrimental effect on wound healing. It therefore seems that sleep deprivation alone does not appear to affect wound healing but in combination with either nicotine or morphine does.

Sleep deprivation in the presence of harmful substances however may affect cellular processes adversely. Naidoo, for example, demonstrated the effect of sleep deprivation on the unfolded protein response in mice. The response was amplified and induced (Naidoo et al. 2005). Gene expression studies by Cirelli et al found that glial proteins such as membrane traffic and cholesterol synthesis were downregulated following sleep loss. Conversely heat shock proteins and mitochondrial and synaptic proteins in the brain were expressed during wakefulness and sleep deprivation (Cirelli et al. 2006).

Oxidative stress following sleep deprivation is difficult to investigate, studies assessing surrogates for oxidative stress such as antioxidant responses did not show any difference in the cerebral cortex or hypothalamus in rats who were sleep deprived, this seems to be converse to the evidence which demonstrated that there was an increase in proteins which may be responsible for cellular activity. Surrogates investigated are superoxide dismutase an enzyme which is expressed as a consequence of oxidative stress. Every et al. investigated this in a paper particularly assessing the glutathione reductase pathway and the enzymes involved, glutathione reductase which converts glutathione disulfide back to glutathione in association with co-factor NADPH. NADPH regeneration requires enzyme glucose-6-phosphate dehydrogenase and 6-phophgluconate dehydrogenase. In addition, the presence of oxidative stress and cell damage was evaluated by measuring levels of aminotransferases. The results confirmed that a reduction in antioxidant levels was present in the liver following sleep deprivation in rats and increased levels of aminotransferases were found in the heart of these animals. These changes were reversed after recovery sleep. Therefore sleep loss not only impacted antioxidant pathways, recovery sleep restored these (Everson et al. 2005). It is worth noting that the levels were measured in liver and heart.

A study reassessing antioxidant response in the brain, demonstrated that here the some of the levels were increased by short term sleep deprivation. Whilst superoxide dismutase did not vary, there were found to be increases in Glutathione peroxidase and total glutathione (Ramanathan et al. 2010).

Whilst most sleep deprivation studies assessed effects on other organs, few studies assessed the effects on either cartilage bone or muscle. In one paper, a clear correlation with muscle mass and sleep deprivation was seen. Rats in this study were subjected to either 96 hrs sleep deprivation or 96 hrs sleep deprivation followed by recover sleep and normal controls, food and water was freely available. Results in this study revealed that muscle mass was decreased in rats who were sleep deprived compared to controls. In the sleep deprived rats, levels of cortisol were vastly increased and levels of testosterone were mildly decreased which were reversed on recovery sleep, therefore, this may be postulated to be the mechanism for muscle breakdown (Mônico-Neto et al. 2017). Human studies on muscle mass following sleep deprivation are few but in one study of levels of IGF-1 in healthy human males showed that there were no changes in this following sleep deprivation but there was an increase in levels in sleep extension compared to habitual hours (Chennaoui et al. 2016).

Sleep deprivation enhances the increase of pro-inflammatory cytokines and in particular TNFα, Il-1β and IL-6 (Krueger et al. 2011). These may be increased due to additional stress of staying awake and concomitant release of cortisol or due to neural factors which regulate sleep/wake states. Whereas recovery sleep will result in a reduction in these cytokines and promote increase in anti-inflammatory cytokines (Besedovsky et al. 2012).

4 Nocturnal Hypoxaemia and Its Impact on Wound Healing

4.1 Understanding How Hypoxaemia in Sleep Affects Tissue Regeneration

Hypoxaemia is defined as a reduction in resting oxygen saturation levels. Nocturnal hypoxaemia is usually caused by repetitive obstructions to the airway in sleep and are termed intermittent hypoxaemia or IH. Sustained hypoxaemia is caused by persistent low levels of oxygen saturation. Many pathways exist for the hypoxia to cause inflammation, one such mechanism is the expression of hypoxia-inducible factor 1 which in turn induces the expression of angiogenesis and erythropoiesis under such conditions. HIF-1 acts through the neutrophil factor kappa B to enhance a neutrophil response and survival (Chen et al. 2019).

Hypoxaemia is a potent driver for a pro-inflammatory state of the organism. Sustained hypoxaemia is known to increase vascular derived endothelial growth factor (VEGF). VEGF levels have been found to stimulate abnormal angiogenesis, which may be implicated in development of atherosclerosis and endothelial damage consequently causing cardiovascular morbidity (Fig. 6). Serum levels of Il-6 and VEGF were assessed in patients with obstructive sleep apnoea by (Burioka et al. 2009). They showed that untreated intermittent hypoxia was associated with higher levels of IL-6 but not VEGF.

However, in a study comparing patients with OSA and controls the baseline VEGF levels were higher (Archontogeorgis et al. 2015). Further the patients with OSA were shown to have a correlation between VEGF levels and severity of OSA as measured by the apnoea/hypopnoea index AHI. Levels of VEGF were shown to decrease on treatment of OSA with CPAP.

Endothelial damage in OSA patients was investigated by Haranczynska et al. in patients with OSA revealed higher soluble intracellular adhesion molecule -1 (sICAM-1) levels and N terminal pro B type natriuretic peptide (NT-proBNP) levels. NT proBNP levels are a marker of right ventricular failure.

Endothelin -1 levels were associated with those with a higher body weight (BMI >35). Endothelin is important in development of atherosclerosis by encouraging monocyte migration and adherence to endothelial surface (Harańczyk et al. 2022).

An in vitro study comparing cells and wound healing from patients with OSA with and without metabolic syndrome and controls found that although levels of VEGF were increased in patients with OSA without metabolic syndrome, wound healing was decreased. It was postulated that other factors may be counteracting the beneficial effects of VEGF on wound healing of the endothelium. In this study C-reactive protein (CRP) was responsible for counteracting the effects of VEGF on wound healing. CRP is also responsible for enhancing monocyte migration which has deleterious effects on endothelium (Briançon-Marjollet et al. 2014).

Intermittent hypoxia may be beneficial to endothelial wound healing in the absence of CRP. In one study mesenchymal stem cells from the bone marrow were

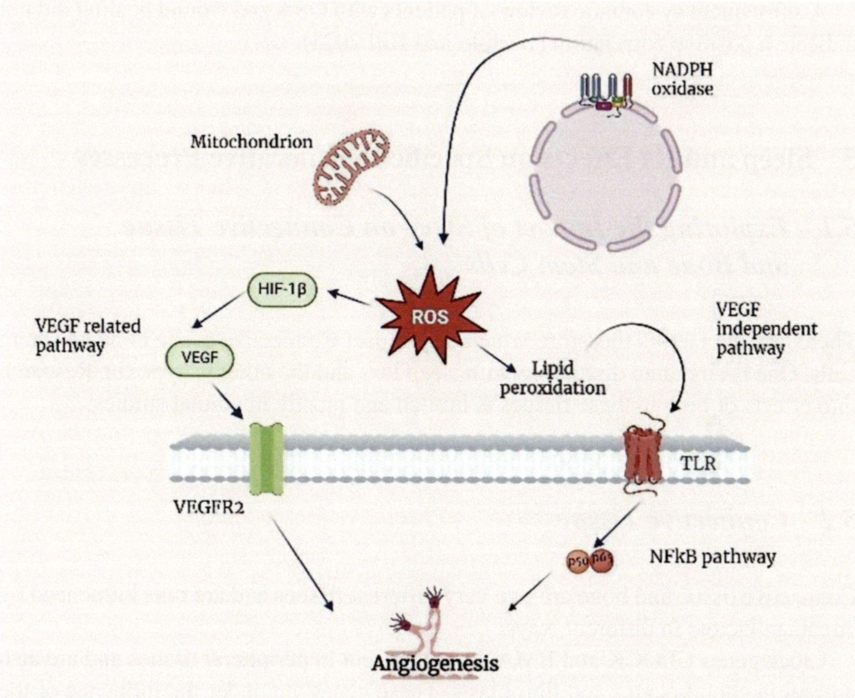

Fig. 6 VEGF can activate angiogenesis via the HIF1α. Sustained hypoxia is a stimulus for this process

shown to more migratory and more adhesive when subjected to the serum form rats subjected to intermittent hypoxia compared with controls (Carreras et al. 2010).

Although many studies have looked at the effects of hypoxia on vascular endothelial system, some studies also assess the effect of hypoxia on wound healing in other tissues.

Avezov et al. investigated the effect of intermittent hypoxia in vitro on wound healing on oral mucosa cell culture. Cells exposed to intermittent hypoxia demonstrated a delay on scratch closure compared with sustained hypoxia or controls. Interestingly, there has been an investigation assessing the osteoarthritis in sleep deprived rats compared with controls. The results demonstrate that the factors VEGF and HIF-1 are expressed in the hypoxia group compared with controls, leading to more risk of osteoarthritis due to induction of abnormal angiogenesis. In fact, the articular surface showed disarrangement of chondrocytes under histological examination (Chen et al. 2019).

Whilst a human study assessing wound healing in diabetic patients with risk of OSA demonstrated poor wound healing compared with those with a STOP-BANG score of <4 (Chapter "Psychology in Musculoskeletal Regeneration"). In addition, a study on surgical wound complications following breast surgery showed an increase in wound complications in patients with OSA.

A subsequent systematic review of patients with OSA and wound healing did not indicate a positive correlation (Bartolo and Hill 2023).

5 Sleep and Its Effects on Specific Regenerative Processes

5.1 Exploring the Impact of Sleep on Connective Tissue and Bone and Stem Cells

There are two factors therefore which can impact Connective tissue bone and stem cells. One is circadian disruption with sleep loss and the other is hypoxia. Research into effects of both in these tissues is limited and mostly in animal studies.

5.2 Connective Tissue

Connective tissue and bone are two very different tissues and are thus influenced by circadian factors in different ways.

Clock genes CLOCK and BMAL 1 are present in peripheral tissues and are also present in chondrocytes and fibroblasts. These genes are under the influence of the master clock in the suprachiasmatic nucleus (SCN). Signals from the master clock are transmitted to align these tissues to the earth cycle through release of hormones especially melatonin and cortisol, however some chondrocytes are in avascular structures. By monitoring Per 2 gene expression Gossan et al demonstrated circadian oscillations which continued in cell cultures for approximately 3 weeks after removal from the rat and these were sensitive to temperature, which is therefore the circadian signal as means of synchrony with the SCN (Gossan et al. 2013). Takarada et al. demonstrated that chondrocytes from rib cartilage demonstrated expression of all clock genes for up to 3 weeks. Period genes Per 1 and Per 2 are involved in chondrogenesis. BMAL -1 was shown to be the promoter gene for chondrogenesis (Fig. 7). Furthermore, rats with absence in BMAl-1 gene had restricted length of bone growth (Takarada et al. 2012).

Disruption of circadian rhythm will have adverse effects on the articular cartilage and bone density. In mice who were phase shifted a reduction in proteoglycan was seen under histological examination compared with control mice and increase in immune cell infiltration was seen along the articular cartilage. Additionally, proinflammatory markers were identified in the articular cartilage of the phase shifted mice compared with controls. The mechanism involves the mitogen activated protein kinase (MAPK) – NFκB pathways (Kc et al. 2015). Dudeck et al's group explored both human and mouse studies to evaluate the contribution of CLOCK genes in the development of osteoarthritis and more importantly the BMAL -1 clock gene. They found that BMAL-1 expression on articular surface cartilage did

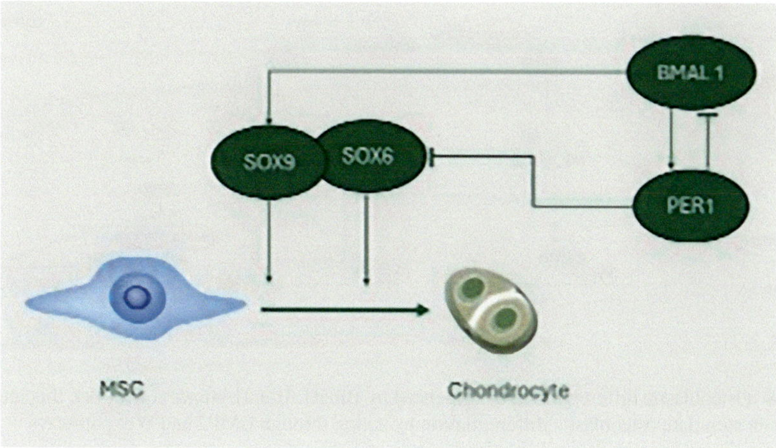

Fig. 7 Bmal1 and Per2 dimer stimulates SOX genes to increase chondrogenesis

decrease in human subjects with osteoarthritis. They then looked at the effects of aging by assessing young mice compared with aged mice and again they found that clock genes did negatively correlate with age leading to a disruption in the chondrocyte circadian rhythm (Dudek et al. 2016).

In later studies, it was found that BMAL-1 may modulate the development of osteoarthritis through the MAPK-NFκB pathway. An experiment on sleep deprived rats (which were succumbed to 18 hrs of paradoxical sleep disruption) assessment of the chondrocyte cell cultures from sacrificed rats showed that chronic sleep disruption resulted in lower levels of BMAL-1 and overexpression of IL-6 and P-ERK with osteophytic changes seen on histology. These pro-inflammatory cytokines were downregulated in rats who then went on to have recovery sleep (Chen et al. 2020).

5.3 Bone

Osteogenic activity is also under circadian influence, and this is so to allow osteoclastic and osteoblastic activity to coordinate with activity and rest cycles of the diurnal animal (Fig. 8).

In Takarada et al's study, examining skeletal tibial bone in BMAL -1 null mice and wild type mice, he showed that BMAL-1 null mice had shorter body length than wild type mice. In a 4-week male the body length decrease was greatest. Also, chondrocyte proteoglycan matrix synthesis was greater in the wild type mice (Takarada et al. 2012). Deficiency of BMAL-1 is associated with reduced numbers of osteoblasts and osteoclasts in vivo and these osteoblasts also differentiate poorly.

The clock proteins CLOCK/BMAL form a heterodimer which interacts with PER and CRY genes to promote protein synthesis or protein breakdown. There are

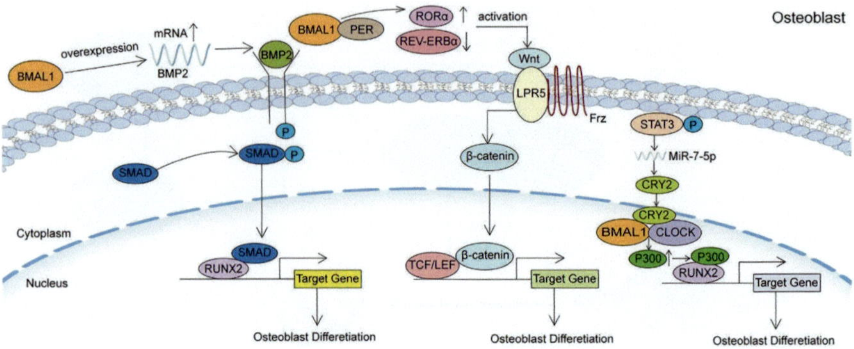

Fig. 8 Osteoblastic differentiation is influenced by Bmal1. Bmal1 interacts with Per, this interaction can stimulate osteoblastic differentiation by acting through BMP2 and Wnt pathways

three subtypes of PER gene, Per 1, Per 2 and Per 3. Of these the most important determinant of circadian rhythmicity is Per 2. By interacting with EBox, it controls the length of the overall circadian cycle. Mutations in Per 2 result in loss of circadian rhythmicity. Additionally, it was demonstrated that Cry 1 and Cry 2 are associated with locomotor rhythmicity in mice. Absence of Cry 1 results in arrhythmic locomotor activity.

Per 2 gene inhibits osteoblasts and affects osteoblast proliferation negatively. Therefore, mice with deficiency in Per 2 hormone appeared to have higher bone formation as shown in the studies (Maronde et al. 2010; Abe et al. 2019). Osteoblasts differentiation or activity, however, was not found in Cry deficient mice. Osteoclastic activity in this group was reduced.

Runt related transcription factors such as Runx2 are involved in bone homeostasis. Expression of Runx2 is also rhythmic and found in other tissues. Runx2 is involved in stimulating osteocalcin in the bone, but it has been shown to be present in the SCN of the mouse and demonstrated rhythmicity. Runx2 was clock controlled with BMAL-1 modulating activity (Fig. 7). This then leads to the hypothesis that ostseocalcin in the bone could be under circadian control.

Overall, there is considerable proof that circadian clock genes positively interact with factors inducing and differentiating osteoblasts in bone.

5.4 *Stem Cells*

Stem cells either originate from embryonic tissue or adult tissue. Whilst embryonic stem cells can differentiate into any tissue, adult stem cells can only differentiate into the tissues they normally arise from. Stem cells have an important function in maintaining the integrity of the organs. Research into circadian influence on stem cell in rats, shows that embryonic stem cells do not express clock genes and have no circadian rhythmicity. As the embryonic stem cells start to differentiate, they start to express clock genes. Meanwhile, adult stem cells express clock genes and are under circadian influence. The Per 2 gene is responsible for enhancing the circulating stem cell migration to cells expressing CXCL2 which is known to attract stem cells. The expression of CXCL2 is highest during the daytime and decreases during the night-time. Hence the stem cell count in plasma is highest at night. This may have important implications for treatments with stem cells.

In adipocyte precursors Per 3 was shown to enhance differentiation, as well as CLOCK and PER 2.

Neural stem cells responded to BMAL-1 and the knockout mice without this gene have been demonstrated to have increased neural cell death. Cry1 and Cry 2 also has an effect on neurospheres by mostly encouraging proliferation and growth.

6 Future Directions in Research and Practice

6.1 Emerging Research Trends

Increasingly mechanisms in which clock genes can influence tissues is being conducted in animal studies alongside research into modifiable factors to improve both quality and quantity of tissue regeneration. Holistic approaches to sleep and micromanagement of cellular biology will form tools for doctors to improve outcomes following tissue injury and healing.

6.2 Future Implications for Regenerative Medicine Practices

Regenerative Medicine is in its infancy. Tissues are complex and affected by many internal and external factors. It is hoped that enhancement of regeneration can be improved through understanding the biology of circadian rhythms and their effects on tissues. Artificial intelligence may play a role in identifying key factors which have the highest impact on enhancing the regenerative processes.

7 Conclusion

7.1 Summarizing the Role of Sleep in Regenerative Medicine

Sleep is a ubiquitous state which occurs in all cellular organisms and has a vital role in resting and repairing the cell. We have described how sleep occurs and specifically how sleep affects the tissues which need to be regenerated. Overall, there are strong influences from clock genes which control circadian rhythmicity. There is a biological advantage to programming regenerative processes to the light/dark cycle. Cells need time to function, and they need time to repair. We have shown how circadian factors influence regeneration. We have also described how these processes are damaged and can be altered in the face of circadian mismatch and hypoxia.

Humans have learnt to break from their light/dark cycles and can live in states with artificial light which profoundly affects the circadian rhythm through the secretion of melatonin.

It is therefore important to assess how we may best be able to take advantage of our biological cycle and protect them from harm and assist regeneration when tissues are injured.

We have shown strong evidence in circadian rhythmicity in cartilage, bone and stem cells. Manipulating factors to improve proliferation and differentiation of these tissues when beneficial is the next step in improving outcomes for patients undergoing surgery.

7.2 *Key Takeaways for Practitioners*

Sleep is part of a circadian rhythm. Sleep has influenced many musculoskeletal regenerative processes on a cellular basis. Sleep also influences the release of hormones timed to the light/dark cycle which promote or demote the formation of musculoskeletal tissue. These processes are adversely affected by loss of sleep or hypoxia because of sleep disorders. More translational research is needed to enhance the understanding of impact of sleep loss and hypoxia on human subjects.

References

Abe T, Sato T, Yoda T, Hoshi K. The period circadian clock 2 gene responds to glucocorticoids and regulates osteogenic capacity. Regen Ther. 2019;11:199–206.

Archontogeorgis K, Nena E, Papanas N, Xanthoudaki M, Hatzizisi O, Kyriazis G, Tsara V, Maltezos E, Froudarakis M, Steiropoulos P. Serum levels of vascular endothelial growth factor and insulin-like growth factor binding Protein-3 in obstructive sleep apnea patients: effect of continuous positive airway pressure treatment. Open Cardiovasc Med J. 2015;9:133–8.

Bae K, Lee C, Sidote D, Chuang KY, Edery I. Circadian regulation of a Drosophila homolog of the mammalian Clock gene: PER and TIM function as positive regulators. Mol Cell Biol. 1998;18(10):6142–51.

Bartolo K, Hill EA. The association between obstructive sleep apnoea and wound healing: a systematic review. Sleep Breath. 2023;27(3):775–87.

Behram Kandemir Y, Sarikcioglu L. Melatonin and its therapeutic actions on peripheral nerve regeneration. Folia Morphol (Warsz). 2015;74(3):283–9.

Benington JH, Heller HC. Restoration of brain energy metabolism as the function of sleep. Prog Neurobiol. 1995;45(4):347–60.

Besedovsky L, Lange T, Born J. Sleep and immune function. Pflugers Arch. 2012;463(1):121–37.

Borbély AA, Neuhaus HU. Sleep-deprivation: effects on sleep and EEG in the rat. J Comp Physiol. 1979;133(1):71–87.

Boyar RM, Kapen S Endocrine concomitants of sleep Proc. V Int. Congr. Endocrinology. Vol. I, 1977
Briançon-Marjollet A, Henri M, Pépin JL, Lemarié E, Lévy P, Tamisier R. Altered in vitro endothelial repair and monocyte migration in obstructive sleep apnea: implication of VEGF and CRP. Sleep. 2014;37(11):1825–32.
Burioka N, Koyanagi S, Fukuoka Y, Okazaki F, Fujioka T, Kusunose N, Endo M, Suyama H, Chikumi H, Ohdo S, Shimizu E. Influence of intermittent hypoxia on the signal transduction pathways to inflammatory response and circadian clock regulation. Life Sci. 2009;85(9–10):372–8.
Carreras A, Rojas M, Tsapikouni T, Montserrat JM, Navajas D, Farré R. Obstructive apneas induce early activation of mesenchymal stem cells and enhancement of endothelial wound healing. Respir Res. 2010;11(1):91.
Chen Y, Zhao B, Zhu Y, Zhao H, Ma C. HIF-1-VEGF-Notch mediates angiogenesis in temporomandibular joint osteoarthritis. Am J Transl Res. 2019;11(5):2969–82.
Chen G, Zhao H, Ma S, Chen L, Wu G, Zhu Y, Zhu J, Ma C, Zhao H. Circadian rhythm protein Bmal1 modulates cartilage gene expression in temporomandibular joint osteoarthritis via the MAPK/ERK pathway. Front Pharmacol. 2020;11:527744.
Chennaoui M, Arnal PJ, Drogou C, Sauvet F, Gomez-Merino D. Sleep extension increases IGF-I concentrations before and during sleep deprivation in healthy young men. Appl Physiol Nutr Metab. 2016;41(9):963–70.
Cirelli C, Gutierrez CM, Tononi G. Extensive and divergent effects of sleep and wakefulness on brain gene expression. Neuron. 2004;41(1):35–43.
Cirelli C, Faraguna U, Tononi G. Changes in brain gene expression after long-term sleep deprivation. J Neurochem. 2006;98(5):1632–45.
Drucker-Colin R, Aguilar-Roblero R, Garcia-Hernandez F, Fernandez-Cancino F, Bermudez Rattoni F. Fetal suprachiasmatic nucleus transplants: diurnal rhythm recovery of lesioned rats. Brain Res. 1984;311(2):353–7.
Dudek M, Gossan N, Yang N, Im HJ, Ruckshanthi JP, Yoshitane H, Li X, Jin D, Wang P, Boudiffa M, Bellantuono I, Fukada Y, Boot-Handford RP, Meng QJ. The chondrocyte clock gene Bmal1 controls cartilage homeostasis and integrity. J Clin Invest. 2016;126(1):365–76.
Dumermuth G, Walz W, Scollo-Lavizzari G, Kleiner B. Spectral analysis of EEG activity in different sleep stages in normal adults. Eur Neurol. 1972;7(5):265–96.
Everson CA, Toth LA. Systemic bacterial invasion induced by sleep deprivation. Am J Physiol Regul Integr Comp Physiol. 2000;278(4):R905–16.
Everson CA, Laatsch CD, Hogg N. Antioxidant defense responses to sleep loss and sleep recovery. Am J Physiol Regul Integr Comp Physiol. 2005;288(2):R374–83.
Feinberg I, Koresko RL, Heller N. EEG sleep patterns as a function of normal and pathological aging in man. J Psychiatr Res. 1967;5(2):107–44.
Fuller CA, Lydic R, Sulzman FM, Albers HE, Tepper B, Moore-Ede MC. Circadian rhythm of body temperature persists after suprachiasmatic lesions in the squirrel monkey. Am J Phys. 1981;241(5):R385–91.
Goichot B, Weibel L, Chapotot F, Gronfier C, Piquard F, Brandenberger G. Effect of the shift of the sleep-wake cycle on three robust endocrine markers of the circadian clock. Am J Physiol Endocrinol Metab. 1998;275(2 38–2):E243–8.
Gossan N, Zeef L, Hensman J, Hughes A, Bateman JF, Rowley L, Little CB, Piggins HD, Rattray M, Boot-Handford RP, Meng QJ. The circadian clock in murine chondrocytes regulates genes controlling key aspects of cartilage homeostasis. Arthritis Rheum. 2013;65(9):2334–45.
Gu JG, Foga IO, Parkinson FE, Geiger JD. Involvement of bidirectional adenosine transporters in the release of L-[3H]adenosine from rat brain synaptosomal preparations. J Neurochem. 1995;64(5):2105–10.
Gümüştekín K, Seven B, Karabulut N, Aktaş O, Gürsan N, Aslan S, Keleş M, Varoglu E, Dane S. Effects of sleep deprivation, nicotine, and selenium on wound healing in rats. Int J Neurosci. 2004;114(11):1433–42.

Harańczyk M, Konieczyńska M, Płazak W. Endothelial dysfunction in obstructive sleep apnea patients. Sleep Breath. 2022;26(1):231–42.
Hurwitz S, Cohen RJ, Williams GH. Diurnal variation of aldosterone and plasma renin activity: timing relation to melatonin and cortisol and consistency after prolonged bed rest. J Appl Physiol. 2004;96(4):1406–14.
Irwin M, Thompson J, Miller C, Gillin JC, Ziegler M. Effects of sleep and sleep deprivation on catecholamine and interleukin-2 levels in humans: clinical implications. J Clin Endocrinol Metab. 1999;84(6):1979–85.
Kc R, Li X, Voigt RM, Ellman MB, Summa KC, Vitaterna MH, Keshavarizian A, Turek FW, Meng QJ, Stein GS, van Wijnen AJ, Chen D, Forsyth CB, Im HJ. Environmental disruption of circadian rhythm predisposes mice to osteoarthritis-like changes in knee joint. J Cell Physiol. 2015;230(9):2174–83.
Krueger JM, Majde JA, Rector DM. Cytokines in immune function and sleep regulation. Handb Clin Neurol. 2011;98:229–40.
Landis CA, Whitney JD. Effects of 72 hours sleep deprivation on wound healing in the rat. Res Nurs Health. 1997;20(3):259–67.
Lee MS, Yin TC, Sung PH, Chiang JY, Sun CK, Yip HK. Melatonin enhances survival and preserves functional integrity of stem cells: a review. J Pineal Res. 2017;62(2)
Legendre R, Piéron H. Résultats de diverses injections de liquides d'animaux insomniques. Bulletin du Muséum national d'histoire naturelle. 1910;16(6):343–6.
Maria S, Swanson MH, Enderby LT, D'Amico F, Enderby B, Samsonraj RM, Dudakovic A, van Wijnen AJ, Witt-Enderby PA. Melatonin-micronutrients Osteopenia Treatment Study (MOTS): a translational study assessing melatonin, strontium (citrate), vitamin D3 and vitamin K2 (MK7) on bone density, bone marker turnover and health related quality of life in postmenopausal osteopenic women following a one-year double-blind RCT and on osteoblast-osteoclast co-cultures. Aging (Albany NY). 2017;9(1):256–85.
Maronde E, Schilling AF, Seitz S, Schinke T, Schmutz I, van der Horst G, Amling M, Albrecht U. The clock genes period 2 and Cryptochrome 2 differentially balance bone formation. PLoS One. 2010;5(7):e11527.
McCarley RW. Neurobiology of REM and NREM sleep.
Moldofsky H. Sleep and the immune system. Int J Immunopharmacol. 1995;17(8):649–54.
Mônico-Neto M, Dáttilo M, Ribeiro DA, Lee KS, de Mello MT, Tufik S, Antunes HKM. REM sleep deprivation impairs muscle regeneration in rats. Growth Factors. 2017;35(1):12–8.
Mostaghimi L, Obermeyer WH, Ballamudi B, Martinez-Gonzalez D, Benca RM. Effects of sleep deprivation on wound healing. J Sleep Res. 2005;14(3):213–9.
Naidoo N, Giang W, Galante RJ, Pack AI. Sleep deprivation induces the unfolded protein response in mouse cerebral cortex. J Neurochem. 2005;92(5):1150–7.
Naidoo N, Casiano V, Cater J, Zimmerman J, Pack AI. A role for the molecular chaperone protein BiP/GRP78 in Drosophila sleep homeostasis. Sleep. 2007;30(5):557–65.
Porkka-Heiskanen T, Strecker RE, Thakkar M, Bjorkum AA, Greene RW, McCarley RW. Adenosine: a mediator of the sleep-inducing effects of prolonged wakefulness. Science. 1997;276(5316):1265–8.
Ramanathan L, Hu S, Frautschy SA, Siegel JM. Short-term total sleep deprivation in the rat increases antioxidant responses in multiple brain regions without impairing spontaneous alternation behavior. Behav Brain Res. 2010;207(2):305–9.
Schmitt LI, Sims RE, Dale N, Haydon PG. Wakefulness affects synaptic and network activity by increasing extracellular astrocyte-derived adenosine. J Neurosci. 2012;32(13):4417–25.
Smith R. Recovery and tissue repair. Br Med Bull. 1985;41(3):295–301.
Takarada T, Kodama A, Hotta S, Mieda M, Shimba S, Hinoi E, Yoneda Y. Clock genes influence gene expression in growth plate and endochondral ossification in mice. J Biol Chem. 2012;287(43):36081–95.
Tune GS. Sleep and wakefulness in normal human adults. Br Med J. 1968;2(5600):269–71.

Vasey C, McBride J, Penta K. Circadian rhythm dysregulation and restoration: the role of melatonin. Nutrients. 2021;13(10):3480.

Vitaterna MHEA. Mutagenesis and mapping of a mouse gene, Clock, essential fo. Science. 1994;264(5159):719.

Zimmermann H. Extracellular metabolism of ATP and other nucleotides. Naunyn Schmiedeberg's Arch Pharmacol. 2000;362(4–5):299–309.

Computer Vision and Deep Learning in Regenerative Medicine: Part 1

Yan Wen and Xujiong Ye

Abstract Artificial intelligence (AI) and computer vision have made great changes in medical image analysis in recent years. In musculoskeletal (MSK) medicine, MRI images provide detailed visualization of soft tissues, bones, and joints, which is important for accurate diagnosis and treatment planning. This research enhances the widely adopted nnU-Net framework by integrating Atrous Spatial Pyramid Pooling (ASPP) and Vision Transformers, resulting in two models: nnAtrousU-Net and nnAtrousTransformer. These improvements contribute to advancements in segmenting complex structures, such as thin and small cartilage, from MRI scans. Experiments were conducted on the OAI-ZIB dataset, including 507 knee MRI scans with manual annotations. Both models outperformed the baseline nnU-Net in key metrics such as Dice Similarity Coefficient (DSC) and average surface distance (ASD). Specifically, nnAtrousU-Net achieved a DSC of 90.46% for femoral cartilage and 86.91% for tibial cartilage, improving the state-of-the-art CNN-3D+SSM approach. The nnAtrousTransformer demonstrated competitive accuracy using half the parameters and one-third of the computational cost compared to nnAtrousU-Net. The proposed architectures offer improved segmentation accuracy, computational efficiency, and adaptability. It provides the solution for enhanced diagnostic and therapeutic applications in healthcare.

Keywords Medical imaging · Cartilage segmentation · Knee MRI · Image analysis · Osteoarthritis

---

Y. Wen (✉)
School of Computer Science, College of Health and Science, University of Lincoln, Lincoln, UK
e-mail: yan@mskdoctors.com

X. Ye (✉)
University of Exeter, Exeter, UK
e-mail: x.ye2@exeter.ac.uk

P. Lee (ed.), *Musculoskeletal Regeneration Medicine*,
https://doi.org/10.1007/978-3-031-77456-0_5

1 Introduction

The advent of Artificial Intelligence (AI) and computer vision in the medical field represents a transformative era in healthcare, improving the efficiency, accuracy, and accessibility of medical services. These technologies have undergone significant evolution over the years, from simple pattern recognition to complex diagnostic tools capable of providing insights into a patient's health that were previously unattainable. This evolution evidences the significance of AI and computer vision in medicine, as they enhance the capabilities of healthcare professionals, streamline diagnostic processes, and improve patient outcomes.

1.1 Evolution and Significance of AI and Computer Vision in Medical Fields

The journey of AI and computer vision in medicine began with simple image analysis and patten recognition, this has progressed to include a wide range of applications, including predictive analytics, disease identification, and automated surgical procedures. Initially, these technologies were leveraged for basic tasks such as the classification of images and pattern identification. However, as AI algorithms became more sophisticated, their application expanded, revolutionizing areas such as radiology, pathology, and ophthalmology. Today, AI and computer vision are integral to developing personalised medicine, enabling the analysis of vast datasets to tailor treatment plans to individual patients' lifestyle and response to previous treatments.

These technologies have the potential to reduce human error, increase the speed and accuracy of diagnoses, and predict disease progression, allowing for early intervention. Furthermore, they facilitate remote monitoring and diagnostics, making healthcare more accessible, especially in underserved areas.

1.2 Specific Relevance to Early Diagnosis and Treatment of Musculoskeletal Diseases

In the context of early diagnosis of musculoskeletal diseases, AI and computer vision hold immense promise. These technologies offer innovative approaches to diagnosing musculoskeletal conditions, assessing tissue damage, and monitoring the healing process. For instance, AI and computer vision driven algorithms can provide detailed analysis of medical images (such as X-rays, MRIs, and CT scans) to detect subtle signs of abnormalities, such as fractures or degenerative diseases in bones, joints, and soft tissue, often before symptoms become apparent to patients, enabling early intervention.

In the treatment and monitoring of musculoskeletal conditions, AI and computer vision offer tools for personalised medicine. By analysing patient data, AI algorithms can predict how individuals might respond to different treatments for musculoskeletal disorders, assisting clinicians in making informed decisions about the most effective interventions.

AI and computer vision also accelerate research in musculoskeletal regeneration. Through the analysis of vast amounts of data, these technologies can uncover new insights into the regenerative processes, identify potential biomarkers for regeneration, and develop new therapeutic targets. Furthermore, computer vision aids in the development of bioengineered tissues and organs by facilitating the analysis of scaffold structures and cell alignments, critical factors in tissue engineering.

The integration of AI and computer vision into the medical field, particularly in the domain of musculoskeletal medicine will enhance diagnostic accuracy, personalising treatment plans, and advancing research, AI and computer vision are paving the way for innovative approaches to medical care that that is more effective, efficient, and patient-centred. As these technologies continue to evolve, their impact on medicine will undoubtedly grow, offering a future where AI and computer vision are at the heart of medical innovation.

2 Overview of Deep Learning Architectures Used for Medical Image Segmentation

Deep learning has revolutionized the field of medical image segmentation, providing powerful tools for accurately delineating anatomical structures and pathological regions in imaging data. This capability is particularly important in musculoskeletal medicine, where accurate segmentation of bones, muscles, cartilage, and other tissues can significantly impact diagnosis, treatment planning, and monitoring of regenerative therapies (Tack et al. 2018; Ambellan et al. 2019). This overview highlights several deep learning models that have been instrumental in advancing medical image segmentation, especially in MSK medicine.

Several deep learning architectures have been adapted and developed specifically for image segmentation in MSK.

2.1 Convolutional Neural Networks (CNNs)

CNNs are the most widely used deep learning architecture for medical image segmentation. They are particularly well-suited for capturing spatial and contextual information in images. Popular CNN architectures used in MSK segmentation include:

U-Net

U-Net (Ronneberger et al. 2015) is a convolutional neural network (CNN) designed specifically for biomedical image segmentation. There are two paths in the architecture, a contracting path (left) and an expansive path (right). The contracting path is also called the encoder, which is a stack of convolutional and max pooling layers for semantic feature extraction. The expansive path is also known as the decoder, which is used to restore the localization precisely using the deconvolutions. To recover the fine-grained information for the output segmentation map in the upsampling, long skip connections are connected from the encoder to the decoder in a symmetrical manner. Figure 1 below is an overview of the model architecture. This progressive upsampling decoder architecture with long skip connections makes it particularly effective for feature extraction of small objects.

Significance in MSK Medicine: U-Net's ability to work with a small amount of training data and its efficiency in segmenting biomedical images make it particularly useful for segmenting musculoskeletal structures from MRI or CT scans, facilitating detailed analysis and treatment planning in MSK regeneration. U-Net can be used for tasks like bone, cartilage, and muscle segmentation (Norman et al. 2018).

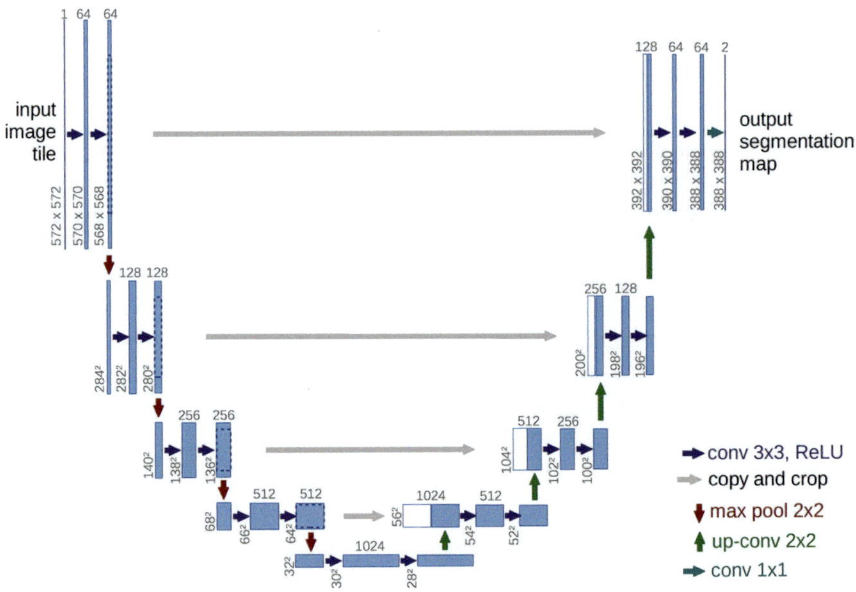

Fig. 1 U-Net architecture (Ronneberger et al. 2015)

Deeplabv3

In most semantic segmentation networks, such as FCN (Long et al. 2015),the segmentation feature maps are coarse due to two main issues, first, the lost information due to the repeated application of max-pooling and downsampling, and second, the lack of constraints on spatial and edge information. To address these problems, the authors at Google Brain (Chen et al. 2017a) several innovations to improve segmentation performance, which includes:

- Atrous Convolutions (Dilated Convolutions): DeepLab v3 utilizes atrous convolutions, which allow for an exponentially expanding receptive field without increasing the number of parameters or computation cost. This enables the model to capture multi-scale information and long-range dependencies, crucial for accurate segmentation of complex structures.
- Atrous Spatial Pyramid Pooling (ASPP): The ASPP module employs parallel atrous convolutions with different rates, effectively capturing objects and context at multiple scales. This helps to segment both large and small structures within the same image].
- Encoder-Decoder Architecture: DeepLab v3 follows an encoder-decoder structure, where the encoder (based on a pre-trained CNN like ResNet) extracts features, and the decoder upsamples and refines the segmentation output, leveraging skip connections from the encoder for better localization.
- Depth-wise Separable Convolutions: To reduce computational complexity and model size, DeepLab v3 incorporates depth-wise separable convolutions, which factorize standard convolutions into depth-wise and point-wise convolutions, resulting in significant parameter and computation savings.

DeepLab v3 has been successfully applied to various medical image segmentation tasks, including MSK applications. Some examples include:

- Bone Segmentation: DeepLab v3 has been used for accurate segmentation of bones from CT or MRI scans, which is crucial for orthopedic surgery planning, fracture detection, and bone disease diagnosis (Heinrich and Hansen 2020).
- Joint Segmentation: The architecture has been adapted for segmenting joints, such as the knee and hip, from medical images. This can aid in the assessment of joint disorders like osteoarthritis and facilitate surgical planning (Tack et al. 2018).
- Muscle Segmentation: DeepLab v3 has been employed for segmenting muscles from MRI or CT data, which can be useful for muscle volume quantification, injury assessment, and planning of musculoskeletal interventions.
- Lesion and Tumour Segmentation: The model has shown promising results in segmenting musculoskeletal lesions and tumours, enabling accurate diagnosis, staging, and treatment monitoring (Yan et al. 2018; Trebeschi et al. 2017)

Figure 2 is a visualization of convolution with 3×3 kernel and dilation rate 2. The dilation ratecontrols the spacing between the pixels in the convolution kernel. From the figure, the atrous convolution have the 7×7 field of view within a 3×3 kernel

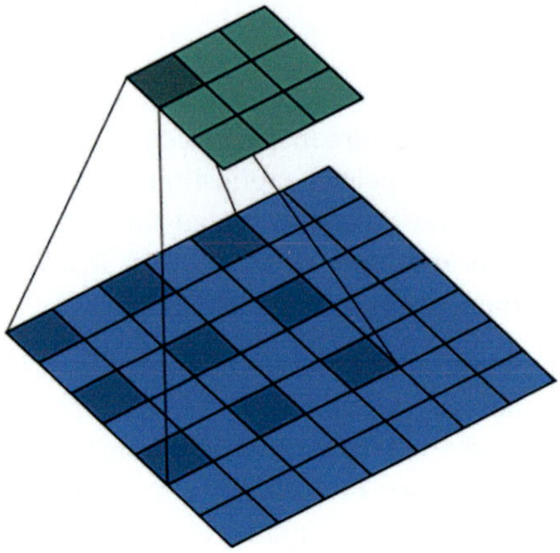

Fig. 2 Dilated convolution. Blue maps are inputs, and cyan maps are outputs (Dumoulin and Visin 2018).

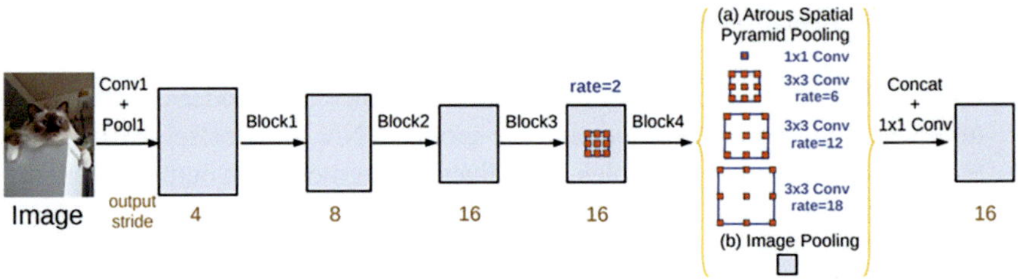

Fig. 3 Atrous Spatial Pyramid Pooling

size. Comparing with the standard convolution, atrous convolution increases the receptive field without increasing the number of parameters.

The Atrous Spatial Pyramid Pooling module is used to extract context information in multiple scales. In the network, the top of the feature map is obtained from the repeat of CNN blocks with atrous convolutions. In the bottleneck, as shown in Fig. 3, one 1×1 convolution and three atrous convolutions with various atrous rates are applied parallelly to effectively captures multi-scale information. Additionally, the image-level features from global average pooling are concatenated with four convolution results and passed through another 1×1 convolution as the final feature map for the decoder.

The strengths of DeepLab v3, including its ability to capture multi-scale information, encoder-decoder architecture, and efficient computation, make it well-suited for segmenting complex anatomical structures in MSK imaging data. However, its performance can be further enhanced by combining it with other techniques, such as attention mechanisms, ensemble methods, or task-specific modifications.

Transformer-Based Architectures

Transformer-based architectures have revolutionized the field of natural language processing (NLP) and have recently made significant strides in computer vision and medical imaging. The Transformer model, introduced by Vaswani et al. in 2017, (Vaswani et al. 2017) replaces traditional recurrent and convolutional layers with a self-attention mechanism that can process sequences of data more effectively and in parallel. Example Transformer Architectures: BERT (Bidirectional Encoder Representations from Transformers) (Devlin et al. 2019), GPT (Generative Pre-trained Transformer) (Radford 2018), Vision Transformer (ViT) and SegFormer.

Segformer (Xie et al. 2021) (see Fig. 4) is a simple, efficient and robust method for semantic segmentation, which comprised a hierarchically structured Transformer encoder. It utilises mix FFN in the Transformer instead of the Positional Embedding to encode the position information. Mix FFN is an extension of the original feed forward network by adding a 3 × 3 deepwise convolutional layer.

Mix-FFN is formulated as:

$$x_{out} = MLP\Big(GELU\Big(Conv_{3\times3}\big(MLP\big(x_{in}\big)\big)\Big)\Big) + x_{in} \; \#\big(\; AUTONUMLGL \;\; \backslash *Arabic \backslash e \; \big)$$

Where x_{in} is the feature from the self-attention algorithm, $Conv_{3\times3}$is a 3 × 3 depth-wise convolution.

In its decoder, with the vast receptive field of self-attention algorithm, the progressive upsampling decoder is replaced by simply yet linear layers. It achieves a state-of-the-art accuracy, as well as a good efficiency by reducing the computational cost and parameters.

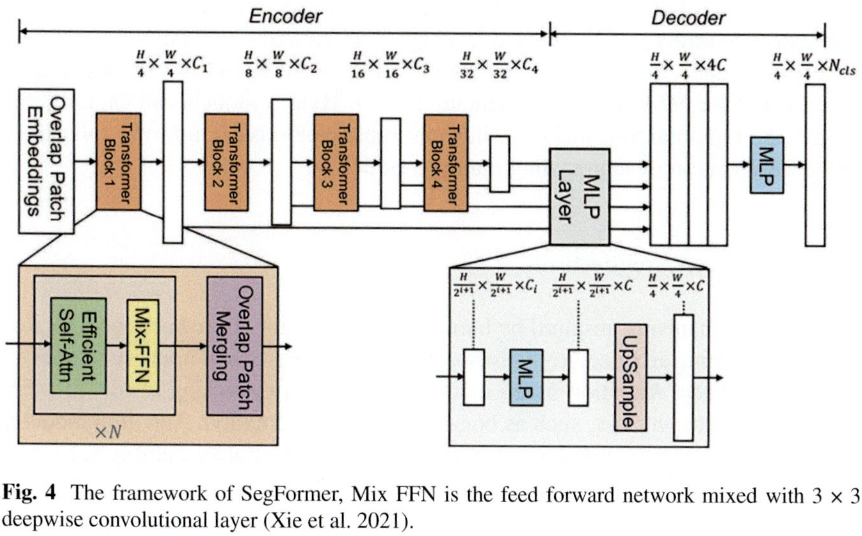

Fig. 4 The framework of SegFormer, Mix FFN is the feed forward network mixed with 3 × 3 deepwise convolutional layer (Xie et al. 2021).

SegFormer's transformer-based architecture and ability to capture global context make it a promising approach for segmenting complex anatomical structures in MSK imaging data.

Recurrent Neural Networks (RNNs)

RNNs, particularly Long Short-Term Memory (LSTM) networks, have been explored for medical image segmentation tasks that involve sequential or temporal data. In MSK, RNNs have been used for segmenting dynamic MRI sequences and analysing gait patterns for musculoskeletal disorder diagnosis. RNNs can capture temporal dependencies and model sequential data, making them suitable for time-series or video data analysis.

Generative Adversarial Networks (GANs)

GANs are a type of deep learning architecture that consists of two competing neural networks: a generator and a discriminator. In the context of MSK image segmentation, GANs have been employed for tasks such as bone and cartilage segmentation (Metcalfe et al. 2012; Duryea et al. 2007), often in combination with other architectures like CNNs. The generator aims to produce realistic segmentations, while the discriminator tries to distinguish between real and generated segmentations, leading to improved segmentation quality.

Hybrid and Ensemble Architectures

Researchers have also explored hybrid and ensemble approaches that combine different deep learning architectures to leverage their strengths and improve segmentation performance. For example, combining CNNs with RNNs or GANs has been investigated for MSK image segmentation tasks. Hybrid models can capture both spatial and temporal information, while ensemble methods can improve robustness and generalization by aggregating multiple models.

Attention-Based Architectures

Attention mechanisms, inspired by human visual attention, have been incorporated into deep learning architectures to focus on relevant features and improve segmentation performance. Attention-based models have shown promising results in MSK image segmentation tasks, such as bone and joint segmentation. Attention modules can highlight important regions or features, enhancing the model's ability to capture relevant information for accurate segmentation.

3 Segmentation of Knee Bone and Cartilage in MRI

Osteoarthritis (OA) is the result losing of articular tissue over time, affecting millions of people worldwide. For example, in the age group over 60 years, 10% of men and 18% of women are affected by osteoarthritis in the world (Glyn-Jones et al. 2015). Statistics research in 2017 indicates the incidence of osteoarthritis per 1000 person-years. Knee osteoarthritis incidence (2.3; 95% CI 2.2 to 2.4) is the highest in the parts of the body (Swain et al. 2020). The typical osteoarthritis-related findings are the degeneration of cartilage, menisci, and ligaments in the knee. In current clinical practice, these structures can be assessed using medical imaging techniques, particularly MRI (Conaghan et al. 2011; Peterfy et al. 2006). In the knee joint, cartilage is an especially concerning region in orthopedists, because it is highly relevant to knee osteoarthritis (Metcalfe et al. 2012). Cartilage is a cover on a bone. In the knee joint, cartilage is located where the tibia, femur and kneecap meet to form the joint. (Fig. 5 Left) is a slice of knee scan in MRI, on the right, it is a ground truth manually annotated by professionals (Ambellan et al. 2019). Unlike small, round targets like lung nodules, knee cartilage spans multiple slices in a 3D MRI scan due to its morphology. Manually quantifying cartilage slice by slice is labor-intensive and resource-demanding, requiring hours for a specialist to segment and annotate. The quality of manual segmentation varies with the operator's experience, leading to interobserver and intra-observer variability (Duryea et al. 2007). Therefore, quantitative measurements have not been widely applied in the clinic yet, because of the lack of a fast, consistent accurate and reliable segmentation approach. In recent years, several semi-automated and automated algorithms have been proposed for knee cartilage segmentation on MRI images.

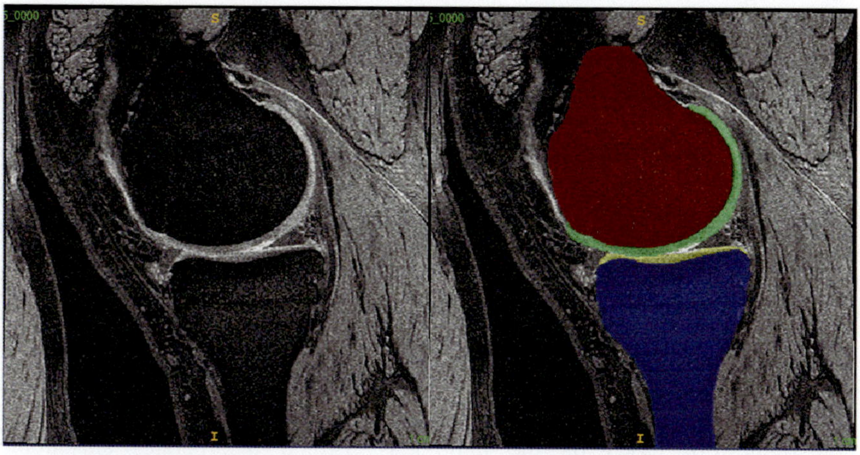

Fig. 5 A slice of knee scan (Left) in MRI with ground truth (Right). Red region: Femoral Bone, Blue region: Tibial Bone, Green region: Femoral Cartilage, Yellow region: Tibial Cartilage

3.1 Graph-Cut Methods

The early research (Shim et al. 2009) related to the semi-automated segmentation method using the graph-cut algorithm (Boykov and Funka-Lea 2006), which is significantly more efficient and reproducible than manual segmentation (mean DSC, 94.3% vs 87.7%). After that, a graphic-cut based fully automatic segmentation for knee bones is proposed by the combination of a content-based image block classification mechanism (Ababneh et al. 2011). However, the graph-based technique does not utilize the knowledge from the built model of the anatomical structure.

3.2 Statistical Modelling Methods

Several statistical modelling-based approaches are used for knee segmentation. In 2010, a fully automatic segmentation based on statistical shape models (SSM) and graph-based optimization was applied to segment the bones and cartilage (Seim et al. 2010). In 2011, High-quality correspondences from MDL-GIR (Minimum Description Length approach to Groupwise Image Registration) (Twining et al. 2004) were proposed to train the Active Appearance Models (Scott et al. 2003) for the segmentation of bone and cartilage. This method shows good generalization performance compared to previous model-based approaches as the testing dataset is independent of the training dataset.

3.3 Registration and Atlas-Based Methods

Image registration is the process of transforming two or more images to align them together. It can be used for object segmentation, by calculating the transformation matrix T_{reg} between manually segmented images X and target images Y. The target segmentation $S(Y)$ can be calculated by $T_{reg}(S(X))$, where $S(X)$ is the segmentation of X by an export. Since image registration is found as a paradigm for medical image segmentation, where the atlas-based approaches are proposed to minimize the registration error by reducing the effect of user variability (Bach Cuadra et al. 2015). The flow is based on building multiple atlases X_i and registering them to the target image Y. The target segmentation $S(Y)$ is computed by label fusion of $S(X_i)$ $i \in \{1 \ldots L\}$, where L is the count of atlases. The initial related research on 19 patients demonstrates (Tamez-Peña et al. 2011) that it is feasible to obtain quantitative data automatically from atlas-based approaches. A novel three-label segmentation framework based on the multi-atlas method incorporates anisotropic regularization to improve the accuracy of cartilage segmentation (Shan et al. 2012, 2014). They compared K Nearest Neighbors (KNN) and Support Vector Machines (SVM), finding similar performance on femoral cartilage, but SVM outperformed

KNN on tibial cartilage. Atlas-based segmentation relies heavily on registration performance, particularly in cases with varying cartilage shapes and losses. This method is computationally intensive, typically doubling the time of a single registration, making it unsuitable for real-time applications.

3.4 Machine Learning Methods

In the past few years, machine learning-based approaches have been widely applied to the segmentation of medical images. At the early stage, the k-nearest neighbour-based (k-NN) classifier is applied to classify tibial, femoral cartilages and background, (Folkesson et al. 2007) they also compared two binary classifiers using a one-versus-all approach and found its higher performance than multiclass classification. To improve the efficiency of k-NN, (Folkesson et al. 2005) a study developed an efficient voxel classification algorithm, whose theory is focused on the cartilage voxels instead of all voxels. As a result, it reduced its computation time from 60 minutes to approximately 10 minutes to make it feasible and reproducible on a standard desktop computer. Based on the success of the k-NN classifier, extended research on a cascaded classification framework is proposed (Prasoon et al. 2012, 2013a). The flow of the algorithm contains two stages. In the first stage, the k-NN classifier is used to generate the initial voxels classes, secondly, the voxels of the front ground (femoral and tibial cartilage) are classified by the support vector machine. The cascaded classification increased the DSC of femoral and tibial cartilage by 4.15% and 3.32% respectively. This is consistent with the findings in the (Folkesson et al. 2007) that a k-NN based segmentation approach cannot segment the region where the two cartilages of femoral and tibial are touching and overlapping. Without prior spatial knowledge, k-NN only uses extracted features by local relations and intensity. Another study (Zhang et al. 2013) combines the support vector machine(SVM) with the discriminative random field (DRF). Compared with the SVM-only method or DRF-only method, the performance of the joint method SVM-DRF is the best of all, whose DSC is 0.864 on femoral cartilage, 0.880 on tibial cartilage and 0.841 on patellar cartilage.

3.5 Deep Learning Methods

While classic machine learning approaches have shown good segmentation performance in knee MRI, deep neural networks have significantly improved segmentation accuracy and feature extraction. Convolutional neural networks (CNNs) have succeeded in multiple vision tasks, and many state-of-the-art methods for medical imaging are now based on deep neural networks. However, deep learning methods, especially 3D CNNs, require substantial computational and memory resources for training. To address this, research (Prasoon et al. 2013b) introduced a 2D triplanar

CNN for voxel classification on 3D MRI, using three 2D CNNs to train and predict on the x-y, y-z, and z-x planes of 3D images. The outputs are combined and fed into a softmax classifier, achieving a DSC of 0.8249, outperforming the state-of-the-art method from 2013 (Folkesson et al. 2007).

Another problem of deep learning for medical image is the shortage of well-annotated datasets. The supervised learning-based method relies on a large-scale and well-annotated dataset. The limited dataset could lead to overfitting easily, which means the model has a low loss on the training dataset while it performs badly on another independent dataset. (Low generalization) (Ying 2019) There are a few solutions to solve overfitting problems. e.g., Dropout (Srivastava et al. 2014), and L1/L2 regularization (Demir-Kavuk et al. 2011, p. 2), another common training strategy for the small dataset is transfer learning (Shin et al. 2016).

A 2D U-Net (Ronneberger et al. 2015) based method with an encoder-decoder framework is proposed for knee cartilage segmentation. Benefiting from the long skip connections, the corresponding features in the encoder are enhanced to preserve sufficient details during the upsampling in the decoder. This method makes CNN very effective in segmentation for dense and small targets. They report the DSC 0.867 (95% CI: 0.859, 0.875) for cartilage. Another method (Liu et al. 2018) concatenates 2D prediction labels from encoder-decoder-based SegNet (Badrinarayanan et al. 2017) into 3D volumes and post-processing by 3D simplex deformable modelling. The boundary of the target labels is smoothed without losing the overall object shape. They also reported the better segmentation performance of SegNet in bone and cartilage compared to the U-Net. The model was ranked at second place in the SKI10 challenge dataset (Zhou et al. 2018). utilized a 3D conditional random field as a post-processing method on 2D slices prediction labels from CNNs for knee segmentation. As a 3D image is split into 2D slices for 2D CNNs, thus, the CNNs cannot consider the inter-slice contextual information. A fully connected 3D conditional random field can rebuild the 3D contextual relationships among voxels to optimize the segmentation mask. The most recent method is based on Mask R-CNN (Felfeliyan et al. 2022). they add an extra decoder block called the ROIAligned block and use the skip connection to connect it to the decoder. They reported their performance of 0.86 for tibial cartilage and 0.82 for femoral cartilage on the OAI-ZIB dataset (Ambellan et al. 2019).

Cartilage thickness is typically thin, with healthy individuals having an average distal medial cartilage thickness of 1.820 mm. In knee osteoarthritis patients, cartilage loss results in even thinner cartilage. This class imbalance and small sample size can lead to overfitting in neural network training. Small errors in quantitative measurements are amplified due to the small cartilage volume, and MRI artifacts or low contrast can complicate classification by convolutional neural networks. To address these issues, a pipeline using 2D CNNs, sub-regional 3D CNNs, and Statistical Shape Models (SSM) has been proposed (Ambellan et al. 2019). Initially, a 2D CNN performs femoral and tibial bone segmentation, refined by a statistical shape model. Then, a 3D CNN segments sub volumes at the bone edges, followed by SSM postprocessing to enhance results. Finally, a 3D CNN segments the femur and tibia cartilage using bone masks. On the OAI ZIB dataset, this method achieved a Dice Similarity Coefficient (DSC) of 0.899 for femur cartilage and 0.856 for tibia cartilage.

3.6 Contributions

This research explores various deep learning methods and frameworks, evaluating them on the OAI-ZIB dataset (Ambellan et al. 2019). The nnU-Net model is enhanced by expanding its auto-configuration search space, particularly in model architecture. Through optimization and new architectures, the nnAtrousU-Net and nnAtrousTransFormer-2 models outperform the current state-of-the-art in knee cartilage segmentation.

Key contributions include:

- Integrating Atrous Spatial Pyramid Pooling (ASPP) (Chen et al. 2017a) into nnU-Net, expanding the field of view without reducing resolution.
- Incorporating Vision Transformer (Dosovitskiy et al. 2021) into nnU-Net, combining convolutional neural networks and Vision Transformer to capture detailed local and global context.
- Developing a new auto-configuration method for patch size calculation to improve model performance by providing larger contextual information.
- Implementing ASPP in 3D convolutional neural networks with optimized atrous rates and enabling Vision Transformer blocks to function in 3D with optimized VRAM settings.

nnU-Net (See Fig. 6 Left) has great success in medical images, the contribution of nnU-Net is the auto-configuration of the training pipeline based on the optimized model. This automated configuration is mainly used for preprocessing and post-processing. In the field of model architecture and training settings (e.g. residual encoder (He et al. 2016), batch normalization (Ioffe and Szegedy 2015), 3-stage convolution, optimizer, loss function) is normally fixed by the author has proposed

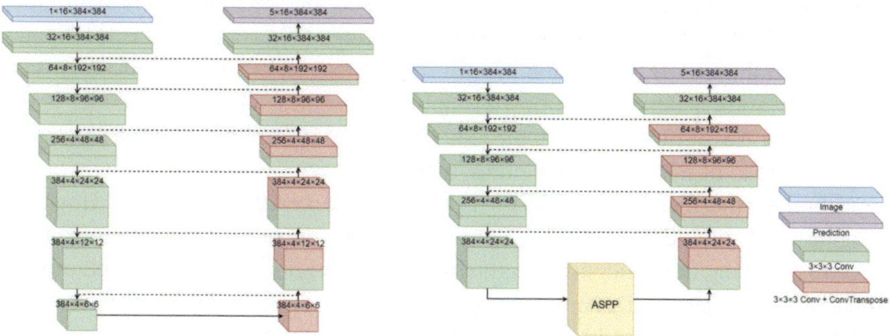

Fig. 6 Comparison of nnU-Net and nnAtrousU-Net

a superior baseline configuration by comparing the results of various settings (nnU-Net 2022). The immutability of the model based on the former U-Net limits the potential improvement of segmentation performance.

3.7 Methods

Overview

The recent new approach proved to work for vision tasks. e.g. Atrous Spatial Pyramid Pooling (Chen et al. 2017b), Vision Transformer [(Dosovitskiy et al. 2021), p. 16]. Inspired by these methods, this research developed an improved nnU-Net by widening the search space, in particular, the architecture of the model. The overall architectures of both nnAtrousU-Net (See Fig. 6 Right) and nnAtrousTransFormer maintain a U shape segmentation network which consists of two paths, a contracting path and an expansive path, which are also called encoder and decoder.

The encoder in the nnAtrousU-Net model consists of the left half of its architecture, with multiple blocks encoding the image into feature representations at various levels. Each block contains two 3×3 convolutional layers. The first convolution layer in each block, except the first, downsamples the feature size using stride 2. The model starts with 32 feature channels, doubling at each downsampling layer until reaching a maximum of 384 channels, constrained by VRAM limitations. Each convolution layer is followed by Instance Normalization (Ulyanov et al. 2017) and Leaky ReLU (Maas et al. n.d.).

In the nnAtrousTransFormer, depending on the number of Vision Transformer blocks in the settings, the convolutional blocks are replaced by Vision Transformer blocks from deep to shallow (From bottom to top). Each Vision Transformer block contains two Transformer Layers. (Fig. 7 Right) In the Transformer layers, there are 6 heads of self-attention with 192 dimensions followed by an MLP (a convolution layer) with 768 dimensions. (Fig. 7 Left).

Maintaining the feature size as large as possible from the encoder can benefit small object segmentation because a small object on the input image (e.g., size of 384×384) will be very tiny and possibly disappear on the high-level feature map (e.g., size of 6×6) after multiple downsampling. (Fig. 6 Left) Atrous Spatial Pyramid Pooling (ASPP) (Chen et al. 2017b) can extract contextual information at multiple various scales without downsampling the scales. Thus, the encoder is followed by an ASPP module to capture the contextual information with a large field of view instead of more downsampling in nnU-Net.

Although, the feature from ASPP is relatively larger than nnU-Net. However, there are still some downsampling operations in the encoder. Thus, the decoder is used for restoring the condensed feature map to the original size of the input image. In the decoder, there is the same number of blocks as the encoder blocks, while the function of each decoder block is opposite to the encoder block. In detail, each decoder block has two 3×3 convolutional layers followed by a deconvolution layer

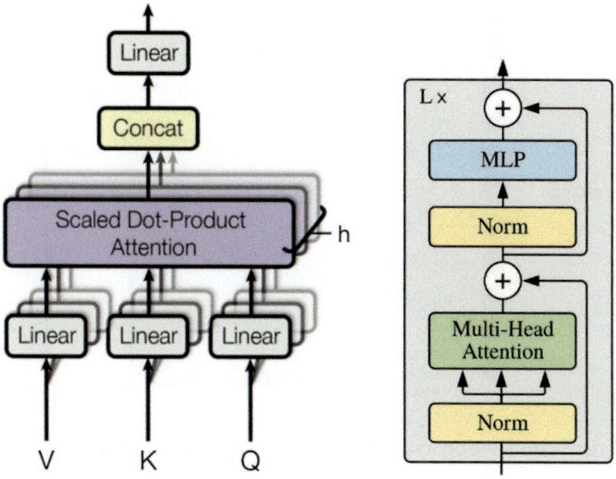

Fig. 7 Multi-Head Attention (Left) and Vision Transformer Block (Right)

corresponding to the downsampling in the encoder by convolution. For every block, feature channels are reduced by 2 times by the first layer of each block. Long skip connections are connected from the encoder to the decoder symmetrically to recover the context information in the higher-resolution feature representations. For the integration of Vision Transformer, because the feature map from the projection of the intermediate Transformer block is the same size as the output from the encoder (Fig. 11 Left), additional deconvolutional layers are applied to the feature map from the projection before concatenating it to symmetrical features in the decoder.

Improved Auto-Configuration

In the automatic experiment planning stage, nnU-Net auto-configures a patch size for segmentation based on VRAM limitations, as working with full 3D image resolution is impractical. The model trains on patches using a random-crop method and predicts full images via a sliding-window approach. Larger patch sizes provide better contextual information, enhancing performance. Therefore, for 3D medical image segmentation, the auto-configured patch size is nearly cubic.

However, 3D medical images often have sparse spacing in one plane, typically the z-plane. For small object segmentation, maintaining contextual information in the denser planes (sagittal) is crucial. Based on this, I propose a new method for auto-configuring patch sizes, prioritizing dense planes.

Initially, the patch size is set to the image size, and the GPU RAM cost is estimated. If the cost exceeds the GPU RAM limit, the patch size is reduced in the sparse plane (z-plane) until it fits within the limit. To prevent the patch size in the sparse plane from becoming too small, a minimum size of 16 is set. If this minimum is reached, size reduction shifts to the other planes.

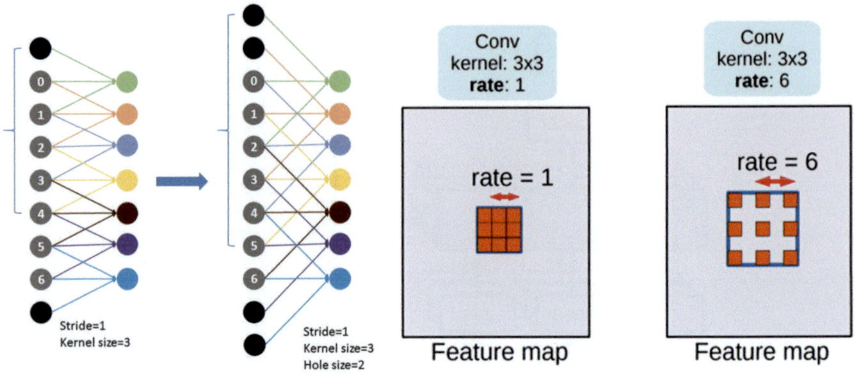

Fig. 8 From standard convolution to atrous convolution (Chen et al. 2017a).

Atrous Convolution

Atrous convolution, (Szegedy et al. 2016) also known as dilated convolution, uses a dilation rate to control the spacing between pixels in the convolution kernel. As shown in Fig. 8, a standard convolution requires a 5 × 5 kernel or two 3 × 3 convolutions to capture a 7 × 7 field of view, whereas an atrous convolution with a dilation rate of 2 achieves the same field of view with a 3 × 3 kernel. This makes atrous convolution more computationally efficient.

Fully convolutional networks are effective for semantic segmentation, (Long et al. 2015) but multiple blocks with downsampling (e.g., max-pooling and striding) significantly reduce the spatial resolution of the feature map. For example, an input size of 16 × 384 × 384 can shrink to 4 × 6 × 6 after six downsampling steps, causing small objects to nearly disappear. Using atrous convolution instead of standard convolution maintains spatial resolution while enlarging the field of view and reducing computational costs.

In the mathematization of atrous convolution

$$y[i]=\sum_{k}x\left[i+r\times(k-1)\right]\times w[k]\#(\ AUTONUMLGL\ \backslash *Arabic\backslash e\ )$$

For each pixel of the location i on the output map y, atrous convolution is calculated from the input map x and filter w. Where r is the dilation rate that represents the spacing between the pixels from the input feature for calculating. The standard convolution's dilation rate is $r = 1$, while $r \geq 2$ is the dilated convolution. By changing the dilation rate value r and kernel size of k, the filter's field-of-view is flexible to adapt to various fields of view for the target contextual information.

nnAtrousU-Net

Improved Atrous Spatial Pyramid Pooling in 3D

Inspired by the success of spatial pyramid pooling (He et al. 2014). The Atrous Spatial Pyramid Pooling (Fig. 9 Right) is four atrous convolutions with various dilation rates which parallelly capture the contextual information on the feature map. The structure of spatial pyramid pooling resamples the features of arbitrary scales, which is an effective and accurate way to classify regions on various scales. Benefiting from the atrous rate, spatial pyramid pooling combining atrous convolution further enlarges the field of view on context information at multi-scales. This structure is applied to multiple approaches to image segmentation and detection. E.g. Pyramid Scene Parsing Net (PSPNet) (Zhao et al. 2017), SPP-Net (He et al. 2014).

Although Atrous Spatial Pyramid Pooling (ASPP) captures context information at multiple scales, downsampling the feature map multiple times reduces its scale. Larger atrous rates result in fewer valid filter weights due to zero padding. To address this, the original ASPP implementation incorporates global context information by adding a parallel path with global average pooling followed by 1×1 convolution, upsampled bilinearly to match the desired spatial dimension. This setup includes one 1×1 convolution and three 3×3 convolutions with dilation rates $r = (6, 12, 18)$.

For 3D feature maps, all 2D convolutions and dilation rates are replaced by 3D equivalents, using trilinear interpolation instead of bilinear. This increases parameters and computational costs to obtain a field of view in three dimensions. Therefore, ASPP must be adapted based on the input's spatial dimension and hardware limitations (VRAM, computational cost).

By reducing the number of convolution stacks and downsampling, ASPP can maintain a similar field of view to the original nnU-Net with less computational cost. For example, with an input size of $16 \times 384 \times 384$, the feature map size reduces to $4 \times 24 \times 24$ after four downsamplings. The original dilation rates $r = (6, 12, 18)$ are too large for such reduced scales, especially for small object segmentation.

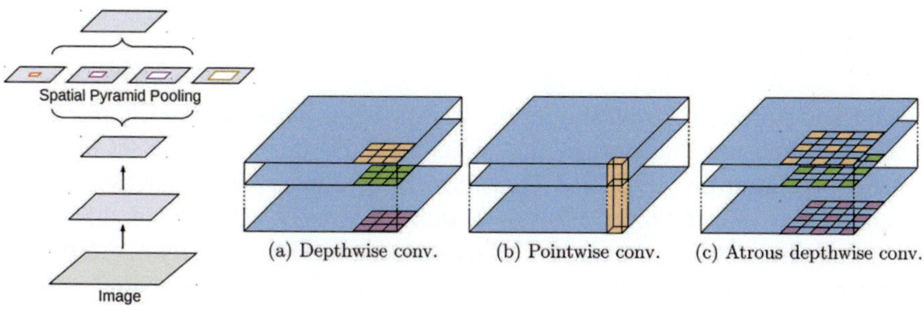

Fig. 9 Atrous Spatial Pyramid Pooling (Left) (Chen et al. 2017a) and Depthwise separable convolution. (Right)

Thus, I propose auto-configured dilation rates based on the downsampling rate in the last convolution stack across three dimensions.

In the mathematization of dilation rates r:

$$r = R \times (1,2,3) \# (\ AUTONUMLGL \ \backslash * Arabic \backslash e\ )$$

Where R is the downsampling rate in the last convolution stack. E.g. downsampling rate of nnAtrousU-Net in the 5th convolution block is 1 × 2 × 2. Thus, four parallel 3 × 3 × 3 convolutions in Atrous Spatial Pyramid Pooling have dilation rates of 1 × 1 × 1, 1 × 2 × 2, 2 × 4 × 4, and 3 × 6 × 6. Their field of view on 4 × 24 × 24 feature maps are 3 × 3 × 3, 3 × 5 × 5, 5 × 9 × 9 and 7 × 13 × 13.

Optimized Atrous Spatial Pyramid Pooling in 3D provide the benefits

- The contextual information can be captured in a large 3D field of view.
- Make the algorithm with fewer convolution layers and downsampling
- Fewer parameters and computational costs.
- Maintain the spatial resolution of the feature responses. (Larger feature map from the encoder)

Depthwise Separable Convolution

Although, replacing the repeat of convolution layers with Atrous Spatial Pyramid Pooling has many benefits including the field of view, computational costs and spatial resolution of the feature. However, the reduction of the model depthwise affects the ability of high-level feature representation for the model. Depthwise separable convolution Fig. 9c is a combination of depthwise convolution Fig. 9a and pointwise convolution Fig. 9b. Specifically, the atrous depthwise convolution first performs an atrous convolution followed by a 1 × 1 × 1 convolution. It makes the model go deeper by adding non-linear transformations to features at a low computational cost.

nnAtrousTransFormer

3D Vision Transformer

Vision Transformer (Dosovitskiy et al. 2021, p. 16) is introduced in Chapter 2.1.3. The flow of Vision Transformer first splits input 2D images into a set of image patches. While in the 3D Vision Transformer, the dimension is increased for the patch splitting. For the input 3D images with the size (H, W, D), input channels C and patch size (P, P, P), the number of patches can be calculated by $N = HWD/P^3$. These patches are flattened into a sequence as token embeddings. Similar to the Transformer's principle (Vaswani et al. 2017), these embeddings are encoded with positional information using the sinusoidal function. The positional encoded sequence is used for capture feature encoding by multiple stacks of Multi-Head Attention block and feed-forward block. The Multi-Head Attention block contains several Scaled Dot-Product Attention independently:

In mathematization of Scaled Dot-Product Attention:

$$Attention(Q,K,V) = Softmax\left(\frac{QK^t}{\sqrt{d_k}}\right)V \#(AUTONUMLGL \ \backslash*Arabic \backslash e)$$

Where $Q = xW_Q$, $K = xW_k$, $V = xW_v$, they respectively represent Query, Key and Value. W is a linear transformation with weights, and x is the positional encoded sequence. d is the dimension of the key vectors, which is used to stabilise gradients. The feature encodings from multiple Scaled Dot-Product Attention are concatenated followed by the feed-forward block to reduce the feature channels as the input size.

Different from the architecture in Transformer, (Vaswani et al. 2017) Vision Transformer performs layer normalization (*LN*) (Ba et al. 2016) before multi-head self-attention (*MSA*) and feed-forward layer(*FFN*) (Wang et al. 2019; Baevski and Auli 2018).

In mathematization:

$$z_l^{'} = MSA\left(LN\left(z_{l-1}\right)\right) + z_{l-1}, \qquad l \in [1, L] \#(AUTONUMLGL \ \backslash*Arabic \backslash e)$$

$$z_l = FFN\left(LN\left(z_l^{'}\right)\right) + z_l^{'}), \qquad l \in [1, L] \#(AUTONUMLGL \ \backslash*Arabic \backslash e)$$

Where L is the number of repeated stacks of Vision Transformer blocks, z_0 is the projected patch embedding in the length of N.

Integration of Vision Transformer

Several works integrate the Vision Transformer into a U-shape encoder-decoder segmentation network. They have obtained superior performance in medical image segmentation. E.g., UNETR (Hatamizadeh et al. 2021), TransUNet (Chen et al. 2021), and nnFormer (Zhou et al. 2022). Their commonalities are the forthputting of the encoder block in Vision Transformer (without MLP head, which is used for classification). To enable the long skip connection from the encoder to the decoder, features from the Vision Transformer blocks are reconstructed by projecting into the symmetrical size of the feature in the decoder. UNETR (Hatamizadeh et al. 2021) (Fig. 10) replaced all convolution layers with Vision Transformer blocks, comprising multi-head self-attention (MSA) and multilayer perceptron (MLP). Its performance set a new state-of-the-art on the BTCV leader board (Multi-Atlas Labeling Beyond the Cranial Vault - Workshop and Challenge 2022). nnFormer (Zhou et al. 2022) is another approach using a full stack of Vision Transformer blocks for the

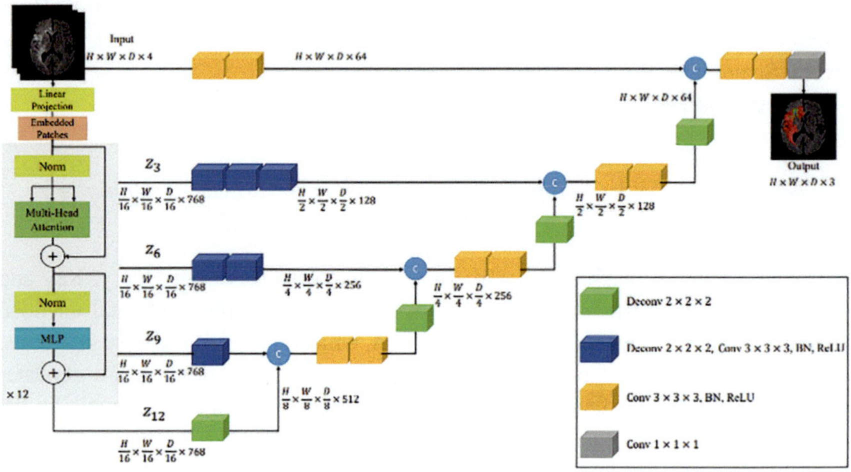

Fig. 10 UNETR (Hatamizadeh et al. 2021)

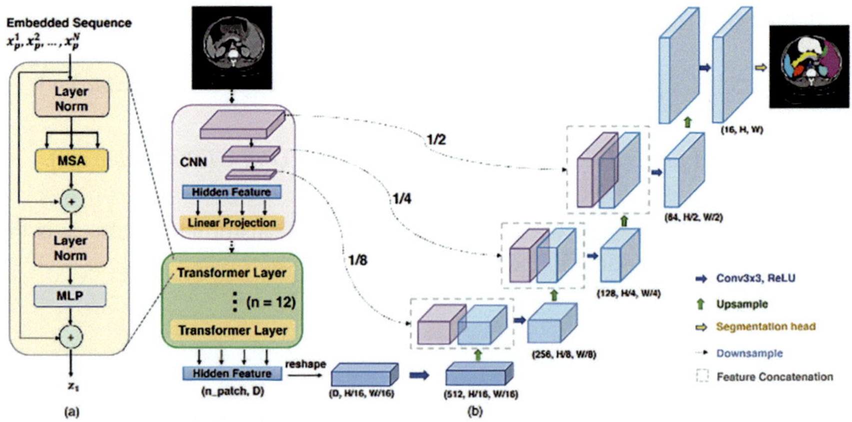

Fig. 11 TransUNet (Chen et al. 2021)

encoder as well as its decoder for semantic feature extraction. However, the research (Chen et al. 2021) indicates that a full Vision Transformer encoder focuses on modelling the global context, resulting in the loss of local localization information, while a convolutional neural network is different to Vision Transformer, the convolution layers are usually calculated on a small field of the view (usually 3 × 3), which is sensitive to the local context. Thus, they have limitations for modelling global relations. TransUNet combines a convolutional neural network and Vision Transformer. The success of TransUNet (Fig. 11), indicates the features of convolutional neural networks can also be embedded into patches. By the combination of convolutional neural networks and Vision Transformer, it balances both detailed high-resolution local context features from the convolutional neural network and the global information encoded by Vision Transformer.

Although, UNETR and TransUNet have achieved great success in specifying medical imaging tasks. When coming up with another medical imaging dataset, especially for which dataset properties vary drastically, transferring these methods from one dataset to another dataset is normally hard to repeat success (Isensee et al. 2021).

Based on these approaches of utilizing the Vision Transformer in the encoder-decoder segmentation network, the ideal approach is inheriting and extending the flexibility of nnU-Net. Thus, we proposed a new approach nnAtrousTransFormer which expands the search space of the base feature encoder approach in nnAtrousU-Net. Specifically, the model can adaptively structure an end-to-end architecture of either a fully CNN or a fully Vision Transformer or Mixed. An option is used to define the amount of Vision Transformer blocks in the model for the replacement of the convolution block. For example, as the overview in Fig. 12, when the setting of 4 Vision Transformer blocks, (Left in Fig. 12) the input images are firstly extracted by a convolutional block with two layers. The captured features by the convolutional block are projected and embedded into patches for the input of the Vision Transformer with 4 blocks. When the setting of 2 Vision Transformer blocks is used, the input images are convoluted into high-level features with two downsampling operations on the scale, followed by two blocks of the Vision Transformer further extracting the global information from the output of 3 convolutional blocks. To get a plain convolutional neural network without Vision Transformer, the setting of the amount of Vision Transformer blocks is 0, which is literally nnAtrousU-Net.

The projected features from the Vision Transformer block are small in scale. The shape difference causes the projected features to not be directly concatenated to the corresponding features to its symmetrical decoder. Thus, a consecutive 3 × 3 × 3 convolutional layer followed by a deconvolutional layer is used for upsampling the transformer output to fit the size of the symmetrical decoder.

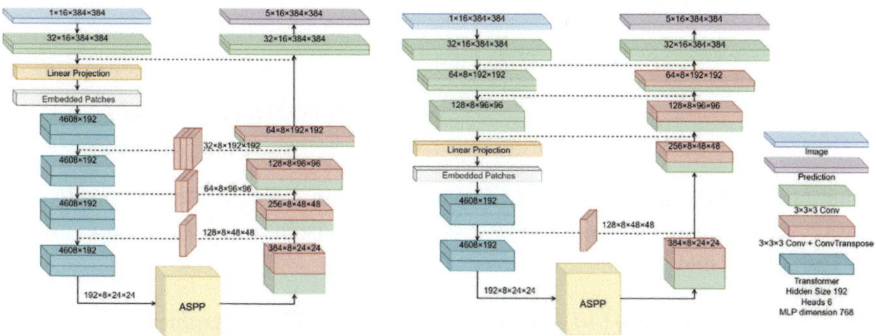

Fig. 12 The overview of nnAtrousTransFormer with different settings. The left diagram is the setting of 4 Vision Transformer blocks, and the right diagram is the setting of 2 Vision Transformer blocks

The Flexibility of Dimension

The nnU-Net is adaptable to diverse datasets, target structures, image types, and properties. Its backend supports various input sizes, modalities, and dimensions in both 2D and 3D. The integrated pipeline handles both 2D and 3D U-Nets, unlike other methods that typically only accept 2D images. E.g., Deeplab (Chen et al. 2017b), Mask R-CNN (He et al. 2018). When coming with 3D medical images, some approaches are converting the image into 2D slices using 2D algorithms (Chen et al. 2021), while some approaches run in a 3D-only approach (e.g. UNETR) (Hatamizadeh et al. 2021) and may not transfer their good performance on 2D medical images. (e.g X-ray) To maintain nnU-Net's flexibility, both nnAtrousU-Net and nnAtrousTrans-Former are implemented in 2D and 3D. Atrous Spatial Pyramid Pooling (ASPP) operates in 2D for both models and extends to 3D in nnAtrousU-Net. The Vision Transformer block in nnAtrousTransFormer also supports 3D patterns, optimized for limited VRAM.

Loss Function

The loss function is a sum of pixel-wise cross entropy and dice loss.

$$l_{CE}(x,y) = \frac{\sum_{n=1}^{N} -loglog \frac{\exp\left(x_{n,y_n}\right)}{\sum_{c=1}^{C} \exp\left(x_{n,c}\right)}}{N} \#(\ AUTONUMLGL\ \backslash *Arabic \backslash e\ )$$

Where x is the prediction from the model, y is the ground truth, C is the number of classes, N is the number of voxels.

$$l_{DSC}(x,y) = \frac{\sum_{n=1}^{N} -\frac{2\left|x_c \cap y_c\right|}{\left|x_c\right| + \left|y_c\right|}}{C} = \frac{\sum_{c=1}^{C} -\frac{2TP_c}{2TP_c + FP_c + FN_c}}{C} \#(\ AUTONUMLGL\ \backslash *Arabic \backslash e\ )$$

Where $|x_c|$ and $|y_c|$ are the number of voxels for class c, TP_c is the true positive for class c, FP_c is false positive for class c, FN_c is the false negative for class c

The sum of two losses:

$$L_{sum} = l_{CE}(x,y) + l_{DSC}(x,y) \#(\ AUTONUMLGL\ \backslash *Arabic \backslash e\ )$$

Deep supervision is applied during training by generating probability maps from intermediate features before each upsampling layer in the decoder. These maps are compared to downsampled ground truth segmentations to calculate loss. The training objective combines the weighted sum of losses from each probability map. $L = w_1 \times L_{sum,1} + w_2 \times L_{sum,2} + \cdots + w_n \times L_{sum,n}$ The weights w are halved along with the decrease in resolution. In mathematization

$$w_n = \frac{1}{2^{n-1}} \#\left(AUTONUMLGL \ \backslash *Arabic \backslash e \right)$$

3.8 *Experiments*

Dataset Introduction

To validate the methods for knee bone and cartilage segmentation in MRI, the experiment uses a publicly available Osteoarthritis Initiative (OAI) dataset from Zuse Institute Berlin (Ambellan et al. 2019). According to the metadata from their publication (See Table 1 below), This dataset includes 507 knee scans with expert manual segmentations. Each scan has 160 slices of size 384 × 384, totaling 81,120 slices. Pixel spacing is 0.3645 mm × 0.3645 mm, with a slice thickness of 0.7 mm. The scans cover rOA (Kellgren and Lawrence/KL) grades (Pai 2022) from 0 (healthy) to 4 (severe), with over half graded 3 or 4. The dataset is split into two folds of 253 and 254 scans, commonly used in published studies.

Measures of Segmentation Accuracy

The method is evaluated on the ground truth by the Dice Similarity Coefficient (DSC), average surface distance (ASD), root mean square surface distance (RSD), maximum surface distance (MSD), volume difference (VD), relative volume difference (RVD), and volume overlap error (VOE).

Table 1 Summary of OAI ZIB dataset used for training and validation

Dataset Name	OAI ZIB
MRI scanner	Siemens 3 T Trio
MRI sequence	DESS
Acquisition plane	sagittal
Image resolution [mm]	0.36 × 0.36 × 0.7
Manual segmentations	Bones and cartilage
Number of subjects	507
Sex (male, female)	(262,245)
Age [years]	61.87 ± 9.33
BMI [kg/m2]	29.27 ± 4.52
rOA grade (0,1,2,3,4)	(60,77,61,151,158)
Timepoints	Baseline

$$DSC = 100 \times \frac{2|X \cap Y|}{|X| + |Y|} \#(\ AUTONUMLGL\ \backslash * Arabic \backslash e\ )$$

$$VOE = 100 \times \left(1 - \frac{DSC}{200 - DSC}\right) \#(\ AUTONUMLGL\ \backslash * Arabic \backslash e\ )$$

$$VD = 100 \times \frac{|X| - |Y|}{|Y|} \#(\ AUTONUMLGL\ \backslash * Arabic \backslash e\ )$$

$$RVD = |VD| \#(\ AUTONUMLGL\ \backslash * Arabic \backslash e\ )$$

Where |*X*| and |*Y*| are the number of voxels of prediction from the model and the ground truth.

Distance from a point *p* to surface *S*:

$$d(p, S') = \|\ p - p_2'\ \| (\ AUTONUMLGL\ \backslash * Arabic \backslash e\ )$$

Where S' is the boundary of the surface *S* and $\|_2$ is the Euclidean Norm (2022).

$$ASD = \frac{\sum_{i=1}^{n_{Y'}} d(Y_i, X') + \sum_{j=1}^{n_{X'}} d(X_j, Y')}{n_{X'} + n_{Y'}} \#(\ AUTONUMLGL\ \backslash * Arabic \backslash e\ )$$

$$RSD = \sqrt{\frac{\sum_{i=1}^{n_{Y'}} d(Y_i, X')^2 + \sum_{j=1}^{n_{X'}} d(X_j, Y')^2}{n_{X'} + n_{Y'}}} \#(\ AUTONUMLGL\ \backslash * Arabic \backslash e\ )$$

$$MSD = \left(d(y, X') + d(x, Y')\right) \#(\ AUTONUMLGL\ \backslash * Arabic \backslash e\ )$$

Where, n_s' are the number of voxels on the boundary S'

Training Details

As shown in Table 2 below, in the training phase, the batch size is set to 2 and the initial learning rate is set to 0.01. The optimizer is Stochastic gradient descent (SGD) with a 0.99 Nesterov momentum factor (Sutskever et al. n.d.) and 3e − 5 weight decay (L2 penalty). The learning rate *Lr* is decayed by the 'poly' learning rate policy throughout the epoch increasing. The networks are trained for 1000 epochs, which have 250 batches in each epoch.

$$Lr = (1 - epoch / epoch_{max})^{0.9} \#(\ AUTONUMLGL\ \backslash * Arabic \backslash e\ )$$

Data augmentation increases training samples by applying transformations to the original images, which helps reduce overfitting, especially with small datasets. It generates new patches from source images using various transformations. To avoid

Table 2 Training Settings

Parameter	Training Settings
Initial learning rate	0.01
Optimizer	SGD
Nesterov momentum	0.99
Weight decay	3e − 5
Learning rate decay	Poly
Total epochs	1000
Batches per epoch	250
Batch Size	2

boundary interpolation issues from spatial augmentations (e.g., rotation, scaling), initial patches are larger than the model input patches. Transformations are applied randomly within specified probabilities and ranges:

- **Rotation**: Probability 0.2, range −30° to 30°
- **Scaling**: Probability 0.2, range 0.7–1.4
- **Gaussian noise**: Probability 0.15, range 0–0.1
- **Gaussian blur**: Probability 0.1, kernel size 0.5–1.5
- **Brightness**: Probability 0.15, range 0.7–1.3
- **Contrast**: Probability 0.15, range 0.65–1.5
- **Low resolution simulation**: Probability 0.125, downsampling factor 1–2
- **Gamma augmentation**: Probability 0.15, range 0.7–1.5
- **Mirroring**: Probability 0.5 on 3 axes

Implementation Details

Software and Hardware Environment

The implementation is based on PyTorch (Paszke et al. 2017) and nnU-Net. The machines have 6 core 12 threading CPUs running on 3.8Ghz (Intel Core i7 6850K and AMD Ryzen 3600X) with 32GB DDR4 3200 RAM. The accelerator of deep learning is NVIDIA RTX 3090 with 24GB VRAM. The software environment is Ubuntu 20.04 with NVIDIA CUDA 11.1.

Deep Learning Framework Comparison

In this experiment, several segmentation approaches from an additional framework are evaluated. In the MMSegmentation (MMSegmentation Contributors 2020), the OAI ZIB dataset (Schock et al. 2020) is converted into 81,120 2D slices and training. Multiple state-of-the-art approaches for nature images are evaluated.

- Swin Transformer+UPerNet (Liu et al. 2021; Xiao et al. 2018): Backbone Swin-B
- SegFormer (Xie et al. 2021) Backbone: MIT-B5

- DeepLabV3+ (Chen et al. 2018) Backbone: R-101-D8

All these methods above have the same settings.

- Initialized the pre-trained model of ImageNet-1K before training.
- Learning rate schedule of 160K iteration
- Input shape of 512 × 512

The other hyperparameters (e.g., data argumentation, optimizer, and loss) are kept by default in the MMSegmentation.

Five baseline approaches from two frameworks are validated with 2-fold cross-validation using five metrics, which are Dice Similarity Coefficient (DSC), average surface distance (ASD), root mean square surface distance (RSD), maximum surface distance (MSD), relative volume difference (RVD), and volume overlap error (VOE)

From the results in Table 3, for bone segmentation, all five baseline approaches from the two frameworks have very competitive performances of 98.6% ± 0.15%. When comparing the segmentation accuracy of the femoral bone, methods on MMSegmentation perform better than nnU-Net, especially since the surface distance-related metric is smaller than nnU-Net. Among them, the best model is DeepLabV3+ whose performance is DSC 98.61 ± 0.29 and ASD 0.18 ± 0.05. However, for the rest of the 3 classes, nnU-Net is better than the methods on MMSegmentation. In particular, the difference in performances for cartilage segmentation is significant. Between 2D and 3D of nnU-Net, 3D nnU-Net has a slightly better performance than 2D nnU-Net on average of 4 classes, while for femoral cartilage segmentation, 2D nnU-Net overperforms 3D nnU-Net. However, 3D nnU-Net has a low standard deviation, which means it has a better generalization compared to 2D nnU-Net.

The comparison result indicates that for knee cartilage segmentation, a variety of U-net frameworks for medical images has more benefits on performance than a general framework for natural image segmentation. In other words, for the segmentation of the knee dataset, the improvement from U-Net to state-of-the-art methods cannot offset the use of an inappropriate training pipeline for medical images. Instead, the data-driven optimization of the training pipeline for medical images is beneficial for the segmentation performance.

The early study (Prasoon et al. 2013b) indicates that 3D CNNs have large memory and computational time requirements compared to 2D CNNs. Nowadays, with the optimized hyperparameters and layers settings, training 3D nnU-Net are feasible with a cost of ~8.5 GB VRAM with a batch size of 2. Without the bottleneck of hardware limitation, the experiment is structured based on a 3D nnU-Net, in the consideration of the following factors:

- Although, 2D nnU-Net and 3D nnU-Net have a slightly better performance on femoral cartilage and tibial cartilage respectively. While 3D nnU-Net is slightly better on average and has a lower standard deviation on the variability of images than 2D nnU-Net.

Table 3 Comparison of approaches from two frameworks

Methods	DSC↑		VOE↓		ASD↓		RSD↓		MSD↓		RVD↓	
Femoral Cartilage												
(3D) nnU-Net	89.78	**±2.43**	18.47	**±3.91**	0.18	**±0.06**	**0.41**	±0.16	5.10	±2.42	4.26	±4.00
(2D) nnU-Net	**90.08**	±3.02	**17.91**	±4.95	**0.18**	±0.07	0.41	±0.17	5.12	±2.39	**3.93**	±3.66
(2D) DeepLabV3+	88.66	±2.68	20.26	±4.28	0.20	±0.06	0.42	±0.16	5.16	±2.39	4.29	±3.98
(2D) SegFormer	88.42	±2.62	20.65	±4.17	0.20	±0.06	0.42	**±0.16**	5.11	±2.41	4.16	**±3.52**
(2D) SwinTransformer	88.64	±2.65	20.31	±4.22	0.20	±0.06	0.42	±0.16	**5.07**	**±2.37**	4.33	±3.66
Tibial Cartilage												
(3D) nnU-Net	**86.53**	**±4.11**	**23.52**	±6.23	**0.21**	**±0.09**	**0.49**	±0.27	4.84	±3.93	9.27	±7.98
(2D) nnU-Net	86.07	±4.31	24.21	±6.49	0.21	±0.09	0.49	**±0.26**	4.84	**±3.92**	**9.10**	**±7.87**
(2D) DeepLabV3+	84.89	±4.16	26.03	±6.10	0.23	±0.10	0.51	±0.28	4.88	±3.99	9.69	±8.48
(2D) SegFormer	84.55	±4.27	26.53	±6.24	0.23	±0.10	0.51	±0.28	5.02	±4.38	9.10	±8.03
(2D) SwinTransformer	84.53	±4.16	26.57	**±6.08**	0.23	±0.10	0.50	±0.27	**4.83**	±4.00	10.57	±9.19
Femoral Bone												
(3D) nnU-Net	98.56	±0.29	2.84	±0.57	0.25	±0.05	0.92	±0.16	13.87	±1.71	0.70	±0.53
(2D) nnU-Net	98.49	±0.31	2.97	±0.59	0.26	±0.06	0.92	±0.18	13.87	**±1.71**	0.72	±0.57
(2D) DeepLabV3+	**98.61**	**±0.29**	**2.73**	**±0.55**	0.18	±0.05	0.37	±0.15	3.28	±2.58	**0.58**	**±0.47**
(2D) SegFormer	98.57	±0.30	2.82	±0.58	0.18	±0.06	0.37	±0.16	3.37	±2.67	0.71	±0.57
(2D) SwinTransformer	98.60	±0.29	2.75	±0.56	**0.18**	**±0.05**	**0.36**	**±0.14**	**3.17**	±2.02	0.62	±0.51
Tibial Bone												
(3D) nnU-Net	**98.74**	±0.32	**2.49**	±0.62	**0.16**	**±0.06**	**0.34**	**±0.11**	**2.86**	±1.25	**0.57**	±0.57
(2D) nnU-Net	98.65	±0.32	2.65	±0.63	0.17	±0.06	0.35	±0.12	2.88	**±1.25**	0.59	**±0.55**

(continued)

Table 3 (continued)

Methods	DSC↑		VOE↓		ASD↓		RSD↓		MSD↓		RVD↓	
(2D) DeepLabV3+	98.54	**±0.29**	2.88	**±0.57**	0.18	±0.06	0.37	±0.19	3.07	±2.46	0.60	±0.57
(2D) SegFormer	98.56	±0.30	2.84	±0.59	0.18	±0.06	0.37	±0.15	3.38	±3.90	0.60	±0.58
(2D) SwinTransformer	98.51	±0.31	2.93	±0.60	0.18	±0.06	0.36	±0.13	3.01	±1.72	0.74	±0.64
Average of four classes												
(3D) nnU-Net	**93.40**	**±1.79**	**11.83**	**±2.83**	0.20	**±0.06**	0.54	±0.18	6.67	±2.33	3.70	±3.27
(2D) nnU-Net	93.32	±1.99	11.94	±3.16	0.21	±0.07	0.54	±0.18	6.68	**±2.32**	**3.59**	**±3.16**
(2D) DeepLabV3+	92.68	±1.85	12.98	±2.87	**0.20**	±0.07	0.42	±0.20	4.10	±2.86	3.79	±3.37
(2D) SegFormer	92.53	±1.87	13.21	±2.89	0.20	±0.07	0.42	±0.19	4.22	±3.34	3.64	±3.18
(2D) SwinTransformer	92.57	±1.85	13.14	±2.86	0.20	±0.07	**0.41**	**±0.17**	**4.02**	±2.53	4.06	±3.50

- The method of this research utilizes Vision Transformer to encode the contextual features globally, as well as Atrous Spatial Pyramid Pooling to capture a large field of view. 3D nnU-Net can capture the contextual relationships in three planes, which could maximize the utilization of Vision Transformer and Atrous Spatial Pyramid Pooling
- A 3D model can efficiently train and infer the input of 3D images with a larger voxel size than the 2D model which works slice by slice.

Architecture Design of nnAtrousU-Net

With the flexibility of nnU-Net, we designed three candidate architectures for Atrous Spatial Pyramid Pooling (ASPP) integration. The main difference between these three architectures is the number of convolution blocks and downsampling strategy, which leads to the differences in model parameters and computational cost. (Table 4) (Fig. 13) nnAtrousU-Net-4 × 24 and nnAtrousU-Net-8 × 24 have 5 blocks of dual convolutional layer, since the 2nd block, dimension is reduced continuously on dense planes. For the sparse planes, nnAtrousU-Net-8 × 24 is reduced on the 2nd block, and an extra reduction is applied for nnAtrousU-Net-4 × 24 of the 4th block. nnAtrousU-Net-4 × 12 has one more block for reducing the sparse planes. In the computational cost, nnAtrousU-Net-4 × 24 is more than nnAtrousU-Net-4 × 24 and nnAtrousU-Net-4 × 12, while nnAtrousU-Net-4 × 12 has the most parameters costing more GPU RAM in the training.

These three candidate models are trained and evaluated on fold 0 of the dataset. The diagram below (Fig. 14) shows the training loss and evaluation loss by epoch for these three models. The diagram shows that these three models have very similar declines in losses by training. Both training and validation loss are rapidly at the

Table 4 Three candidate architectures of nnAtrousU-Net

Properties	nnAtrousU-Net-4 × 24 Fig. 13 Left	nnAtrousU-Net-4 × 12 Fig. 13 Middle	nnAtrousU-Net-8 × 24 Fig. 13 Right
Convolution blocks	5	6	5
Patch size	16 × 384 × 384	16 × 384 × 384	16 × 384 × 384
Downsampling per stride	[2, 2, 2], [1, 2, 2], [2, 2, 2], [1, 2, 2]	[2, 2, 2], [1, 2, 2],[2, 2, 2], [1, 2, 2], [1, 2, 2]	[2, 2, 2], [1, 2, 2], [1, 2, 2], [1, 2, 2]
Downsampling per axis	[2, 4, 4]	[2, 5, 5]	[1, 4, 4]
Downsampled feature size	4 × 24 × 24	4 × 12 × 12	8 × 24 × 24
Atrous dilation rate	[1, 2, 2], [2, 4, 4], [3, 6, 6]	[1, 2, 2], [2, 4, 4], [3, 6, 6]	[1, 2, 2], [2, 4, 4], [3, 6, 6]
Flops	766 G	763.98 G	896.01 G
Params	62.49 M	83 M	62.36 M
Training VRAM cost	8071 M	9001 M	9063 M

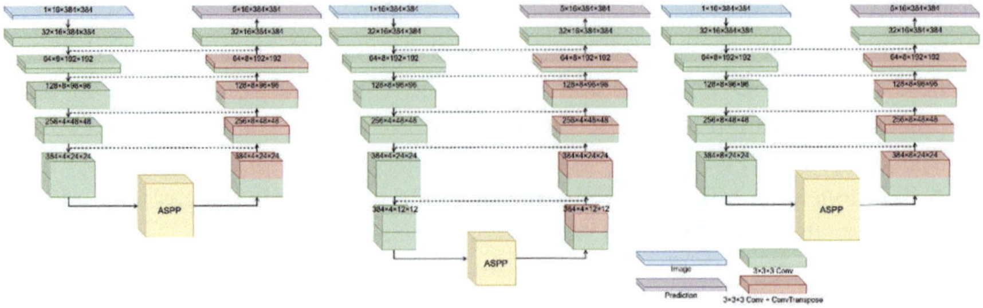

Fig. 13 Three candidates of nnAtrousU-Nets

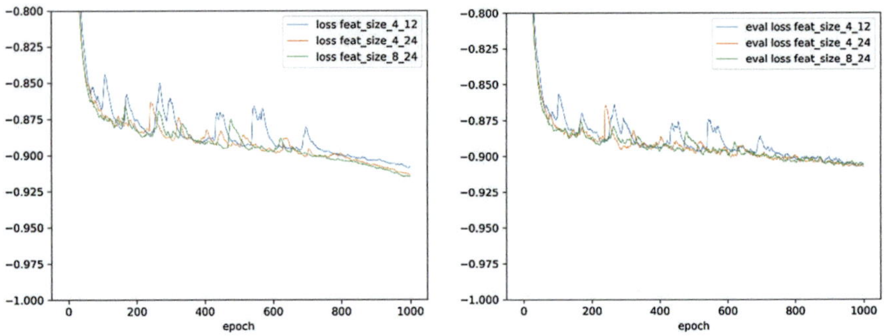

Fig. 14 Training loss and evaluation loss by epoch

first 200 epochs. Because of the momentum in the optimizer as well as the small batch size, there are some spikes in the loss between 100 epochs and 700 epochs. With the continuous decay in the learning rate, the spikes have disappeared since 700 epochs. Among these models, nnAtrousU-Net-4 × 24 is less impacted by spikes than others. Since the epoch of 800, nnAtrousU-Net-8 × 24 started overfitting slightly. All three models have a very close evaluation loss in the ending.

The evaluation results in Table 5 show that while the accuracy differences for femoral cartilage and bones are minimal, nnAtrousU-Net-4 × 24 performs better on tibial cartilage, leading to a slightly better overall performance. Due to its low computational cost and parameter count, nnAtrousU-Net-4 × 24 is chosen as the final setting for nnAtrousU-Net.

Figure 15 compares nnU-Net and nnAtrousU-Net-4 × 24. The nnAtrousU-Net shows a faster decrease in loss than the default nnU-Net. Early on, training loss is higher than evaluation loss due to regularization techniques like data augmentation, which introduce noise and make training data more challenging. Despite these challenges, nnAtrousU-Net maintains consistency between training and validation losses and ultimately achieves a lower loss than nnU-Net.

Table 5 Evaluation of three candidate nnAtrousU-Net architectures

Methods	DSC↑		VOE↓		ASD↓		RSD↓		MSD↓		RVD↓	
Femoral Cartilage												
nnAtrousU-Net-4 × 24	**90.53**	±2.62	**17.20**	±4.31	**0.17**	±0.05	0.39	±0.14	4.97	±2.33	3.65	±3.43
nnAtrousU-Net-4 × 12	90.52	**±2.57**	17.22	**±4.22**	0.17	**±0.05**	**0.39**	**±0.14**	4.98	±2.34	**3.64**	**±3.40**
nnAtrousU-Net-8 × 24	90.50	±2.60	17.25	±4.27	0.17	±0.05	0.39	±0.14	**4.95**	**±2.31**	3.75	±3.59
Tibial Cartilage												
nnAtrousU-Net-4 × 24	**86.93**	**±4.35**	**22.87**	**±6.58**	**0.20**	**±0.09**	**0.46**	±0.22	**4.47**	±1.88	**9.32**	**±7.67**
nnAtrousU-Net-4 × 12	86.80	±4.44	23.07	±6.69	0.20	±0.09	0.46	**±0.22**	4.50	**±1.83**	9.50	±7.70
nnAtrousU-Net-8 × 24	86.78	±4.38	23.10	±6.60	0.20	±0.09	0.47	±0.23	4.52	±1.91	9.36	±7.91
Femoral Bone												
nnAtrousU-Net-4 × 24	**98.79**	**±0.27**	**2.39**	**±0.53**	**0.16**	**±0.05**	**0.34**	**±0.12**	2.95	±1.87	**0.57**	**±0.43**
nnAtrousU-Net-4 × 12	98.79	±0.28	2.40	±0.54	0.16	±0.05	0.34	±0.14	2.93	±1.88	0.57	±0.43
nnAtrousU-Net-8 × 24	98.79	±0.27	2.40	±0.54	0.16	±0.05	0.34	±0.14	**2.91**	**±1.84**	0.58	±0.43
Tibial Bone												
nnAtrousU-Net-4 × 24	98.78	**±0.34**	2.41	**±0.66**	**0.15**	**±0.06**	**0.32**	**±0.12**	2.59	**±0.98**	0.59	±0.64
nnAtrousU-Net-4 × 12	98.77	±0.35	2.42	±0.67	0.15	±0.06	0.32	±0.12	**2.58**	±1.03	0.59	**±0.63**
nnAtrousU-Net-8 × 24	**98.78**	±0.34	**2.41**	±0.67	0.15	±0.06	0.32	±0.12	2.59	±0.99	**0.58**	±0.63
Average of four classes												
nnAtrousU-Net-4 × 24	**93.76**	**±1.89**	**11.22**	±3.02	**0.17**	**±0.06**	**0.38**	**±0.15**	3.74	±1.76	**3.53**	**±3.04**
nnAtrousU-Net-4 × 12	93.72	±1.91	11.28	±3.03	0.17	±0.06	0.38	±0.15	3.75	±1.77	3.58	±3.04
nnAtrousU-Net-8 × 24	93.71	±1.90	11.29	**±3.02**	0.17	±0.07	0.38	±0.16	**3.74**	**±1.76**	3.57	±3.14

Architecture Design of nnAtrousTransFormer

The flexibility of the nnAtrousTransFormer allows defining the number of convolutional blocks replaced by Vision Transformer blocks. By setting 2 and 4 for nnAtrousU-Net, there are two models are obtained, which are nnAtrousTransFormer-2 and nnAtrousTransFormer-4. The following table (Table 6) shows the detailed properties of two nnAtrousTransFormers.

The challenge of training on nnAtrousTransFormer is the VRAM cost during training. It is caused by the increased feature channels from convolutional blocks as

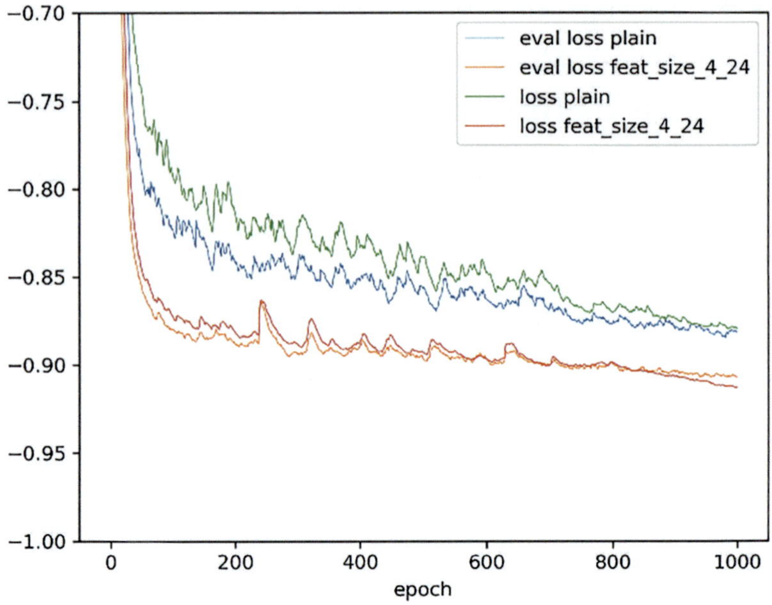

Fig. 15 Comparison of nnU-Net and nnAtrousU-Net-4 × 24

Table 6 Architectures of nnAtrousTransFormer-2 and nnAtrousTransFormer-4

Properties	nnAtrousTransFormer-2 Fig. 12 Left	nnAtrousTransFormer-4 Fig. 12 Right
Count of Convolutional blocks	3	1
Count of Transformer blocks	2	4
Patch size	16 × 384 × 384	16 × 384 × 384
Downsampling per stride	[2, 2, 2], [1, 2, 2], [1, 2, 2], [1, 2, 2]	[2, 2, 2], [1, 2, 2], [1, 2, 2], [1, 2, 2]
Input size of Vision Transformer	8 × 96 × 96	16 × 384 × 384
Input feature channels	128	32
Transformer patch size	1 × 4 × 4	2 × 16 × 16
Hidden size	192	192
Heads	6	6
MLP dimensions	768	768
Downsampling per axis	[1, 4, 4]	[1, 4, 4]
Downsampled feature size	8 × 24 × 24	8 × 24 × 24
Flops	215.96 G	246.73 G
Params	16.24 M	19.61 M
VRAM cost in Training	15,521 M	23,767 M

well as one more dimension of the input and layers. Compared to the 2D Vision Transformer (ViT-Base) (Dosovitskiy et al. 2021), Hidden size *D*, MLP size as well as heads are reduced. The patch size for position embedding is calculated as (Input size of Vision Transformer) divided by (Downsampled feature size). nnAtrousTransFormer-2 has 3 blocks of a convolutional neural network followed by two blocks of Vision Transformer. The input image is downsampled by two times on the dense plane and once on the sparse plane. As a result, the input size of the Vision Transformer is 8 × 96 × 96. nnAtrousTransFormer-4 has two more Vision Transformers. Because the first block of the convolution neural network has no downsampling, Vision Transformer blocks in nnAtrousTransFormer-4 are calculated on the full scale of the input image. In computational cost and amount of model parameters, nnAtrousTransFormer-4 is more than nnAtrousTransFormer-2.

Figure 16 shows that nnAtrousTransformer-2 has a slightly better decline rate of loss than nnAtrousTransformer-4. The reason for the differences is that nnAtrousTransformer-2 have more blocks of the convolutional neural network, its encoder detailed high-resolution local context features followed by the Vision Transformer for global information encoding, while nnAtrousTransformer-4 receives the full scale of features with less local context features from the convolutional neural network. This finding is consistent with previous research on Vision Transformers (Dosovitskiy et al. 2021; Chen et al. 2021). These two architectures are evaluated on the 2 folds of the knee bone and cartilage dataset. (Table 7) The overall performance shows that nnAtrousTransFormer-2 is better. Finally, the structure of nnAtrousTransFormer-2 is applied as the final nnAtrousTransFormer model.

Figure 17 is the comparison between nnU-Net, nnAtrousU-Net and nnAtrousTransFormer on fold 0 of training loss. The figure shows all my proposed approaches have a faster decline of loss than nnU-Net, resulting in a lower loss than nnU-Net. In the ending epoch, nnAtrousU-Net obtained the lowest loss, followed by nnAtrousTransFormer.

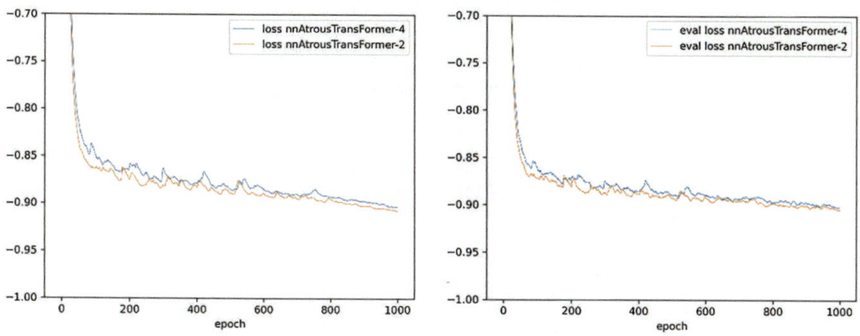

Fig. 16 Comparison of nnAtrousTransFormer

Table 7 Comparison between nnAtrousTransFormer-4 and nnAtrousTransFormer-2

Methods	DSC↑		VOE↓		ASD↓		RSD↓		MSD↓		RVD↓	
Femoral Cartilage												
nnAtrousTransFormer-4	89.85	**±2.46**	18.34	**±3.99**	0.18	**±0.06**	0.41	±0.16	5.09	±2.39	4.28	±4.04
nnAtrousTransFormer-2	**90.08**	±2.63	**17.94**	±4.29	**0.18**	±0.06	**0.40**	**±0.15**	**5.07**	**±2.31**	**4.05**	**±3.83**
Tibial Cartilage												
nnAtrousTransFormer-4	86.38	**±4.16**	23.75	**±6.26**	0.21	**±0.10**	0.49	**±0.28**	4.84	±3.94	9.69	±8.31
nnAtrousTransFormer-2	**86.57**	±4.18	**23.44**	±6.32	**0.21**	±0.10	**0.49**	±0.29	**4.78**	**±3.93**	**9.31**	**±7.88**
Femoral Bone												
nnAtrousTransFormer-4	98.73	±0.31	2.50	±0.59	0.16	±0.05	0.35	**±0.13**	**3.04**	**±1.72**	0.62	±0.51
nnAtrousTransFormer-2	**98.76**	**±0.30**	**2.45**	**±0.58**	**0.16**	**±0.05**	**0.34**	±0.13	3.04	±1.78	**0.60**	**±0.50**
Tibial Bone												
nnAtrousTransFormer-4	98.75	±0.32	2.47	±0.63	0.16	±0.06	0.33	±0.11	2.81	±1.44	**0.58**	±0.57
nnAtrousTransFormer-2	**98.76**	**±0.32**	**2.44**	**±0.62**	**0.15**	**±0.06**	**0.33**	**±0.11**	**2.71**	**±1.14**	0.58	**±0.57**
Average of four classes												
nnAtrousTransFormer-4	93.43	**±1.81**	11.76	**±2.87**	0.18	**±0.07**	0.40	**±0.17**	3.94	±2.37	3.79	±3.36
nnAtrousTransFormer-2	**93.54**	±1.86	**11.57**	±2.95	**0.17**	±0.07	**0.39**	±0.17	**3.90**	**±2.29**	**3.63**	**±3.19**

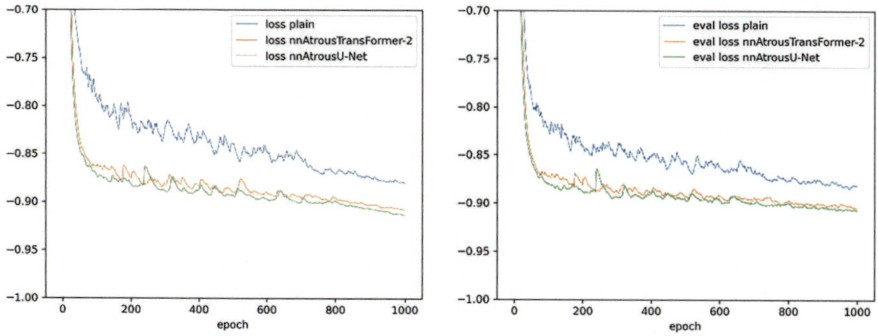

Fig. 17 Comparison of nnU-Net, nnAtrousU-Net and nnAtrousTransFormer

Experimental Results

Based on the architecture design of nnAtrousU-Net and nnAtrousTransFormer, I further evaluated the best model of these two approaches with a two-fold cross-validation. The accuracy evaluations of my two methods as well as nnU-Net are summarized in Table 8. From the table below, in any of the four classes, the proposed nnAtrousU-Net nnAtrousTransFormer approach both overperforms the nnU-Net. nnAtrousU-Net.

The Dice of the best model nnAtrousU-Net increased for Femoral Cartilage (90.46 ± 2.86%), Tibial Cartilage (86.91 ± 4.15%), Femoral Bone (98.79 ± 0.28%) compared to the dice score of nnU-Net for Femoral Cartilage (90.46 ± 2.86%), Tibial Cartilage (86.91 ± 4.15%) and Femoral Bone (98.56 ± 0.29%). Another published research has improved the nnU-Net as well and reports their implementation of plain nnU-Net (Liang et al. 2022). Caused by differences in settings (e.g., optimized patch size, in my implementation), the performances of the baseline nnU-Net are different. While my improvement on the nnU-Net exceeds the improvement of nnU-Net+PCAM.

nnAtrousTransFormer has shown a competitive performance for Femoral Cartilage (90.08 ± 2.63%), Femoral Bone (98.76 ± 0.30%). Although, replacing the two deep convolution blocks with Vision Transformer blocks does not surpass the nnAtrousU-Net. nnAtrousTransFormer has shown its superiority in parameters and computational cost. Specifically, nnAtrousTransFormer has a size of 19.61 M parameters, (See Table 9 below) which are 4 times smaller than nnAtrousU-Net (62.49 M) and 2 times smaller than nnU-Net. (38.84 M) It is even further smaller and faster than 2D-based segmentation methods. In computational cost, nnAtrousTransFormer is 246.73 GFlops, which is 2.5 ~ 3.5 times fewer than nnAtrousTransFormer and nnU-Net. Compared with the 2D approach, its computational cost is competitive. Because 2D-based method is not an end-to-end method, the GFlops of the 2D methods are measured by inferring one single slice. While nnAtrousTransFormer can predict with 16 slices at a similar computational cost. In other words, 19.6 times faster than SwinTransformer-B, 3.5 times faster than SegFormer MIT-B5, and 16.5 times faster than DeepLabV3+ R101. Compared with the accuracy,

Table 8 Quantitative analysis of my approach with nnU-Net

Methods	DSC↑		VOE↓		ASD↓		RSD↓		MSD↓		RVD↓	
Femoral Cartilage												
nnAtrousU-Net	**90.46**	±2.86	**17.30**	±4.71	**0.17**	±0.06	**0.39**	**±0.15**	**4.99**	**±2.30**	**3.99**	**±3.73**
nnAtrousTransFormer	90.08	±2.63	17.94	±4.29	0.18	±0.06	0.40	±0.15	5.07	±2.31	4.05	±3.83
nnU-Net	89.78	**±2.43**	18.47	**±3.91**	0.18	**±0.06**	0.41	±0.16	5.10	±2.42	4.26	=4.00
nnU-Net + PCAM (Liang et al. 2022)	89.35	±2.69	19.14	±4.32	0.24	±0.32		×		×		×
nnU-Net (Liang et al. 2022)	89.03	±2.73	19.21	±4.37	0.26	±0.32		×		×		×
Tibial Cartilage												
nnAtrousU-Net	**86.91**	±4.15	**22.92**	±6.36	**0.20**	±0.10	**0.47**	±0.28	**4.69**	±3.93	**9.17**	**±7.74**
nnAtrousTransFormer	86.57	±4.18	23.44	±6.32	0.21	±0.10	0.49	±0.29	4.78	±3.93	9.31	±7.88
nnU-Net	86.53	**±4.11**	23.52	**±6.23**	0.21	**±0.09**	0.49	**±0.27**	4.84	**±3.93**	9.27	±7.98
nnU-Net+PCAM (Liang et al. 2022)	86.14	±4.43	24.08	±6.63	0.22	±0.12		×		×		×
nnU-Net (Liang et al. 2022)	86.00	±4.52	24.29	±6.73	0.21	±0.11		×		×		×
Femoral Bone												
nnAtrousU-Net	**98.79**	**±0.28**	**2.39**	**±0.55**	**0.15**	**±0.05**	**0.33**	**±0.11**	**2.88**	**±1.67**	**0.58**	**±0.46**
nnAtrousTransFormer	98.76	±0.30	2.45	±0.58	0.16	±0.05	0.34	±0.13	3.04	±1.78	0.60	±0.50
nnU-Net	98.56	±0.29	2.84	±0.57	0.25	±0.05	0.92	±0.16	13.87	±1.71	0.70	±0.53
Tibial Bone												

nnAtrousU-Net	**98.79**	**±0.32**	**2.38**	**±0.62**	**0.15**	**±0.06**	**0.32**	**±0.10**	**2.64**	**±1.14**	**0.56**	**±0.55**
nnAtrousTransFormer	98.76	±0.32	2.44	±0.62	0.15	±0.06	0.33	±0.11	2.71	±1.14	0.58	±0.57
nnU-Net	98.74	±0.32	2.49	±0.62	0.16	±0.06	0.34	±0.11	2.86	±1.25	0.57	±0.57
Average of four classes												
nnAtrousU-Net	**93.74**	±1.90	**11.25**	±3.06	**0.17**	**±0.06**	**0.38**	**±0.16**	**3.80**	**±2.26**	**3.58**	**±3.12**
nnAtrousTransFormer	93.54	±1.86	11.57	±2.95	0.17	±0.07	0.39	±0.17	3.90	±2.29	3.63	±3.19
nnU-Net	93.40	**±1.79**	11.83	**±2.83**	0.20	±0.06	0.54	±0.18	6.67	±2.33	3.70	±3.27

Table 9 Computational cost and parameters and its mean performance on cartilage

Method	GFlops	Params (M)	DSC↑		VOE↓		ASD↓	
			Mean Cartilage					
(2D) DeepLabV3+ R101	253.93	62.57	86.78	±3.42	23.15	±5.19	0.22	±0.08
(2D) SegFormer MIT-B5	**53.16**	81.97	86.49	±3.45	23.59	±5.20	0.22	±0.08
(2D) SwinTransformer-B	302.22	121.30	86.59	±3.40	23.44	±5.15	0.22	±0.08
nnAtrousU-Net	766.00	62.49	**88.68**	±3.51	**20.11**	±5.54	**0.18**	±0.08
nnAtrousTransFormer	246.73	**19.61**	88.33	±3.41	20.69	±5.31	0.19	±0.08
nnU-Net	630.59	38.84	88.15	**±3.27**	20.99	**±5.07**	0.19	**±0.07**

nnAtrousTransFormer utilizes only half parameters and computational cost to overperform nnU-Net.

The parameters in nnAtrousTransFormer are reduced because the Transformer patch size, hidden size, Heads and MLP dimensions are greatly reduced from the Vision Transformer-Base. Two blocks (two layers each) of Vision Transformer, only have 1.77 M params. Replacing the complex 3D high-cost convolutional block in the deep of the network with lite Vision Transformer followed by an Atrous Spatial Pyramid Pooling. It can provide a good balance between computational cost and performance.

The latest related research developed their method based on the nnU-Net, by adding a clustering-based self-attention module into the nnU-Net (Liang et al. 2022). The performance of their approach nnU-Net+PCAM is improved based on their implementation of nnU-Net. For the comparison of their method, they evaluated three additional models, which are V-Net without any modification (Milletari et al. 2016), Generative Adversarial Network(GAN) based approach (Tan et al. 2019) and ACFNet (Zhang et al. 2019). They have an extremely competitive performance to CNN-3D+SSM (Ambellan et al. 2019), while CNN-3D+SSM used additional independent data for training the SSM model (Schock et al. 2020). They proposed a pipeline including 2D-based pre-segmentation, 3D shape regression and subsequent patch-based 3D segmentation in the final. In the DSC performance, they overperformed the previous state-of-the-art method (CNN-3D+SSM), but they use 80% and 20% for training and evaluation, which is different from all settings in the paper of the dataset (Ambellan et al. 2019). Improved MaskRCNN (iMask RCNN) (Felfeliyan et al. 2022) train on the 80%/20% as well, by adding a long skip connection from additional ROI Aligned features in large scale to the decoder, it greatly improved the performance of Mask RCNN (He et al. 2018) on this dataset.

Table 10 Summarizes the results of the most recent research for the OAI-ZIB dataset. From the table, we could find that the proposed nnAtrousU-Net has achieved the state-of-the-art performance among the published methods.

Figure 18 is the visualization of two slices in the axial plane (Top) and sagittal plane (Bottom) by ITK-SNAP (Yushkevich et al. 2006). All the predictions from the three models have a good consistency with the annotation by the specialist. In the axial plane, the inference of Femoral Cartilage (Green) from nnAtrousU-Net is very consistent with the ground truth, comparing the two other models. In the sagittal plane on another scan, the inference of Tibial Cartilage from nnAtrousU-Net is closest to ground truth among the three models, especially, in thickness and shape.

Table 10 Quantitative comparison among segmentation methods

Methods	DSC↑		VOE↓		ASD↓		Data Split Train/Valid
Femoral Cartilage							
Our nnAtrousU-Net	**90.46**	±2.86	**17.30**	±4.71	0.17	±0.06	253/254
Our nnAtrousTransFormer	90.08	±2.63	17.94	±4.29	0.18	±0.06	253/254
(3D) nnUNet (Isensee et al. 2021)	89.78	±2.43	18.47	**±3.91**	0.18	**±0.06**	253/254
(2D) nnUNet (Isensee et al. 2021)	90.08	±3.02	17.91	±4.95	0.18	±0.07	253/254
(2D) DeepLabV3+ (Chen et al. 2018)	88.66	±2.68	20.26	±4.28	0.20	±0.06	253/254
(2D) SegFormer (Xie et al. 2021)	88.42	±2.62	20.65	±4.17	0.20	±0.06	253/254
(2D) SwinTransformer (Liu et al. 2021)	88.64	±2.65	20.31	±4.22	0.20	±0.06	253/254
CNN-3D+SSM (Ambellan et al. 2019)	89.90	±3.60	18.10	±5.90	**0.16**	±0.07	253/254 + 60
(3D) nnUNet (Liang et al. 2022)	89.03	±2.73	19.21	±4.37	0.26	±0.32	253/254
nnU-Net+PCAM (Liang et al. 2022)	89.35	±2.69	19.14	±4.32	0.24	±0.32	253/254
V-Net (Milletari et al. 2016; Liang et al. 2022)	87.71	±2.77	21.79	±4.28	0.23	±0.10	253/254
(Tan et al. 2019; Liang et al. 2022)	87.86	±2.84	21.53	±4.40	0.24	±0.12	253/254
ACFNet (Zhang et al. 2019; Liang et al. 2022)	87.91	±2.82	21.45	±4.38	0.22	±0.11	253/254
(Schock et al. 2020)	89.90			×		×	80%/20%
Mask RCNN (He et al. 2018)	71.	±3.		×		×	80%/20%
iMask RCNN	86.	±2.		×		×	80%/20%
Tibial Cartilage							
nnAtrousU-Net	**86.91**	±4.15	**22.92**	±6.36	**0.20**	±0.10	253/254
nnAtrousTransFormer-2	86.57	±4.18	23.44	±6.32	0.21	±0.10	253/254
(3D) nnUNet (Isensee et al. 2021)	86.53	±4.11	23.52	±6.23	0.21	±0.09	253/254
(2D) nnUNet (Isensee et al. 2021)	86.07	±4.31	24.21	±6.49	0.21	±0.09	253/254
(2D) DeepLabV3+ (Chen et al. 2018)	84.89	±4.16	26.03	±6.10	0.23	±0.10	253/254
(2D) SegFormer (Xie et al. 2021)	84.55	±4.27	26.53	±6.24	0.23	±0.10	253/254
(2D) SwinTransformer (Liu et al. 2021)	84.53	±4.16	26.57	±6.08	0.23	±0.10	253/254
CNN-3D+SSM (Ambellan et al. 2019)	85.60	±4.54	24.90	±6.79	0.23	±0.12	253/254 + 60
(3D) nnUNet original (Liang et al. 2022)	86.00	±4.52	24.29	±6.73	0.21	±0.11	253/254
nnU-Net+PCAM (Liang et al. 2022)	86.14	±4.43	24.08	±6.63	0.22	±0.12	253/254
V-Net (Milletari et al. 2016)	84.11	±3.93	27.23	**±5.79**	0.23	±0.11	253/254
Tan et al. (2019)	84.08	±4.04	27.25	±5.94	0.26	±0.14	253/254
ACFNet (Zhang et al. 2019)	84.57	±4.16	26.52	±6.20	0.24	**±0.09**	253/254
(Schock et al. 2020)	85.90			×		×	80%/20%
Mask RCNN (He et al. 2018)	81.	±3.		×		×	80%/20%
(Felfeliyan et al. 2022)	82.	±4.		×		×	80%/20%

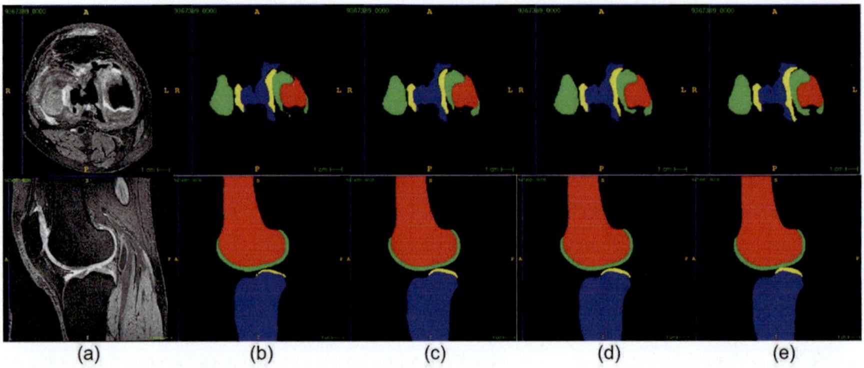

Fig. 18 Qualitative analysis of segmentation results. (**a**) is one slice of in 3D MRI image. (**b**) is ground truth generated by professionals. (**c**) is the inference from nnAtrousU-Net. (**d**) is the inference from nnAtrousTransForme.r (**e**) is the inference from nnU-Net. Red represents Femoral Bone, Green represents Femoral Cartilage, Yellow represents Tibial Cartilage and Blue represents Tibial Bone

This research enhances the superior nnU-Net by expanding its universal medical image segmentation approach with new search spaces. By inheriting nnU-Net's automated configuration pipeline, various medical datasets can be trained out-of-the-box, benefiting from state-of-the-art algorithms. On the knee OAI-ZIB dataset, a new strategy for input patch size calculation was added, allowing for larger image scales on dense planes and providing detailed information under limited VRAM, resulting in advanced accuracy. The proposed methods, nnAtrousU-Net and nnAtrousTransformer, integrate Atrous Spatial Pyramid Pooling and Vision Transformer into nnU-Net, achieving accuracy improvements on the OAI-ZIB dataset over the previously optimized nnU-Net and outperforming the current best CNN-3D+SSM (Ambellan et al. 2019). For small object segmentation, nnAtrousU-Net exceeds CNN-3D+SSM by 0.935% in cartilage segmentation. nnAtrousTransformer combines detailed high-resolution local context features from CNN with global information from Vision Transformer, using only half the parameters and one-third the computational cost, achieving accuracy close to nnAtrousU-Net (Fig. 19).

4 Conclusion

This research enhances the superior nnU-Net by expanding its universal medical image segmentation approach with new search spaces. By inheriting nnU-Net's automated configuration pipeline, various medical datasets can be trained out-of-the-box, benefiting from state-of-the-art algorithms. On the knee OAI-ZIB dataset, a new strategy for input patch size calculation was added, allowing for larger image scales on dense planes and providing detailed information under limited VRAM, resulting in advanced accuracy. The proposed methods, nnAtrousU-Net and

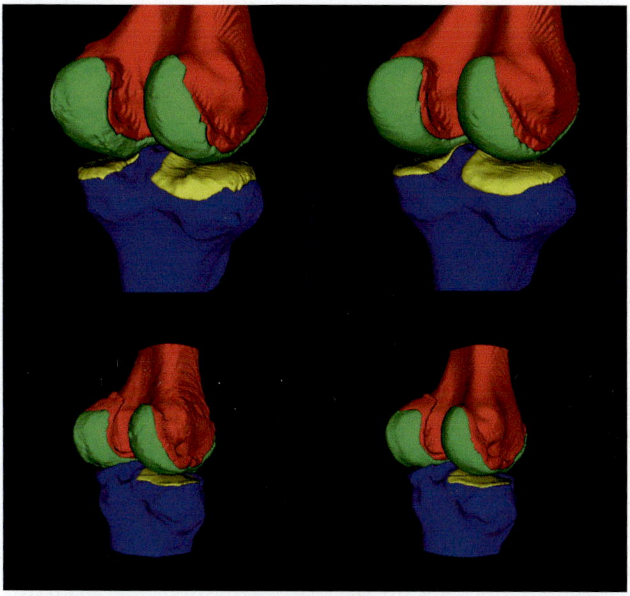

Fig. 19 Visualization of knee bone and cartilage. The Left figures are the ground truth, and the Right figures are the model inferences from nnAtrousU-Net

nnAtrousTransformer, integrate Atrous Spatial Pyramid Pooling and Vision Transformer into nnU-Net, achieving accuracy improvements on the OAI-ZIB dataset over the previously optimized nnU-Net and outperforming the current best CNN-3D+SSM (Ambellan et al. 2019). For small object segmentation, nnAtrousU-Net exceeds CNN-3D+SSM by 0.935% in cartilage segmentation. nnAtrousTransformer combines detailed high-resolution local context features from CNN with global information from Vision Transformer, using only half the parameters and one-third the computational cost, achieving accuracy close to nnAtrousU-Net.

References

Ababneh SY, Prescott JW, Gurcan MN. Automatic graph-cut based segmentation of bones from knee magnetic resonance images for osteoarthritis research. Med Image Anal. 2011;15(4):438–48. https://doi.org/10.1016/j.media.2011.01.007.

Ambellan F, Tack A, Ehlke M, Zachow S. Automated segmentation of knee bone and cartilage combining statistical shape knowledge and convolutional neural networks: data from the Osteoarthritis Initiative. Med Image Anal. 2019;52:109–18. https://doi.org/10.1016/j.media.2018.11.009.

Ba JL, Kiros JR, Hinton GE. Layer normalization, 2016, July 21, arXiv: arXiv:1607.06450. https://doi.org/10.48550/arXiv.1607.06450.

Bach Cuadra M, Duay V, Thiran J-P. Atlas-based segmentation. In: Paragios N, Duncan J, Ayache N, editors. Handbook of biomedical imaging: methodologies and clinical research. Boston: Springer US; 2015. p. 221–44. https://doi.org/10.1007/978-0-387-09749-7_12.

Badrinarayanan V, Kendall A, Cipolla R. SegNet: a deep convolutional encoder-decoder architecture for image segmentation. IEEE Trans Pattern Anal Mach Intell. 2017;39(12):2481–95. https://doi.org/10.1109/TPAMI.2016.2644615.

Baevski A, Auli M. Adaptive input representations for neural language modeling. 2018, Sept. https://doi.org/10.48550/arXiv.1809.10853.

Boykov Y, Funka-Lea G. Graph cuts and efficient N-D image segmentation. Int J Comput Vis. 2006;70(2):109–31. https://doi.org/10.1007/s11263-006-7934-5.

Chen L-C, Papandreou G, Schroff F, Adam H. Rethinking Atrous convolution for semantic image segmentation, 2017a, Dec 5, arXiv: arXiv:1706.05587. Accessed 20 Aug 2022. [Online]. Available: http://arxiv.org/abs/1706.05587.

Chen LC, Papandreou G, Kokkinos I, Murphy K, Yuille AL. DeepLab: semantic image segmentation with deep convolutional nets, Atrous convolution, and fully connected CRFs, 2017b, May 11, arXiv: arXiv:1606.00915. Accessed 22 Aug 2022. [Online]. Available: http://arxiv.org/abs/1606.00915.

Chen LC, Zhu Y, Papandreou G, Schroff F, Adam H. Encoder-decoder with Atrous separable convolution for semantic image segmentation, 2018, Aug 22, arXiv: arXiv:1802.02611. Accessed 26 Aug, 2022. [Online]. Available: http://arxiv.org/abs/1802.02611.

Chen J, et al. TransUNet: transformers make strong encoders for medical image segmentation, 2021, Feb 8, arXiv: arXiv:2102.04306. Accessed 22 Aug 2022. [Online]. Available: http://arxiv.org/abs/2102.04306.

Conaghan PG, Hunter DJ, Maillefert JF, Reichmann WM, Losina E. Summary and recommendations of the OARSI FDA osteoarthritis Assessment of Structural Change Working Group. Osteoarthr Cartil. 2011;19(5):606–10. https://doi.org/10.1016/j.joca.2011.02.018.

Demir-Kavuk O, Kamada M, Akutsu T, Knapp E-W. Prediction using step-wise L1, L2 regularization and feature selection for small data sets with large number of features. BMC Bioinform. 2011;12:412. https://doi.org/10.1186/1471-2105-12-412.

Devlin J, Chang MW, Lee K, Toutanova K. BERT: Pre-training of deep bidirectional transformers for language understanding, 2019, May 24, arXiv: arXiv:1810.04805. https://doi.org/10.48550/arXiv.1810.04805.

Dosovitskiy A, et al. An image is worth 16x16 words: transformers for image recognition at scale, 2021, Jun 3. arXiv: arXiv:2010.11929. Accessed: Aug. 14, 2022. [Online]. Available: http://arxiv.org/abs/2010.11929.

Dumoulin V, Visin F. A guide to convolution arithmetic for deep learning, 11 Jan 2018, arXiv: arXiv:1603.07285. https://doi.org/10.48550/arXiv.1603.07285.

Duryea J, et al. Novel fast semi-automated software to segment cartilage for knee MR acquisitions. Osteoarthr Cartil. 2007;15(5):487–92. https://doi.org/10.1016/j.joca.2006.11.002.

'Euclidean Norm - an overview | ScienceDirect Topics. Accessed 25 Aug 2022. [Online]. Available: https://www.sciencedirect.com/topics/computer-science/euclidean-norm.

Felfeliyan B, Hareendranathan A, Kuntze G, Jaremko JL, Ronsky JL. Improved-mask R-CNN: towards an accurate generic MSK MRI instance segmentation platform (data from the Osteoarthritis initiative). Comput Med Imaging Graph. 2022;97:102056. https://doi.org/10.1016/j.compmedimag.2022.102056.

Folkesson J, Olsen OF, Pettersen P, Dam E, Christiansen C. Combining binary classifiers for automatic cartilage segmentation in knee MRI. In: Liu Y, Jiang T, Zhang C, editors. Computer vision for biomedical image applications. Lecture notes in computer science. Berlin/Heidelberg: Springer; 2005. p. 230–9. https://doi.org/10.1007/11569541_24.

Folkesson J, Dam EB, Olsen OF, Pettersen PC, Christiansen C. Segmenting articular cartilage automatically using a voxel classification approach. IEEE Trans Med Imaging. 2007;26:106–15.

Glyn-Jones S, et al. Osteoarthritis. Lancet. 2015;386(9991):376–87. https://doi.org/10.1016/S0140-6736(14)60802-3.

Hatamizadeh A, et al. UNETR: transformers for 3D medical image segmentation, 2021, Oct 9, arXiv: arXiv:2103.10504. Accessed 22 Aug 2022. [Online]. Available: http://arxiv.org/abs/2103.10504.

He K, Zhang X, Ren S, Sun J. Spatial pyramid pooling in deep convolutional networks for visual recognition. 2014;8691:346–61. https://doi.org/10.1007/978-3-319-10578-9_23.

He K, Zhang X, Ren S, Sun J. Deep residual learning for image recognition. In: 2016 IEEE conference on computer vision and pattern recognition (CVPR). Las Vegas, NV, USA: IEEE; 2016. p. 770–8. https://doi.org/10.1109/CVPR.2016.90.

He K, Gkioxari G, Dollár P, Girshick R. Mask R-CNN, 2018, Jan 24, arXiv: arXiv:1703.06870. https://doi.org/10.48550/arXiv.1703.06870.

Heinrich MP, Hansen L. Highly accurate and memory efficient unsupervised learning-based discrete CT registration using 2.5D displacement search. In: Martel AL, Abolmaesumi P, Stoyanov D, Mateus D, Zuluaga MA, Zhou SK, Racoceanu D, Joskowicz L, editors. Medical image computing and computer assisted intervention—MICCAI 2020, in lecture notes in computer science, vol. 12263. Cham: Springer International Publishing; 2020. p. 190–200. https://doi.org/10.1007/978-3-030-59716-0_19.

Ioffe S, Szegedy C. Batch normalization: accelerating deep network training by reducing internal covariate shift. In: Proceedings of the 32nd international conference on international conference on machine learning - volume 37, in ICML'15. Lille, France: JMLR.org; 2015. p. 448–56.

Isensee F, Jaeger PF, Kohl SAA, Petersen J, Maier-Hein KH. nnU-Net: a self-configuring method for deep learning-based biomedical image segmentation. Nat Methods. 2021;18(2):Art. no. 2. https://doi.org/10.1038/s41592-020-01008-z.

Liang D, Liu J, Wang K, Luo G, Wang W, Li S. Position-prior clustering-based self-attention module for Knee Cartilage segmentation, 2022, Jun 21, arXiv: arXiv:2206.10286. Accessed 26 Aug 2022. [Online]. Available: http://arxiv.org/abs/2206.10286.

Liu F, Zhou Z, Jang H, Samsonov A, Zhao G, Kijowski R. Deep convolutional neural network and 3D deformable approach for tissue segmentation in musculoskeletal magnetic resonance imaging. Magn Reson Med. 2018;79(4):2379–91. https://doi.org/10.1002/mrm.26841.

Liu Z, et al. Swin transformer: hierarchical vision transformer using shifted windows, 2021, Aug 17, arXiv: arXiv:2103.14030. https://doi.org/10.48550/arXiv.2103.14030.

Long J, Shelhamer E, Darrell T. Fully convolutional networks for semantic segmentation. In: 2015 IEEE conference on computer vision and pattern recognition (CVPR); 2015. p. 3431–40. https://doi.org/10.1109/CVPR.2015.7298965.

Maas AL, Hannun AY, Ng AY. Rectifier nonlinearities improve Neural Network Acoustic Models, p. 6.

Metcalfe AJ, Andersson MLE, Goodfellow R, Thorstensson CA. Is knee osteoarthritis a symmetrical disease? Analysis of a 12 year prospective cohort study. BMC Musculoskelet Disord. 2012;13:153. https://doi.org/10.1186/1471-2474-13-153.

Milletari F, Navab N, Ahmadi SA. V-Net: fully convolutional neural networks for volumetric medical image segmentation, 2016, Jun 15, arXiv: arXiv:1606.04797. https://doi.org/10.48550/arXiv.1606.04797.

MMSegmentation Contributors, OpenMMLab semantic segmentation toolbox and benchmark, 2020, July. Python. Accessed 20 Aug 2022. [Online]. Available: https://github.com/open-mmlab/mmsegmentation.

Multi-Atlas Labeling Beyond the Cranial Vault - Workshop and Challenge - syn3193805 - Wiki'. Accessed 22 Aug 22, 2022. [Online]. Available: https://www.synapse.org/#!Synapse:syn3193805/wiki/89480.

nnU-Net. Python. MIC-DKFZ, 2022, Aug 21. Accessed 21 Aug 2022. [Online]. Available: https://github.com/MIC-DKFZ/nnUNet/blob/d0ba941a8a8996dca3ac36afc34988bafbc575a1/nnunet/experiment_planning/alternative_experiment_planning/readme.md.

Norman B, Pedoia V, Majumdar S. Use of 2D U-Net convolutional neural networks for automated cartilage and Meniscus segmentation of knee MR imaging data to determine Relaxometry and morphometry. Radiology. 2018;288(1):177–85. https://doi.org/10.1148/radiol.2018172322.

Pai V. Kellgren and Lawrence system for classification of osteoarthritis | Radiology Reference Article | Radiopaedia.org, Radiopaedia. Accessed 25 Aug 2022. [Online]. Available:

https://radiopaedia.org/articles/kellgren-and-lawrence-system-for-classification-of-osteoarthritis?lang=gb.

Paszke A, et al. Automatic differentiation in PyTorch, 2017, Oct. Accessed 26 Aug 2022. [Online]. Available: https://openreview.net/forum?id=BJJsrmfCZ.

Peterfy C, Woodworth T, Altman R. Workshop for consensus on osteoarthritis imaging: MRI of the knee. Osteoarthr Cartil. 2006;14:44–5. https://doi.org/10.1016/j.joca.2006.02.025.

Prasoon A, Igel C, Loog M, Lauze F, Dam E, Nielsen M. Cascaded classifier for large-scale data applied to automatic segmentation of articular cartilage. In: Medical imaging 2012: image processing. SPIE; 2012. p. 1371–9. https://doi.org/10.1117/12.910809.

Prasoon A, Igel C, Loog M, Lauze F, Dam E, Nielsen M. Femoral cartilage segmentation in Knee MRI scans using two stage voxel classification. In: Conf Proc Annu Int Conf IEEE Eng Med Biol Soc IEEE Eng Med Biol Soc Conf, vol. 2013; 2013a. p. 5469–72. https://doi.org/10.1109/EMBC.2013.6610787.

Prasoon A, Petersen K, Igel C, Lauze F, Dam E, Nielsen M. Deep feature learning for Knee Cartilage segmentation using a triplanar convolutional neural network. In: Mori K, Sakuma I, Sato Y, Barillot C, Navab N, editors. Medical image computing and computer-assisted intervention—MICCAI 2013. 讲义的. Berlin/Heidelberg: Springer; 2013b. p. 246–53. https://doi.org/10.1007/978-3-642-40763-5_31.

Radford A. Improving language understanding by generative pre-training. 2018, Accessed 18 Sep 2024. [Online]. Available: https://hayate-lab.com/wp-content/uploads/2023/05/43372bfa750340059ad87ac8e538c53b.pdf.

Ronneberger O, Fischer P, Brox T. U-Net: convolutional networks for biomedical image segmentation. In: Navab N, Hornegger J, Wells WM, Frangi AF, editors. Medical image computing and computer-assisted intervention—MICCAI 2015, in lecture notes in computer science. Cham: Springer International Publishing; 2015. p. 234–41. https://doi.org/10.1007/978-3-319-24574-4_28.

Schock J, et al. A method for semantic knee bone and cartilage segmentation with deep 3D shape fitting using data from the Osteoarthritis initiative. In: Reuter M, Wachinger C, Lombaert H, Paniagua B, Goksel O, Rekik I, editors. Shape in medical imaging, in lecture notes in computer science. Cham: Springer International Publishing; 2020. p. 85–94. https://doi.org/10.1007/978-3-030-61056-2_7.

Scott IM, Cootes TF, Taylor CJ. Improving appearance model matching using local image structure. In: Taylor C, Noble JA, editors. Information processing in medical imaging, in lecture notes in computer science, vol. 2732. Berlin/Heidelberg: Springer Berlin Heidelberg; 2003. p. 258–69. https://doi.org/10.1007/978-3-540-45087-0_22.

Seim H, Kainmueller D, Lamecker H, Bindernagel M, Malinowski J, Zachow S. Model-based auto-segmentation of knee bones and cartilage in MRI data. Proc Med Image Anal. 2010;

Shan L, Charles C, Niethammer M. Automatic multi-atlas-based cartilage segmentation from knee MR images. In: Proc. IEEE Int. Symp. Biomed. Imaging Nano Macro IEEE Int. Symp. Biomed. Imaging, vol. 2012; 2012. p. 1028–31. https://doi.org/10.1109/ISBI.2012.6235733.

Shan L, Zach C, Charles C, Niethammer M. Automatic atlas-based three-label cartilage segmentation from MR knee images. Med Image Anal. 2014;18(7):1233–46. https://doi.org/10.1016/j.media.2014.05.008.

Shim H, Chang S, Tao C, Wang J-H, Kwoh CK, Bae KT. Knee Cartilage: efficient and reproducible segmentation on high-spatial-resolution MR images with the semiautomated Graph-Cut Algorithm Method1. Radiology. 2009;251 https://doi.org/10.1148/radiol.2512081332.

Shin H-C, et al. Deep convolutional neural networks for computer-aided detection: CNN architectures, dataset characteristics and transfer learning. IEEE Trans Med Imaging. 2016;35(5):1285–98. https://doi.org/10.1109/TMI.2016.2528162.

Srivastava N, Hinton G, Krizhevsky A, Sutskever I, Salakhutdinov R. Dropout: a simple way to prevent neural networks from overfitting. J Mach Learn Res. 2014;15(56):1929–58.

Sutskever I, Martens J, Dahl G, Hinton G. On the importance of initialization and momentum in deep learning, p. 14.

Swain S, et al. Trends in incidence and prevalence of osteoarthritis in the United Kingdom: findings from the Clinical Practice Research Datalink (CPRD). Osteoarthr Cartil. 2020;28(6):792–801. https://doi.org/10.1016/j.joca.2020.03.004.

Szegedy C, Vanhoucke V, Ioffe S, Shlens J, Wojna Z. Rethinking the inception architecture for computer vision. In: 2016 IEEE conference on computer vision and pattern recognition (CVPR). Las Vegas: IEEE; 2016, June. p. 2818–26. https://doi.org/10.1109/CVPR.2016.308.

Tack A, Mukhopadhyay A, Zachow S. Knee menisci segmentation using convolutional neural networks: data from the Osteoarthritis Initiative. Osteoarthr Cartil. 2018;26(5):680–8. https://doi.org/10.1016/j.joca.2018.02.907.

Tamez-Peña J, González P, Farber J, Baum K, Schreyer E, Totterman S. Atlas based method for the automated segmentation and quantification of knee features: data from the osteoarthritis initiative. In: 2011 IEEE international symposium on biomedical imaging: from nano to macro; 2011. p. 1484–7. https://doi.org/10.1109/ISBI.2011.5872681.

Tan C, Yan Z, Zhang S, Li K, Metaxas DN. Collaborative multi-agent learning for MR knee articular cartilage segmentation, 2019, Aug 12, arXiv: arXiv:1908.04469. Accessed 26 Aug 2022. [Online]. Available: http://arxiv.org/abs/1908.04469.

Trebeschi S, et al. Deep learning for fully-automated localization and segmentation of rectal cancer on multiparametric MR. Sci Rep. 2017;7(1):5301. https://doi.org/10.1038/s41598-017-05728-9.

Twining CJ, Marsland S, Taylor C. Groupwise non-rigid registration: the minimum description length approach. In: Procedings of the British machine vision conference 2004. Kingston: British Machine Vision Association; 2004. p. 44.1–44.10. https://doi.org/10.5244/C.18.44.

Ulyanov D, Vedaldi A, Lempitsky V. Instance normalization: the missing ingredient for fast stylization, 2017, Nov 6, arXiv: arXiv:1607.08022. Accessed 23 Aug 2022. [Online]. Available: http://arxiv.org/abs/1607.08022.

Vaswani A, et al. Attention is all you need. In: Advances in neural information processing systems. Curran Associates, Inc; 2017. Accessed: Aug. 14, 2022. [Online]. Available: https://proceedings.neurips.cc/paper/2017/hash/3f5ee243547dee91fbd053c1c4a845aa-Abstract.html.

Wang Q, et al. Learning deep transformer models for machine translation. In: Proceedings of the 57th annual meeting of the Association for Computational Linguistics. Florence, Italy: Association for Computational Linguistics; 2019. p. 1810–22. https://doi.org/10.18653/v1/P19-1176.

Xiao T, Liu Y, Zhou B, Jiang Y, Sun J. Unified perceptual parsing for scene understanding. 2018, July 26, arXiv: arXiv:1807.10221. Accessed 20 Aug 2022. [Online]. Available: http://arxiv.org/abs/1807.10221.

Xie E, Wang W, Yu Z, Anandkumar A, Alvarez JM, Luo P. SegFormer: simple and efficient design for semantic segmentation with transformers, 2021, Oct 28, arXiv: arXiv:2105.15203. Accessed 20 Aug 2022. [Online]. Available: http://arxiv.org/abs/2105.15203.

Yan K, Wang X, Lu L, Summers RM. DeepLesion: automated mining of large-scale lesion annotations and universal lesion detection with deep learning. J Med Imaging. 2018;5(3) https://doi.org/10.1117/1.JMI.5.3.036501.

Ying X. An overview of overfitting and its solutions. J Phys Conf Ser. 2019;1168:022022. https://doi.org/10.1088/1742-6596/1168/2/022022.

Yushkevich PA, et al. User-guided 3D active contour segmentation of anatomical structures: significantly improved efficiency and reliability. NeuroImage. 2006;31(3):1116–28. https://doi.org/10.1016/j.neuroimage.2006.01.015.

Zhang K, Lu W, Marziliano P. Automatic knee cartilage segmentation from multi-contrast MR images using support vector machine classification with spatial dependencies. Magn Reson Imaging. 2013;31(10):1731–43. https://doi.org/10.1016/j.mri.2013.06.005.

Zhang F, et al. ACFNet: attentional class feature network for semantic segmentation, 2019, Oct 17, arXiv: arXiv:1909.09408. Accessed 26 Aug 2022. [Online]. Available: http://arxiv.org/abs/1909.09408.

Zhao H, Shi J, Qi X, Wang X, Jia J. Pyramid Scene parsing network, 2017, April 27, arXiv: arXiv:1612.01105. Accessed 24 Aug 2022. [Online]. Available: http://arxiv.org/abs/1612.01105.

Zhou Z, Zhao G, Kijowski R, Liu F. Deep convolutional neural network for segmentation of knee joint anatomy. Magn Reson Med. 2018;80(6):2759–70. https://doi.org/10.1002/mrm.27229.

Zhou HY, Guo J, Zhang Y, Yu L, Wang L, Yu Y. nnFormer: interleaved transformer for volumetric segmentation, 2022, Feb 4, arXiv: arXiv:2109.03201. Accessed 25 Aug, 2022. [Online]. Available: http://arxiv.org/abs/2109.03201.

Computer Vision and Deep Learning in Regenerative Medicine: Part 2

Kai Armstrong and Xujiong Ye

Abstract The clinical environment all over the world has been changing rapidly, with a higher demand for healthcare availability without an increase in available resources to account for this demand. Thus, computational tools are required to help alleviate this burden and allow for patients to not only be seen faster but provide more objective data in the decision-making process. Orthopaedics is a particularly challenging field to implement novel computational tools given the subjective nature of many of the currently used diagnostic workflows. Additionally, much of the research in this field has been focused on biomedical image analysis which requires the patient to have radiographic imaging, which is an expensive and time-consuming task. In this chapter we propose the use of marker-less motion capture techniques to allow the collection of clinically relevant motion data from patients. This motion data can be used to initially quantify the severity of a problem and thus lift the load on the more expensive services and those with long waiting lists. In this work we showed that marker-less motion capture is sensitive enough to distinguish between a patient with knee pain and a patient without knee pain, this was done using a diagnostic block to remove the effects of pain and allow us to see the knee's physiological capabilities. This motion data has also been further utilised in a biomarker identification pipeline to outline actions and kinematic metrics that are clinically relevant to both diagnosis and rehabilitative progress tracking.

Keywords Health informatics · Clinical biomechanics · Biomarkers · Computer vision · Machine learning

---

K. Armstrong (✉)
University of Lincoln, Lincoln, UK
e-mail: karmstrong@lincoln.ac.uk

X. Ye
University of Exeter, Exeter, UK

P. Lee (ed.), *Musculoskeletal Regeneration Medicine*,
https://doi.org/10.1007/978-3-031-77456-0_6

1 Clinical Introduction

Although MRI remains the gold standard for diagnosing musculoskeletal (MSK) conditions, its high cost and maintenance requirements pose significant challenges for clinics and hospitals, especially in low- and middle-income countries (Jalloul et al. 2023). Consequently, there is a pressing need for low-cost, portable techniques to diagnose and monitor patients effectively. One promising alternative is the analysis of human motion, which involves quantifying a person's movements and using this data to inform clinical decision-making, including diagnosis and treatment (Simon 2004). Recently, several studies have evaluated the effectiveness of human pose estimation (HPE) in clinical environments, highlighting its potential applications and current limitations (Stenum et al. 2021; Chen et al. 2018; Armstrong et al. 2022). HPE can be used to assess gait, posture, and functional movements, providing valuable insights for diagnosing conditions such as arthritis, Parkinson's disease, and recovery progress post-surgery (Lv et al. 2022; Li et al. 2018). Despite its promise, the integration of HPE into clinical practice requires further validation and standardisation to address challenges related to accuracy, reliability, and the diversity of patient populations.

2 Human Pose Estimation

HPE is a method of estimating the positions of keypoints on the human body based on either an image or a video, these keypoints are typically in the form of joints in the human body (Wang et al. 2021). Pose estimation models are typically trained on large annotated datasets and are structured in a variety of ways depending on the purpose, multi-person pose estimation falls into either top-down or bottom-up methods while single-person pose estimation uses either a regression-based approach or body part detection methods (Andriluka et al. 2014; Zheng et al. 2023).

Top-down HPE methods first use a detector to identify each person in the image and create a bounding box, followed by a joint detection pipeline typically utilising CNN architecture (Dai et al. 2017) for feature extraction trained using large-scale image datasets such as ImageNet (Deng et al. 2009) or COCO (Lin et al. 2014). An example of this pipeline is shown in Fig. 1, which outlines the pipeline from the initial human detection and cropping followed by the pose estimation for each person (Chen et al. 2023).

Bottom-up architecture on the other hand has two primary steps: first is the feature extraction typically using a neural network or transformer to predict the body joint candidates, followed by the assembly of these joint candidates into a skeleton for each person in the image or video (Zheng et al. 2023). This pipeline can be seen in Fig. 2, showing a method developed by Bao et al. (2023) who presented Pose ResNet 3D which performs a bottom-up methodology of HPE which works for both 2D and 3D data (Bao et al. 2023).

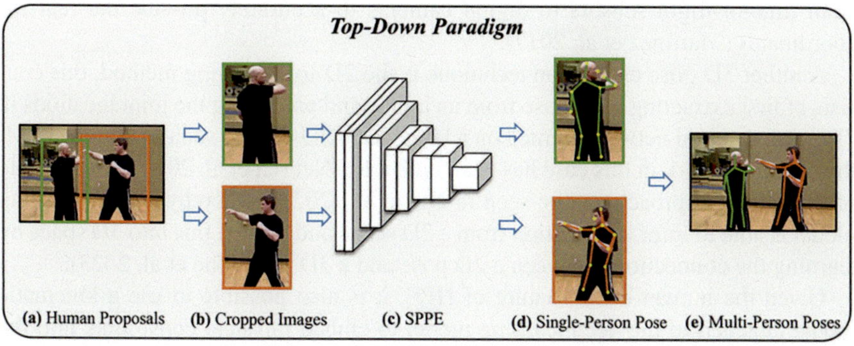

Fig. 1 Top-down human pose estimation pipeline, showing the initial bounding box detection followed by the keypoint detection for each bounding box by neural network approach (Andriluka et al. 2014)

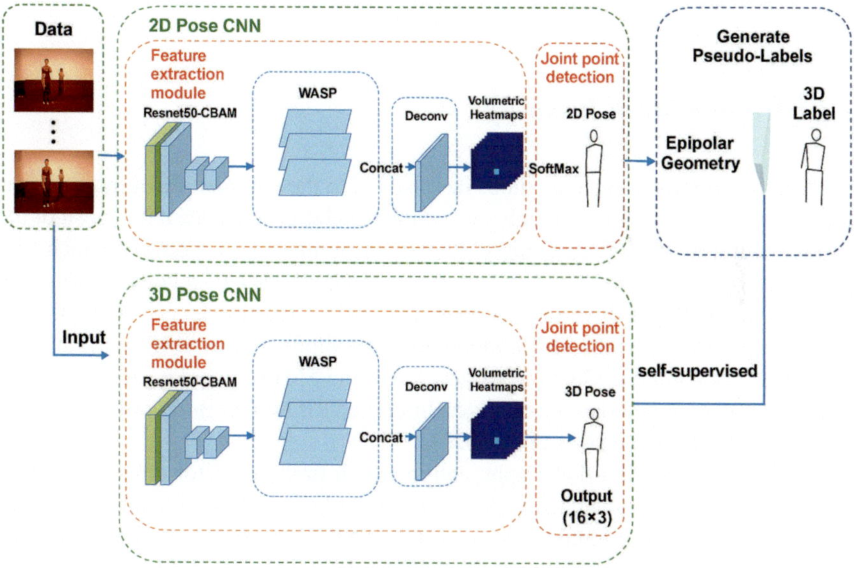

Fig. 2 Bottom-up approach to human pose estimation showing the feature extraction architecture to produce a heatmap of predicted joint locations followed by a probabilistic model to put each of these keypoints into an assembled skeleton (Bao et al. 2023)

The above methods are empirically 2D pose estimation methods, while 3D pose estimation is more computationally challenging, this area has made significant strides recently. One of these methods is the direct estimation, this technique can predict the 3D pose of a person directly from a 2D image without any 2D pose intermediate steps. This can be done using a deep CNN network trained on a large-scale 3D human pose dataset such as Human3.6M (Ionescu et al. 2013), a dataset which contains 3.6 million different articulated 3D poses captured with 15 sensors ranging from time-of-flight sensors to digital cameras to accurately provide the real 3D coordinates (Martinez et al. 2017).

Another 3D pose estimation technique is the 2D-to-3D lifting method, this consists of first extracting a 2D pose from an image and predicting the joint locations in 3D using a neural network trained on a large-scale 3D dataset such as Human3.6M; the neural network in this case has been both a ResNet (Ci et al. 2019). An example of this lifting approach can be seen in Nie et al. (2023), whereby a deep learning model is able to infer information from a 2D scene and project this into 3D space by learning the connections between a 2D pose and a 3D pose (Nie et al. 2023).

Given the human-bound nature of HPE, it is also possible to use a kinematic model attached to the deep learning model to embed physical constraints into the prediction which is important for 3D pose estimation using videos as input. This can be done in a few different ways, one of the more prominent of these methods is the application of a temporal attention mechanism that ensures the joint movement is consistent with the pose sequence rather than relying on just the image alone (Liu et al. 2020).

2.1 Human Mesh Recovery

Each of the aforementioned methods has also led to the culmination of more powerful techniques that extract not only an accurate skeleton sequence but a mesh of the surface features of a person; this is known as human mesh recovery which was pioneered by the skinned multi-person linear model (SMPL)—a body system with 23 joints and 6890 mesh vertices (Loper et al. 2015). This model can be applied on either images to extract the SMPL model using methods such as SMPLify (Pavlakos et al. 2019) or can be applied on a video where the SMPL model is created at each time point using methods such as Video Inference for human Body pose and shape Estimation (VIBE) (Kocabas et al. 2020).

VIBE also uses a top-down approach in which first a detector model is used to identify where in a frame each person is located, then a CNN with gated recurrent units (GRUs) create a temporally consistent SMPL sequence which is validated with a motion discriminator generative adversarial network (GAN) trained using a large-scale dataset known as AMASS (Kocabas et al. 2020; Mahmood et al. 2019).

2.2 Application to the Clinical Environment

These technologies have shown a lot of potential for a portable solution for clinical motion analysis, given the environmental conditions present in clinical environments it is not always possible to have a dedicated space required for gait analysis and classical biomechanics analyses with tools such as optical tracking mocap with many both tripod-based cameras and ceiling mounted cameras surrounding the data capture environment (Moro et al. 2022).

Studies have shown the potential for RGB-D cameras for a markerless motion capture system with minimal error when focusing on the knee joint kinematics (Pfister et al. 2014). Given this, experiments have since been run to test if markerless motion capture methods can be used for measuring the intervention success when treating knee disorders.

This technique has been proven to be useful when using gait analysis, the understanding of how a person walks based on a set of gait parameters such as cadence, stride length and step width (Viswakumar et al. 2019). The HPE methodology of OpenPose (Wei et al. 2016) is utilised to extract the gait cycle, focusing on knee flexion/extension and hip abduction/adduction from heel strike onwards. This method has been validated in both bright and dim clinical conditions (D'Antonio et al. 2020).

To best show the capabilities in clinical practice this method of marker-less HPE has been applied on a pool of patients with diagnosed knee osteoarthritis (OA) related joint pain before and after receiving a local anaesthetic injection to the knee, this shows the physical capabilities of the patients by performing actions without the limitation of pain (Kim et al. 2018).

3 Methods

3.1 Human Pose and Human Mesh Recovery

Azure Kinect

In this study, the process began with the setup and configuration of the Azure Kinect device to record depth frames and sensor data during a session. The recorded data was saved to files for later processing, where we utilised the Azure Kinect DK as a post-processing device to analyse the recorded videos and extract pose data.

In the post-processing phase, we initialised the body tracking system using the recorded data files instead of live sensor input. The body tracker was created with the necessary sensor calibration information. The recorded frames were sequentially read and enqueued into the body tracker, which processed each frame to extract body tracking data. This data included the detection and tracking of human poses frame by frame.

The body tracking results were stored, detailing the poses and movements detected in the recorded video. This approach allowed for detailed analysis and review of the body tracking data after the recording session was completed, enabling thorough post-event examination and application of the pose data for various analytical purposes. Finally, the system was properly shut down, ensuring all resources were released and the processed data was saved for further use. The benefits of using a depth sensor such as the Azure Kinect for capturing more detailed

Fig. 3 Demonstration of Azure Sensor showing how the combination of RGB and near-IR can be combined to provide a depth understanding to the capture environment

environment data can be seen in Fig. 3, by combining the RGB video output with the near-IR time of flight data, it is possible to create a more depth-aware digital twin reducing the possibility of occlusion caused by lack of scene understanding.

VIBE

In this study, we also employed the VIBE model developed by Kocabas et al. (2020) to extract 3D mesh sequences from video footage. Initially, video sequences were input into the VIBE model, which utilises a temporal neural network to estimate 3D human body poses and shapes for each frame (Kocabas et al. 2020). The network incorporates recurrent connections to ensure temporal consistency and smooth transitions between frames. Training of the VIBE model involved a combination of 2D keypoint detection, 3D pose estimation, and adversarial learning, facilitating the generation of realistic and coherent human body reconstructions (Kocabas et al. 2020). By processing entire video sequences, the method effectively reduced noise and inaccuracies typically associated with single-frame estimations. The resulting output was a sequence of 3D meshes representing the detailed and temporally consistent human motion captured in the videos. An example frame of this can be seen in Fig. 4, showing the locations of the joints in the SMPL-X body system along with the mesh vertices showing the surface information of a human (Pavlakos et al. 2019).

Fig. 4 Visualisation of the SMPL-X body model (Pavlakos et al. 2019) showing the morphable mesh of a human overlayed on top of a human performing a jumping exercise. This mesh was created using the VIBE mesh reconstruction tool (Kocabas et al. 2020)

3.2 Clinical Study

Participants

Twenty participants (10 men and 10 women) aged over 55 years and clinically diagnosed with knee pain were recruited for this study. Inclusion criteria required participants to have experienced chronic knee pain for at least 6 months prior to the study. Participants with other musculoskeletal disorders or severe comorbidities that could affect their gait were excluded. Informed consent was obtained from all participants.

Intervention

Each participant received a diagnostic local anaesthetic block to temporarily relieve knee pain. This procedure involved injecting a local anaesthetic into the knee joint under sterile conditions to confirm that the pain originated from the knee.

Data Collection

Participants were recorded using an Azure Kinect RGB-D camera positioned 3 metres away, with the data processed through the Azure Kinect Skeleton Tracking SDK, which has previously been validated against motion capture (Tölgyessy et al. 2021; Kurillo et al. 2022), and the VIBE model (Kocabas et al. 2020) to generate 3D skeleton and mesh systems. The actions performed by participants included a sit-to-stand, where individuals moved from a sitting position to a standing position and back to sitting, and a squat, performed to the maximum depth without any pain. These actions were selected for their commonality in daily activities. Participants who were unable to perform at least three repetitions of both actions were excluded from the study.

Inference and Analysis

Inference was conducted on a local PC with Nvidia GTX 1080ti, capable of running the VIBE model at an inference speed of 40 seconds per 10 seconds of video, while Azure Kinect Skeleton Tracking provided real-time performance at 24 frames per second.

Data Analysis

Pose and mesh sequences captured before and after the treatment were systematically compared using a range of metrics and statistical tests. Joint angles were calculated utilising the cosine formula (Eq. 1), which allowed for precise determination of knee flexion/extension and hip abduction/adduction. These joint angles were then summarised using two key metrics: smoothness (Eq. 2), which quantified the consistency of movement, and impulse (Eq. 3), which measured the cumulative absolute rotational acceleration of the joints. These metrics provided a comprehensive overview of the changes in movement dynamics induced by the treatment. The derived joint angles and their summary metrics were subjected to statistical tests to assess the significance of the observed changes, thereby ensuring a robust evaluation of the treatment's impact on joint mechanics.

$$\cos(A) = \frac{b^2 + c^2 - a^2}{2bc} \tag{1}$$

Where a is the distance between points B and C, b is the distance between points A and C, and c is the distance between points A and B.

$$\sigma_m = \sqrt{\frac{\sum_{i=1}^{n} (m_i - \underline{m})^2}{n-1}} \tag{2}$$

Where n is the number of data points, m_i is the gradient at point i and $\underline{m}$ is the mean of all gradients.

$$J = \int |\alpha| \cdot dt \tag{3}$$

Where J is the approximation of the impulse using the cumulative absolute rotational acceleration of a single joint. Using $|\alpha|$ as the absolute rotational acceleration and *dt* is the change in time.

Feature Extraction

Principal Component Analysis (PCA) was utilised to identify significant joints for various tasks, enabling the extraction of relevant features from the dataset. Implemented via Scikit-learn using linear dimensionality reduction with single value decomposition (SVD) (Pedregosa et al. 2011; Zhang et al. 2007), PCA commenced with the preprocessing of data to ensure uniformity and relevance through normalisation and feature scaling. PCA effectively reduced the dimensionality of the dataset while preserving its essential characteristics by transforming it into principal components, thereby facilitating the interpretation and visualisation of data patterns. The SVD method decomposed the dataset matrix into three matrices: U, Σ, and V^T, which identified the most significant features. By separating the decomposition into two components, each corresponding to a specific action, PCA allowed for a targeted analysis of joint movements pertinent to each task, thereby determining the most representative features.

Comparison and Correlation

To compare the data obtained from the Azure Kinect and the SMPL models, a comprehensive statistical analysis was conducted. Initially, linear regression analysis was applied to examine the relationship between the joint angle data from both systems, ensuring that the differences in data acquisition methods did not affect the overall trends. The Pearson correlation coefficient was then calculated to quantify the degree of correlation between the two datasets, providing a measure of the linear relationship and consistency between the Azure Kinect and SMPL-derived joint angles. Summary metrics, specifically the smoothness and impulse metrics as defined in Eqs. 2 and 3, were computed for both pre- and post-injection conditions. These metrics were used to create a dataset comprising the relevant joint angles. A paired t-test was subsequently performed on this dataset to determine the presence of statistically significant changes in joint angles following the diagnostic local anaesthetic block. This rigorous approach ensured that the analysis accounted for individual variability and provided a robust comparison of the joint kinematics before and after the intervention.

Table 1 Summary statistics of the pearson correlation coefficient for each participant in the squat action showing the mean correlation coefficient, standard deviation, minimum, lower quartile, median, upper quartile, and the maximum

	Left Flexion	Right Flexion
mean	0.9	0.92
std	0.18	0.08
min	−0.18	0.61
25%	0.91	0.93
50%	0.95	0.95
75%	0.97	0.97
max	0.99	0.99

Table 2 Summary statistics of the pearson correlation coefficient for each participant in the sit-to-stand action showing the mean correlation coefficient, standard deviation, minimum, lower quartile, median, upper quartile, and the maximum

	Left Flexion	Right Flexion
mean	0.95	0.95
std	0.04	0.04
min	0.75	0.86
25%	0.94	0.94
50%	0.97	0.97
75%	0.98	0.98
max	1	1

4 Results

4.1 Linear Regression Analysis and Correlation Analysis

The initial results of the correlation analysis between the Azure Kinect skeleton tracking and the SMPL-based tracking using the VIBE model show a strong correlation between the two data extraction methodologies. Tables 1 and 2 show that for both the left and right knee flexion for both the squat and the sit-to-stand actions, the two data extraction modalities are strongly correlated with a pearson mean correlation coefficient >= 0.9. This is also more exaggerated in the sit-to-stand action where the knee flexion for both left and right knees had a standard deviation of just 0.04 which is greater than the standard deviation of the squat: 0.18 for the left knee and 0.08 for the right knee. The sit-to-stand action also has a maximum value of 1 which ultimately means that for at least one repeat of both actions the mean left and right knee flexion was identical for the whole action.

These summary statistics are also visualised in Fig. 5 where the box plots represent the median, upper quartile, lower quartile, minimum, and maximum values. These plots help accentuate the small variance between the upper and lower quartiles for both actions for both knee flexion values.

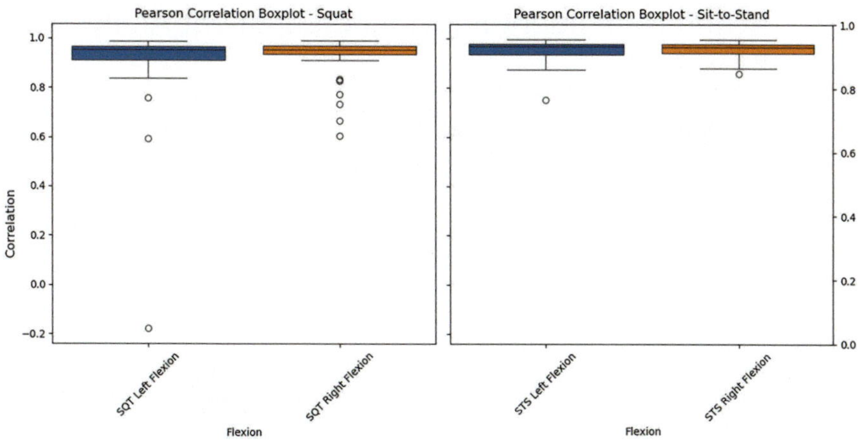

Fig. 5 Boxplots showing the pearson correlation coefficient for both the squat and the sit-to-stand actions where the boxes represent the pearson coefficient of each participant represented as the mean, median, upper quartile, lower quartile, minimum and maximum values

4.2 *Principal Component Analysis and Biomarker Identification*

The principal component analysis performed for each person, one example shown in Fig. 6, is a method of reducing the dimensionality of a set of data into feature space. This plot shows the clear separation between the squat and sit-to-stand data, by taking the furthest points we can identify which data is more representative of each action. After the PCA was performed for each participant, the counts of these most representative data were visualised in the histograms in Figs. 7 and 8 whereby every time one of these features was found in the top 10 most representative features the sum of these times were calculated. This identifies likely candidates for representative biomarkers for knee pain, where the left and right knee flexion is important for both actions. However, the minimum knee flexion was shown to be more important in the squat than in the sit-to-stand. The sit-to-stand action was also found to feature the elbow flexion and arm abduction rather than lower limb kinematics alone.

4.3 *Statistical Testing*

To determine the statistical significance, the paired t-test was used to compare the pre-injection and post-injection data for the entire population of participants. These results can be seen in Tables 3 and 4 with the first value being the t-value and the second being the p-value. This shows that for the squat action there were only two metrics that were statistically significant with a p-value <0.05 which were the left and right maximum knee flexion curve's smoothness. However, in the sit-to-stand

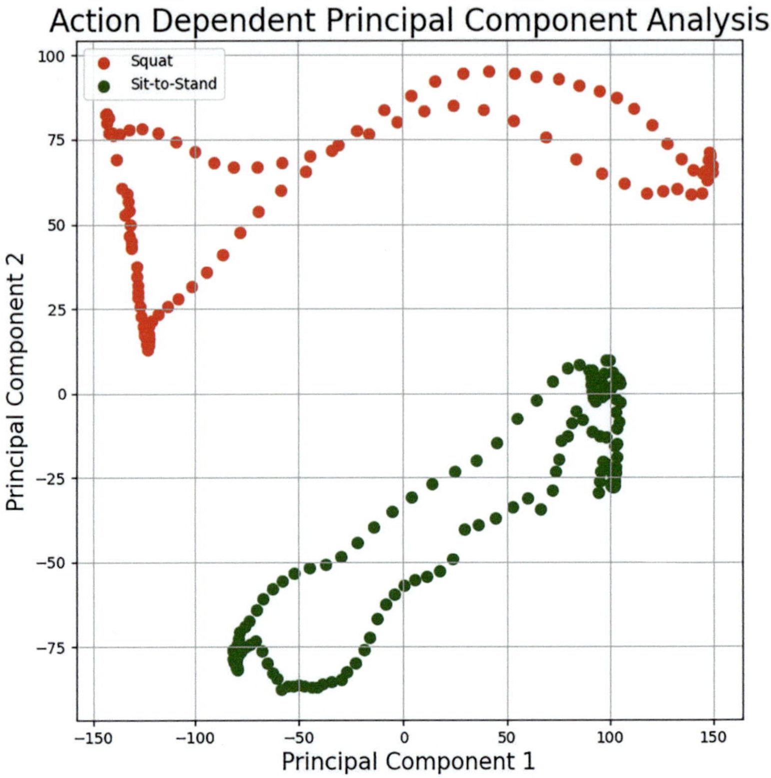

Fig. 6 Plot showing the clear divide between the squat action metrics and the sit-to-stand action metrics

there were considerably more significant metrics with a p-value <0.05 such as the cumulative acceleration of the left and right maximum elbow flexion and the smoothness of all displayed metrics.

4.4 *Clinical Trial Results*

Taking the results of the statistical testing further, we can take those statistically significant metrics and use these as biomarkers for knee pain. By plotting the percentage change between the pre- and post-injection time points we can identify whether the statistically significant change mentioned previously is also clinically relevant. Both Figs. 9 and 10 show that for each statistically significant biomarker the majority of points have at least a 10% increase after the injection. There are some anomalies that have been identified, this is where there is either a worsening or a change considered very large, this will be investigated further.

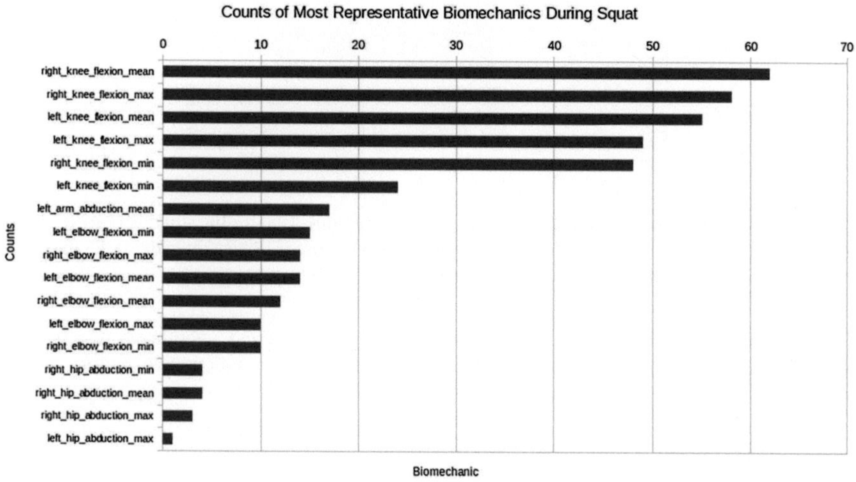

Fig. 7 Histogram showing the most commonly identified metrics for the squat action, with counts of the times each metric appeared in the 10 most representative features from the PCA

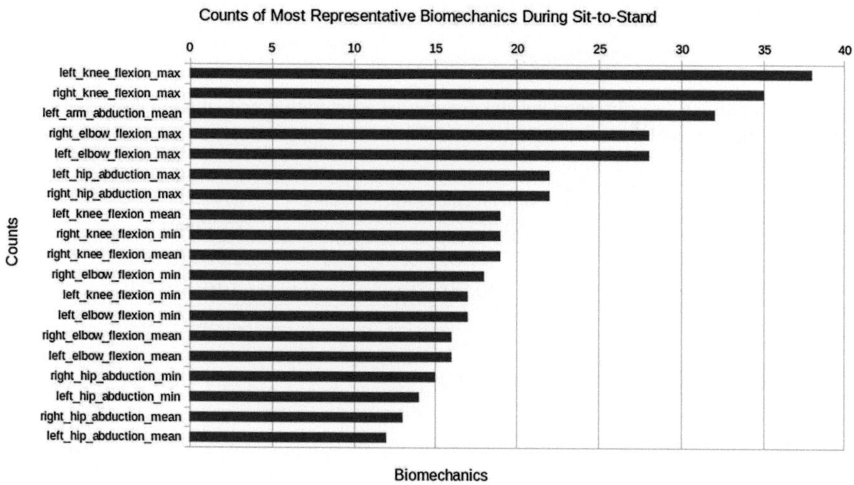

Fig. 8 Histogram showing the most commonly identified metrics for the sit-to-stand action, with counts of the times each metric appeared in the 10 most representative features from the PCA

5 Discussion

5.1 Correlation

When interpreting the results of these methods, it is imperative to understand their implications and significance within the context of clinical applications. Firstly, when considering the correlation between the Azure Kinect Skeleton tracking and

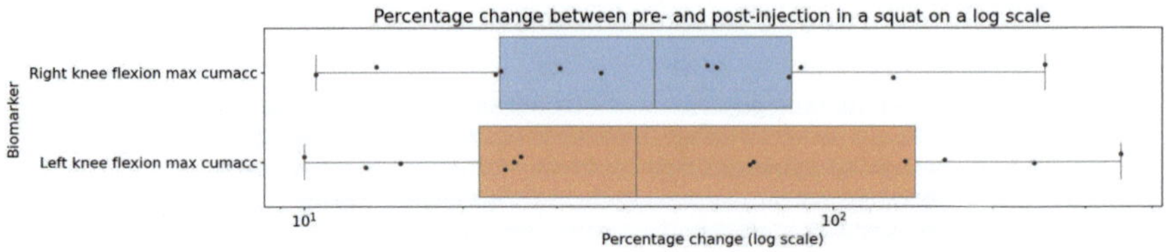

Fig. 9 Box plot showing the change between pre- and post-injection for the squat action on a logarithmic scale. The two boxes represent the minimum, maximum, lower quartile, median, upper quartile and mean for the selected clinical metrics

Table 3 Paired t-test results showing the cumulative acceleration and smoothness results when comparing the pre- and post- injection samples for the most representative metrics of the squat action

Metric (squat)	Cum. Acceleration—t-value (p-value)	Smoothness—t-value (p-value)
Right Knee Flexion (mean)	1.786 (0.090)	1.203 (0.244)
Right Knee Flexion (max)	0.126 (0.901)	2.324 (0.031)
Left Knee Flexion (mean)	1.907 (0.072)	1.196 (0.246
Left Knee Flexion (max)	0.483 (0.635)	2.52 (0.021)
Right Knee Flexion (min)	0.203 (0.841)	1.385 (0.182

Table 4 Paired t-test results showing the cumulative acceleration and smoothness results when comparing the pre- and post- injection samples for the most representative metrics of the sit-to-stand action

Metric (sit-to-stand)	Cum. Acceleration—t-value (p-value)	Smoothness—t-value (p-value)
Left Knee Flexion (max)	0.772 (0.450)	2.976 (0.008)
Right Knee Flexion (max)	0.558 (0.584)	2.401 (0.027)
Left Arm Abduction (mean)	0.675 (0.508)	3.586 (0.002)
Right Elbow Flexion (max)	2.451 (0.024)	3.592 (0.002)
Left Elbow Flexion (max)	2.364 (0.029)	2.604 (0.017)

the SMPL-based extraction methods of the VIBE model this will build the foundations of the development of a new clinical workflow. Tables 1 and 2, along with Fig. 5 show a strong correlation between the two data extraction methodologies; thus the desired workflow for clinical applications will shape the method. One of the requirements for this clinical workflow is the initial cost. Although the Azure Kinect is now discontinued, its price of $399 was still considered by many a low-cost solution for a depth camera (Kurillo et al. 2022). Whereas the SMPL-based methods do not require any specific hardware and can therefore be run with either a smartphone camera or a computer webcam which many people already have.

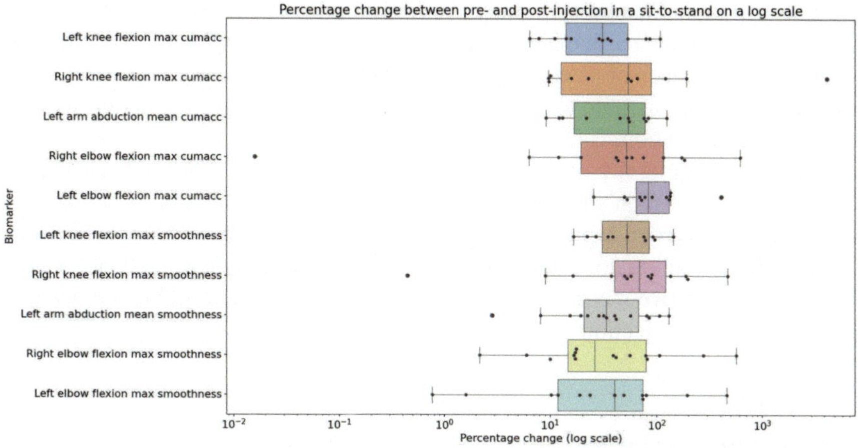

Fig. 10 Box plot showing the change between pre- and post-injection for the squat action on a logarithmic scale. The boxes represent the minimum, maximum, lower quartile, median, upper quartile and mean for the selected clinical metrics

Therefore, when moving forward with the experiments only the data from the SMPL-based extraction have been used as this will link more closely to what would be used in a clinical environment.

5.2 PCA

Dimensionality reduction by PCA is an effective method to extract biomarkers, this has previously been done for many clinical applications especially in the analysis of cancer biomarkers (Filella et al. 2018; Velonas et al. 2013). Thus, the aim is to use the PCA to identify features that have the potential to be clinically relevant biomarkers of knee pain. In the results of Figs. 9 and 10 one can see that there are many viable candidates for further analysis into their clinical and statistical significance. The results of the PCA do not necessarily indicate that these are the most important metrics, however, do imply that these are important in the distinguishing between the sit-to-stand action and the squat and thus infer meaning to the data.

5.3 Statistical Testing

To determine whether there is any statistical significance between the pre- and post-injection the paired t-test is used as a population study. The results in Tables 3 and 4 show that the right and left knee flexion maximum values are statistically significant, this is likely because the participants could perform a deeper squat without the presence of pain. Interestingly, only the smoothness was found to be significant which means that the participants could perform a smoother action without the presence of pain. The lack of a cumulative acceleration significant result could be due to the participants being able to stay in the squat longer therefore the action will not have more acceleration.

On the other hand, two of the significant results shown in Table 4 indicate the left and right elbow flexion's maximum values cumulative acceleration is statistically significant; this can be explained by the participants no longer needing to use their arms to lift themselves up. Table 4 also indicates that each of the smoothness values for the representative metrics were found to have a significant difference between the pre- and post- injection.

5.4 Clinical Trial Results

To analyse the effectiveness of these statistically significant metrics as biomarkers we can analyse the change between the pre- and post-injection time points. The results in Figs. 7 and 8 indicate that each of the measured variables show an overall improvement, albeit with a few outliers anomalous results. These anomalous results would need to be investigated further to truly know the cause. However, one potential cause could be as a result of self-occlusion during the action causing some joints to be unseen by the camera resulting in an inconsistent joint curve which is an issue for many pose estimation methods (Radwan et al. 2013).

5.5 Next Steps

Moving forward, the next steps to consider are associated with the development of a new technology that can be easily applied to clinical applications; this must use the advantages and disadvantages of each mentioned methodology. Firstly, the matter of barrier to entry, one of the most necessary requirements of the system is a cheap initial set-up cost as this will allow use in less developed areas. Another of these barriers to entry is the lack of mobility in the developing countries, this means that a portable system will allow people who might have gone unseen to be diagnosed (Baart and Taaka 2017).

This technology also drives another form of condition detection through the continuous monitoring allowed by CCTV cameras, the combination of RGB-based human pose estimation with the ability for pose estimation of crowds this has the potential for people to be diagnosed based on their gait alone (Bouchrika et al. 2011). This has previously been done to identify individuals based on their gait patterns using CCTV, thus the potential to take this one step further and allow for early detection of musculoskeletal conditions (Zhang et al. 2020).

6 Conclusion

Given that the amount of data is increasing at a substantial rate, this brings new potential for using motion data to differentiate different knee disorders. This has been reviewed in the past and it is widely accepted that different musculoskeletal and neurological disorders significantly affect the knee biomechanics, but some disorders have very similar effects on movement patterns and thus would require large amounts of data to distinguish (Armstrong et al. 2024). However, as it stands, marker-less motion capture technologies are still capable of determining knee pain in patients and are highly effective at monitoring the progression of treatment and recovery.

References

Andriluka M, Pishchulin L, Gehler P, Schiele B. 2d human pose estimation: new benchmark and state of the art analysis. In: Proceedings of the IEEE conference on computer vision and pattern recognition; 2014. p. 3686–93.

Armstrong K, Wen Y, Zhang L, Ye X, Lee P. Novel clinical applications of marker-less motion capture as a low-cost human motion analysis method in the detection and treatment of knee osteoarthritis. J Arthritis. 2022;11(1):001–5.

Armstrong K, Zhang L, Wen Y, Willmott AP, Lee P, Ye X. A marker-less human motion analysis system for motion-based biomarker identification and quantification in knee disorders. Front Digital Health. 2024;6:1324511.

Baart J, Taaka F. Barriers to healthcare services for people with disabilities in developing countries: a literature review. Disabil CBR Inclus Dev. 2017;28(4):26–40.

Bao W, Ma Z, Liang D, Yang X, Niu T. Pose ResNet: 3D human pose estimation based on self-supervision. Sensors. 2023;23(6):3057.

Bouchrika I, Goffredo M, Carter J, Nixon M. On using gait in forensic biometrics. J Forensic Sci. 2011;56(4):882–9.

Chen K, Gabriel P, Alasfour A, Gong C, Doyle WK, Devinsky O, et al. Patient-specific pose estimation in clinical environments. IEEE J Transl Eng Health Med. 2018;6:1–11.

Chen H, Feng R, Wu S, Xu H, Zhou F, Liu Z. 2D human pose estimation: a survey. Multimedia Systems. 2023;29(5):3115–38.

Ci H, Wang C, Ma X, Wang Y. Optimizing network structure for 3d human pose estimation. In: Proceedings of the IEEE/CVF international conference on computer vision; 2019. p. 2262–71.

D'Antonio E, Taborri J, Palermo E, Rossi S, Patanè F. A markerless system for gait analysis based on OpenPose library. In: 2020 IEEE international instrumentation and measurement technology conference (I2MTC). IEEE; 2020, May. p. 1–6.

Dai J, Qi H, Xiong Y, Li Y, Zhang G, Hu H, Wei Y. Deformable convolutional networks. In: Proceedings of the IEEE international conference on computer vision; 2017. p. 764–73.

Deng J, Dong W, Socher R, Li LJ, Li K, Fei-Fei L. Imagenet: a large-scale hierarchical image database. In: 2009 IEEE conference on computer vision and pattern recognition. IEEE; 2009, June. p. 248–55.

Filella X, Fernández-Galan E, Fernandez Bonifacio R, Foj L. Emerging biomarkers in the diagnosis of prostate cancer. Pharmacogenom Personal Med. 2018;11:83–94.

Ionescu C, Papava D, Olaru V, Sminchisescu C. Human3. 6m: large scale datasets and predictive methods for 3d human sensing in natural environments. IEEE Trans Pattern Anal Mach Intell. 2013;36(7):1325–39.

Jalloul M, Miranda-Schaeubinger M, Noor AM, Stein JM, Amiruddin R, Derbew HM, Mollura DJ. MRI scarcity in low-and middle-income countries. NMR Biomed. 2023;36(12):e5022.
Kim DH, Choi SS, Yoon SH, So-Hee L, Seo DK, Lee IG, et al. Ultrasound-guided genicular nerve block for knee osteoarthritis: a double-blind, randomized controlled trial of local anesthetic alone or in combination with corticosteroid. Pain Physician. 2018;21(1):41.
Kocabas M, Athanasiou N, Black MJ. Vibe: video inference for human body pose and shape estimation. In: Proceedings of the IEEE/CVF conference on computer vision and pattern recognition; 2020. p. 5253–63.
Kurillo G, Hemingway EC, Heng ML, Cheng L. Evaluating the accuracy of the azure kinect and kinect v2. Sensors. 2022;22(7):2469.
Li MH, Mestre TA, Fox SH, Taati B. Vision-based assessment of parkinsonism and levodopa-induced dyskinesia with pose estimation. J Neuroeng Rehabil. 2018;15:1–13.
Lin TY, Maire M, Belongie S, Hays J, Perona P, Ramanan D, et al. Microsoft coco: common objects in context. In: Computer vision–ECCV 2014: 13th European conference, Zurich, Switzerland, September 6–12, 2014, proceedings, part V 13. Springer International Publishing; 2014. p. 740–55.
Liu R, Shen J, Wang H, Chen C, Cheung SC, Asari V. Attention mechanism exploits temporal contexts: real-time 3d human pose reconstruction. In: Proceedings of the IEEE/CVF conference on computer vision and pattern recognition; 2020. p. 5064–73.
Loper M, Mahmood N, Romero J, Pons-Moll G, Black MJ. SMPL: A skinned multi-person linear model. In: Seminal graphics papers: pushing the boundaries, vol. 2; 2015. p. 851–66.
Lv X, Ta N, Chen T, Zhao J, Wei H. Analysis of gait characteristics of patients with knee arthritis based on human posture estimation. Biomed Res Int. 2022;2022:7020804.
Mahmood N, Ghorbani N, Troje NF, Pons-Moll G, Black MJ. AMASS: archive of motion capture as surface shapes. In: Proceedings of the IEEE/CVF international conference on computer vision; 2019. p. 5442–51.
Martinez J, Hossain R, Romero J, Little JJ. A simple yet effective baseline for 3d human pose estimation. In: Proceedings of the IEEE international conference on computer vision; 2017. p. 2640–9.
Moro M, Marchesi G, Hesse F, Odone F, Casadio M. Markerless vs. marker-based gait analysis: a proof of concept study. Sensors. 2022;22(5):2011.
Nie Q, Liu Z, Liu Y. Lifting 2d human pose to 3d with domain adapted 3d body concept. Int J Comput Vis. 2023;131(5):1250–68.
Pavlakos G, Choutas V, Ghorbani N, Bolkart T, Osman AA, Tzionas D, Black MJ. Expressive body capture: 3d hands, face, and body from a single image. In: Proceedings of the IEEE/CVF conference on computer vision and pattern recognition; 2019. p. 10975–85.
Pedregosa F, Varoquaux G, Gramfort A, Michel V, Thirion B, Grisel O, et al. Scikit-learn: machine learning in Python. J Machine Learn Res. 2011;12:2825–30.
Pfister A, West AM, Bronner S, Noah JA. Comparative abilities of Microsoft Kinect and Vicon 3D motion capture for gait analysis. J Med Eng Technol. 2014;38(5):274–80.
Radwan I, Dhall A, Goecke R. Monocular image 3D human pose estimation under self-occlusion. In: Proceedings of the IEEE international conference on computer vision; 2013. p. 1888–95.
Simon SR. Quantification of human motion: gait analysis—benefits and limitations to its application to clinical problems. J Biomech. 2004;37(12):1869–80.
Stenum J, Cherry-Allen KM, Pyles CO, Reetzke RD, Vignos MF, Roemmich RT. Applications of pose estimation in human health and performance across the lifespan. Sensors. 2021;21(21):7315.
Tölgyessy M, Dekan M, Chovanec Ľ. Skeleton tracking accuracy and precision evaluation of Kinect V1, Kinect V2, and the azure kinect. Appl Sci. 2021;11(12):5756.
Velonas VM, Woo HH, dos Remedios CG, Assinder SJ. Current status of biomarkers for prostate cancer. Int J Mol Sci. 2013;14(6):11034–60.

Viswakumar A, Rajagopalan V, Ray T, Parimi C. Human gait analysis using OpenPose. In: 2019 fifth international conference on image information processing (ICIIP). IEEE; 2019, November. p. 310–4.

Wang J, Tan S, Zhen X, Xu S, Zheng F, He Z, Shao L. Deep 3D human pose estimation: a review. Comput Vis Image Underst. 2021;210:103225.

Wei S-E, Ramakrishna V, Kanade T, Sheikh Y. Convolutional pose machines. CVPR; 2016.

Zhang L, Marron JS, Shen H, Zhu Z. Singular value decomposition and its visualization. J Comput Graph Stat. 2007;16(4):833–54.

Zhang L, Liu G, Han B, Wang Z, Yan Y, Ma J, Wei P. Knee joint biomechanics in physiological conditions and how pathologies can affect it: a systematic review. Appl Bionics Biomech. 2020;2020:1–22.

Zheng C, Wu W, Chen C, Yang T, Zhu S, Shen J, et al. Deep learning-based human pose estimation: a survey. ACM Comput Surv. 2023;56(1):1–37.

Multi-Omic Approaches for the Evaluation of the Synovial Joint Fluid

Ozan Kaplan, Batuhan Aktaş, Emine Koç, Mustafa Çelebier, and Feza Korkusuz

Abstract The incidence and prevalence of OA have notably increased over the past few decades, particularly in low and middle-income countries, posing significant challenges to the public health systems worldwide. There is an increasing need for innovative diagnostic, monitoring the condition and therapeutic approaches as OA adversely impacts the quality of life in and aging population. Exercise and sports related joint injuries are furthermore increasing. The initial intensity of the injury during exercise and sport is a major determinant of the damage to the joint structures. Even appropriate and timely treatment may cause the early onset of secondary OA in this age group with heightened social and physical expectations. Synovial joint fluid (SJF) emerges as a promising source for monitoring the inflammatory, degenerative and regenerative molecules of the joint microenvironment. By analyzing SJF proteomic and metabolomic profiles, researchers may identify specific biomarkers offering insights into the disease progression and facilitating early diagnosis. In this chapter, we overviewed the published metabolomics and proteomics studies performed on SJF that may reflect the dynamic composition of local pathophysiological processes. Multi-omic approaches may aid the uncovering of novel biomarkers for diagnostic, prognostic and therapeutic purposes. This approach may guide a better definition of the epigenetics, phenotypes and clinical, medicinal and molecular endotypes of OA in the near future. Mapping of various biomedical domains may leverage the comprehensive insight into the disease and/or condition by proteomics and metabolomics may serve as a cutting-edge diagnostic tool in future research and clinical settings.

Keywords Osteoarthritis · Proteomics · Metabolomics · OA biomarkers

---

O. Kaplan · E. Koç · M. Çelebier
Department of Analytical Chemistry, Faculty of Pharmacy, Hacettepe University, Ankara, Türkiye
e-mail: celebier@hacettepe.edu.tr

B. Aktaş · F. Korkusuz (✉)
Department of Sports Medicine, Faculty of Medicine, Hacettepe University, Ankara, Türkiye
e-mail: feza.korkusuz@hacettepe.edu.tr

P. Lee (ed.), *Musculoskeletal Regeneration Medicine*,
https://doi.org/10.1007/978-3-031-77456-0_7

1 Introduction

Osteoarthritis (OA) incidence increased about 9% during the last 28 years (Quicke et al. 2022). The prevalence of the disease is about 16% (Cui et al. 2020) and it is higher in low and middle income countries (Yahaya et al. 2021). Knee joint OA was diagnosed in about 654 million individuals in 2022 in 40+ years old and the incidence was published as 203 per ten thousand people (Cui et al. 2020). In Europe, medicinal consumption due to OA is about 47% (Kingsbury et al. 2014). OA is a disease and/or condition that decreases the quality of life in an aging society that has higher social and physical expectancies more than previous generations. The prevalence of the disease is therefore expected to increase from 26,6% to 29,5% in the year of 2032 (Turkiewicz et al. 2014). Pain, loss of function and deformity in the joint are the most common findings. Swelling may occur during the inflammatory phase of this multi-dimensional and multi-genetic disease.

Synovial tissue, which is a major immune modulator (Korkusuz et al. 2006) of joins regulates the synthesis of the synovial joint fluid (SJF) under normal and pathological conditions. SJF is a specialized and viscous liquid confined within the diarthrodial joint space including the knees, hips and wrists. It acts as an ultrafiltrate of blood plasma. Lubricating the joint space minimizes cartilage friction and abrasion, absorbs shock and facilitates the transport of nutrients and waste into and out from the tissues (Freemont 1996). The concentration and composition of SJF within the synovial space is critical. Deficiency and/or loss of homeostasis may lead to articular cartilage degeneration and degradation. It is primarily composed of hyaluronic acid, lubricin, plasma proteins like albumin and globulin and as well as cells including white blood cells and synoviocytes. The SJF is produced by fibroblast-like type B synovial cells of the synovium. The composition and volume of SJF are dynamically regulated in response to trauma, disease or infection (Prekasan and Saju 2016).

SJF analysis is randomly carried out for identifying underlying causes of joint conditions. Trauma and diseases such as arthritis and joint hemorrhage can be classified through the examination of SJF including macroscopic, microscopic and biochemical analysis (Lotz and Loeser 2012). Changes in SJF composition can also be used to monitor treatment efficiency. Analyzing the SJF also aids for biomarker development, offering potential for early diagnosis and patient stratification, leading to personalized treatment approaches. Understanding the roles of proteins, lipids and metabolites additionally could lead to new therapeutic targets, potentially for modifying the progression of OA.

In OA (Fig. 1), the body releases more inflammation-causing proteins such as tumour necrosis factor-alpha (TNFa), interleukin (IL)-1β, IL-6, IL-8, IL-15, IL-17, IL-18 and IL-21, which can lead to increase in pain (Sokolove and Lepus 2013; Wojdasiewicz et al. 2014). There is also an increase in the production of Prostoglandin E2 (PGE2) and nitric oxide. The synovium and its' lining cells produce specifically ICAM-1 and VCAM-1 adhesion molecules (Wang et al. 2020; Wu et al. 2022). These molecules may lead to more cells gathering at the inflamed site that can worsen the condition. There is additionally a rise in the activity of certain enzymes

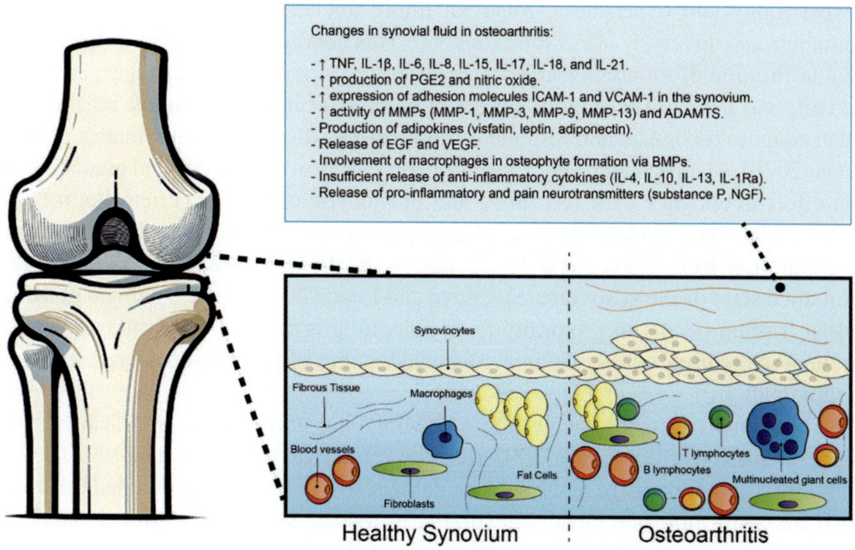

Fig. 1 The figure illustrates the profound changes occurring in the synovium and SJF due to OA, juxtaposing the normal state against the diseased condition. It highlights the thickened synovial lining (synovial hypertrophy and hyperplasia) and increased cell numbers, including the infiltration of monocytes, macrophages, activated B cells and T cells. It additionally shows an increase in pro-inflammatory cytokines (TNF, IL-1b, etc.), PGE2, nitric oxide, adhesion molecules (ICAM-1, VCAM-1), enzymes breaking down joint cartilage (MMPs, ADAMTS) and the release of growth factors (EGF, VEGF) that stimulate neovascularization. Macrophages clustering to form multinucleated giant cells are also represented

such as matrix metalloproteinases (MMP)-1, MMP-3, MMP-9, MMP-13 and a disintegrin and metalloproteinase with thrombospondin motifs (ADAMTS). These may cause degradation and degeneration of the articular joint cartilage (Watt et al. 2020). Fat cells in the joint area release substances like nicotinamide phosphoribosyl-transferase, leptin and adiponectin. These can also affect inflammation and the functioning of the joint (Han et al. 2020). Growth factors that can lead to the formation of new blood vessels and potentially affect the healing or damage processes of the articular joint cartilage. Although the joint space receives no blood vessels, mediators such as epidermal growth factor (EGF) and vascular endothelial growth factor (VEGF) are released in higher amounts in pathological conditions (Sokolove and Lepus 2013). Macrophages function in the growth of osteophytes that are painful. The growth of osteophytes is regulated by bone and cartilage-derived morphogenetic proteins (BMPs). There is a reduced release of cytokines like IL-4, IL-10, IL-13 and IL-1Ra that usually help control the inflammation (Fernandes et al. 2020). There is furthermore a release of pro-inflammatory and pain neurotransmitters such as substance P (Korkusuz et al. 2006) and nerve growth factor (NGF), contributing to the pain experience in OA (Yu et al. 2022). Significant alterations occur within the synovial tissue in the progression of joint diseases. A notable change is synovial hypertrophy and hyperplasia. Synovial lining thickens due to the increased cell numbers and adversely affect joint function. This condition is often accompanied by the infiltration of mononuclear cells, including monocytes, macrophages, activated B cells and T cells. These infiltrates not only indicate an active immune reaction but also adapt to recognize and respond to fragments of the extracellular matrix (Zhang et al. 2020). Macrophages may cluster together to form multinucleated giant cells in an effort to enhance their resorptive and phagocytic capability (Prieto-Potin et al. 2015). There is a marked increase in angiogenesis in the synovium alongside these cellular transformations. New blood vessels in the tissue may contribute to the inflamed state of the synovium (Sokolove and Lepus 2013). This is defined as synovitis. It often occurs in proximity to degenerating articular joint cartilage or after trauma while highlighting the interconnected nature of joint tissue degradation and inflammation.

The Human Metabolome Database (HMDB) has underwent a significant development since its inception in 2007 with the release of Version 1. The HMDB was initially a comprehensive and freely available resource containing detailed information on small molecule metabolites found in the human body. Version 1 provided data on approximately 2500 metabolites, including their chemical structures, physico-chemical properties, biological roles and associated pathways. Over the years, HMDB exponentially grew with advancements in metabolomics technologies and increased research interest in understanding the human metabolome. The subsequent versions of HMDB, including versions 2.0, 3.0 and beyond, witnessed substantial expansions in both the number and depth of metabolite entries. By 2022, HMDB has evolved into a vast repository housing information on over 114,000 metabolites, encompassing endogenous human metabolites, drugs, food components and environmental chemicals. Each metabolite entry is meticulously curated, providing comprehensive details such as chemical names, synonyms, chemical structures, mass spectra, biofluid concentrations, metabolic pathways, physiological roles and associated diseases. The HMDB furthermore integrated data from various metabolomics studies, bioinformatics analyses and literature sources, ensuring its accuracy, reliability and relevance to diverse research endeavors spanning metabolomics, systems biology, pharmacology and clinical medicine. The growth and development of the HMDB underscored its pivotal role as a valuable resource for elucidating the complex metabolic landscape of the human body and its implications for health, disease and therapeutics (Wishart et al. 2022).

The development of liquid chromatography–mass spectrometry (LC-MS) based proteomics stimulated transformation in the field of biology and biomedical research. Emerging in the late twentieth century, LC-MS proteomics initially faced numerous challenges, including limited sensitivity, dynamic range and throughput. LC-MS proteomics has evolved into a powerful and versatile tool for comprehensive characterization and quantification of proteins on a global scale through continuous advancements in instrumentation, methodology and bioinformatics. Key milestones in its development include improvements in mass spectrometer resolution, sensitivity and speed, enabling the detection and analysis of low-abundance proteins and post-translational modifications (PTMs) with unprecedented accuracy and depth. Innovations in chromatographic separation techniques, such as nanoLC and ultra-high-performance liquid chromatography (UHPLC) moreover, have enhanced peptide separation and throughput, leading to improved proteome coverage and depth of analysis. Advancements in data acquisition strategies, including data-dependent acquisition (DDA) and data-independent acquisition (DIA), have enabled efficient and comprehensive proteome profiling and quantification concurrently. The integration of advanced bioinformatics tools and algorithms for data processing, protein identification and quantification furthermore has streamlined data analysis workflows and enhanced the reliability and interpretability of LC-MS proteomics data. Today, LC-MS based proteomics has become a valuable tool for addressing a wide range of biological questions, including protein expression profiling, biomarker discovery, PTM analysis and elucidation of protein-protein interactions, paving the way for groundbreaking discoveries in basic and translational research, as well as personalized medicine and diagnostics. LC-MS proteomics holds immense promise for unraveling the complexities of the proteome and its role in health and disease research (Seger and Salzmann 2020; Rozanova et al. 2021; Noor et al. 2021) as technological innovations and interdisciplinary collaborations enhance.

Combining proteomics and metabolomics data holds immense potential for discovering novel biomarkers with diagnostic, prognostic and therapeutic significance across various fields of biomedical research. Researchers can gain a more comprehensive understanding of the molecular mechanisms underlying physiological and pathological processes by integrating information on protein expression levels and metabolic profiles within biological systems. Proteomics offers insights into the dynamic regulation of cellular pathways and the identification of protein biomarkers associated with specific diseases or biological states. Metabolomics on the other hand provides a snapshot of the small molecule metabolites involved in cellular metabolism, reflecting changes in metabolic pathways and cellular responses to perturbations. Researchers can identify potential biomarker signatures for OA by correlating changes in protein expression with alterations in metabolite levels. Integration of proteomics and metabolomics data moreover enables the identification of interconnected molecular networks and pathways implicated in disease pathogenesis, facilitating the discovery of novel therapeutic targets and personalized treatment strategies. The synergistic combination of proteomics and metabolomics holds promise for advancing precision medicine, biomarker-driven diagnostics and the development of targeted therapies tailored to individual patient needs (Çelebier and Ercan 2020; Wörheide et al. 2021).

OA is a slowly developing and debilitating disease with limited non-specific biomarkers for its early detection. There is a need to identify specific molecules and biomarkers that will enable early detection of this disease to improve treatment approaches. This chapter aims to provide a comprehensive overview of the current landscape of SJF metabolomics and proteomics research, synthesizing existing literature to highlight both the advancements and challenges in the field. By merging diverse studies, it explains the potential of multi-omic approaches in elucidating the molecular keystones of joint diseases and covering the way for innovative diagnostic and therapeutic strategies.

Despite intensive efforts, there remains a possibility of missed studies during our literature search. Additionally, older studies employing less advanced analytical techniques may yield limited results compared to more contemporary approaches. The inherent heterogeneity in study populations and disease groups further complicates data interpretation. Despite these limitations, our work stands out as one of the earliest attempts to integrate metabolomics and proteomics data for SJF analysis. Leveraging this combined dataset, we conducted a comprehensive pathway analysis, shedding new light on the molecular mechanisms underlying joint diseases.

The changes in the proteome level of SJF, as outlined in the introduction, underscore the multifaceted nature of OA pathogenesis and its intricate interplay with inflammatory mediators, cellular transformations, and angiogenic processes within the joint microenvironment. Table 1 present the proteins found to be changed in SJF as a result of OA.

Barreto et al. (2019) reported a study to uncover the molecular mechanisms underlying primary OA by investigating protein expression dynamics in SJF from two different types of OA joints. Using label-free quantitative liquid chromatography mass spectrometry, proteomic analysis revealed differential expression of 40 proteins, including clusterin, paraoxonase/arylesterase 1 (PON1) and transthyretin. Validation through enzyme-linked immunosorbent assays and droplet digital PCR confirmed these findings. Functional assays on primary chondrocytes demonstrated that SJF from knee OA patients induced a significant pro-inflammatory response, cell death, and hypertrophy, whereas SJF from carpometacarpal-I (CMC-I) OA joints showed primarily lower protein expression levels and different effects on chondrocytes. This suggests that molecular mechanisms driving OA vary between different joints, providing new insights into primary OA pathogenesis.

In a study published in 2021 (Li et al. 2021), SJF samples were collected from patients undergoing diagnostic hip or knee aspiration for suspected periprosthetic joint infection (PJI), diagnosed based on modified Musculoskeletal Infection Society criteria. LC-MS/MS analysis of 117 samples (51 PJI, 66 non-PJI) presented that particularly MNDA and PRTN3, can be used as a potent biomarker for PJI diagnosis.

In another study published in 2021, Seny et al. (2022) performed a proteomic investigation utilizing liquid chromatography-mass spectrometry (LC-MS/MS) for synovial inflammation across various arthritis joint diseases, analyzing synovial biopsies from patients with OA, chronic pyrophosphate arthropathy (CPPA), and rheumatoid arthritis (RA). The study identified and quantified 1.871 proteins, revealing differential expression of ten proteins (LAP3, MANF, LCP1, CTSZ, PTPRC, DNAJB11, EML4, SCARA5, EIF3K, C1orf123) across the disease groups. Increased expression of seven proteins correlated with histological inflammatory scores in RA, indicating their potential role in inflammation. Immunohistochemistry validated these findings.

The reported data in the literature (Huang et al. 2022) for proteomics analysis targeting 1.678 credible proteins in SJF-derived exosomes in inflammatory arthritis

Table 1 Proteomics Profiling of SJF in OA

Model species	Fluid sampled	Fluid Analysis Method	Sample size(s)	Metabolites/ Metabolic Pathways of Importance	Chromatographic Conditions	References
Human	Synovial fluid	UPLC/ LC-MS	CMC OA: 12 PIA: 10		Column: BEH-130 C18 (1.7 × 75 × 9 250 mm) Mobile phase: A (0.1% formic acid) Solution B (0.1% formic acid/ acetonitrile)	Barreto et al. (2019)
Human	Synovial fluid	LC-MS/ MS	PJI: 51 Non-PJI: 66	Myeloid nuclear differentiation antigen (MNDA) Polymorphonuclear leukocyte serine protease 3 (PRTN3)	Column: C18 (1.9 μm, 120 Å) Mobile phase (acetonitrile and 0.1% formic acid)	Li et al. (2021)
Human	Knee synovial biopsies	LC-MS/ MS	OA: 9 CPPA: 7 RA: 8	1871 10 proteins (LAP3, MANF, LCP1, CTSZ, PTPRC, DNAJB11, EML4, SCARA5, EIF3K, C1orf123	Mobile phase A (0.1% formic acid, B = acetonitrile, 0.1% formic acid) Positive ESI	de Seny et al. (2022)
Human	Synovial fluid	LC-MS/ MS	Gout: 42 AxSpA: 10 OA: 18 RA: 30		Column: C18 (15 cm × 75) Mobile phase: A (2% acetonitrile in HPLC water) and B (90% acetonitrile in HPLC water).	Huang et al. (2022)
Horses	Carpal synovial fluid	LC-MS	OA: 6 HC: 6	Increased O-glycosylation and sialylation, along with decreased fucosylation	Column: ZIC pHILIC (5 μm, 2.1 × 150 mm) Mobile phase (A) 10 mm ammonium acetate in Water pH = 9.8 and (B) acetonitrile. Negative and positive ESI	Noordwijk et al. (2022)
Horse	Synovial fluid	Q-TOF LC-MS/ MS)		26 differentially abundant proteins compared to non-septic arthritis	Column: C18 (75 μm × 150 mm) Mobile phase A (0.1% formic acid in water) B (0.1% formic acid in acetonitrile). Positive ESI	Koziy et al. (2022)

(continued)

Table 1 (continued)

Model species	Fluid sampled	Fluid Analysis Method	Sample size(s)	Metabolites/ Metabolic Pathways of Importance	Chromatographic Conditions	References
Human	Synovial fluid	Nano-LC-MS/MS	ACL: 29	CRIP1, S100A11, PLS3, POSTN, and VIM, CHI3L2 (YKL-39), TNFAIP6/TSG6, DEFA1, SPP1, and CILP	Column: C18 (2.5 μL75 μXm × 50 cm) Mobile phase: A (1% FA) containing 5% B (100% MeCN, 1% FA) The data was acquired in data-dependent acquisition mode.	Rai et al. (2023)
Human	Synovial fluid	LC-MS/MS	OA: 12	SLPI, C8, CLU, FN1, RARRES2, MATN3, MMP3, TNC, IGF2, AHSG, FN1, CFB, KNG and C8	Column: C18 (EASY-Spray PepMap RSLC C18, 75 μm × 50 cm, 2 μm packing material, 100 Å) Mobile phase: Mobile phase A (0.1% formic acid in water) B (0.1% formic acid in water: acetonitrile 80:20 v/v).	Timur et al. (2021)

MCP Metacarpophalangeal, *CIA* collagen-induced arthritis, *PJI* periprosthetic joint infection, *ACL* acute traumatic tear, *CPPA* chronic pyrophosphate arthropathy, *RA* rheumatoid arthritis, *CMC* carpometacarpal OA, *PIA* primary idiopathic OA. *AS* ankylosing spondylitis, *SLE* systemic lupus erythematosus, *SSc* systemic sclerosis, *ESI* electrospray ionisation, *HC* healthy control

presented that SJF-derived exosomes across RA, axial spondyloarthritis (axSpA), gout, and OA patients were different in proteome level. 69 differentially expressed proteins identified in gout, 84 in axSpA, and 184 in RA, compared to OA, axSpA, and gout simultaneously. Unique protein expressions in each condition, such as lysozyme C and protein S100-A9 in gout, RNA-binding protein 8A and protein transport protein Sec24C in axSpA, and pregnancy zone protein and stromelysin-1 in RA, were identified, with associated biological pathways.

In a recent study Noordwijk et al. (2022) reported an in vivo metabolomics and glycomics study aimed to identify differentially expressed metabolic and glycosylation pathways in SJF from healthy horses and those with naturally occurring carpal OA, a common issue in equine athletes. High-resolution LC-MS analysis of carpal SJF from 12 horses (6 healthy, 6 OA) revealed 84 identifiable metabolites with significant differences between groups, implicating pathways related to inflammation, oxidative stress and collagen metabolism. Lectin microarrays and lubricin sandwich ELISA conducted on SJF from 40 horses (20 healthy, 20 OA) identified distinct glycosylation patterns, including increased O-glycosylation and sialylation, along with decreased fucosylation in OA.

Koziy et al. reported (2022) a study to discover potential biomarkers for eradicating joint infection in horses with septic arthritis, a life-threatening condition. Proteomics analysis of SJF and plasma from horses with experimental SA, non-septic lipopolysaccharide-induced arthritis, and controls was conducted. Comparisons before and after infection eradication revealed 26 differentially abundant proteins, predominantly upregulated in SA SJF compared to other groups and post-eradication samples. These proteins, primarily of cellular origin, were involved in processes such as ubiquitination, signal transduction and apoptosis. The significant fold differences in their abundance, exceeding expected levels based solely on cell count variations, suggest their potential as biomarkers. Notably, neutrophil- and cell-associated proteins exhibited variable abundance, indicative of possible biomarkers for SA eradication in horses. This study underscores the importance of identifying biomarkers for SA diagnosis and treatment monitoring in equine health.

In another study (Rai et al. 2023), the proteomic profile of knee SJF over time following anterior cruciate ligament (ACL) injury was examined. SJF was collected from patients shortly after injury evaluation (aspiration 1) and at the time of surgery (aspiration 2). Analysis of 58 SJF samples from 29 patients revealed alterations in the levels of 130 proteins over time, with notable increases in catabolic/inflammatory proteins (e.g., CRIP1, S100A11, PLS3, POSTN, VIM) and decreases in chondroprotective proteins (e.g., CHI3L2, TNFAIP6/TSG6, DEFA1, SPP1, CILP) in aspiration 2. These findings suggest an initial disruption of joint homeostasis following ACL tears, potentially contributing to OA development.

In a comparative proteomics study (Timur et al. 2021), tissue-dependent proteins and their specific abundance in OA SJF were investigated. Tissue samples including articular joint cartilage, synovium, Hoffa's fat pad and meniscus were collected alongside SJF from knee arthroplasty patients and analyzed using LC-MS/MS for protein identification and quantification. By comparing the proteomic profiles of OA and non-OA conditions, proteins released from each tissue and those with altered abundance in OA were identified. The study aimed to elucidate OA pathogenesis given its complex nature affecting the entire joint. Key findings highlighted tissue-dependent proteins such as secretory leukocyte protease inhibitor (SLPI), complement component 8 (C8), clusterin (CLU), fibronectin 1, retinoic acid receptor responder 2 (RARRES2), matrilin 3 (MATN3), matrix metallopeptidase 3 (MMP3) and tenascin C (TNC), indicating various roles in inflammation, tissue remodeling, adipogenesis, extracellular matrix changes and joint degeneration. Additionally, specific proteins like insulin-like growth factor 2 (IGF2), alpha-2-HS-glycoprotein (AHSG), complement factor beta (CFB), and kininogen (KNG) were identified in SJF, suggesting their involvement in tissue repair, immune response and pain mediation in OA.

In another study, eight healthy controls (HC) and OA patients were divided into two groups, OA-obese (n = 10; body mass index (BMI) ≥30 kg/m^2) and non-OA-obese (n = 10; BMI <30 kg/m^2). This study identified 509 proteins, 279 of which were measured. Pair-wise comparisons using PCA revealed 8 proteins differentially regulated between OA-obese and OA-non-obese controls. sPLS-DA also distinguished OA from controls (9 proteins) and OA-obese from OA-non-obese (23 proteins). Component 2 proteins have been linked to obesity. Component 1 proteins upregulated by both OA groups compared to controls included CRTAC1, which has been identified as a strong OA biomarker candidate. FBN1, VDBP, and potentially SERPINF1 were also highlighted as potential biomarkers. The results were important in terms of the role of obesity in the development of OA (Tardif et al. 2022).

Changes in lipid metabolism attracted attention when OA biomarker studies were examined. SJF metabolites were analyzed distinctive markers such as phospholipid species, fatty acid metabolites, acylcarnitine species, ursolic acid and sulfonic acid derivatives were linked to the differentiation between OA and HC. Phosphatidic acid, lysophosphatidic acid, phosphatidylglycerol and lysylphosphatidylglycerol species, which play a role in signal transduction and membrane traffic, increased in HC versus OA (Carlson et al. 2018).

In another metabolomics study conducted according to OA stages, LC-MS analysis of SJF obtained from patients with early OA and late OA. The metabolomic profile of OA stages was found to be different from each other. A total of 43 metabolites were detected in the study but phosphatidylcholine 20: 0/18:2 and 18:0/20:2, sphingomyelin and ceramide, inosine 5′-phosphate, adenosine thiamine diphosphate and diadenosine 5′,5′-diphosphate were the ones of significance. All except ceramide (d18:1/20:0) were enriched in the OA group. These changes are associated with the impaired metabolome of OA (Bocsa et al. 2022).

In a separate investigation utilizing UPLC-MS analysis of rabbit SJF, different phospholipid species, tryptophan and tryptophan metabolites (including kynurenic acid), acylcarnitine species, phenylalanine and phenylalanine metabolites, lysine, arginine and histidine were associated with OA (Hu et al. 2020).

OA and RA exhibit distinct clinicopathological features. To investigate this, global metabolomic changes in the SJF of 19 patients with RA and OA were evaluated. Various statistical analyses, including partial least squares-discriminant analysis, hierarchical clustering and univariate analysis, were conducted to differentiate between RA and OA metabolites. A total of 121 metabolites were identified in the SJF samples. The partial least squares-discriminant analysis revealed a clear separation between RA and OA samples. Twenty-eight key metabolites, such as hypoxanthine, xanthine, adenosine, citrulline, histidine and tryptophan, were identified as discriminatory markers distinguishing RA from OA metabolism, predominantly associated with purine and amino acid pathways (Kim et al. 2022).

In another study, changes in the lysophosphatidylcholine, phosphatidylcholine, ether-linked phosphatidylethanolamine and sphingomyelin subclasses were identified in association with RA activity (Koh et al. 2022).

Metabolic changes in SJF among RA patients were investigated utilizing gas chromatography/time-of-flight mass spectrometry (GC/TOF MS) to assess alterations in the SJF metabolome corresponding to RA disease activity levels. Samples were obtained from 48 RA patients and disease activity was gauged using Disease Activity Score in joint-28 (DAS28) for RA (Jensen Hansen et al. 2017). Analysis revealed a notable correlation between DAS28-Erythrocyte Sedimentation Rate (ESR (3)) values and the abundance of 12 metabolites. Specifically, glycocyamine and indole-3-lactate exhibited a positive correlation with DAS28-ESR (3) values, while β-alanine, asparagine, citrate, cyano-L-alanine, leucine, nicotinamide, citrulline, methionine, oxoproline and salicylaldehyde showed a negative correlation.

This study, fifteen pathways were identified as significantly associated with RA disease activity, indicating that higher disease activity levels were linked to more pronounced disruptions in amino acid metabolic processes (Ahn et al. 2020).

OA and autoimmune-induced RA were analyzed in another study where the metabolic profile of SJF and infrapatellar fat pad was analyzed. Non-targeted metabolite profiling was performed for five non-inflammatory trauma controls, 10 primary OA (pOA) patients and 10 seropositive RA patients with high-resolution mass spectrometry-based techniques and metabolites were matched with known metabolite identities (Nieminen et al. 2022).

In a study comparing the metabolomic profiles of early OA (grades I-II), and late OA (grades III-IV) OA subtypes in SJF with HC, the metabolomic profiles of healthy, early OA and late OA donors were found to be different from each other. Pathways differentially activated across these groups include structural degradation, glycerophospholipid metabolism, inflammation, central energy metabolism, oxidative stress and vitamin metabolism. These results provided insight into pathogenesis in human SJF, represented novel biomarkers and lead to personalized treatments in OA subgroup patients (Carlson et al. 2019).

Another study measured lipid species using electrospray ionization tandem mass spectrometry. SJF from nine control, 18 RA, 17 eOA, and 13 IOA patients were investigated. Lipidomic profile of osteoarthritis subtypes eOA and IOA were compared to normal SJF. Twenty-seven phospholipid species differed between RA and eOA SJF, and 64 phospholipid species differed between IOA SJF. The data provided the basis for studying the biosynthesis and function of lipid species in health and the most common joint diseases (Kosinska et al. 2014).

To differentiate between septic and non-septic arthritis in SJF, metabolomic analyses using proton H^1NMR spectroscopy were conducted on samples from patients with septic arthritis, crystal arthropathy, various forms of inflammatory arthritis or OA. A total of 35 distinct metabolites were identified. The metabolic profile of septic arthritis exhibited distinguishable characteristics compared to other conditions, while no significant differences were observed between OA and crystal-associated arthritis, RA, or spondyloarthritis. The changed metabolites were as follows; hippuric acid, phenylalanine, tyrosine, D-glucose, valine, TCA cycle metabolites, carnitine species, glutamine, lactate, alanine and lysine signatures present on H^1 NMR of SJF. Recognizing the distinct metabolism in septic arthritis offered crucial insights for effectively distinguishing between septic and non-septic forms of arthritis (Hügle et al. 2012).

Metabolomic analysis of SJF obtained from symptomatic chronic knee OA patients and normal human cadaver knee joints was evaluated and the samples were analyzed using NMR and GC-MS. The metabolic profile of OA patients was found to be significantly altered in 11 metabolites that are crucial for distinguishing them from controls. Citrate increased, malate, acylcarnitine species, creatine decreased in the control group versus the OA group (Mickiewicz et al. 2015).

The research aimed to explore variances in metabolites within the SJF of OA hips and knees using metabolic profiling via NMR spectroscopy. Twenty-four SJF samples, comprising 12 from hips and 12 from knees, were gathered from individuals undergoing hip or knee arthroplasty due to end-stage OA. The analysis revealed notably elevated levels of four specific metabolites in the knee group compared to the hip group, namely N-acetylated molecules, glycosaminoglycans, citrate and glutamine (Akhbari et al. 2019).

OA can result from an individual's overall systemic health or be influenced by factors such as age, gender, BMI, diabetes, depression and cardiovascular disease can cause changes in the metabolome in OA. Previous studies show that injury type and gender affect metabolic phenotypes in SJF. In a study comparing the cellular mechanisms and repair pathways in the SJF of male and female participants with various injuries, significant differences were observed both between genders and injury types (Welhaven et al. 2023).

In the examination of the interconnection between OA and metabolic syndrome, metabolomic analysis was conducted on the SJF and plasma of OA patients with and without diabetes, in comparison to HC. The analysis revealed the presence of phosphatidylcholine acyl-alkyl C34:3 and phosphatidylcholine acyl-alkyl C36:3 in OA and diabetic patients, indicating an association with phosphatidylcholine metabolism. The plasmatic lysophosphatidylcholine/ phosphatidylcholine ratio also emerged as a potential novel biomarker for predicting advanced knee OA. The study also confirmed that the phosphatidylcholine metabolism was associated with both OA and diabetes mellitus (Zhang et al. 2016).

Based on the results evaluated in the scope of this chapter, a comprehensive list detailing metabolites documented to undergo alterations in SJF due to OA, alongside their corresponding Kyoto Encyclopedia of Genes and Genomes codes (KEGG), is provided in Table 2. When a metabolic pathway analyzes conducted using Reactome (https://reactome.org/), a metabolic pathway analysis tool, the mainly altered pathways in SJF were found to be associated with amino acid transport across the plasma membrane, disorders of transmembrane transporters, metabolism of amino acids and derivatives, phospholipid metabolism and tRNA aminoacylation.

Table 2 List of metabolites altered in SJF as a result of OA

Metabolite name	KEGG ID	Analytical Method
Glucosamine	**C00329**	LC-MS
Galactosamine	**C01132**	LC-MS
Phosphatidylserine	**C02737**	LC-MS
Phosphatidylcholine	**C00157**	LC-MS
Phosphatidylethanolamine	**C00350**	LC-MS
Kynurenic acid	**C01717**	LC-MS
O-acylcarnitine	**C02301**	LC-MS
L-phenylalanine	**C00079**	LC-MS
L-lysine	**C00047**	LC-MS
L-arginine	**C00062**	LC-MS
L-histidine	**C00135**	LC-MS
Tryptophan	**C00078**	LC-MS
Arachidonic acid	**C0021.9**	LC-MS
Acylcarnitine	**C02301**	LC-MS
Hydroxyproline	**C01157**	LC-MS
L-glutamine	**C00064**	LC-MS
Oxiglutatione	**C00127**	LC-MS
L-isoleucine	**C00407**	LC-MS
Citrate	**C00158**	H1-NMR, GC–MS
Malate	**C00149**	H1-NMR, GC–MS
Creatine	**C00300**	H1-NMR, GC–MS
D-serine	**C00740**	H1-NMR, GC–MS
L-asparagine	**C00152**	H1-NMR
Uridine triphosphate	**C00075**	H1-NMR
Isobutyrate	**C02632**	H1-NMR
Hydroxyproline	**C01157**	H1-NMR
D-glucose	**C00031**	H1-NMR
Phosphatidic acid	**C00416**	LC-MS
Lysophosphatidic acid	**C00681**	LC-MS
Hippuric acid	**C01586**	LC-MS

2 Conclusion

SJF holds great importance as a potential source of biomarkers due to its proximity to the joint microenvironment and its dynamic composition reflecting local pathophysiological processes. As the lubricant and shock absorber of joints, SJF serves as a reservoir for proteins, metabolites and other molecules derived from articular tissues, including cartilage, synovium, and surrounding structures. Importantly, SJF composition can be altered in various joint diseases, including OA, RA, and septic arthritis, reflecting disease-specific changes such as inflammation, tissue damage, and metabolic dysregulation. Furthermore, SJF can be obtained through minimally invasive procedures such as joint aspiration, making it an accessible and relatively non-invasive source for biomarker discovery. By analyzing the proteomic, metabolomic and molecular profiles of SJF, researchers can identify disease-specific biomarkers that offer insights into disease pathogenesis, facilitate early diagnosis, monitor disease progression, and guide personalized treatment strategies. Thus, SJF represents a promising avenue for the development of biomarkers with clinical utility in the management of joint diseases and musculoskeletal disorders.

Analyzing the diverse findings from studies investigating SJF in various joint diseases provides valuable insights into the metabolic pathways implicated in these conditions. In primary OA, proteomic analysis revealed differential expression of proteins like clusterin, paraoxonase/arylesterase 1 (PON1), and transthyretin, suggesting a pro-inflammatory response, cell death, and hypertrophy induced by SJF from knee OA patients. Conversely, SJF from CMC-I OA joints exhibited primarily lower protein expression levels and different effects on chondrocytes, indicating joint-specific variations in OA pathogenesis. Similarly, in PJI, proteomic analysis identified MNDA and PRTN3 as potential biomarkers for PJI diagnosis, underscoring the importance of protein markers in infectious joint diseases. Proteomic investigations across various arthritis joint diseases, including OA, chronic pyrophosphate arthropathy (CPPA), and RA, revealed differential expression of proteins associated with inflammation, suggesting their role in disease pathogenesis. Additionally, analysis of SJF-derived exosomes in inflammatory arthritis identified unique protein expressions and associated biological pathways, highlighting disease-specific alterations in the proteome. Metabolomic studies further elucidated distinct metabolic profiles in OA and RA SJF, indicating significant alterations in pathways such as lipid metabolism, amino acid metabolism, and energy metabolism. Metabolic profiling also distinguished between different stages of OA, providing insights into disease progression and potential biomarkers for personalized treatment strategies. Furthermore, the impact of demographic characteristics and comorbidities, including age, gender, BMI, and diabetes, on SJF metabolome underscores the multifactorial nature of joint diseases. In conclusion, combining proteomics and metabolomics data from SJF offers a comprehensive understanding of the molecular mechanisms underlying joint diseases, facilitating the discovery of novel biomarkers and therapeutic targets for improved diagnosis and treatment outcomes.

Acknowledgments The Scientific and Technological Research Council of Türkiye (TÜBİTAK) granted the "Development of a new Immobilized Metal Affinity Chromatography (IMAC) Sorbent for Phosphopeptide Enrichment for Monitoring the Diagnosis, Course, and Treatment of Osteoarthritis and Implementing the Omic Mapping of Synovial Fluid in Proteomic, Metabolomic and Lipidomic Studies" project (#223S509) in relevance to the EU COST CA21110 – "Building an open European Network on OsteoArthritis research (NetwOArk)" Action. This study is partially based on the PhD Thesis of Serhat Aladağ, PhD, entitled "Development of Monodisperse-Mesoporous SiO2 Microspheres as a New Metal Oxide Affinity Chromatography (MOAC) Sorbent for Phospholipid Enrichment in Synovial Joint Fluid" and authors thnak to Serhat Aladağ PhD for his contributions. Feza Korkusuz is an active members of the Turkish Academy of Sciences (TUBA).

References

Ahn JK, Kim J, Cheong YE, Kim KH, Cha H-S. Variation in the synovial fluid metabolome according to disease activity of rheumatoid arthritis. Clin Exp Rheumatol. 2020;38(3):500–7.

Akhbari P, Jaggard MK, Boulangé CL, Vaghela U, Graça G, Bhattacharya R, et al. Differences in the composition of hip and knee synovial fluid in osteoarthritis: a nuclear magnetic resonance (NMR) spectroscopy study of metabolic profiles. Osteoarthr Cartil. 2019;27(12):1768–77.

Barreto G, Soliymani R, Baumann M, Waris E, Eklund KK, Zenobi-Wong M, et al. Functional analysis of synovial fluid from osteoarthritic knee and carpometacarpal joints unravels different molecular profiles. Rheumatol (United Kingdom). 2019;58(5):897–907.

Bocsa D-C, Socaciu C, Iancu SD, Pelea MA, Gutiu RI, Leopold N, et al. Stage related metabolic profile of the synovial fluid in patients with acute flares of knee osteoarthritis. Med Pharm reports. 2022;95(4):438–45.

Carlson AK, Rawle RA, Wallace CW, Brooks EG, Adams E, Greenwood MC, Olmer M, Lotz MK, Bothner B, June RK. Characterization of synovial fluid metabolomic phenotypes of cartilage morphological changes associated with osteoarthritis. Osteoarthritis Cartilage. 2019;27(8):1174–1184. https://doi.org/10.1016/j.joca.2019.04.007. Epub 2019 Apr 25. PMID: 31028882; PMCID: PMC6646055

Carlson AK, Rawle RA, Adams E, Greenwood MC, Bothner B, June RK. Application of global metabolomic profiling of synovial fluid for osteoarthritis biomarkers. Biochem Biophys Res Commun [Internet]. 2018;499(2):182–8. Available from: https://www.sciencedirect.com/science/article/pii/S0006291X18306090.

Çelebier M, Ercan A. Recent approaches to integrate MultiOmics data on system biology. Curr Anal Chem. 2020;16:1243.

Cui A, Li H, Wang D, Zhong J, Chen Y, Lu H. Global, regional prevalence, incidence and risk factors of knee osteoarthritis in population-based studies. EClinicalMedicine. 2020;29–30:100587.

de Seny D, Baiwir D, Bianchi E, Cobraiville G, Deroyer C, Poulet C, et al. New proteins contributing to immune cell infiltration and pannus formation of synovial membrane from arthritis diseases. Int J Mol Sci. 2022;23(1):1–18.

Fernandes TL, Gomoll AH, Lattermann C, Hernandez AJ, Bueno DF, Amano MT. Macrophage: a potential target on cartilage regeneration. Front Immunol. 2020;11:111.

Freemont AJ. Microscopic analysis of synovial fluid—the perfect diagnostic test? Ann Rheum Dis. 1996;55(10):695–7.

Han D-F, Li Y, Xu H-Y, Li R-H, Zhao D. An update on the emerging role of Visfatin in the pathogenesis of osteoarthritis and pharmacological intervention. Evid Based Complement Alternat Med. 2020;2020:8303570.

Hu Y, Wu Q, Qiao Y, Zhang P, Dai W, Tao H, et al. Disturbances in metabolic pathways and the identification of a potential biomarker panel for early cartilage degeneration in a Rabbit anterior cruciate ligament transection model. Cartilage [Internet]. 2020;13(2_suppl):1376S–1387S. Available from: https://doi.org/10.1177/1947603520921434.

Huang Y, Liu Y, Huang Q, Sun S, Ji Z, Huang L, et al. TMT-based quantitative proteomics analysis of synovial fluid-derived exosomes in inflammatory arthritis. Front Immunol. 2022;13(March):1–14.

Hügle T, Kovacs H, Heijnen IAFM, Daikeler T, Baisch U, Hicks JM, et al. Synovial fluid metabolomics in different forms of arthritis assessed by nuclear magnetic resonance spectroscopy. Clin Exp Rheumatol [Internet]. 2012;30(2):240–5. Available from: https://www.scopus.com/inward/record.uri?eid=2-s2.0-84863737272&partnerID=40&md5=38a6eda6325ccadf4419e11e4cae5964.

Jensen Hansen IM, Asmussen Andreasen R, van Bui Hansen MN, Emamifar A. The reliability of disease activity score in 28 joints-C-reactive protein might be overestimated in a subgroup of rheumatoid arthritis patients, when the score is solely based on subjective parameters: a cross-sectional, exploratory study. J Clin Rheumatol Pract reports Rheum Musculoskelet Dis. 2017;23(2):102–6.

Kim S, Hwang J, Kim J, Lee S-H, Cheong YE, Lee S, et al. Metabolic discrimination of synovial fluid between rheumatoid arthritis and osteoarthritis using gas chromatography/time-of-flight mass spectrometry. Metabolomics. 2022;18(7):48.

Kingsbury SR, Gross HJ, Isherwood G, Conaghan PG. Osteoarthritis in Europe: impact on health status, work productivity and use of pharmacotherapies in five European countries. Rheumatology (Oxford). 2014;53(5):937–47.

Koh JH, Yoon SJ, Kim M, Cho S, Lim J, Park Y, et al. Lipidome profile predictive of disease evolution and activity in rheumatoid arthritis. Exp Mol Med [Internet]. 2022;54(2):143–55. Available from: https://doi.org/10.1038/s12276-022-00725-z.

Korkusuz F, Korkusuz P, Ozkul A. Identification of substance P receptors on fibroblast-like cells derived from the periosteum: an in vitro cell culture study Periost kökenli fibroblast benzeri hücreler üzerinde P maddesi al›c›lar›n›n gösterilmesi: ‹n vitro hücre kültürü çal›flmas›. In 2006. Available from: https://api.semanticscholar.org/CorpusID:73602747.

Kosinska MK, Liebisch G, Lochnit G, Wilhelm J, Klein H, Kaesser U, et al. Sphingolipids in human synovial fluid - a lipidomic study. PLoS One [Internet]. 2014;9(3):1–11. Available from: https://doi.org/10.1371/journal.pone.0091769.

Koziy RV, Bracamonte JL, Yoshimura S, Chumala P, Simko E, Katselis GS. Discovery proteomics for the detection of putative markers for eradication of infection in an experimental model of equine septic arthritis using LC-MS/MS. J Proteome. 2022;261:104571.

Li R, Song L, Quan Q, Liu MW, Chai W, Lu Q, et al. Detecting periprosthetic joint infection by using mass spectrometry. J Bone Jt Surg. 2021;103(20):1917–26.

Lotz M, Loeser RF. Effects of aging on articular cartilage homeostasis. Bone. 2012 Aug;51(2):241–8.

Mickiewicz B, Heard BJ, Chau JK, Chung M, Hart DA, Shrive NG, et al. Metabolic profiling of synovial fluid in a unilateral ovine model of anterior cruciate ligament reconstruction of the knee suggests biomarkers for early osteoarthritis. J Orthop Res [Internet]. 2015;33(1):71–7. Available from: https://onlinelibrary.wiley.com/doi/abs/10.1002/jor.22743.

Nieminen P, Hämäläinen W, Savinainen J, Lehtonen M, Lehtiniemi S, Rinta-Paavola J, et al. Metabolomics of synovial fluid and infrapatellar fat pad in patients with osteoarthritis or rheumatoid arthritis. Inflammation [Internet]. 2022;45(3):1101–17. Available from: https://doi.org/10.1007/s10753-021-01604-x.

Noor Z, Ahn SB, Baker MS, Ranganathan S, Mohamedali A. Mass spectrometry-based protein identification in proteomics-a review. Brief Bioinform. 2021;22(2):1620–38.

Noordwijk KJ, Qin R, Diaz-Rubio ME, Zhang S, Su J, Mahal LK, et al. Metabolism and global protein glycosylation are differentially expressed in healthy and osteoarthritic equine carpal synovial fluid. Equine Vet J. 2022;54(2):323–33.

Prekasan D, Saju KK. Review of the tribological characteristics of synovial fluid. Proc Technol [Internet]. 2016;25:1170–4. Available from: https://doi.org/10.1016/j.protcy.2016.08.235.

Prieto-Potin I, Largo R, Roman-Blas JA, Herrero-Beaumont G, Walsh DA. Characterization of multinucleated giant cells in synovium and subchondral bone in knee osteoarthritis and rheumatoid arthritis. BMC Musculoskelet Disord. 2015;16:226.

Quicke JG, Conaghan PG, Corp N, Peat G. Osteoarthritis year in review 2021: epidemiology & therapy. Osteoarthr Cartil. 2022;30(2):196–206.

Rai MF, Cai L, Zhang Q, Townsend RR, Brophy RH. Synovial fluid proteomics from serial aspirations of ACL-injured knees identifies candidate biomarkers. Am J Sports Med. 2023;51(7):1733–42.

Rozanova S, Barkovits K, Nikolov M, Schmidt C, Urlaub H, Marcus K. Quantitative mass spectrometry-based proteomics: an overview. Methods Mol Biol. 2021;2228:85–116.

Seger C, Salzmann L. After another decade: LC-MS/MS became routine in clinical diagnostics. Clin Biochem. 2020;82:2–11.

Sokolove J, Lepus CM. Role of inflammation in the pathogenesis of osteoarthritis: latest findings and interpretations. Ther Adv Musculoskelet Dis. 2013;5(2):77–94.

Tardif G, Paré F, Gotti C, Roux-Dalvai F, Droit A, Zhai G, et al. Mass spectrometry-based proteomics identify novel serum osteoarthritis biomarkers. Arthritis Res Ther [Internet]. 2022;24(1):1–16. Available from: https://doi.org/10.1186/s13075-022-02801-1.

Timur UT, Jahr H, Anderson J, Green DC, Emans PJ, Smagul A, et al. Identification of tissue-dependent proteins in knee OA synovial fluid. Osteoarthr Cartil. 2021;29(1):124–33.

Turkiewicz A, Petersson IF, Björk J, Hawker G, Dahlberg LE, Lohmander LS, et al. Current and future impact of osteoarthritis on health care: a population-based study with projections to year 2032. Osteoarthr Cartil [Internet]. 2014;22(11):1826–32. Available from: https://www.sciencedirect.com/science/article/pii/S1063458414011881.

Wang S, Jiao J, Dong X, Yang F. Effect of total knee replacement combined with Yunnan Baiyao on the expression of serum IL-6, TNF-α, VCAM-1 and MMP-9 in patients with knee osteoarthritis. Panminerva Med. 2020;

Watt FE, Hamid B, Garriga C, Judge A, Hrusecka R, Custers RJH, et al. The molecular profile of synovial fluid changes upon joint distraction and is associated with clinical response in knee osteoarthritis. Osteoarthr Cartil. 2020;28(3):324–33.

Welhaven HD, Welfley AH, Pershad P, Satalich J, O'Connell R, Bothner B, et al. Metabolic phenotypes reflect patient sex and injury status: a cross-sectional analysis of human synovial fluid. Osteoarthr Cartil. 2023;31:S342.

Wishart DS, Guo A, Oler E, Wang F, Anjum A, Peters H, et al. HMDB 5.0: the human metabolome database for 2022. Nucleic Acids Res. 2022;50(D1):D622–31.

Wojdasiewicz P, Poniatowski ŁA, Szukiewicz D. The role of inflammatory and anti-inflammatory cytokines in the pathogenesis of osteoarthritis. Mediat Inflamm. 2014;2014:561459.

Wörheide MA, Krumsiek J, Kastenmüller G, Arnold M. Multi-omics integration in biomedical research - a metabolomics-centric review. Anal Chim Acta. 2021;1141:144–62.

Wu T-J, Chang SL-Y, Lin C-Y, Lai C-Y, He X-Y, Tsai C-H, et al. IL-17 facilitates VCAM-1 production and monocyte adhesion in osteoarthritis synovial fibroblasts by suppressing miR-5701 synthesis. Int J Mol Sci. 2022;23(12):6804.

Yahaya I, Wright T, Babatunde OO, Corp N, Helliwell T, Dikomitis L, et al. Prevalence of osteoarthritis in lower middle- and low-income countries: a systematic review and meta-analysis. Rheumatol Int. 2021;41(7):1221–31.

Yu H, Huang T, Lu WW, Tong L, Chen D. Osteoarthritis pain. Int J Mol Sci. 2022;23:9.

Zhang W, Sun G, Likhodii S, Aref-Eshghi E, Harper PE, Randell E, et al. Metabolomic analysis of human synovial fluid and plasma reveals that phosphatidylcholine metabolism is associated with both osteoarthritis and diabetes mellitus. Metabolomics [Internet]. 2016;12(2):24. Available from: https://doi.org/10.1007/s11306-015-0937-x.

Zhang H, Cai D, Bai X. Macrophages regulate the progression of osteoarthritis. Osteoarthr Cartil. 2020;28(5):555–61.

Clinical Applications of Musculoskeletal Regeneration Medicine: From Theory to Practice

Paul Lee, Tanvi Verma, and Kar Hao Teoh

Abstract This chapter explores the current landscape and prospects of musculoskeletal (MSK) regenerative medicine, focusing on the principles, clinical applications, and innovative approaches that are transforming patient care. It begins by outlining the foundational aspects of regenerative therapies, including the role of stem cells, platelet-rich plasma (PRP), and the extracellular matrix (ECM) in tissue repair. The chapter then delves into the practical application of these therapies, highlighting their effectiveness in treating conditions such as osteoarthritis and tendon injuries. Emerging technologies like induced pluripotent stem cells (iPSCs), photo-biomodulation therapy (PBMT), and 3D bioprinting are discussed as promising tools that could revolutionise MSK regeneration. Additionally, the chapter addresses the challenges of translating research into clinical practice, including regulatory and ethical considerations, and emphasises the importance of personalised medicine in optimising therapeutic outcomes. Looking ahead, we are in need for sustainable, long-term strategies to ensure the widespread adoption of regenerative therapies. Overall, the chapter provides a comprehensive overview of the advancements and future directions in MSK regenerative medicine, underscoring its potential to significantly improve patient outcomes in the coming years.

Keywords Stem cells · Platelet-Rich Plasma (PRP) · Regenerative therapies · Extracellular Matrix (ECM) · Photo-biomodulation

P. Lee (✉)
College of Health and Science, University of Lincoln, Lincoln, UK

London Cartilage Clinic, London, UK

MSK Doctors, ICRS Center of Excellence for Cartilage Regeneration & Joint Preservation, Grantham, UK
e-mail: leep@mskdoctors.com

T. Verma · K. H. Teoh
MSK Doctors, ICRS Center of Excellence for Cartilage Regeneration & Joint Preservation, Grantham, UK
e-mail: tanvi@mskdoctors.com

P. Lee (ed.), *Musculoskeletal Regeneration Medicine*,
https://doi.org/10.1007/978-3-031-77456-0_8

1 Introduction

Musculoskeletal (MSK) regeneration medicine has become a cornerstone in modern clinical practice, focusing on innovative biological therapies that aim to repair and regenerate damaged tissues such as bone, cartilage, muscle, tendons, and ligaments. The increasing incidence of musculoskeletal disorders—driven by factors like aging populations, sports injuries, and sedentary lifestyles—has led to a growing demand for treatments that go beyond symptom management, offering long-term solutions to tissue degeneration and injury.

Traditionally, treatments for musculoskeletal disorders have ranged from physical therapy and pain management to surgical interventions. However, these approaches often fail to fully restore the function and integrity of damaged tissues. As a result, there has been a significant shift towards regenerative medicine, which seeks to leverage the body's natural healing processes to repair tissues at the cellular and molecular levels.

1.1 Overview of Musculoskeletal (MSK) Regeneration Medicine

Musculoskeletal regeneration medicine encompasses a wide array of strategies aimed at restoring the function of damaged tissues through biological means. These strategies include the use of autologous platelet-rich plasma (PRP), mesenchymal stem cells (MSCs), and stromal vascular fraction (SVF) derived from adipose tissue. Each of these therapies plays a unique role in promoting tissue repair and regeneration, making them invaluable tools in the treatment of conditions such as osteoarthritis, tendon injuries, and chronic musculoskeletal pain.

Platelet-Rich Plasma (PRP) PRP therapy involves concentrating platelets from the patient's own blood and re-injecting them into the damaged tissue. The high concentration of growth factors and cytokines in PRP promotes cellular proliferation, tissue remodeling, and angiogenesis, all of which are crucial for the healing process. PRP has been extensively used in the treatment of tendinopathies, ligament injuries, and degenerative joint diseases, showing promising results in both pain reduction and functional improvement (Andia and Maffulli 2013). This is an important method for doctors to manipulate the chemical environment within the body to enhance regeneration.

Mesenchymal Stem Cells (MSCs) MSCs are multipotent cells capable of differentiating into various cell types, including osteoblasts, chondrocytes, and myocytes, which are essential for bone, cartilage, and muscle repair. Currently they are clinically sourced primarily from bone marrow and adipose tissue, MSCs cannot contribute to tissue regeneration by differentiating into the required cell types clinically but it modulate the immune response and reduce inflammation. This makes MSCs particularly effective in treating conditions like osteoarthritis.

Stromal Vascular Fraction (SVF) SVF is a component of adipose tissue that contains a heterogeneous mix of cells, including MSCs, endothelial cells, and immune cells. The use of SVF in regenerative medicine is gaining traction due to its relative ease of extraction and its potent regenerative capabilities. SVF has shown promise in the treatment of cartilage lesions and osteoarthritis, where it helps restore tissue structure and function by promoting the formation of new cartilage and reducing inflammation.

1.2 Integration of Regenerative Principles in Clinical Practice

The successful application of musculoskeletal regeneration therapies in clinical practice requires a deep understanding of the underlying principles that govern tissue repair and regeneration. These principles are grounded in the disciplines of physics, biology, chemistry, and even sleep medicine, each contributing to the overall effectiveness of regenerative treatments.

Physics The physical forces acting on tissues play a crucial role in their health and regeneration. Concepts such as mechanotransduction—the process by which cells convert mechanical stimuli into biochemical signals—are fundamental to understanding how therapies like PRP and MSCs can be optimised for better clinical outcomes. For example, mechanical loading through physical activity is known to enhance the efficacy of regenerative treatments by promoting tissue remodeling and strengthening.

Biology At the heart of musculoskeletal regeneration is the biological response to injury. The cascade of cellular events that follow tissue damage—ranging from inflammation to cell proliferation and matrix remodelling—are orchestrated by various signalling molecules, including growth factors and cytokines. Understanding these biological processes is critical for the effective application of regenerative therapies, ensuring that they align with the body's natural healing mechanisms.

Chemistry The chemical environment within tissues, particularly the extracellular matrix (ECM), is integral to tissue repair. The ECM not only provides structural support but also regulates cell behaviour through molecular interactions. Chemical factors such as ion concentrations, nutrient availability, and the activity of ensymes like matrix metalloproteinases (MMPs) influence the success of regenerative therapies. For instance, maintaining an optimal chemical environment can enhance the integration of MSCs into the host tissue, promoting more effective regeneration.

Sleep Recent research has highlighted the importance of sleep and circadian rhythms in tissue regeneration. Sleep influences hormone levels, such as growth hormone, which are critical for tissue repair. Poor sleep quality can impair the body's ability to heal, reducing the effectiveness of regenerative therapies. Therefore, optimising sleep patterns and addressing sleep disorders are increasingly being recognised as important components of comprehensive musculoskeletal regeneration strategies.

1.3 Objectives and Scope of the Chapter

This chapter aims to provide a comprehensive overview of the clinical applications of musculoskeletal regeneration medicine, with a particular focus on integrating the principles of physics, biology, chemistry, and sleep into clinical practice. It will explore the mechanisms of action behind key regenerative therapies, such as PRP and MSCs, and examine their efficacy in treating various musculoskeletal conditions. Additionally, the chapter will address the challenges associated with these therapies, including the need for standardisation and the variability in clinical outcomes. By bridging the gap between theory and practice, this chapter seeks to offer valuable insights for clinicians and researchers working in the field of musculoskeletal regeneration medicine.

2 Foundations of Regenerative Medicine

Regenerative medicine, particularly in the context of musculoskeletal (MSK) disorders, builds upon a deep understanding of the body's natural processes for healing and repair. To effectively leverage these processes in clinical practice, it is essential to understand the fundamental principles that govern tissue regeneration. These principles are rooted in the complex interplay between biological, chemical, and physical factors, each playing a critical role in determining the success of regenerative therapies. This section will delve into the basic mechanisms of tissue repair, the role of key biological components such as stem cells and growth factors, and the integration of these elements into clinical applications.

2.1 Basic Principles of Regenerative Medicine

Overview of Tissue Repair Mechanisms Tissue repair is a complex, multi-phase process that begins immediately after injury. The initial inflammatory phase involves the recruitment of immune cells to the injury site, which helps to clear debris and prevent infection. This phase is followed by the proliferative phase, where new tissue is formed through the proliferation of cells and the synthesis of extracellular matrix (ECM) components. The final phase, remodelling, involves the maturation and reorganisation of the newly formed tissue, restoring its structure and function.

- **Inflammatory Phase:**
  - The first response to tissue injury is inflammation, which is critical for initiating the healing process. Inflammation is characterised by the release of pro-inflammatory cytokines and the activation of immune cells, such as macrophages. These cells play a key role in clearing the injury site of dead cells and debris, as well as in secreting growth factors that stimulate the subsequent phases of healing.
- **Proliferative Phase:**
  - During the proliferative phase, new tissue begins to form. This involves the proliferation of various cell types, including fibroblasts, which are responsible for producing collagen and other ECM components. In musculoskeletal tissues, specific cells such as chondrocytes in cartilage and osteoblasts in bone also proliferate during this phase to contribute to tissue repair.
- **Remodeling Phase:**
  - The final phase of tissue repair is remodeling, where the newly formed tissue undergoes maturation and reorganisation. This process can take months to years, depending on the extent of the injury and the tissue involved. During remodeling, collagen fibers are realigned along lines of mechanical stress, enhancing the strength and functionality of the tissue.

Role of Stem Cells, Growth Factors, and Extracellular Matrix (ECM) Central to the regenerative process are stem cells, growth factors, and the extracellular matrix, which together orchestrate tissue repair and regeneration.

- **Stem Cells:**
  - Stem cells are undifferentiated cells with the potential to differentiate into various cell types needed for tissue repair. In musculoskeletal regeneration, mesenchymal stem cells (MSCs) are particularly important. In the lab, these cells can differentiate into osteoblasts, chondrocytes, and myocytes, which are essential for bone, cartilage, and muscle repair, respectively. But in clinical practice these MSCs does not differentiate, but they can secrete a range of bioactive molecules that modulate the immune response, reduce inflammation, and promote tissue regeneration.
- **Growth Factors:**
  - Growth factors are signaling proteins that play a crucial role in regulating cellular processes such as proliferation, differentiation, and migration. In regenerative medicine, growth factors like platelet-derived growth factor (PDGF),

transforming growth factor-beta (TGF-β), and vascular endothelial growth factor (VEGF) are commonly used to enhance the healing process. These factors are often delivered through therapies like platelet-rich plasma (PRP) to boost the body's natural healing response, it is a great way to modify chemistry.

- **Extracellular Matrix (ECM):**
  - The ECM is a complex network of proteins, glycoproteins, and proteoglycans that provides structural support to tissues and regulates cell behavior. In tissue regeneration, the ECM not only serves as a scaffold for new tissue growth but also plays a key role in signaling cellular activities through its interaction with cell surface receptors. The composition and organisation of the ECM are critical for the successful integration and function of regenerated tissues.

2.2 Regenerative Principles: Integrating Physics, Biology, Chemistry, and Sleep

Physics in Regeneration

- **Mechanical Forces and Tissue Response:**
  - The physical forces acting on tissues—such as tension, compression, and shear—are vital for maintaining tissue health and promoting regeneration. Mechanotransduction is the process by which cells sense and respond to these mechanical stimuli, converting them into biochemical signals that influence cell behavior and tissue repair. For example, in bone tissue, mechanical loading enhances osteoblast activity, leading to increased bone formation and strength. Similarly, in cartilage, compressive forces promote chondrocyte function and the production of cartilage-specific extracellular matrix components.
- **The Role of Mechanotransduction in Cell Behavior and Tissue Repair:**
  - Mechanotransduction is particularly important in musculoskeletal tissues, where mechanical loading is a natural part of daily function. Cells such as osteocytes in bone and chondrocytes in cartilage are equipped with mechanoreceptors that detect mechanical stress and translate it into cellular responses that promote tissue repair and maintenance. This principle is leveraged in regenerative therapies, where controlled mechanical loading is used to enhance the effectiveness of treatments like stem cell therapy and tissue-engineered scaffolds.

Biology in Regeneration

- **Cellular Responses to Injury and the Healing Cascade:**
  - Biological responses to tissue injury involve a well-coordinated sequence of events that ensure proper healing. The healing cascade begins with an inflam-

matory response, followed by the proliferation of cells needed to rebuild the tissue, and concludes with the remodeling of the new tissue to restore its original structure and function. Understanding these biological processes is crucial for developing regenerative therapies that align with the body's natural healing mechanisms, thereby enhancing their efficacy.

- **The Importance of Cellular Environment in Promoting Regeneration:**
  - The microenvironment surrounding cells, often referred to as the cellular niche, plays a critical role in determining the success of tissue regeneration. This environment includes not only the ECM but also the presence of signaling molecules, oxygen levels, and other cells that influence stem cell behavior and differentiation. Optimising the cellular environment is a key strategy in regenerative medicine, as it can enhance the regenerative capacity of stem cells and improve the integration of engineered tissues with the host tissue.

Chemistry in Regeneration

- **Molecular Interactions within the ECM:**
  - The extracellular matrix is not merely a structural support; it is also a dynamic environment where chemical interactions play a crucial role in regulating cell behavior. The binding of growth factors to ECM components, the crosslinking of collagen fibers, and the activity of ensymes like matrix metalloproteinases (MMPs) all contribute to the remodeling and stabilisation of the regenerating tissue. These molecular interactions are critical for ensuring that the newly formed tissue is functional and well-integrated with the surrounding tissue.
- **Role of Ion Balance, Ensymatic Activity, and Nutrient Availability in Tissue Repair:**
  - Chemical factors such as ion concentrations, ensymatic activity, and nutrient availability are essential for tissue repair. For example, calcium and phosphate ions are crucial for bone mineralisation, while ensymes like collagenases are necessary for remodeling the ECM during tissue regeneration. Adequate nutrient supply, including oxygen and glucose, is also vital for supporting the metabolic needs of proliferating cells during the healing process. Understanding these chemical principles allows for the development of regenerative therapies that provide the necessary biochemical support for effective tissue repair.

Sleep and Regeneration

- **The Role of Circadian Rhythms in Cellular Repair Processes:**
  - Sleep and circadian rhythms have a profound impact on tissue regeneration. During sleep, the body undergoes various physiological processes that are critical for tissue repair, including the release of growth hormone, which stimulates cell proliferation and protein synthesis. Disruptions in sleep patterns can negatively affect these processes, leading to impaired healing and suboptimal outcomes in regenerative therapies. Ensuring that patients have ade-

quate and quality sleep is therefore an important consideration in the overall strategy for musculoskeletal regeneration.

- **Impact of Sleep Quality on Hormonal Balance and Tissue Healing:**
  - Hormonal balance is crucial for tissue repair, and sleep plays a key role in maintaining this balance. For instance, cortisol, a stress hormone, is regulated by sleep and has a catabolic effect on tissues, meaning it can break down proteins and hinder tissue repair if not properly balanced. Conversely, growth hormone and testosterone, which are anabolic hormones that promote tissue growth and repair, are released in higher quantities during sleep. Poor sleep quality can disrupt the release of these hormones, leading to slower recovery times and reduced effectiveness of regenerative treatments.

3 Autologous Platelet-Rich Plasma (PRP) in Clinical Practice

Platelet-rich plasma (PRP) offers a minimally invasive approach to enhance tissue repair and regeneration. PRP therapy leverages the body's natural healing processes by concentrating and delivering growth factors directly to areas of injury or degeneration (Foster et al. 2009).

3.1 Mechanisms of PRP in Tissue Healing

The therapeutic potential of PRP is rooted in its high concentration of growth factors, cytokines, and other bioactive molecules that are essential for initiating and regulating tissue repair processes (Eppley et al. 2004). PRP is derived from autologous blood, meaning it is sourced from the patient themselves, thereby minimising the risk of immune reactions or transmission of infectious diseases (Marx 2004).

Growth Factors and Cytokines in PRP PRP is rich in a variety of growth factors, including platelet-derived growth factor (PDGF), transforming growth factor-beta (TGF-β), vascular endothelial growth factor (VEGF), and insulin-like growth factor-1 (IGF-1). These growth factors play a pivotal role in promoting cellular proliferation, angiogenesis, and extracellular matrix (ECM) synthesis, all of which are critical for effective tissue repair (Sánchez et al. 2009; Mishra and Pavelko 2006)

- **PDGF:** Stimulates the proliferation of mesenchymal stem cells (MSCs) and fibroblasts, which are essential for the formation of new connective tissue.
- **TGF-β:** Regulates cell differentiation and the production of ECM components such as collagen, thereby contributing to the structural integrity of the regenerating tissue (Murray et al. 2007)

- **VEGF:** Promotes the formation of new blood vessels (angiogenesis), which is crucial for delivering nutrients and oxygen to the healing tissue.
- **IGF-1:** Enhances protein synthesis and cell proliferation, facilitating the repair of damaged tissues such as tendons and cartilage.

PRP Preparation and Activation The preparation of PRP involves drawing blood from the patient and centrifuging it to separate the plasma and platelets from the other blood components (Marx 2004). The resultant PRP can be either leukocyte-rich or leukocyte-poor, depending on the presence of white blood cells, which can influence the inflammatory response at the site of injection (de Vos et al. 2010).

Activation

It can be activated by agents such as thrombin or calcium chloride, which trigger the release of growth factors stored in the platelets' alpha granules (Fufa et al. 2008). This process mimics the body's natural response to injury, where platelets are activated to release cytokines and growth factors that initiate tissue repair. When thrombin or calcium chloride is added, it prompts the platelets to degranulate, releasing vital proteins like platelet-derived growth factor (PDGF), transforming growth factor-beta (TGF-β), and vascular endothelial growth factor (VEGF). These proteins are crucial for tissue regeneration, as they promote cellular proliferation, angiogenesis, and collagen synthesise of Growth Factors.

- **PDGF**: This growth factor plays a significant role in promoting angiogenesis (formation of new blood vessels) and collagen synthesis, both of which are essential for tissue repair. PDGF also enhances the recruitment of stem cells and stimulates cellular processes that aid in wound healing which is vital for chondrocyte proliferation and differentiation, making it a key player in cartilage repair. It also modulates the immune response, reducing inflammation at the injury site, which is crucial for preventing chronic pain and promoting healing.
- **VEGF**: Promotes angiogenesis, improving oxygen and nutrient supply to the healing tissue. This is particularly beneficial for cartilage repair, as cartilage is avascular.

The variability in PRP preparation methods—including differences in centrifugation speed, time, and the use of activators—leads to significant differences in PRP composition, which can affect its therapeutic efficacy (Dragoo et al. 2012). This has been a point of contention in clinical practice, as there is no standardised protocol for PRP preparation, making it difficult to compare outcomes across different studies.

3.2 Clinical Applications of PRP

PRP has been widely applied in the treatment of various musculoskeletal conditions, with particular success in managing osteoarthritis, tendinopathies, and sports injuries (Mishra and Pavelko 2006). The efficacy of PRP in these contexts is largely attributed to its ability to modulate inflammation, promote tissue regeneration, and reduce pain.

PRP in Ligament Injuries PRP has also been extensively used in the treatment of ligament injuries, where its ability to enhance the healing of soft tissues is particularly valuable (Jo et al. 2015). Ligaments typically have a limited blood supply, which slows their healing process. The application of PRP can accelerate this process by providing a concentrated dose of growth factors that stimulate cell proliferation, collagen synthesis, and angiogenesis.

One of the most important effects is the enhancement of collagen production, a fundamental component for the structural integrity of tendons ligaments. PRP accelerates collagen synthesis, leading to the formation of stronger and more organised collagen fibres, which results in faster and more robust tissue repair (Murray et al. 2007). In addition, PRP boosts angiogenesis through the action of VEGF, promoting the formation of new blood vessels. This improves the delivery of oxygen and nutrients to the damaged tissue, creating a more favourable environment for healing (Anitua et al. 2006). Furthermore, TGF-β plays a pivotal role in modulating the inflammatory response, helping to reduce pain, prevent chronic inflammation, and promote a more efficient healing process. Together, these mechanisms make PRP a valuable treatment option for tendon and ligament injuries, offering not only accelerated healing but also improved tissue quality and a reduced risk of re-injury.

Clinical evidence supports the use of PRP in treating conditions such as lateral epicondylitis (tennis elbow), Achilles tendinopathy, and rotator cuff injuries. Patients receiving PRP injections often report reduced pain and improved function, with some studies suggesting that PRP may also reduce the risk of re-injury compared to conventional treatments (Filardo et al. 2014).

3.3 *Challenges and Future Directions on PRP in MSK Regeneration*

Despite the promising results associated with PRP therapy, several challenges remain that need to be addressed to optimise its clinical use.

Standardisation of PRP Protocols One of the major challenges in PRP therapy is the lack of standardised preparation and administration protocols. The variability in PRP composition due to different preparation techniques makes it difficult to predict and compare clinical outcomes (Dragoo et al. 2012). Efforts are needed to establish guidelines for PRP preparation and application, which would help to ensure consistent and reliable results across different clinical settings.

Long-Term Efficacy and Outcomes While short-term benefits of PRP therapy, such as pain reduction and improved function, are well-documented, there is still limited evidence on its long-term efficacy (Jo et al. 2015). Some studies suggest that the effects of PRP may diminish over time, necessitating repeated injections to maintain the therapeutic benefits (Filardo et al. 2014). Long-term follow-up studies are required to determine the sustainability of PRP's effects and to assess whether it can alter the course of chronic degenerative conditions like osteoarthritis.

PRP in Combination Therapies The future of PRP therapy may lie in its combination with other regenerative approaches, such as stem cell therapy or hyaluronic acid injections (Taylor et al. 2011). Preliminary studies indicate that combining PRP with other treatments may enhance its efficacy by providing a synergistic effect, improving tissue repair and regeneration beyond what PRP alone can achieve. Further research is needed to explore these combination therapies and to establish protocols that maximise their potential.

1. **PRP + Hyaluronic Acid (HA)**
   This combination offers a synergistic effect. HA provides lubrication and acts as a scaffold, while PRP delivers growth factors that promote tissue healing. Studies suggest that HA enhances PRP's ability to regenerate cartilage, as it supports cell migration and proliferation (Foster et al. 2009). In some trials, PRP combined with HA resulted in better cartilage regeneration and symptom relief compared to PRP or HA alone. The mechanical properties of HA support the localised retention of PRP's growth factors at the injection site, which improves tissue regeneration over time (Anitua et al. 2006).
2. **PRP + Mesenchymal Stem Cells (MSC)**
   MSCs' have the potential for cartilage repair and PRP's can enhance cell proliferation. MSCs are effective in regenerating cartilage, and PRP enhances the differentiation of these cells into chondrocytes, promoting collagen and extracellular matrix production (Murray et al. 2007). This accelerates tissue repair, reduces inflammation, and improves outcomes in conditions like osteoarthritis.
3. **Hydrogel (IPAAG)**
   Hydrogels such as IPAAG (Injectable Polyacrylamide Gel) are used as scaffolds in regenerative medicine. When combined with PRP, these hydrogels provide a 3D matrix that enhances the stability of the injected PRP, allowing for better localised tissue regeneration (Anitua et al. 2006). The hydrogel offers mechanical support while PRP releases growth factors that stimulate cell proliferation and tissue healing. In some animal models, PRP combined with hydrogel significantly improved cartilage repair, showing enhanced structural and functional recovery (Mishra and Pavelko 2006).
4. HA + MSCs has shown promise for cartilage repair due to HA's supportive properties and MSCs' regenerative potential. HA helps MSCs remain localised at the injury site and provides a conducive environment for MSC differentiation into chondrocytes (Murray et al. 2007). In studies, this combination improved cartilage regeneration and pain relief in osteoarthritis patients. HA's viscoelastic properties not only lubricate the joint but also help maintain MSC viability and function (Anitua et al. 2006).

MSC therapy can be introduced first, allowing the cells to initiate the cartilage repair process. After a 6-week period, a hydrogel scaffold (like IPAAG) can be applied to stabilise the regenerating tissue and further enhance the environment for healing. Three months later, PRP injections serve as a "top-up" to stimulate further tissue regeneration by delivering concentrated growth factors to boost cellular activity and collagen production. This staged approach provides time for each phase of healing to progress effectively before introducing the next treatment (Murray et al. 2007).

4 Mesenchymal Stem Cells (MSCs) in Regenerative Medicine

4.1 Biological Properties of MSCs

MSCs are defined by their ability to self-renew and differentiate into multiple cell lineages, including osteoblasts (bone cells), chondrocytes (cartilage cells), and myocytes (muscle cells). These multipotent capabilities and immunomodulatory properties are used to repair or replace damaged tissues.

Sources of MSCs MSCs can be isolated from various tissues, with bone marrow and adipose tissue being the most common sources used in clinical practice. However, the process of harvesting BM-MSCs is invasive and yields a relatively low number of cells, which has led to increasing interest in alternative sources such as adipose-derived MSCs (AD-MSCs) (Anitua et al. 2006). Adipose tissue, commonly obtained through liposuction, offers a more accessible and abundant source of MSCs with high regenerative potential.

Immunomodulatory and Paracrine Functions In addition to their differentiation capabilities, MSCs exert significant immunomodulatory effects, making them suitable for treating inflammatory conditions like osteoarthritis. MSCs can modulate the immune response by inhibiting the proliferation of T cells, promoting the formation of regulatory T cells, and shifting macrophages from a pro-inflammatory (M1) to an anti-inflammatory (M2) phenotype (Jo et al. 2015). These effects are largely mediated through the secretion of bioactive molecules, including cytokines, growth factors, and exosomes, which create a pro-regenerative microenvironment conducive to tissue healing .

4.2 Clinical Applications of MSCs

MSCs in Osteoarthritis (OA) Treatment Osteoarthritis is one of the most common conditions treated with MSCs. Intra-articular injections of MSCs have been shown to reduce pain, improve joint function, and potentially slow the progression of OA (Sánchez et al. 2009). The therapeutic effects of MSCs in OA are believed to result from their ability to modulate the inflammatory environment within the joint, promote the repair of cartilage, and inhibit further degradation of the extracellular matrix. Studies have shown that MSCs can enhance the production of cartilage matrix proteins like collagen and aggrecan, which are essential for maintaining cartilage integrity.

Cartilage Repair Cartilage repair is another critical area where MSCs have shown significant promise. Cartilage is notoriously difficult to repair due to its avascular nature, which limits the natural healing response. MSCs can be used to overcome this limitation by promoting the regeneration of cartilage tissue. In some clinical trials, MSCs have been combined with scaffolds or growth factors to enhance their chondrogenic potential, leading to the formation of cartilage tissue that closely resembles native hyaline cartilage.

Bone Regeneration Bone regeneration using MSCs is particularly relevant in the treatment of fractures, bone defects, and conditions like osteoporosis. MSCs contribute to bone repair by differentiating into osteoblasts, the cells responsible for new bone formation. Additionally, MSCs secrete factors that promote angiogenesis, which is crucial for providing the necessary blood supply to the regenerating bone. Clinical studies have demonstrated that MSCs can improve the healing of fractures and even aid in the repair of large bone defects when combined with appropriate scaffolds or growth factors.

Stromal Vascular Fraction (SVF) and its Clinical Potential Stromal vascular fraction (SVF), derived from adipose tissue, is a rich source of MSCs and other regenerative cells. SVF has gained attention for its use in treating musculoskeletal conditions, particularly osteoarthritis. SVF contains not only MSCs but also endothelial cells, pericytes, and immune cells, all of which contribute to tissue repair. Clinical studies have shown that SVF can enhance the healing of cartilage lesions and improve joint function in patients with OA, making it a promising alternative to more invasive treatments.

4.3 Enhancing MSC Therapy

Preconditioning Strategies Preconditioning MSCs before their application can enhance their therapeutic potential. This can involve exposing the cells to specific conditions, such as hypoxia or inflammatory cytokines, to "prime" them for better performance in the target tissue. Preconditioned MSCs have been shown to exhibit enhanced survival, migratory capacity, and secretion of regenerative factors, making them more effective in promoting tissue repair.

Combination with Growth Factors and Cytokines Combining MSCs with growth factors or cytokines can amplify their regenerative effects. For example, adding platelet-rich plasma (PRP) to MSC therapy has been shown to boost the production of cartilage matrix components and enhance the overall effectiveness of the treatment. Similarly, integrating MSCs with bioactive scaffolds that release growth factors can provide sustained support for tissue regeneration, leading to better clinical outcomes.

Use of Scaffolds The use of biocompatible scaffolds in MSC therapy provides structural support for the regenerating tissue and can improve the integration of MSCs into the damaged area. Scaffolds can be designed to mimic the natural extracellular matrix, providing cues that guide MSC differentiation and tissue formation. In cartilage repair, scaffolds loaded with MSCs have been shown to produce cartilage with mechanical properties similar to native tissue, offering a promising approach for treating joint injuries and defects.

5 Innovative Approaches in MSK Regeneration

5.1 Induced Pluripotent Stem Cells (iPSCs) in Tendon Regeneration

Induced pluripotent stem cells (iPSCs) have the ability to differentiate into any cell type, including those necessary for tendon and cartilage repair. iPSCs are derived from adult somatic cells, such as skin fibroblasts, which are reprogrammed back into a pluripotent state through the introduction of specific transcription factors (e.g., OCT4, SOX2, KLF4, and c-MYC). This process allows iPSCs to gain similar properties to embryonic stem cells without the ethical concerns associated with the latter.

Tendon Regeneration with iPSCs In tendon regeneration, induced pluripotent stem cells (iPSCs) can differentiate into tenocytes, the specialised cells responsible for producing the extracellular matrix (ECM) that supports tendon structure and function. The process includes the differentiation of iPSCs into tenocyte-like cells, the secretion of specific cytokines and growth factors, and the integration of these cells into the damaged tendon tissue to promote repair.

Differentiation of iPSCs into Tenocytes
iPSCs are first induced to differentiate into mesodermal progenitors, which are the precursors to tenocytes. This differentiation is primarily driven by the activation of specific transcription factors, such as Scleraxis (Scx) (Yoshimoto et al. 2022). Additionally, the TGF-β (transforming growth factor-beta) pathway induces the expression of Scx and other tenocyte-specific markers like type I collagen and tenomodulin to promote tenogenic differentiation (Havis et al. 2014).

Cytokines and Growth Factors Involved in Tendon Regeneration

- **TGF-β:** Reduces the infiltration of inflammatory cells by modulating the inflammatory response and promotes tenogenic differentiation.
- **IGF-1 (Insulin-like Growth Factor 1):** Stimulates the proliferation and migration of tenocytes, as well as enhances collagen synthesis.
- **PDGF (Platelet-Derived Growth Factor):** Enhances the proliferation of tenocytes and fibroblasts and promotes angiogenesis, which improves blood supply to the healing tendon.

Integration and Functional Restoration

Once the iPSCs have differentiated into tenocytes and are introduced into the damaged tendon tissue—either through direct injection or via scaffolds—they begin to integrate into the existing tissue. The newly formed tenocytes align along the axis of the tendon, producing type I collagen necessary for the tensile strength of the tissue.

The remodelling process, of interaction between the iPSC-derived tenocytes and the native tendon cells, is influenced by mechanotransduction pathways, where mechanical forces are converted into biochemical signals that further promote tissue repair. Activation of mechanosensitive ion channels like TRPV4 and integrins on the surface of tenocytes initiates signalling cascades that enhance the synthesis and organisation of collagen fibres.

Mechanosensitive Ion Channels

TRPV4 (Transient Receptor Potential Vanilloid 4) is a mechanosensitive ion channel that is activated by mechanical forces such as stretching, compression, or shear stress. When tenocytes in a regenerating tendon experience these mechanical forces, TRPV4 channels open, allowing the influx of calcium ions (Ca^{2+}) into the cell (Muramatsu et al. 2007). The increase in intracellular calcium concentration serves as a secondary messenger triggering intracellular signaling pathways (Cao et al. 2018).

- **Calcium Signaling and Collagen Synthesis:**
  - The influx of Ca^{2+} through TRPV4 channels activates the calcium/calmodulin-dependent protein kinase (CaMK) pathway, regulating gene expression for extracellular matrix (ECM) production. This activation leads to the upregulation of collagen type I and III genes.
  - Additionally, calcium signaling also activates the MAPK/ERK pathway, promoting the proliferation and differentiation of tenocytes.

Integrins and Their Role in Mechanotransduction

Integrins are transmembrane receptors that mediate the interaction between tenocytes and the extracellular matrix. They are crucial for transmitting mechanical signals from the ECM into the cell, a process known as mechanotransduction. When mechanical stress is applied to the tendon, integrins cluster and bind to ECM proteins such as fibronectin and collagen (Hynes 2002). This binding initiates the formation of focal adhesion complexes, which are multi-protein structures that link integrins to the actin cytoskeleton inside the cell.

- **Focal Adhesion Kinase (FAK) Activation:**
  - The clustering of integrins activates focal adhesion kinase (FAK), a key enzyme in mechanotransduction. FAK activates the RhoA/ROCK pathway phosphorylating the Src family kinases. This pathway is necessary for the reorganisation of the actin cytoskeleton, which is necessary for the proper alignment and orientation of newly synthesised collagen fibers.
  - FAK signaling also interacts with the MAPK/ERK pathway, increasing the expression of genes encoding for collagen for ECM production.
- **Integrin-Mediated Signaling and ECM Remodeling:**
  - The mechanical signals transmitted by integrins regulate the activity of matrix metalloproteinases (MMPs), enzymes that degrade and remodel the ECM. It removes the damaged matrix components and replaces them with newly synthesised collagen.

Synergistic Effects of TRPV4 and Integrin Activation
The activation of TRPV4 and integrins works synergistically to ensure efficient tendon regeneration. The calcium influx through TRPV4 channels amplifies the integrin-mediated signaling pathways, creating a robust response that enhances the synthesis and organisation of collagen fibers in the proper alignment and direction of mechanical load.

Challenges and Future Directions Despite their promise, the use of iPSCs in clinical applications faces several challenges, including the risk of teratoma formation, the need for precise control over differentiation pathways, and the technical complexities of generating iPSCs. Future research is focused on overcoming these hurdles through improved reprogramming techniques, the development of safer and more efficient differentiation protocols, and the exploration of combination therapies that integrate iPSCs with other regenerative approaches.

5.2 *Photobiomodulation Therapy*

Photobiomodulation therapy (PBMT) is an emerging non-invasive treatment that uses specific wavelengths of light to promote tissue repair and reduce inflammation. Specifically, the therapy targets cytochrome c oxidase, a key enzyme in the electron transport chain located within the mitochondria (Hamblin 2006). Upon absorption of light, typically in the red to near-infrared spectrum (600–1000 nm), cytochrome c oxidase becomes more efficient in its role of transferring electrons. This enhanced electron transfer activity increases the mitochondrial membrane potential, which in turn accelerates the production of adenosine triphosphate (ATP) (Karu 1999).

Mechanisms of Action PBMT operates by delivering light energy to the mitochondria within cells, which enhances ATP production and activates signalling pathways that lead to increased cell proliferation, migration, and the synthesis of ECM components.

Activation of Cellular Signaling Pathways
The increase in ATP levels and low levels of reactive oxygen species (ROS) trigger tissue regeneration through the following pathways:

- **NF-κB and MAPK Pathways:** The NF-κB pathway (nuclear factor kappa-light-chain-enhancer of activated B cells) activates the production of various growth factors that promote tissue healing. The MAPK pathway (mitogen-activated protein kinase) helps regulate cell proliferation and apoptosis.
- **Akt/PI3K Pathway: Akt/PI3K pathway** (phosphoinositide 3-kinase) ensures that cells in stressed or injured tissues do not die off prematurely, thereby supporting the overall healing process.

Modulation of Inflammatory Responses
- **Reduction of Pro-Inflammatory Cytokines:** PBMT reduces the levels of harmful inflammatory molecules TNF-α (tumor necrosis factor-alpha), IL-1β (interleukin-1 beta), and IL-6, which are typically elevated in injured tissues and contribute to pain and swelling.
- **Increase in Anti-Inflammatory Cytokines:** PBMT boosts the production of beneficial cytokines like IL-10 and TGF-β. These molecules help resolve inflammation and encourage the repair process.

Enhanced Cell Proliferation and Migration
With increased ATP and the activation of signaling pathways, cells are better equipped to proliferate and migrate to where they are needed. Fibroblasts are cells responsible for producing the extracellular matrix, including collagen providing structural support to tissues. PBMT also promotes the proliferation and movement of endothelial cells lining blood vessels causing angiogenesis.

Synthesis and Remodeling of the Extracellular Matrix (ECM)
Finally, PBMT supports the synthesis and proper organisation of the extracellular matrix. PBMT helps regulate the activity of MMPs, ensuring that they remove only the damaged matrix while leaving the newly synthesised matrix intact, which is crucial for effective tissue remodeling.

Clinical Applications in MSK Disorders PBMT has been applied in the treatment of tendinopathies, muscle strains, and osteoarthritis. Clinical studies have shown that PBMT can reduce pain and inflammation in these conditions, leading to improved function and faster recovery times. The therapy is particularly beneficial when used in conjunction with other regenerative treatments, such as PRP or MSC therapy, as it can enhance the effects of these therapies by creating a more favourable environment for tissue repair.

5.3 Combination Therapies

PBMT and Stem Cell Therapy Combining PBMT with stem cell therapy, such as MSCs or iPSCs, offers another promising approach. PBMT can enhance the regenerative potential of stem cells by promoting their proliferation, migration, and differentiation (Houreld 2010). This combination has been shown to improve outcomes in tissue regeneration, particularly in challenging cases such as chronic tendinopathies and advanced osteoarthritis.

6 Translating Research into Clinical Practice

The translation of musculoskeletal (MSK) regenerative research into clinical practice involves multiple challenges, including bridging the gap between laboratory findings and real-world applications, navigating regulatory and ethical considerations, and evaluating clinical outcomes through rigorous studies.

6.1 Bridging the Gap Between Research and Application

Translating regenerative medicine research into clinical practice requires overcoming significant barriers. These include the variability in preclinical models, the complexity of human biology, and the need for large-scale, controlled clinical trials to demonstrate efficacy and safety.

Preclinical to Clinical Transition
Preclinical studies, often conducted in animal models, provide foundational data on the safety and potential efficacy of regenerative therapies. However, differences between animal models and human physiology can lead to discrepancies in outcomes when these therapies are tested in human clinical trials. For example, the regenerative capabilities observed in animal models may not fully replicate in humans due to differences in immune responses and tissue characteristics.

Clinical Trials and Evidence-Based Medicine
The transition from research to clinical application necessitates well-designed clinical trials that adhere to the principles of evidence-based medicine. These trials must be large enough to provide statistically significant results, yet focused enough to assess specific outcomes related to the therapy being tested. The challenges include recruiting sufficient participants, ensuring adherence to trial protocols, and managing the long-term follow-up required to assess the durability of regenerative treatments.

Bridging Research and Practice

Bridging this gap also involves the integration of new therapies into existing treatment paradigms. This requires collaboration between researchers, clinicians, and policymakers to develop guidelines that support the use of regenerative therapies in a safe and effective manner. Education and training for healthcare providers are also essential to ensure that they are well-equipped to implement these therapies in clinical settings.

6.2 *Regulatory and Ethical Considerations*

The field of regenerative medicine is heavily regulated due to the complex nature of the therapies involved, which often include living cells and tissues. Navigating these regulatory frameworks is crucial for the successful translation of research into clinical practice.

Regulatory Frameworks

Regenerative therapies must comply with stringent regulatory requirements before they can be used in clinical settings. In many countries, this involves approval from agencies like the U.S. Food and Drug Administration (FDA) or the European Medicines Agency (EMA). These agencies require extensive preclinical data, as well as evidence from clinical trials, to ensure that new therapies are both safe and effective.

Ethical Considerations

Ethical issues are particularly prominent in regenerative medicine, especially when it comes to the use of stem cells. Concerns include the source of stem cells (e.g., embryonic stem cells vs. adult stem cells), the potential for tumor formation, and the long-term effects of altering cellular processes. There is also a need to ensure informed consent from patients, particularly in early-phase clinical trials where the risks and benefits may not be fully understood.

Regulatory Challenges

A major regulatory challenge is the classification of regenerative therapies. These therapies often do not fit neatly into existing categories of drugs or medical devices, leading to delays in approval processes. Additionally, the need for post-market surveillance to monitor long-term safety adds another layer of complexity to the regulatory landscape.

6.3 *Case Studies of Successful Clinical Applications*

Real-world examples of regenerative therapies provide insights into the challenges and successes of translating research into practice.

Platelet-Rich Plasma (PRP) in Sports Medicine
PRP has been successfully integrated into clinical practice, particularly in sports medicine, where it is used to treat tendon injuries, muscle strains, and joint disorders. Despite its widespread use, PRP therapy's effectiveness varies depending on the preparation methods and the specific condition being treated, highlighting the need for standardisation and further research.

Mesenchymal Stem Cells (MSCs) in Osteoarthritis
MSC therapy has shown promise in treating osteoarthritis by promoting cartilage repair and modulating the inflammatory environment within joints. Clinical trials have demonstrated improvements in pain and function, though the variability in outcomes underscores the importance of refining patient selection criteria and optimising MSC delivery methods.

Photobiomodulation Therapy (PBMT) in Chronic Pain Management
PBMT has been used effectively to manage chronic pain associated with musculoskeletal disorders. Its non-invasive nature and ability to modulate inflammation make it an attractive option for patients who are not candidates for surgery. The successful application of PBMT in clinical practice illustrates the potential for integrating innovative therapies into existing treatment frameworks.

7 Future Perspectives and Innovations

The field of musculoskeletal (MSK) regenerative medicine is on the cusp of a transformation driven by emerging technologies, personalised medicine, and considerations of long-term sustainability. These advancements promise to enhance therapeutic outcomes, address current limitations, and pave the way for more accessible and effective treatments.

7.1 Emerging Technologies in Regenerative Medicine

Biomaterials and 3D Bioprinting
Biomaterials are critical in tissue engineering, providing scaffolds that support cell growth and tissue formation. Recent innovations have focused on developing biomimetic scaffolds that mimic the natural extracellular matrix (ECM) more closely. These scaffolds are designed to enhance cell attachment, proliferation, and differentiation, thereby improving tissue regeneration outcomes. Multiphasic scaffolds are being developed to regenerate both cartilage and subchondral bone simultaneously, addressing the complexity of osteochondral defects.

3D bioprinting allows for the precise fabrication of complex tissue structures layer by layer. This technology enables the creation of customised tissue grafts that match the patient's anatomical and biological needs, which is particularly beneficial for MSK applications. For instance, 3D bioprinting can produce cartilage or bone tissues with embedded vascular networks, ensuring adequate nutrient supply and integration with the host tissue.

Gene Editing and CRISPR-Cas9
Gene editing technologies, particularly CRISPR-Cas9, are revolutionising regenerative medicine by allowing precise modifications to the genome. In the context of MSK regeneration, gene editing can be used to enhance the regenerative capacity of cells or to correct genetic defects that contribute to degenerative conditions. For example, CRISPR-Cas9 can be employed to modify the expression of genes involved in collagen production, potentially improving the quality of regenerated cartilage or bone.

7.2 *Personalised Medicine in MSK Regeneration*

Genomics and Patient Profiling Advances in genomics have made it possible to identify genetic markers associated with musculoskeletal diseases, such as osteoarthritis or osteoporosis. By analysing a patient's genetic profile, clinicians can predict how they might respond to specific regenerative therapies. This genetic profiling helps in customising treatment plans that are more likely to be effective for the individual patient.

Tailored Therapeutic Approaches Personalised medicine also involves customising the type, dosage, and delivery method of regenerative therapies based on patient-specific factors. For instance, the concentration of growth factors in platelet-rich plasma (PRP) can be adjusted according to the patient's healing requirements. Similarly, the selection of donor cells for therapies like stem cell transplantation can be optimised to reduce the risk of immune rejection and improve therapeutic efficacy.

Integration of Lifestyle Factors Beyond genetics, lifestyle factors such as diet, physical activity, and sleep patterns significantly influence the success of regenerative treatments. Personalised medicine considers these factors when designing rehabilitation programs and other supportive therapies. For example, optimising a patient's sleep patterns could enhance hormonal balance, thereby improving tissue healing and overall recovery.

7.3 *Long-Term Implications and Sustainability*

Evaluating Long-Term Success Long-term studies are essential to assess the durability and safety of regenerative treatments.

Barriers to Widespread Adoption Despite their potential, several challenges must be overcome to ensure the widespread adoption of regenerative therapies. These include the high costs associated with advanced treatments, the need for specialised training for healthcare providers, and the scalability of production processes. Addressing these challenges requires the development of cost-effective manufacturing methods, as well as comprehensive training programs to equip healthcare providers with the necessary skills.

Sustainability and Ethical Considerations Sustainability in regenerative medicine also involves ethical considerations, such as the responsible use of gene editing and the long-term impact of introducing bioengineered tissues into the human body. Ensuring that these technologies are developed and implemented in a way that respects patient autonomy and promotes equitable access is vital for the future of the field. Additionally, regulatory frameworks must evolve to keep pace with technological advancements, ensuring that new treatments are both safe and effective.

8 Conclusion

Musculoskeletal (MSK) regenerative medicine is rapidly evolving, driven by advances in stem cell biology, biomaterials, gene editing, and personalised medicine. The integration of these innovative approaches into clinical practice has the potential to significantly improve outcomes for patients suffering from a wide range of musculoskeletal disorders, from osteoarthritis to tendon injuries.

8.1 Summary of Key Points

Foundations of Regenerative Medicine The principles of MSK regeneration are deeply rooted in the biology of stem cells, the chemistry of extracellular matrix (ECM) interactions, and the physics of mechanotransduction. Understanding these principles is crucial for the development of effective regenerative therapies that align with the body's natural healing processes.

Clinical Applications of PRP and MSCs Platelet-rich plasma (PRP) and mesenchymal stem cells (MSCs) have become central to regenerative treatments, particularly in managing conditions like osteoarthritis, tendon injuries, and chronic pain. While PRP leverages the body's own growth factors to promote healing, MSCs offer a versatile tool capable of differentiating into various cell types necessary for tissue repair.

Innovative Approaches Emerging technologies, such as induced pluripotent stem cells (iPSCs), photobiomodulation therapy (PBMT), and combination therapies, are expanding the horisons of MSK regeneration. These approaches offer promising alternatives for cases where traditional therapies may fall short, particularly in complex or chronic conditions.

Translation to Clinical Practice Bridging the gap between research and clinical practice involves navigating regulatory and ethical challenges, standardising treatment protocols, and ensuring long-term efficacy through rigorous clinical trials. Successful case studies demonstrate the potential for regenerative therapies to significantly improve patient outcomes, but widespread adoption will require continued efforts in education, cost management, and regulatory compliance.

References

Andia I, Maffulli N. Platelet-rich plasma for managing pain and inflammation in osteoarthritis. Nat Rev Rheumatol. 2013;9(12):721–30. https://doi.org/10.1038/nrrheum.2013.141.

Anitua E, Sanchez M, Nurden AT, Nurden P, Orive G, Andía I. New insights into and novel applications for platelet-rich fibrin therapies. Trends Biotechnol. 2006;24(5):227–34. https://doi.org/10.1016/j.tibtech.2006.02.010.

Cao S, Anishkin A, Zinkevich NS, Nishijima Y, Korishettar A, Wang Z, Fang J, Wilcox DA, Zhang DX. Transient receptor potential vanilloid 4 (TRPV4) activation by arachidonic acid requires protein kinase A-mediated phosphorylation. J Biol Chem. 2018;293(14):5307–22. https://doi.org/10.1074/jbc.M117.811075. Epub 2018 Feb 8. PMID: 29462784; PMCID: PMC5892583.

de Vos RJ, Weir A, van Schie HT, Bierma-Zeinstra SM, Verhaar JA, Tol JL. Platelet-rich plasma injection for chronic Achilles tendinopathy: a randomized controlled trial. JAMA. 2010;303(2):144–9. https://doi.org/10.1001/jama.2009.1986.

Dragoo JL, Braun HJ, Durham JL, Ridley BA, Odegaard JI, Luong R. Comparison of the acute inflammatory response of two commercial platelet-rich plasma systems in healthy rabbit tendons. Am J Sports Med. 2012;40(6):1274–81. https://doi.org/10.1177/0363546512442334.

Eppley BL, Woodell JE, Higgins J. Platelet quantification and growth factor analysis from platelet-rich plasma: implications for wound healing. Plast Reconstr Surg. 2004;114(6):1502–8. https://doi.org/10.1097/01.PRS.0000138251.07040.51.

Filardo G, Di Matteo B, Kon E, Di Martino A, Tesei G, Marcacci M. Platelet-rich plasma for the treatment of patellar tendinopathy: clinical and imaging findings at medium-term follow-up. Int Orthop. 2014;37(8):1583–9. https://doi.org/10.1007/s00264-013-1921-8.

Foster TE, Puskas BL, Mandelbaum BR, Gerhardt MB, Rodeo SA. Platelet-rich plasma: from basic science to clinical applications. Am J Sports Med. 2009;37(11):2259–72. https://doi.org/10.1177/0363546509349921.

Fufa D, Shealy B, Jacobson M, Kevy S, Murray MM. Activation of platelet-rich plasma using soluble type I collagen. J Oral Maxillofac Surg. 2008;66(4):684–90. https://doi.org/10.1016/j.joms.2007.10.011.

Hamblin MR, Demidova TN. Mechanisms of low-level light therapy. Proc SPIE. 2006;6140:614001. https://doi.org/10.1117/12.646294.

Havis E, Bonnin MA, Olivera-Martinez I, Nazaret N, Ruggiu M, Weibel J, Durand C, Guerquin MJ, Bonod-Bidaud C, Ruggiero F, Schweitzer R, Duprez D. Transcriptomic analysis of mouse limb tendon cells during development. Development. 2014;141(19):3683–96. https://doi.org/10.1242/dev.108654. Epub 2014 Sep 5. PMID: 25249460.

Houreld N, Abrahamse H. Low-intensity laser irradiation stimulates wound healing in diabetic wounded fibroblast cells (WS1). Diabetes Technol Ther. 2010;12(12):971–8. https://doi.org/10.1089/dia.2010.0039. PMID: 21128844.
Hynes RO. Integrins: bidirectional, allosteric signaling machines. Cell. 2002;110(6):673–87. https://doi.org/10.1016/S0092-8674(02)00971-6.
Jo CH, Kim JE, Yoon KS, Lee JH. Does platelet-rich plasma accelerate recovery after rotator cuff repair? A prospective cohort study. Am J Sports Med. 2015;43(8):1831–7. https://doi.org/10.1177/0363546515587084.
Karu T. Primary and secondary mechanisms of action of visible to near-IR radiation on cells. J Photochem Photobiol B. 1999;49(1):1–17. https://doi.org/10.1016/S1011-1344(98)00219-X. PMID: 10365442.
Marx RE. Platelet-rich plasma: evidence to support its use. J Oral Maxillofac Surg. 2004;62(4):489–96. https://doi.org/10.1016/j.joms.2003.12.003.
Mishra A, Pavelko T. Treatment of chronic elbow tendinosis with buffered platelet-rich plasma. Am J Sports Med. 2006;34(11):1774–8. https://doi.org/10.1177/0363546506288850.
Muramatsu S, Wakabayashi M, Ohno T, Amano K, Ooishi R, Sugahara T, Shiojiri S, Tashiro K, Suzuki Y, Nishimura R, Kuhara S, Sugano S, Yoneda T, Matsuda A. Functional gene screening system identified TRPV4 as a regulator of chondrogenic differentiation. J Biol Chem. 2007;282(44):32158–67. https://doi.org/10.1074/jbc.M706158200. Epub 2007 Sep 5. PMID: 17804410.
Murray MM, Spindler KP, Ballard P, Welch TP, Zurakowski D, Nanney LB. Enhanced histologic repair in a central wound in the anterior cruciate ligament with a collagen-platelet-rich plasma scaffold. J Orthop Res. 2007;25(8):1007–17. https://doi.org/10.1002/jor.20391.
Sánchez M, Anitua E, Orive G, Andía I. Platelet-rich therapies in the treatment of orthopaedic sport injuries. Sports Med. 2009;39(5):345–54. https://doi.org/10.2165/00007256-200939050-00002.
Taylor DW, Petrera M, Hendry M, Theodoropoulos JS. A systematic review of the use of platelet-rich plasma in sports medicine as a new treatment for tendon and ligament injuries. Clin J Sport Med. 2011;21(4):344–52. https://doi.org/10.1097/JSM.0b013e31821e1da3.
Yoshimoto Y, Uezumi A, Tanaka K, Yu X, Kurosawa T, Yambe S, Maehara K, Ohkawa Y, Sotomaru Y, Shukunami C. Tenogenic induction from induced pluripotent stem cells unveils the trajectory towards tenocyte differentiation. Front Cell Dev Biol. 2022;10:780038. https://doi.org/10.3389/fcell.2022.780038.

Musculoskeletal Injection Therapy: Part 1

Feza Korkusuz, Elif Üstün, Emine Çiftçi, Bilge Başak Fidan, and Paul Lee

Abstract There is no miracle or magic in medicine. Most injection therapies are used to treat sport injuries as well as the symptoms of degenerative diseases and disorders such as osteoarthritis (OA). Some injection therapies such as corticosteroids (CS) and hyaluronans (HA) are well documented for their safety and potential side effects. Some such as hydrolyzed collagens (HC) and hydrogels (HG) however are experimented in a limited number of patients and their follow-up outcomes are time limited. Still, they are broadly used to relieve pain and increase function in musculoskeletal conditions. The European Society for Sports Traumatology, Knee Surgery and Arthroscopy (ESSKA)-Orthobiologic Initiative (ORBIT) recently supported the safety and clinical efficacy of platelet rich plasma for mild to moderate knee OA. CS are strong anti-inflammatory agents that can be used in inflamed joints however as they have degenerative effects on articular joint cartilage and they are only recommended to patients that are in the waiting list for a joint replacement surgery. There is a critical time window between two injections and/or surgery. In case of CS injections, there is an increased risk of infection if the duration between injection and surgery is less than 12 months. HA is debatable for their effectiveness. About 80% of our patients re-visit the outpatient clinic in about 12 to 24 months for their regular injection. The formulations of HA are improving rapidly. Combinations are frequent however the best treatment option is not well established yet. Their

F. Korkusuz (✉)
Hacettepe University, Faculty of Medicine, Department of Sports Medicine, Altındağ, Ankara, Türkiye
e-mail: feza.korkusuz@hacettepe.edu.tr; https://avesis.hacettepe.edu.tr/feza.korkusuz

E. Üstün
Hacettepe University Faculty of Pharmacy, Ankara, Türkiye

E. Çiftçi
Department of Sports Medicine, Hacettepe University Faculty of Medicine, Ankara, Türkiye

B. B. Fidan
Department of Biomedical Engineering, Hacettepe University, Ankara, Türkiye

P. Lee
College of Health and Science, University of Lincoln, Lincoln, UK

P. Lee (ed.), *Musculoskeletal Regeneration Medicine*,
https://doi.org/10.1007/978-3-031-77456-0_9

long-lasting beneficial effect is also hardly reported in randomized controlled studies. Recent randomized clinical studies revealed that there is a place for a special formulation of HA in the treatment of degenerative and stable meniscus tears. Selected patients who refrain from forceful rotational knee activities may benefit from these injections. Concentrated HG collagens are recently introduced into injection therapies for musculoskeletal acute or chronic injuries. Relief of pain and increase in function is observed in most patients. HG on the other hand significantly decrease swelling and pain in knee OA patients. They do not recover deformity however they open a window for patients that cannot undergo joint replacement surgery. Platelet rich plasma (PRP), total stromal cells from adipose tissues (TOST) and extracellular vesicles (EV) have the potential to regenerate tissues. In this chapter, we are willing to give insight into orthobiologics and injection therapies that are frequently used in musculoskeletal conditions by including our expert opinions.

Keywords Osteoarthritis · Orthobiologics · Hyaluronan · Platelet rich plasma · Hydrogels · Peptides · Collagens · Total stromal cells from adipose tissues

1 Introduction

2 Osteoarthritis and Its Therapies

OA is a heterogeneous and complex disease with symptoms of pain, loss of function and deformity. A cascade with intervening periods of inflammation, degeneration and regeneration in and around the articular joint structures is obvious (Fig. 1). The broad and general term of OA does not determine all of the states of this chronic and disabling disease. We are aware that not all of our OA patients are the same. There are subtypes of OA determined by their genomic, proteomic and metabolomic profiles. We meet patients at a young age with secondary OA after sport injuries. An early joint replacement surgery may need revision in the future. Slowing down the degeneration process in such patients using orthobiologics is a preferred method. On the other hand, there are senior patients with too many comorbidities and the anesthesiology departments do not allow joint replacement surgery. In such cases non-surgical treatment methods should be chosen wisely. Diagnosis and progression of OA is conventionally carried out by history, physical examination and imaging. Comorbidities of OA are however frequently overlooked. The unmet challenges of OA include the early diagnosis of the onset and progression of OA, typing and staging of the disease and the monitoring of the response to treatment. Only a few applications and therapeutics can currently decrease OA symptoms and cure is seldom observed as the disease progresses with regression. Injection therapies, which are recently also named as orthobiologics, are comparatively new applications to

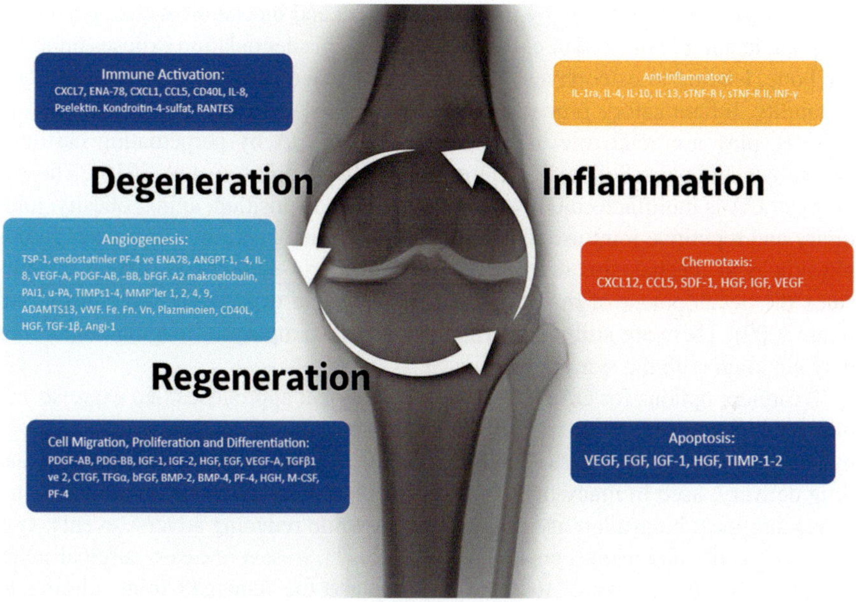

Fig. 1 The intervening cascade of inflammation, degeneration and regeneration in OA

overcome symptoms of OA. Most clinical outcomes such as decrease in pain and increase in function are usually recorded by semi-quantitative questionnaires. There is therefore an obvious need for the quantification of regeneration, if any, in OA related joint damage after such therapies. The average time to physiotherapy and/or joint replacement surgery in most European and associated countries is about 4–6 months after decision (The Economist Intelligence Unit n.d.). Injection therapies on the other hand are mostly applied in the outpatient setting to overcome OA related symptoms. The most common injections that are used to control and decrease OA symptoms are CS, HA, HC, HG and cellular injections. Cellular injections include PRP, TOST and EV. Although covered under the name of cellular injections, PRP and EV are cellular particles effective in decreasing inflammation and stimulating regeneration. Quantification of regeneration of the articular joint structures is extensively studied currently. Injection therapies are used alone or in combination as peri- and/or intra-articular injections. Although broadly applied in musculoskeletal injuries and OA, the science behind injection therapies is still evolving.

The degenerative breakdown of articular joint cartilage, subchondral bone and synovial tissue in OA determines the pathophysiology. It is the most prevalent form of arthritis, affecting millions of individuals worldwide, particularly those over the age of 60. The knee, hip and hand joints are most commonly affected by OA, leading to pain, stiffness, reduced mobility and significant impairment in the quality of life (Hunter and Bierma-Zeinstra 2019; Glyn-Jones et al. 2015).

OA is theoretically driven by mechanical stress and biochemical changes within the joint, leading to the loss of cartilage integrity. This degradation exposes underlying bone, leading to osteophyte formation, joint space narrowing and further joint instability. Inflammatory processes, although less pronounced than in rheumatoid arthritis, play a crucial role in the progression of OA by perpetuating cartilage breakdown and pain (Sellam and Berenbaum 2010; Robinson et al. 2016). The etiology of OA is multifactorial, involving genetic predisposition, aging, obesity, joint injury and repetitive joint use. Despite its widespread occurrence, the exact pathophysiological mechanisms driving OA remain poorly understood, which complicates the development of targeted therapies (Bijlsma et al. 2011; Abramson and Attur 2009). There are still unknowns for OA. In example the imaging sometimes may not align with the symptoms (Fig. 2).

Treatment options for OA range from non-surgical approaches like exercise and nonsteroidal anti-inflammatory drugs (NSAID) to more advanced therapies such as hyaluronic acid injections, PRP, cellular therapies and HG. (Table 1) Intra-articular drug delivery, used in many of these therapies, provides direct access to the joint, increasing local bioavailability of therapeutics while reducing adverse events, systemic exposure, and overall costs (Cao et al. 2021). In severe cases, surgical interventions may be necessary to restore or replace the damaged joint, aiming to improve function and quality of life (Brumat et al. 2022; Concoff et al. 2019). The

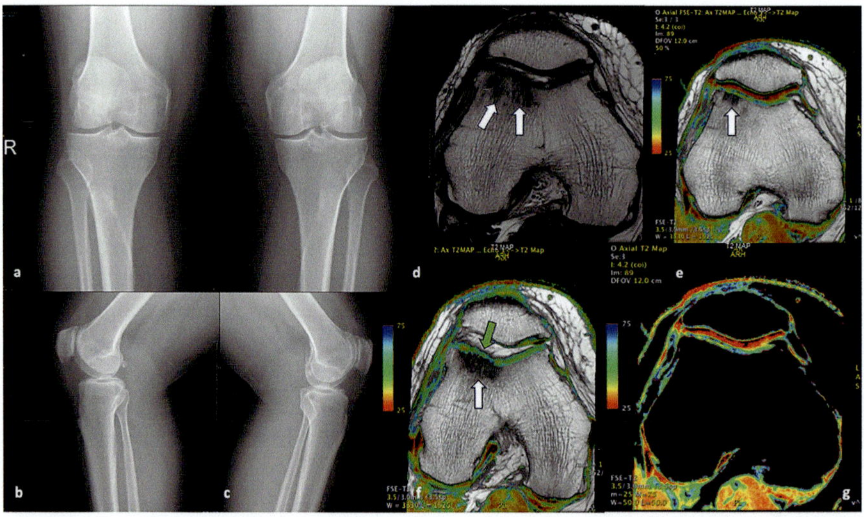

Fig. 2 X-ray (**a**, **b**, **c**), MR (**d**) and qMR (**e**, **f**, **g**) images of a 61y old male asymptomatic individual with an intense background of rowing training. HT and sub-DM are existing co-morbidities. X-ray images (**a**, **b**, **c**) can be graded as KL-0 or-1. White arrows (**d**) mark subchondral bone lesions that are not observed in X-ray pictures. There are also patellar osteophytes that indicate the presence of early degeneration. In T2 mapping (**e**), using a 3 T MR, the patellar cartilage that appears in red is normal and the articular joint cartilage of the femoral condyle that is colored in green indicates mild degeneration. Volumetric analysis and mapping software in qMR images (**f**, **g**) may therefore not predict progression of OA

Table 1 Knee OA treatment algorithm modified from the OARSI guidelines

1. Non-Surgical Exercise Weight management NSAIDs Unloading braces	**4. Cellular** PRPs Activated PRPs Stromal Vascular Fraction Mesenchymal Stem Cells Extracellular Vesicles	**5. Surgical** Micro or Nano-Fx Mosaicplasty Allografts MACI High tibial osteotomy Subchondral bioplasty Partial or total joint replacement surgery
2. **Disease Modifying** Oral GAG & CS IA Hyaluronan IA Corticosteroid IA Peptide		
3.Hydrogel Non-degradable polymeric hydrogels		

current orthobiologic treatments for OA include a variety of advanced therapies such as hyaluronic acid, PRP, stromal vascular fraction (SVF), peptides, and HG. These treatments are designed to enhance the body's ability to heal and repair joint tissues, potentially altering the disease course rather than merely alleviating symptoms (Moreno-Garcia and Rodriguez-Merchan 2022; Delanois et al. 2022).

3 Orthobiologics

Orthobiologics are therapies derived from natural biological substances, used to treat acute injuries and chronic conditions affecting muscles, bones, joints and soft tissues such as ligaments, tendons, and cartilage. The human body has a self-repair and regeneration capacity. This is also the key to homeostasis in organs and tissues. These treatments utilize cells from your own body or a donor to enhance recovery after an injury or minimize damage from degenerative conditions like OA. Orthobiologics are particularly effective in alleviating pain and other symptoms associated with early OA, potentially delaying the need for surgical intervention (Moreno-Garcia and Rodriguez-Merchan 2022). They are also employed in treating tendinopathies, such as lateral epicondylitis and serve as supportive measures in the healing process following surgical repairs of specific tendons, including those of the rotator cuff (Nelson et al. 2024; Konopitski and Malige 2022). Research on orthobiologics indicated that they are safe with minimum complications (Budhiparama et al. 2024; Lattermann et al. 2022; Sutton et al. 2023). Recently the ESSKA Orbit Consensus Report provided an informative road map to this issue (ESSKA 2022).

3.1 Corticosteroids

Intra-articular CS injections, first introduced by Dr. Hollander and colleagues in the 1950s for treating rheumatoid arthritis, have since become a widely recognized method for managing pain in various joint conditions (Hollander et al. 1951). The injection of CS into a localized painful area, such as the articular joint, synovial sheath, canal, epidural space, bursa or peritendinous region, is intended to specifically target and treat a particular joint disorder (Condé et al. 2023). The pain symptoms caused by the inflammatory processes involved in the pathogenesis of OA have included CS as an analgesic technique in this field. Clinical and epidemiological studies have established a correlation between inflammation in the knee joints of individuals with OA and the progression of cartilage damage. The potential for utilizing the immunosuppressive properties of CS to suppress this inflammatory process has sparked interest in their ability to slow the progression of knee OA. Studies addressing this hypothesis have prompted a more detailed investigation into the mechanisms of action of CS from multiple perspectives, contributing to a deeper understanding of their therapeutic potential and limitations.

CS, synthetic derivatives of steroid hormones secreted by the adrenal glands, are especially known for their anti-inflammatory properties. Intra-articular CS injections interrupt the inflammatory cascade, leading to the inhibition of the accumulation of inflammatory cells and mediators such as leukocytes, neutrophils, prostaglandins and interleukin-1. The inflammatory response is alleviated by limiting capillary permeability and preventing vasodilation and leukocyte migration. The immunomodulatory effects induced by steroids help reduce pain and local inflammation but also carry certain risks.

The alteration in gene expression induced by intra-articular CS injections is thought to lead to undesirable side effects, such as the inhibition of chondrocyte anabolic activity, decreased local collagen synthesis and additional joint damage. Systematic reviews, while highlighting some inconsistencies in in vitro and in vivo studies characterize intra-articular CS as an unfavorable treatment option for their effects on metabolism and tissue in normal human joints (Tokawa et al. 2023). A particularly noteworthy study, conducted in 2017, was a prospective, double-blind, stratified, block-randomized, placebo-controlled clinical trial that yielded striking findings. In that trial, a group of 70 individuals with symptomatic knee OA who received intra-articular CS injections every 3 months for 2 years experienced a significant increase in cartilage volume loss compared to a control group that received saline injections with the same frequency (McAlindon et al. 2017). Additionally, no notable differences in knee pain were reported between the two groups. A more recent study indicates that intra-articular CS injections may increase the risk of knee arthroplasty in patients with symptomatic knee OA or those at risk of developing it (Wijn et al. 2020). Additionally, it has been estimated that approximately 30% of patients receive intra-articular CS injections before total knee arthroplasty (TKA), and as the interval between the injection and primary TKA decreases, the risk of postoperative infection increases. However, no increase in infection risk has been observed when the interval between the injection and TKA exceeds 6–7 months (Bedard et al. 2017). In 2023, a retrospective evaluation of 682 hips with primary OA that underwent intra-articular CS injections was published, showing different findings. The evaluation concluded that the risk of periprosthetic joint infections (PJI) following total hip arthroplasty (THA) was minimal, even among patients who had received multiple injections within 3 months prior to surgery. Additionally, the study calculated the rate of rapidly progressive osteoarthritis (RPIO) to be 0.6%, a figure significantly lower than the previously reported RPIO rates of 2.8% to 21% following intra-articular CS injections in other studies (Streck et al. 2023). The need for a solid evidence base to clarify major uncertainties among the results of various studies is growing in many fields. This is particularly important because intra-articular CS injections are a common practice among athletes. The treatment remains surrounded by deficiencies and uncertainties despite their widespread use. To the best of our knowledge, the first study to examine the use of CS injections for knee pain in the triathlete population was conducted in 2023 (Norman et al. 2023).

The study reported that many triathletes described the injections as beneficial both in the short and long term. A large proportion of the triathletes were furthermore able to return to sport within 2 weeks after their injection. While this timeframe appears short compared to triathletes who opted for alternative treatments, the long-term effects may result in adverse outcomes. These outcomes may include joint cartilage damage, bone mass reduction, fascia or tendon rupture or general joint degeneration. Additionally, the fact that triathletes who received CS injections were generally older than those who pursued alternative treatments could imply that side effects may manifest more quickly, with a heightened risk of further joint damage. Therefore, athletes, especially older individuals, should be made aware of the potential side effects and counseling should be provided. Care should also be taken to avoid administering CS injections immediately after injury or just before returning to sport (Hainline et al. 2017).

In conclusion, it is evident that intra-articular CS remain controversial in terms of both efficacy and safety. National and international guidelines also reflect the lack of consensus on CS use in OA (Pavone et al. 2021). Considering the rising prevalence of OA, an increase in CS injections can be anticipated. At this point, optimizing treatment through high-quality research will be crucial. It will be essential to conduct large-scale, long-term studies that closely examine critical factors such as the types of CS, complications that may arise depending on the injection site, and the duration of their effects. In our clinical approach, we combine CS with HA in patients only once that have joint effusion and those who will receive joint replacement surgery after 12 months.

3.2 Hyaluronan (HA)

HA, also known as hyaluronic acid, is a glycosaminoglycan that is a vital component of the extracellular matrix in connective tissues, including articular joint cartilage. In the context of joint health, HA plays a critical role in maintaining the viscoelastic properties of the tissue, which is essential for lubrication (Abatangelo et al. 2020; Migliore and Procopio 2015).

About 75% of the wet weight of articular joint cartilage is synovial fluid. HA is the main link protein on which chondrocytes, chondroitin sulphate and glycoseaminoglycan are attached. The loss of HA from the articular joint cartilage means deterioration of lubrication. Injection of HA into the articular joint space slows down or stops the loss of HA and modifies the progression of the disease. In OA, the concentration and molecular weight of HA in the synovial fluid are often reduced, leading to compromised joint lubrication and increased mechanical stress on the cartilage. This exacerbates cartilage degradation and contributes to the symptomatic pain and stiffness experienced by OA patients (Ghosh and Guidolin 2002; Altman et al. 2015; Goldberg and Buckwalter 2005). Intra-articular injections of HA aim to restore the viscoelastic properties of the synovial fluid, providing symptomatic relief and potentially slowing disease progression. Its therapeutic effects include restoring viscoelastic properties, reducing inflammation by interacting with CD44 receptors on cell surfaces, and decreasing the production of pro-inflammatory cytokines like IL-1β and tumor necrosis factor-alpha (TNF-α) (Altman et al. 2015, 2019; Marinho et al. 2021). HA also provides a protective coating over the cartilage, shielding it from mechanical damage and enzymatic degradation, potentially inhibiting enzymes like matrix metalloproteinases that contribute to cartilage breakdown (Takahashi et al. 1999). Additionally, HA offers pain relief by enhancing lubrication and reducing pain receptor sensitivity through both mechanical and biochemical pathways (Altman et al. 2019).

Meta-analyses of randomized controlled trials have shown that HA injections can provide significant pain relief and functional improvement in patients with knee OA (Altman et al. 2015; Li et al. 2020; Colen et al. 2012). The therapeutic benefits, however, tend to be moderate and vary depending on the severity of the disease, the specific formulation of HA used, and the frequency of administration. Recent studies have suggested that high-molecular-weight HA may be more effective than low-molecular-weight formulations in improving clinical outcomes (Shewale et al. 2017; Bannuru et al. 2019). Furthermore, some evidence supports the long-term benefits of repeated HA injections, although the optimal dosing regimen remains to be established (McAlindon et al. 2014). The best dose and the cross-linking of the molecule issue is still debatable. Recent advancements in HA therapy for OA have focused on enhancing the efficacy and duration of its effects. Combination therapies, where HA is used alongside other orthobiologics, such as PRP or CS, are being explored to achieve synergistic benefits (Delanois et al. 2022). Additionally, novel formulations, such as cross-linked HA or HA-loaded HG, are under investigation for their potential to provide prolonged relief and better patient outcomes (Pérez et al. 2021). One recent case who had SARS-Cov2 infection experienced increase of pain and swelling after intra-articular sodium hyaluronate hexadecylamide injection for a degenerated meniscal lesion.

Platelet-Rich Plasma (PRP)

For over three decades, PRP treatments have been employed across various medical disciplines for a wide range of indications. Recently, there has been a growing interest in the potential of autologous PRP in regenerative medicine, particularly due to its ability to enhance the body's natural healing processes, promote tissue repair and regeneration. This interest is especially notable in the treatment of OA, tendinopathies and among other musculoskeletal conditions, as PRP's ability to promote healing and tissue regeneration continues to garner attention (Everts et al. 2020; Gupta et al. 2023). Platelets, also known as thrombocytes, are small, disk-shaped cell fragments in the blood that are crucial for blood clotting and wound healing. They also play important roles in repairing damaged tissue. They are produced in the bone marrow from larger cells called megakaryocytes, which release platelets into the bloodstream. A normal platelet count ranges from 150,000 to 450,000 platelets per microliter of blood. The primary function of platelets is to stop bleeding by forming clots in blood vessels. When a blood vessel is injured, platelets are quickly attracted to the site of injury. They adhere to the exposed collagen in the vessel wall, become activated, and release granules that contain substances essential for clot formation, such as ADP (adenosine diphosphate), serotonin, and thromboxane A2. These substances further recruit and activate more platelets to the site, forming a "platelet plug." This plug is then stabilized by a mesh of fibrin strands, forming a blood clot that prevents further bleeding and allows for tissue repair.

PRP is an autologous blood product that is highly concentrated with platelets, which play a crucial role in tissue repair and regeneration due to their rich content of growth factors and cytokines. PRP is derived from the patient's own blood through a process called centrifugation, which separates the blood into different components, including red blood cells, platelet-poor plasma, and platelet-rich plasma. The resulting PRP has a platelet concentration typically 4–5 times higher than that found in normal blood, making it particularly effective in enhancing the body's natural healing processes. PRP has shown promise in reducing pain and improving joint function in OA patients by promoting cartilage repair and reducing inflammation. Recent meta-analyses and randomized controlled trials (RCTs) have confirmed the efficacy of PRP injections in improving symptoms in knee OA patients, with benefits often sustained over several months. Studies moreover suggest that multiple PRP injections may provide better long-term outcomes compared to a single injection.

Function and Mechanism of Action

PRP functions primarily through the high concentration of bioactive molecules contained within the platelets. These molecules include growth factors and cytokines that play a vital role in promoting tissue repair, regeneration, and pain relief. PRP's therapeutic effects are primarily driven by the high concentration of bioactive molecules within the platelets. The key mechanisms include:

1. **Growth Factors:** PRP is rich in growth factors such as platelet-derived growth factor (PDGF), transforming growth factor-beta (TGF-β), and vascular endothelial growth factor (VEGF). These growth factors are crucial for promoting cell proliferation, differentiation, and extracellular matrix synthesis, which enhance tissue repair and regeneration by stimulating the production of new cells and blood vessels, contributing to faster healing (Chiou et al. 2018; Shen et al. 2015).
2. **Stimulation of Cartilage Repair**: Specific growth factors in PRP, including PDGF, TGF-β, and insulin-like growth factor are pivotal in promoting chondrocyte proliferation and extracellular matrix synthesis, which are crucial for cartilage repair and regeneration (Russo et al. 2016).
3. **Anti-inflammatory Effects:** PRP helps modulate the inflammatory response by reducing the production of pro-inflammatory cytokines, such as interleukin-1 (IL-1) and TNF-α. This reduction in inflammation is beneficial in mitigating tissue damage and chronic inflammatory conditions, including OA, thereby protecting cartilage from further degradation (Andia et al. 2015).

1. **Tissue Regeneration:** PRP supports the regeneration of damaged tissues by providing a concentrated source of platelets that release growth factors and cytokines. These factors enhance essential cellular activities for tissue repair, including chemotaxis, proliferation, and differentiation of cells such as fibroblasts and mesenchymal stem cells.
2. **Pain Relief:** PRP reduces inflammation and promotes tissue healing, which decreases nociceptive signals from the affected area, thereby providing pain relief. This effect is particularly valuable in managing conditions like OA, where joint pain is a significant symptom.

Types of PRP

PRP can be classified based on its composition, particularly the concentration of platelets, leukocytes and fibrin. The various types of PRP used in clinical practice include:

1. **Pure Platelet-Rich Plasma (P-PRP) / Leukocyte-Poor PRP (LP-PRP):** P-PRP or LP-PRP is characterized by a high concentration of platelets with minimal or no leukocytes. It is preferred when the goal is to reduce inflammation, making it ideal for treating chronic conditions or sensitive joints where excessive inflammation could be detrimental

2. **Leukocyte-Rich PRP (LR-PRP):** This type contains a high concentration of both platelets and leukocytes. The presence of leukocytes can enhance the immune response, potentially improving healing, but it also carries a higher risk of provoking an inflammatory reaction. LR-PRP may be particularly useful in cases where a more vigorous immune response is desired, such as in acute injuries
3. **Platelet-Rich Fibrin (PRF):** PRF includes fibrin along with platelets and leukocytes, forming a fibrin matrix that serves as a scaffold for tissue repair. This matrix supports the sustained release of growth factors, aiding in the migration of cells involved in healing processes
4. **Leukocyte-Poor PRP (LP-PRP):** A more refined version of P-PRP, LP-PRP minimizes leukocyte content to further reduce the risk of inflammatory responses. It is particularly useful for intra-articular applications where minimizing inflammation is crucial
5. **Activated PRP:** In some protocols, PRP is activated before injection using agents like calcium chloride or thrombin to initiate the immediate release of growth factors from the platelets. Activated PRP is believed to provide a more rapid therapeutic effect, which can be beneficial in certain clinical scenarios (Fig. 3).

The clinical efficacy of PRP in the treatment of OA has been demonstrated in numerous studies, showing significant improvements in pain relief and joint function (Gupta et al. 2023; Coulange Zavarro et al. 2022). It is widely debated as to which preparation is better for which structures of the musculoskeletal system. The choice between LR-PRP and LP-PRP often depends on the patient's specific

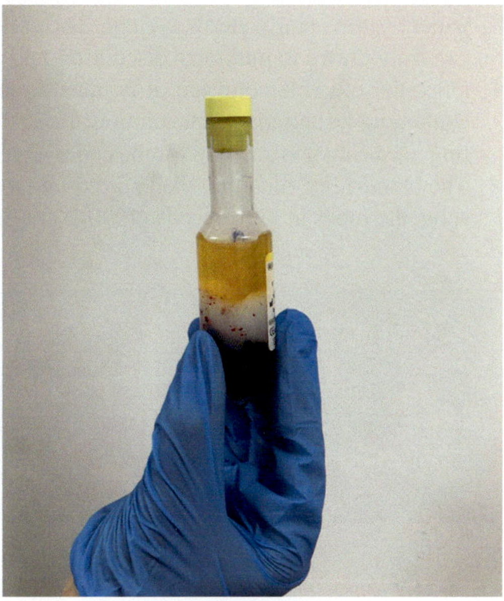

Fig. 3 PRP Activated with the beads system

condition and the desired therapeutic outcomes. For instance, LP-PRP might be preferred for patients with chronic conditions or those who are more sensitive to inflammation, while LR-PRP could be more suitable for patients with acute injuries requiring a stronger immune response (Gupta et al. 2023; Yusof and Mohamed Ali 2014). PRP therapy is generally considered safe, with almost mild or no adverse effects. Deep massage after periarticular injection while local anesthesia is present decreases localized pain or swelling at the injection site. Equimosis can be observed in patients under anti-coagulant therapy. The inclusion of leukocytes in LR-PRP can increase the risk of a more pronounced inflammatory response, while LP-PRP tends to have a milder side effect profile, making it more suitable for patients with sensitive joints or those prone to inflammation. Research and application of PRP in OA treatment are rapidly evolving. Current trends include the exploration of combination therapies, such as PRP with HA or stem cells, to enhance therapeutic outcomes (Russo et al. 2016). Advances in PRP preparation methods and the development of standardized classification systems are critical for ensuring consistent and optimal results across different clinical settings.

PRP represents a promising and versatile treatment option for OA, offering a potential disease-modifying approach that goes beyond mere symptom relief. As research continues to refine PRP formulations and protocols, it is expected that PRP will play an increasingly important role in the management of OA and other musculoskeletal disorders.

The PRP preparation involves several steps (Fig. 4) to ensure that the PRP is concentrated with growth factors and cytokines essential for healing. Below is a step-by-step guide to the PRP preparation process, accompanied by images to illustrate each step.

- In the first step, a small amount of blood, typically 30–60 ml, is drawn from the patient's arm using a sterile syringe. The collection site is prepared, and blood is carefully drawn to minimize discomfort and avoid contamination.
- Once the blood is collected, it is transferred into a specialized collection tube containing an anticoagulant solution. This solution prevents the blood from clotting, ensuring it is suitable for the centrifugation process that follows.
- The blood-filled tubes are then placed into a centrifuge machine. The centrifuge spins the tubes at high speeds (1000-3000 g), separating the blood components

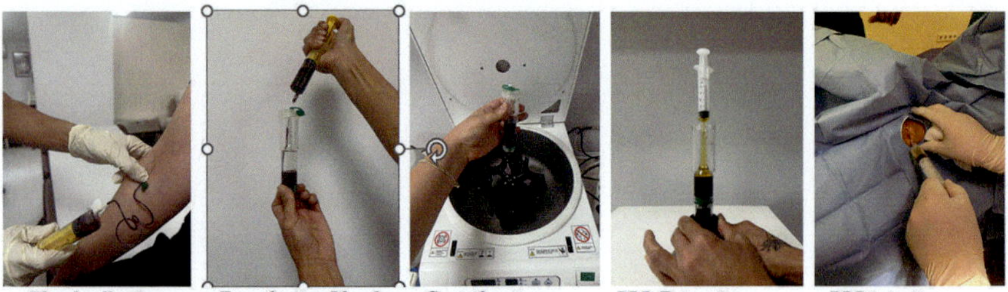
Blood collection Transferring Blood Centrifugation PRP Extraction PRP injection

Fig. 4 PRP preparation steps

based on their density. The heavier components, such as red blood cells, settle at the bottom, while the lighter components, including PRP, rise to the top.
- After centrifugation, the separated layers of blood can be clearly seen in the tube. The top layer, which is PRP, is carefully extracted using a sterile syringe. This layer is rich in growth factors and cytokines, which are key to the therapeutic properties of PRP.
- The extracted PRP is now ready for injection. It is stored in a sterile syringe and prepared under aseptic conditions to ensure safety and efficacy. The PRP can then be injected into the target area, such as joints, tendons, or other tissues, to promote healing and regeneration.

PRP Preparation Method

One of the main challenges in PRP therapy is the lack of standardization in the preparation and content of PRP. For example, variations in platelet concentration, leukocyte content, and preparation techniques can affect therapeutic outcomes. Therefore, establishing more standardized protocols is essential for achieving consistency and clinical success in PRP applications. A standardized approach would lead to more predictable and effective results across different patient population (Chahla et al. 2017).

3.3 Peptides

Peptide-based therapies are emerging as promising treatments for OA, targeting specific molecular pathways involved in inflammation, cartilage degradation, and tissue repair. Peptides, as short chains of amino acids, function as essential regulators in the body, influencing hormone regulation, cell signaling, and intercellular communication. These bioactive molecules act as key modulators, making them attractive for OA management (Kesiktas et al. 2020; Martínez-Puig et al. 2023). Peptides used in OA treatment operate through various mechanisms. One of their primary actions is reducing inflammation. Peptides like Thymosin Beta-4 (Tβ4) exhibit strong anti-inflammatory properties by inhibiting pro-inflammatory cytokines and enzymes like matrix metalloproteinases (MMPs), which are responsible for cartilage breakdown (Chahla et al. 2017; Martínez-Puig et al. 2023). Moreover, peptides such as BMP-7 and Growth Differentiation Factor 5 (GDF-5) stimulate cartilage regeneration by promoting chondrocyte activity, aiding in the repair and maintenance of damaged cartilage (Kesiktas et al. 2020). In addition to their effects on inflammation and cartilage regeneration, peptides can provide pain relief by modulating pain pathways, offering a natural analgesic effect without the risks associated with opioid medications (Kesiktas et al. 2020; Lin et al. 2023). Some peptides, like Parathyroid Hormone-related Peptide (PTHrP), also contribute to the health of subchondral bone, the bone layer beneath the cartilage, thus supporting joint integrity and potentially slowing OA progression. This multifaceted approach makes peptides a unique therapeutic option for OA (Kesiktas et al. 2020).

Recent research highlights peptides' efficacy in OA treatment. A 2023 study demonstrated that Thymosin Beta-4 injections significantly reduced pain and improved joint function in patients with knee OA (Lin et al. 2023). In another study from 2022, daily collagen peptide supplementation improved cartilage quality, as seen through MRI scans, and reduced markers of cartilage degradation (Martínez-Puig et al. 2023). Additionally, a 2021 study by Edwin Lee and Blake Padgett evaluated the effects of BPC-157 injections in knee pain management. The results showed significant improvement in 91.6% of patients receiving only BPC-157, further emphasizing the peptide's potential in OA therapy (Lee and Padgett 2021).

Peptides represent a novel approach to managing OA by targeting inflammation, supporting cartilage repair, and relieving pain. Peptides show great potential for cartilage regeneration by promoting hyaline cartilage formation. Their versatility and low toxicity make them valuable candidates. Some peptides independently drive chondrogenesis, while others work with growth factors. However, more controlled in vivo studies are needed to confirm their clinical efficacy (Ajeeb et al. 2022). Their safety profile, short half-life, and effectiveness across multiple biological pathways make them a highly promising therapeutic option for OA management.

3.4 *Hydrolyzed Collagen*

Collagen is a protein that plays a pivotal role in maintaining the structural integrity of articular cartilage. It is primarily composed of three alpha chains that form a triple helix, consisting of glycine (33%), proline, and hydroxyproline (22%) amino acids (primary structure) (Gelse et al. 2003). In addition to providing elasticity and strength, collagen also serves as a ligand for the OSCAR (Osteoclast-associated Receptor). The potential to modulate the activity of this receptor, which is associated with bone homeostasis, opens new avenues for the treatment of OA (Nedeva et al. 2021). There has been a significant increase in the number of studies investigating the effects of collagen derivatives on OA between 2000 and 2020.

The use of collagen as an oral treatment is supported by studies demonstrating a significant safety and efficacy profile in managing various joint diseases (Santos et al. 2023). The majority of studies examining collagen from various sources at different doses have reported that collagen exhibits chondroprotective activity (Honvo et al. 2020). It is also indicated that HC provides better absorption and bioavailability compared to undenatured native collagen (Felim et al. 2022). HC, or collagen hydrolysate, is produced through the enzymatic hydrolysis of natural collagen. Through various hydrolysis processes, collagen peptides with very low molecular weights, as low as 3 kDa, are derived from natural collagen, which initially has a molecular weight of approximately 300 kDa. In addition to not requiring a stepwise or complex extraction procedure, the extraction of HC is considered satisfactory in terms of biological efficacy. It is important to emphasize that the solubility and functional activity of HC are dependent on the type and degree of hydrolysis, as well as the specific enzymes used throughout the process (León-López et al. 2019). Collagen injections for OA appear to offer the advantage of reliable efficacy with negligible side effects. A "solution of low molecular weight peptides derived from hydrolyzed collagen" has indicated that no side effects have been reported in relation to the injection of low molecular weight peptides. However, this product is contraindicated in the presence of joint or peri-articular infections, hemarthrosis, erythema, or psoriatic lesions near the injection sites.

This product is a medical device indicated for the direct reinforcement of the extracellular matrix of connective tissues, which is naturally rich in collagen fibers. Its formulation is also recommended for improving articular mobility and managing pain symptoms in joints and surrounding connective structures (muscle, tendon, ligament, menisci) caused by overload and postural defects. Additionally, it is indicated after joint surgeries (meniscectomy, ligament reconstruction, articular cartilage repair) to accelerate functional recovery, improve postoperative comfort, and counteract the physiological aging and/or degeneration of connective and epithelial tissues (from PEPTYS).

Several authors highlight the cost-effective nature of HC, along with its biocompatibility, tolerable side effects, and chondroprotective properties. The authors stated that the reduced cost of collagen compared to low- or high-molecular weight HA could allow a greater use of intra-articular collagen as an alternative therapeutic option to HA for OA, resulting in an NSAID intake reduction, as well as social cost reduction due to working days lost and caregivers' time off work. Despite its promising therapeutic and supportive potential, it should be noted that the use of HC for intra-articular injections in OA is still in its early stages. To the best of our knowledge, the first study that examines the use of intra-articular collagen injections for OA in humans was conducted in 2023. The short-term and small-scale nature of other studies in this area presents a significant gap (Honvo et al. 2020). In most of the clinical trials, treatment durations varied between 3 and 6 months, with the shortest being 1.4 months and the longest 11 months.

All of these parameters might, in fact, be relevant for future meta-analyses, particularly in the investigation of possible sources of inconsistent results from clinical trials.

In summary, current evidence suggests that HC may represent a viable option for the management of OA patients. However, further advancements in both laboratory and clinical research are essential to draw definitive conclusions.

3.5 Hydrogels

HG have emerged as an innovative approach in managing OA, offering the potential to alleviate pain, provide structural support, and promote tissue regeneration. HG are biomaterials designed to mimic the natural extracellular matrix of cartilage. As a three-dimensional polymer network, resembling the extracellular matrix, it facilitates cellular proliferation and supports cartilage healing. Their highly absorbent nature allows them to hold water, enhancing lubrication and reducing joint friction, which is particularly beneficial in the context of OA. When injected, HG integrates with the synovial membrane, providing structural support and enhancing the elasticity of the joint capsule. This scaffold-like structure promotes the proliferation of synovial cells, facilitating the production of synovial fluid and improving joint lubrication.

Polyacrylamide hydrogels (PAAG), a cross-linked HG, is non-biodegradable and remains structurally stable after injection. Unlike biodegradable materials, PAAG remains intact over time, offering continuous support to the damaged joint. This enhances the elasticity of the synovium but also forms a protective barrier against inflammatory cytokines and enzymes that degrade cartilage, effectively slowing the progression of OA. The anti-inflammatory effects further contribute to pain relief and improved joint mobility in patients.

PAAG combines with the synovial tissue, providing structural stability, reducing inflammation, and acting as a barrier against cartilage-degrading cytokines and enzymes. The material's ability to enhance synovial elasticity and promote cellular proliferation in the synovium makes it effective in long-term OA treatment (Vaca-González et al. 2023).

It supports cellular proliferation by creating a microenvironment conducive to tissue growth while reducing the influx of inflammatory cytokines that contribute to cartilage degradation in the treatment of advanced OA (Bliddal et al. 2021, 2024; Atwal et al. 2023). This environment also shields the joint from inflammatory mediators, which can otherwise accelerate cartilage degradation.

Studies have shown that HG injections can significantly reduce pain and improve joint function by disrupting the cycle of inflammation commonly seen in OA. By distancing inflammatory cells from the affected joint surfaces, HG helps to prevent further degradation of cartilage and subchondral bone.

The **15-Step** injection protocol outlines the systematic approach utilized by focusing on precision and safety during musculoskeletal injections, particularly for treatments involving Arthrosamid and MSK-P1 Mix.

1. Before proceeding, ensure that the patient is free from active infections. Blood clotting must be controlled, with an INR (International Normalized Ratio) below 4. There is no requirement to stop any medications before the procedure.
2. Obtain informed consent from the patient.
3. Position and prepare the patient by ensuring the area is exposed with no obstruction.
4. Administer IV antibiotics to prevent infection:
    - Augmentin 1.2 g or
    - Cefuroxime 750 mg (as an alternative if the patient has allergies or contraindications to Augmentin).
5. Use an ultrasound scanner (USS) to identify and mark the injection site, which may be: Supra-patellar, medial, lateral or all three regions. Clean the skin with

Hydrex 0.5% Chlorhexidine or Chloraprep. Apply a sterile drape around the area, either with a hole drape or a square drape to isolate the injection site.

6. Use Local Anaesthetic (LA) to numb the area and reduce discomfort: Inject 2% Lidocaine mixed with 0.1 ml adrenaline (to a total of 5 ml per injection site) using: 25G orange needle initially, followed by a 23G blue needle.
7. While waiting for the local anaesthetic to take effect: Prepare Arthrosamid: Open all Arthrosamid boxes. Unwrap the packages and untwist the tops of all six syringes.

   Set up the MSK-P1 Mix: Combine the following for use during the procedure:

   (a) 4 ml of 0.25% Marcaine,
   (b) 20 mg of Kenalog (half a vial),
   (c) 4 ml of 1% lidocaine,
   (d) 0.1 ml of adrenaline.

8. Change into sterile gloves and reclean the injection area with Hydrex 0.5% Chlorhexidine or Chloraprep, ensuring thorough sterility before proceeding.
9. Clean the ultrasound scanner and apply sterile gel to the probe. Then place a sterile cover over the scanner to avoid contamination during the procedure.
10. Using a Pink 18G needle, inject 5–8 ml of MSK-P1 Mix under ultrasound guidance (USS). After the initial injection, leave the needle in place to facilitate the transition to the next step.
11. Switch the syringe attached to the same needle for the Arthrosamid injection. Inject the Arthrosamid slowly, using the same needle track to ensure the accurate placement of the medication. Repeat the same process for each marked injection site.
12. After the Arthrosamid injection, flush the needle with 2 ml of MSK-P1 Mix to clear the needle and ensure complete delivery of the medication.
13. Withdraw the needle carefully and immediately apply pressure to the injection site to prevent hematoma formation. Use an appropriate sterile dressing to cover the site.
14. Instruct the patient to take NSAIDs (Non-Steroidal Anti-Inflammatory Drugs) for 7 days to reduce inflammation and lower the risk of infection. This is an important step in managing post-injection discomfort and preventing complications.
15. Upon discharge, provide the patient with a discharge letter detailing post-procedure care instructions, including any follow-up appointments or additional recommendations.

Despite their many benefits, HG like PAAG have potential side effects, although they are generally well-tolerated. Commonly reported adverse effects include mild swelling, temporary joint stiffness, and localized pain at the injection site. These reactions are typically short-lived and resolve without serious complications. More significant adverse events, such as severe inflammation or infection, are rare. In terms of their long-term safety profile, PAAG has shown minimal risk, making it an attractive option for OA management. However, careful patient selection is important to minimize any potential complications, especially in high-risk individuals (Fig. 5).

HG therapy, particularly with PAAG, offers long-lasting benefits with minimal intervention. In many cases, a single injection can significantly improve pain and joint function for up to 12 months, with follow-up studies confirming its effects for as long as 52 weeks. This contrasts with other treatments that often require repeated administrations over a shorter time frame. The sustained impact of HG is attributed to their stable, non-biodegradable structure, which allows them to remain effective within the joint environment over an extended period.

Research shows that HG injections help distance inflammatory cells, effectively breaking the cycle of inflammation that exacerbates OA symptoms. This function is vital for reducing pain and improving mobility in OA patients (Bliddal et al. 2024).

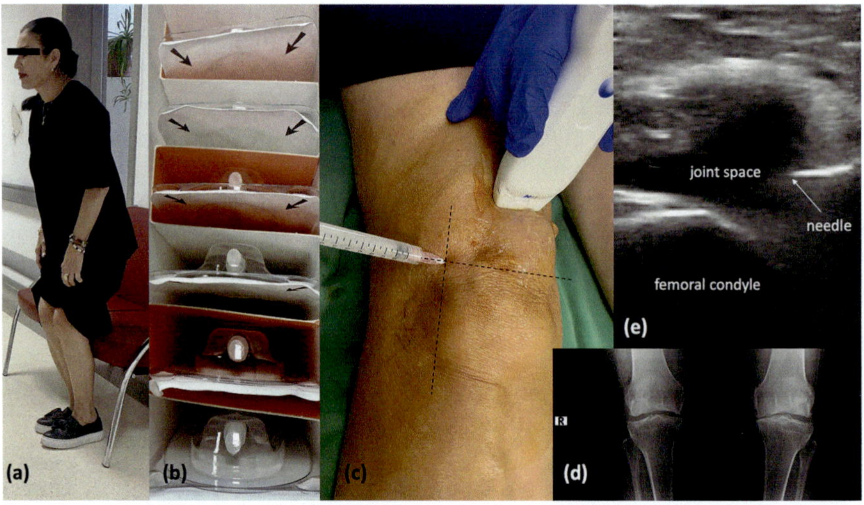

Fig. 5 Before HG injection, (**a**) patients undergo repeated (x3) sit and stand tests that are evaluated by the motion analysis MAI technology. The HG (**b**) comes as six separate 1.0 ml single-use disposable injections. The superior-lateral portal is anesthetized by 2.5 ml 2% prilocaine hydrochloride injection 10 min after dermal sanitation using 10% polyvinyl iodine (ALG, Istanbul, Turkey). (**c**) A 18G needle is used to reach the joint space. Synovial fluid, if any, is removed before the HG injection. Removal of the synovial fluid proves that the needle is in the joint space. X-ray pictures (**d**) of this 55y female patient revealed bilateral KL grade 3 to 4 secondary knee OA due to partial meniscectomy 17 years ago. Ultrasound (**e**) recorded the tip of the needle and injection of the HG into the joint space. 1.0 mg of second-generation cephalosporin is administered IV during the procedure to prevent infection. Small joint spaces as in females require the IV administration of 20 mg tenoxicam administration

Total Stromal Cells from Adipose Tissues (TOST)

SVF is derived from adipose tissue and contains a mixture of cells, including mesenchymal stem cells, which have the potential to differentiate into cartilage cells. When injected into the joint, SVF may contribute to the repair of damaged cartilage by promoting regeneration and modulating the inflammatory environment within the joint (Fig. 6).

The main idea behind TOST is the transfer of pericytes together with adipocytes that have a stem cell potential into the damaged tissue together with their environment. These cells are pro-angiogenic and have a chondrogenic potential. The optimum number of cells are frequently studied however recent work suggest that not only the cell number but also the environment is critical for their implementation. As a general rule, enzymatical separation of the cells is recently avoided due to potential complications. TOST can be combined with PRP for the homing of cells in the applied tissues. The age and sex of the patient as well as co-morbidities including obesity and diabetes can affect the outcome of cellular efficiency. Combining TOST and PRP is a good option in modifying the course of degeneration in young patients who have secondary OA mostly due to trauma.

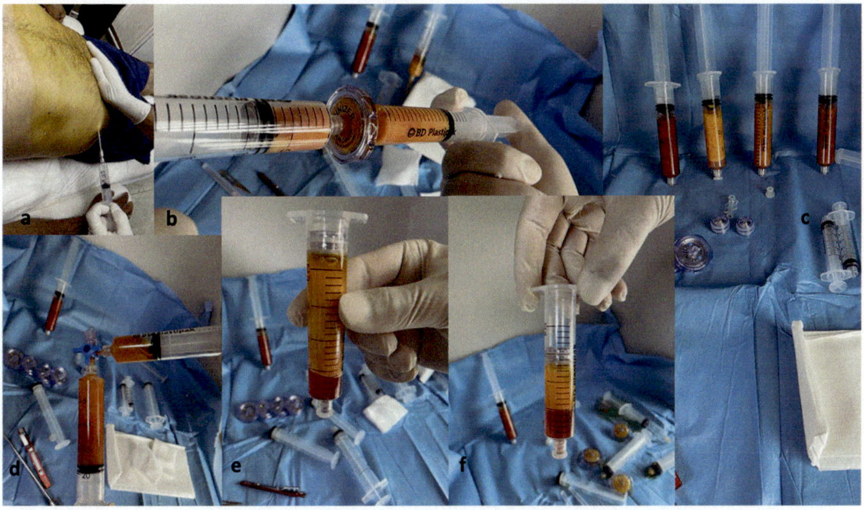

Fig. 6 TOST preparation steps: (**a**) liposuction, (**b**) gradual mechanical nano-fat extraction, (**c**) lipid fractions, (**d**) preparing for centrifugation, (**e**) first and (**f**) second centrifugation

4 Discussion

Joint degeneration, particularly OA, is a prevalent condition associated with significant morbidity and recently mortality. Injection therapies have gained attention as treatment modalities even the pathophysiology is less identified. These therapies aim to decrease pain, restore joint function and regenerate damaged tissues. Among the most common injectable treatments are CS, HA, HC, HG, PRP, TOST, EV. Each of these interventions brings unique mechanisms of action, efficacy profiles, and potential drawbacks.

CS are one of the earliest injectable therapies used for joint pain management, particularly in OA. They exhibit potent anti-inflammatory effects by inhibiting the phospholipase A2 enzyme and reducing the synthesis of pro-inflammatory mediators such as prostaglandins and leukotrienes. CS injections provide rapid relief of pain and swelling in the clinical setting. Therapeutic effects of CS however are typically short-lived, lasting from a few weeks to a couple of months. Repeated injections have been associated with degenerative effects, including cartilage damage, subchondral bone loss and a paradoxical increase in inflammation in some cases. While CS are effective for short-term symptom relief, they are not considered as regenerative or long-term therapeutics in joint degeneration. They are mostly limited to inflammatory conditions and for patients in the waiting list of joint replacement.

HA is a key component of articular joint cartilage that plays a crucial role in maintaining joint viscosity and facilitating gliding of the joint that is also named as lubrication. In OA, the quality and quantity of HA in the joint are reduced, contributing to stiffness and pain. HA injections, also known as viscosupplementation aim to restore synovial fluid viscoelasticity, thereby improving joint lubrication and potentially reducing pain. HA injections have been found to provide moderate pain relief, particularly in the knee joint. Benefits however are not immediate and usually manifest over several weeks, with the effects lasting for up to 6 months. While HA injections are well-tolerated and have a good safety profile, their regenerative capacity is limited. They are primarily used as a palliative treatment rather than a disease-modifying therapy. Combining them with CS is also a recent alternative.

Collagens are major structural proteins of the articular joint cartilage that is essential for maintaining cartilage integrity. Collagen injections, particularly those derived from bovine or porcine sources, are gaining interest as potential therapeutic agents for joint degeneration. These injections aim to provide the necessary building blocks for cartilage repair and regeneration. Preclinical studies suggest that collagen can promote chondrocyte activity and reduce cartilage degradation while their usage is still in its early stages. Collagen's ability to modulate inflammation and contribute to extracellular matrix remodeling makes it a promising candidate for further investigation. However, clinical data on the efficacy of collagen injections in large patient populations are currently limited.

HG, composed of highly hydrated polymer networks, have gained attention as potential delivery systems for joint therapeutics. Their biocompatibility and ability to mimic the extracellular matrix environment make them ideal scaffolds for cartilage repair. HG can be used as carriers for bioactive molecules, such as growth factors or stem cells, facilitating the targeted delivery of regenerative agents. In the context of OA and joint degeneration, HG have been investigated for their ability to promote cartilage regeneration and improve joint function. Early studies suggest that HG injections, particularly when combined with cells or growth factors, can enhance cartilage repair and reduce OA symptoms. Nevertheless, more robust clinical trials are needed to establish the long-term efficacy and safety of hydrogel-based therapies.

PRP injections have gained widespread popularity in recent years due to their autologous nature and potential for tissue regeneration. PRP is derived from the patient's blood and contains a high concentration of platelets, which are rich in growth factors, cytokines, and other bioactive molecules. These factors are thought to promote tissue repair, reduce inflammation and enhance healing in degenerative joint conditions. PRP has shown promise in the treatment of various musculoskeletal conditions, including OA, tendinopathies, and ligament injuries. In OA, PRP injections have been reported to reduce pain and improve joint function, particularly in early to moderate stages of the disease. However, variability in PRP preparation methods and the lack of standardized protocols have led to inconsistent results across studies. Research continues to explore the optimal PRP formulations and injection protocols that can maximize therapeutic efficacy.

TOST refers to a heterogeneous cell population derived from adipose tissue, containing adipose-derived stem cells (ADSCs), endothelial cells, pericytes, and immune cells. TOST injections are an emerging treatment modality for joint degeneration due to their regenerative potential. ADSCs within the TOST can differentiate into various cell types, including chondrocytes, making them attractive candidates for cartilage repair. Intra-articular TOST injections have demonstrated promising results in preclinical and early-phase clinical studies, showing improvements in pain relief, cartilage regeneration, and joint function. Immunomodulatory properties of TOST may help reduce inflammation in the joint, contributing to symptom relief. However, the use of TOST in clinical practice is still under investigation. Further studies are needed to determine optimal dosing, safety, and long-term efficacy.

EV, including exosomes and microvesicles, are small membrane-bound particles released by cells that carry proteins, lipids, nucleic acids, and other bioactive molecules. In recent years, EV have emerged as potential mediators of intercellular communication and tissue repair. In the context of joint degeneration, EV derived from stem cells or chondrocytes have shown promise in promoting cartilage regeneration and modulating inflammation.

EV offer several advantages over cell-based therapies, including reduced risk of immune rejection and tumor formation. Additionally, EV can be engineered to deliver specific therapeutic molecules, such as anti-inflammatory cytokines or growth factors, directly to the joint. Although research on EV injections for OA is still in its infancy, early studies indicate that they may provide significant benefits in terms of pain reduction, cartilage repair, and slowing disease progression.

5 Conclusion

Injection therapies also named as orthobiologics represent a rapidly evolving field in the management of OA. By harnessing the body's natural healing processes, these therapies offer the potential to not only alleviate symptoms but also to modify the underlying disease processes of OA. Nevertheless, numerous questions arising during the application of orthobiologics remain unanswered (Rodeo 2023). Continued research is essential to fully understand the mechanisms of action, optimize treatment protocols, and determine the long-term benefits of orthobiologic therapies in OA management. The landscape of injectable therapies for joint degeneration is rapidly evolving, with a wide range of biological treatments being explored. While CS and HA have long been in clinical practice, emerging treatments such as PRP, TOST, and EVs offer the potential for not only symptom relief but also joint regeneration. HG and collagen-based injections represent exciting avenues for scaffold-based cartilage repair. Despite their promise, many of these therapies are still in the experimental stage. Long-term patient data on their efficacy and safety are needed. As the field advances, the development of standardized protocols, along with a better understanding of the underlying mechanisms of joint degeneration, will be crucial for optimizing the use of these therapies and achieving better patient outcomes. Future research will likely focus on combination therapies, personalized medicine approaches, and the integration of injectable treatments with other modalities, such as physical therapy and pharmacological interventions, to maximize therapeutic benefits.

Acknowledgement Authors thank to İbrahim Vargel MD, PhD and plastic and reconstructive surgeon, for the harvesting and preparation of the TOST in this case study. Feza Korkusuz MD is an active member of the Turkish Academy of Sciences (TUBA). The Scientific and Technological Research Council of Türkiye (TÜBİTAK) funded this project (#222S093).

References

Abatangelo G, Vindigni V, Avruscio G, Pandis L, Brun P. Hyaluronic acid: redefining its role. Cells. 2020;9(7):1743.

Abramson SB, Attur M. Developments in the scientific understanding of osteoarthritis. Arthritis Res Ther. 2009;11:1–9.

Ajeeb B, Acar H, Detamore MS. Chondroinductive peptides for cartilage regeneration. Tissue Eng Part B Rev. 2022;28(4):745–65.

Altman R, Manjoo A, Fierlinger A, Niazi F, Nicholls M. The mechanism of action for hyaluronic acid treatment in the osteoarthritic knee: a systematic review. BMC Musculoskelet Disord. 2015;16:1–10.

Altman R, Bedi A, Manjoo A, Niazi F, Shaw P, Mease P. Anti-inflammatory effects of intra-articular hyaluronic acid: a systematic review. Cartilage. 2019;10(1):43–52.

Andia I, Rubio-Azpeitia E, Maffulli N. Platelet-rich plasma modulates the secretion of inflammatory/angiogenic proteins by inflamed tenocytes. Clin Orthop Relat Res. 2015;473:1624–34.

Atwal A, Dale TP, Snow M, Forsyth NR, Davoodi P. Injectable hydrogels: an emerging therapeutic strategy for cartilage regeneration. Adv Colloid Interf Sci. 2023;321:103030.

Bannuru RR, Osani M, Vaysbrot E, Arden N, Bennell K, Bierma-Zeinstra S, et al. OARSI guidelines for the non-surgical management of knee, hip, and polyarticular osteoarthritis. Osteoarthr Cartil. 2019;27(11):1578–89.

Bedard NA, Pugely AJ, Elkins JM, Duchman KR, Westermann RW, Liu SS, et al. The John N. Insall award: do intraarticular injections increase the risk of infection after TKA? Clin Orthop Relat Res. 2017;475:45–52.

Bijlsma JW, Berenbaum F, Lafeber FP. Osteoarthritis: an update with relevance for clinical practice. Lancet. 2011;377(9783):2115–26.

Bliddal H, Overgaard A, Hartkopp A, Beier J, Conaghan P, Henriksen M. Polyacrylamide hydrogel injection for knee osteoarthritis: a 6 months prospective study. J Orthop Res Ther. 2021;6(2):S278.

Bliddal H, Beier J, Hartkopp A, Conaghan PG, Henriksen M. Effectiveness and safety of polyacrylamide hydrogel injection for knee osteoarthritis: results from a 12-month follow up of an open-label study. J Orthop Surg Res. 2024;19(1):274.

Brumat P, Kunšič O, Novak S, Slokar U, Pšenica J, Topolovec M, et al. The surgical treatment of osteoarthritis. Life. 2022;12(7):982.

Budhiparama NC, Putramega D, Lumban-Gaol I. Orthobiologics in knee osteoarthritis, dream or reality? Arch Orthop Trauma Surg. 2024;1-10:3937–46.

Cao Y, Ma Y, Tao Y, Lin W, Wang P. Intra-articular drug delivery for osteoarthritis treatment. Pharmaceutics. 2021;13(12):2166.

Chahla J, Cinque ME, Piuzzi NS, Mannava S, Geeslin AG, Murray IR, et al. A call for standardization in platelet-rich plasma preparation protocols and composition reporting: a systematic review of the clinical orthopaedic literature. JBJS. 2017;99(20):1769–79.

Chiou C-S, Wu C-M, Dubey NK, Lo W-C, Tsai F-C, Tung TDX, et al. Mechanistic insight into hyaluronic acid and platelet-rich plasma-mediated anti-inflammatory and anti-apoptotic activities in osteoarthritic mice. Aging (Albany NY). 2018;10(12):4152–65.

Colen S, Van Den Bekerom MP, Mulier M, Haverkamp D. Hyaluronic acid in the treatment of knee osteoarthritis: a systematic review and meta-analysis with emphasis on the efficacy of different products. BioDrugs. 2012;26:257–68.

Concoff A, Rosen J, Fu F, Bhandari M, Boyer K, Karlsson J, et al. A comparison of treatment effects for nonsurgical therapies and the minimum clinically important difference in knee osteoarthritis: a systematic review. JBJS Rev. 2019;7(8):e5.

Condé K, Adjakou F, Barry A, Touré M, Wouaseha RLF, Cissé FA. Assessment of the effectiveness of corticosteroid infiltrations in the Ignace Deen University Hospital Center in Conakry. Cureus. 2023;15(11):e49555.

Coulange Zavarro A, De Girolamo L, Laver L, Sánchez M, Tischer T, Filardo G, et al. The top 100 most cited articles on platelet-rich plasma use in regenerative medicine—a bibliometric analysis—from the ESSKA orthobiologic initiative. Bioengineering. 2022;9(10):580.

Delanois RE, Sax OC, Chen Z, Cohen JM, Callahan DM, Mont MA. Biologic therapies for the treatment of knee osteoarthritis: an updated systematic review. J Arthroplast. 2022;37(12):2480–506.

ESSKA Orthobiologic initiative (ORBIT): consensus process on blood derived products [press release]. 2022.

Everts P, Onishi K, Jayaram P, Lana JF, Mautner K. Platelet-rich plasma: new performance understandings and therapeutic considerations in 2020. Int J Mol Sci. 2020;21(20):7794.

Felim J, Chen C-K, Tsou D, Kuo H-P, Kong Z-L. Effect of different collagen on anterior cruciate ligament transection and medial meniscectomy-induced osteoarthritis male rats. Front Bioeng Biotechnol. 2022;10:917474.

Gelse K, Pöschl E, Aigner T. Collagens–structure, function, and biosynthesis. Adv Drug Deliv Rev. 2003;55(12):1531–46.

Ghosh P, Guidolin D, editors. Potential mechanism of action of intra-articular hyaluronan therapy in osteoarthritis: are the effects molecular weight dependent? Seminars in arthritis and rheumatism. Elsevier; 2002.

Glyn-Jones S, Palmer A, Agricola R, Price AJ, Vincent TL, Weinans H, Carr AJ. Osteoarthritis. Lancet. 2015;386(9991):376–87.

Goldberg V, Buckwalter J. Hyaluronans in the treatment of osteoarthritis of the knee: evidence for disease-modifying activity. Osteoarthr Cartil. 2005;13(3):216–24.
Gupta A, Potty AG, Maffulli N. Allogenic platelet-rich plasma for treatment of knee and hip osteoarthritis. Front Pain Res. 2023;4:1216190.
Hainline B, Derman W, Vernec A, Budgett R, Deie M, Dvořák J, et al. International Olympic Committee consensus statement on pain management in elite athletes. Br J Sports Med. 2017;51(17):1245–58.
Hollander JL, Brown EM, Jessar RA, Brown CY. Hydrocortisone and cortisone injected into arthritic joints: comparative effects of and use of hydrocortisone as a local antiarthritic agent. JAMA J Am Med Assoc. 1951;147(17):1629–35.
Honvo G, Lengelé L, Charles A, Reginster J-Y, Bruyère O. Role of collagen derivatives in osteoarthritis and cartilage repair: a systematic scoping review with evidence mapping. Rheumatol Ther. 2020;7:703–40.
Hunter D, Bierma-Zeinstra S. Osteoarthritis. Lancet. 2019;393:1745–59.
Kesiktas FN, Dernek B, Sen EI, Albayrak HN, Aydin T, Yildiz M. Comparison of the short-term results of single-dose intra-articular peptide with hyaluronic acid and platelet-rich plasma injections in knee osteoarthritis: a randomized study. Clin Rheumatol. 2020;39:3057–64.
Konopitski A, Malige A. Review of ortho-biologics in rotator cuff repair. Shoulder surgery for RC pathology, arthropathy and tumors. IntechOpen; 2022.
Lattermann C, Leite CB, Frisbie DD, Schlegel TS, Bramlage LR, Koch T, et al. Orthobiologics in orthopedic applications: a report from the TMI Havemeyer meeting on orthobiologics. J Cartil Jt Preserv. 2022;2(3):100055.
Lee E, Padgett B. Intra-articular injection of BPC 157 for multiple types of knee pain. Altern Ther Health Med. 2021;27(4):8–13.
León-López A, Morales-Peñaloza A, Martínez-Juárez VM, Vargas-Torres A, Zeugolis DI, Aguirre-Álvarez G. Hydrolyzed collagen—sources and applications. Molecules. 2019;24(22):4031.
Li C, Cao Z, Li W, Liu R, Chen Y, Song Y, et al. A review on the wide range applications of hyaluronic acid as a promising rejuvenating biomacromolecule in the treatments of bone related diseases. Int J Biol Macromol. 2020;165:1264–75.
Lin C-R, Tsai SHL, Huang K-Y, Tsai P-A, Chou H, Chang S-H. Analgesic efficacy of collagen peptide in knee osteoarthritis: a meta-analysis of randomized controlled trials. J Orthop Surg Res. 2023;18(1):694.
Marinho A, Nunes C, Reis S. Hyaluronic acid: a key ingredient in the therapy of inflammation. Biomolecules. 2021;11(10):1518.
Martínez-Puig D, Costa-Larrión E, Rubio-Rodríguez N, Gálvez-Martín P. Collagen supplementation for joint health: the link between composition and scientific knowledge. Nutrients. 2023;15(6):1332.
McAlindon TE, Bannuru RR, Sullivan M, Arden N, Berenbaum F, Bierma-Zeinstra S, et al. OARSI guidelines for the non-surgical management of knee osteoarthritis. Osteoarthr Cartil. 2014;22(3):363–88.
McAlindon TE, LaValley MP, Harvey WF, Price LL, Driban JB, Zhang M, et al. Effect of intra-articular triamcinolone vs saline on knee cartilage volume and pain in patients with knee osteoarthritis: a randomized clinical trial. JAMA. 2017;317(19):1967–75.
Migliore A, Procopio S. Effectiveness and utility of hyaluronic acid in osteoarthritis. Clin Cases Miner Bone Metab. 2015;12(1):31.
Moreno-Garcia A, Rodriguez-Merchan EC. Orthobiologics: current role in orthopedic surgery and traumatology. Arch Bone Joint Surg. 2022;10(7):536.
Nedeva IR, Vitale M, Elson A, Hoyland JA, Bella J. Role of OSCAR signaling in osteoclastogenesis and bone disease. Front Cell Deve Biol. 2021;9:641162.
Nelson PA, George T, Bowen E, Sheean AJ, Bedi A. An update on Orthobiologics: cautious optimism. Am J Sports Med. 2024;52(1):242–57.
Norman MB, Norman ER, Langer GH, Allen MR, Meller L, Vitale KC. Return to sport using corticosteroid injections for knee pain in triathletes. Cureus. 2023;15(6):e39985.

Pavone V, Vescio A, Turchetta M, Giardina SMC, Culmone A, Testa G. Injection-based management of osteoarthritis of the knee: a systematic review of guidelines. Front Pharmacol. 2021;12:661805.

Pérez LA, Hernández R, Alonso JM, Pérez-González R, Sáez-Martínez V. Hyaluronic acid hydrogels crosslinked in physiological conditions: synthesis and biomedical applications. Biomedicine. 2021;9(9):1113.

Robinson WH, Lepus CM, Wang Q, Raghu H, Mao R, Lindstrom TM, et al. Low-grade inflammation as a key mediator of the pathogenesis of osteoarthritis. Nat Rev Rheumatol. 2016;12(10):580–92.

Rodeo SA. Orthobiologics: current status in 2023 and future outlook. JAAOS-J Am Acad Orthop Surg. 2023;31(12):604–13.

Russo F, D'Este M, Vadalà G, Cattani C, Papalia R, Alini M, et al. Platelet rich plasma and hyaluronic acid blend for the treatment of osteoarthritis: rheological and biological evaluation. PLoS One. 2016;11(6):e0157048.

Santos DR, Xavier DP, Ataíde LAP, Bentes LGB, Lemos RS, Giubilei DB, et al. The effects of hydrolyzed collagen and collagen peptide in the treatment of superficial chondral lesions: an experimental study. Rev Bras Ortop. 2023;58(1):72–8.

Sellam J, Berenbaum F. The role of synovitis in pathophysiology and clinical symptoms of osteoarthritis. Nat Rev Rheumatol. 2010;6(11):625–35.

Shen J, Gao Q, Zhang Y, He Y. Autologous platelet-rich plasma promotes proliferation and chondrogenic differentiation of adipose-derived stem cells. Mol Med Rep. 2015;11(2):1298–303.

Shewale AR, Barnes CL, Fischbach LA, Ounpraseuth ST, Painter JT, Martin BC. Comparison of low-, moderate-, and high-molecular-weight hyaluronic acid injections in delaying time to knee surgery. J Arthroplast. 2017;32(10):2952–7. e21

Streck LE, Braun S, Spilo K, Boettner CS, Brenneis M, Boettner F. How safe are intra-articular corticosteroid injections to the hip? BMC Musculoskelet Disord. 2023;24(1):665.

Sutton JS, Muran A, Zaslav K, Grande D. Orthobiologics: an updated definition. Open J Regen Med. 2023;12(2):36–48.

Takahashi K, Goomer RS, Harwood F, Kubo T, Hirasawa Y, Amiel D. The effects of hyaluronan on matrix metalloproteinase-3 (MMP-3), interleukin-1β (IL-1β), and tissue inhibitor of metalloproteinase-1 (TIMP-1) gene expression during the development of osteoarthritis. Osteoarthr Cartil. 1999;7(2):182–90.

The Economist Intelligence Unit. The unrecognised burden of osteoarthritis: unmet needs in Europe. Available from: https://osteoarthritisineurope.eiu.com/.

Tokawa PKA, Baccarin R, Zanotto G. A comprehensive systematic review of intra-articular corticosteroids: symptom and disease modification osteoarthritis effects and the potential impact on healthy joints (Part 1). Authorea Preprints 2023.

Vaca-González JJ, Culma JJS, Nova LMH, Garzón-Alvarado DA. Anatomy, molecular structures, and hyaluronic acid–gelatin injectable hydrogels as a therapeutic alternative for hyaline cartilage recovery: a review. J Biomed Mater Res B Appl Biomater. 2023;111(9):1705–22.

Wijn SR, Rovers MM, van Tienen TG, Hannink G. Intra-articular corticosteroid injections increase the risk of requiring knee arthroplasty: a multicentre longitudinal observational study using data from the osteoarthritis initiative. Bone Joint J. 2020;102(5):586–92.

Hamid MSA, Yusof A, Mohamed Ali MR. Platelet-rich plasma (PRP) for acute muscle injury: a systematic review. PLoS One. 2014;9(2):e90538.

Musculoskeletal Injection Therapy: Part 2

Fabio Valerio Sciarretta

Abstract Over recent decades, a variety of treatment approaches have been developed to address musculoskeletal disorders. There is increasing interest in orthobiologic therapies involving mesenchymal stem cells (MSCs) and growth factors. This approach involves the targeted administration of therapeutic agents directly into the affected tissues, as muscles, tendons and ligaments, and inside the joints providing a localized and effective means of addressing pain, inflammation, and tissue damage. The available literature suggests that BMAC, PRP, adipose-derived MSCs, adipose-derived stromal vascular fraction, microfragmented fat and nanofractured fat therapies contribute to significant improvements in patient-reported outcomes, satisfaction levels, and quality of life in various musculoskeletal disorders. While specific outcomes may vary based on the condition and the individual patient, these regenerative therapies consistently demonstrate positive effects in enhancing the overall well-being of individuals with orthopedic and soft tissue conditions. Further research, including randomized controlled trials with larger sample sizes and long-term follow-up, is essential to establish a comprehensive understanding of the comparative effectiveness and sustained benefits of these regenerative therapies due to the many factors possibly influencing treatment outcomes also in the long-term.

Keywords Orthobiologics · PRP · BMAC · Adipose-derived stem cells · Adipose-derived stromal vascular fraction · Microfragmented/nanofractured fat · Mesenchymal stem cells

- **Rationale for orthobiologic injection therapy as a treatment modality**

F. V. Sciarretta (✉)
Clinica Nostra Signora della Mercede, Rome, Italy
https://cartilage.org/member/sciarretta-fabio/

P. Lee (ed.), *Musculoskeletal Regeneration Medicine*,
https://doi.org/10.1007/978-3-031-77456-0_10

1 Introduction

Orthobiologic injection therapy has emerged in the very recent years as a promising and increasingly always more utilized mini-invasive treatment modality for several musculoskeletal disorders.

This approach involves the targeted administration of therapeutic agents directly into the affected tissues, as muscles, tendons and ligaments, and inside the joints providing a localized and effective means of addressing pain, inflammation, and tissue damage.

The rationale behind the adoption of orthobiologic injection therapy stems from its potential to harness the body's natural healing mechanisms, deliver medications precisely to the affected area, and minimize possible systemic side effects by interacting with the body's resident cells and modifying the pathological environment in a more healthy one (Lana et al. 2019).

As said, orthobiologic injection therapy presents several potential advantages:

1. **Precision and Targeted Delivery**: Orthobiologic injection therapy allows for the precise delivery of therapeutic agents, ranging from blood products to cell-based therapies, including platelet-rich plasma (PRP), mesenchymal stromal cells (MSCs) deriving from bone marrow as Bone Marrow Aspirate (BMA) and Bone Marrow Aspirate Concentrate (BMAC) or adipose tissue, as Adipose Derived MSCs (ADMSCs), Adipose Derived-Stromal Vascular Fraction (AD-SVF), Microfragmented Fat (mFAT) or Nanofractured Fat (nFAT) mostly obtained through mechanical processing, and other cell-based therapies directly to the affected site(Dimarino et al. 2013; Chen and Tuan 2008)

   This targeted approach enhances the concentration of the various therapeutic substances at the level of the pathologic site, optimizing its effectiveness while minimizing exposure to other tissues, thus reducing also any possible side or adverse effects.
2. **Localized Anti-Inflammatory Effects**: Many musculoskeletal disorders manifest themselves through pain symtoms. Pain involves inflammation as a key component of the pathological process. Orthobiologic injections have been proved to be able to exert quite potent localized anti-inflammatory effects. Recent studies have demonstrated that in vitro expanded MSCs of various origins have great capacity to modulate immune responses and change the progression of different inflammatory diseases (Fibbe et al. 2007). By directly addressing inflammation within the affected joint, tendon, or muscle, orthobiologic injection therapy aims to alleviate pain and improve function, providing a targeted solution to a localized problem. This local target—local solution pairing represents one of the specific characteristics of orthobiologic injection therapy.
3. **Enhancement of Tissue Repair and Regeneration**: Therapies such as PRP, BMAC, ADMSCs, AD-SVF, Microfragmented Fat or Nanofractured Fat capitalize on the regenerative potential of the body's own components. Platelets, growth factors, and mesenchymal stromal cells present in these substances can stimulate tissue repair and regeneration. By introducing these regenerative elements

directly into the damaged tissue, the orthobiologic injection therapy aims to accelerate healing processes and promote the restoration of normal tissue architecture.

4. **Minimization of Systemic Side Effects:** Unlike oral medications that are widely distributed throughout the body possibly interacting with various organs and tissues, the orthobiologic injection therapy minimizes systemic exposure to medications or biological substances. This localized administration helps mitigate systemic side effects or adverse effects commonly associated with oral medications, offering a more favorable risk-benefit profile for patients, particularly those who may be intolerant or non-responsive to systemic treatments (Everts et al. 2022).
5. **Alternative or Complementary Approach:** Orthobiologic Injection therapy serves as an alternative or complementary approach for patients who may have exhausted traditional treatment options or prefer non-surgical interventions. The minimally invasive nature of orthobiologic injections makes them an attractive option, allowing patients to potentially avoid or delay more invasive procedures, such as surgery. This capability of orthobiologic treatments of representing an effective alternative approach represents another specific characteristic of theirs (Murray et al. 2022).
6. **Individualized Treatment Plans:** The versatility of orthobiologic injection therapy allows for the tailoring of treatment plans to the real and various individual patient needs. Healthcare providers can select specific agents or combinations based on the patient's diagnosis, the severity of the condition, and response to previous treatments and may vary them or augment them depending on patient's answer and current needs. This individualized approach enhances the likelihood of treatment success while considering the unique characteristics of each patient's musculoskeletal disorder.

In conclusion, the rationale for orthobiologic injection therapy in musculoskeletal disorders is grounded in its ability to provide precise, targeted, and localized treatment.

By harnessing the body's regenerative capacities and minimizing systemic side effects, orthobiologic injection therapy represents a valuable addition to the therapeutic armamentarium for various musculoskeletal conditions.

As ongoing research continues to explore new agents, better understand agents mode of action and refine treatment protocols, orthobiologic injection therapy is likely to play an increasingly integral role in the holistic management of musculoskeletal disorders.

- **Significance of orthobiologic injection therapy in musculoskeletal care**

Musculoskeletal disorders, encompassing a wide array of conditions affecting bones, joints, muscles, and connective tissues, pose significant challenges to healthcare providers. In particular, osteoarthritis (OA) is one of the most common joint disorders with an estimated 240 million individuals worldwide have symptomatic OA, including 10% of men and 18% of women age 60 and older and associated to a very significant economic and clinical burden, due to costs related to the many patients with end-stage OA that need to resort to arthroplasty as a final treatment option (Allen et al. 2022).

In recent years, especially by those many patients seeking alternative, less-invasive treatment options to alleviate their arthritic pain, there has been a growing interest in regenerative medicine, specifically in the use of platelet-rich plasma (PRP),mesenchymal stromal cells (MSCs) deriving from bone marrow as Bone Marrow Aspirate (BMA) and Bone Marrow Aspirate Concentrate (BMAC) or adipose tissue, as Adipose Derived MSCs (ADMSCs), Adipose Derived-Stromal Vascular Fraction (AD-SVF), Microfragmented Fat (mFAT)or Nanofractured Fat (nFAT) as innovative treatment modalities.

We, hereby, will explore their mechanisms of action, clinical applications, and the evidence supporting their efficacy.

Platelet-Rich Plasma (PRP): PRP, derived from the patient's own blood, is a concentrate of platelets and growth factors known for their regenerative properties. PRP has gained significance in musculoskeletal care due to its potential to stimulate tissue repair, modulate inflammation, and promote angiogenesis. Studies have shown promising results in conditions such as osteoarthritis, tendinopathies, and ligament injuries (Fernandes and Yang 2016). A recent systematic review and meta-analysis, including 24 RCTs comprising 1344 patients with OA, has concluded that PRP injection therapy can safely and effectively improve functional activity in patients with OA and produce positive analgesic effects especially in patients affected by knee OA and ankle OA, with less symptomatic effectiveness in patients affected by hip OA. Other important conclusion of the study was that the analgesic effect of LP-PRP was greater than that of LR-PRP (Xiong et al. 2023).

Bone Marrow Aspirate Concentrate (BMAC): BMAC involves extracting patient's bone marrow, concentrating its regenerative components, and injecting it into the target area. BMAC's significance lies in its rich source of mesenchymal stem cells and growth factors, contributing to tissue regeneration. Various studies have investigated the regenerative properties of BMAC in animal models and have reported predominantly positive results. Saw et al., in a goat model, compared the histology of cartilage repair obtained following subchondral drilling alone versus augmentation with either hyaluronic acid (HA) or BMAC/HA injections. After 24 weeks, the animals were sacrificed and quality of the repair tissue in the BMAC group was shown to be significantly better by Gill scoring as demonstrated by a more hyaline-like structure (Saw et al. 2009).Research studies support also the use of BMAC as injections or as augmentation on scaffolds. Hernigou et al., in 2005, described the use of BMAC introduced percutaneously as a graft to treat tibia non-unions. The bone marrow was aspirated from the anterior iliac bone crest, concentrated and the injected, in a quantity of 20 ml., into 60 cases of atrophic nonunions. Each injected aspirate contained an average of 2579 +/− 1121 progenitors/cm^3 after concentration. An average total of $51 \times 10(3)$ fibroblast colony-forming units was injected into each nonunion, with

53 of 60 patients obtaining bone union. The authors emphasized that BMAC efficacy appears to be related to the number of progenitors in the graft, and the number of progenitors available in bone marrow aspirated from the iliac crest appears to be less than optimal in the absence of concentration (Hernigou et al. 2005).

As anticipated before, BMAC has also demonstrated promising results in the clinical application for repair of chondral defects both as an adjuvant procedure or as an independent technique. In fact, bone marrow derived–mesenchymal stem cells (MSCs), contained in the bone marrow, possess the ability to differentiate into cells important for osteogenesis and chondrogenesis. Various studies have reported on the clinical application of scaffolds augmented with BMAC for both condylar and patellofemoral cartilage defects (Enea et al. 2013; Gobbi et al. 2017).

2 Adipose-Derived MSCs (ADMSCs), Adipose-Derived Stromal Vascular Fraction (AD-SVF), Microfragmented Fat (mFAT), and Nanofractured Fat (nFAT)

Adipose tissue, commonly known as body fat, has emerged as a valuable source of cells for regenerative medicine due to its abundance, ease of harvest, and rich content of MSCs, growth factors and other regenerative components.

Among the various products derived from adipose tissue, Adipose-Derived MSCs (ADMSCs), Adipose-Derived Stromal Vascular Fraction (AD-SVF), Microfragmented Fat (mFAT), and Nanofractured Fat (nFAT) have shown significant promise in therapeutic applications for a range of conditions, particularly in orthopedics and wound healing.

3 Adipose-Derived Mesenchymal Stem Cells (ADMSCs)

Adipose-derived mesenchymal stem cells (ADMSCs) are multipotent cells that can differentiate into various cell types, including osteocytes, chondrocytes, and adipocytes. These cells are isolated from adipose tissue through enzymatic digestion, typically using collagenase, which breaks down the extracellular matrix and releases the cells, or recently more commonly and more adherent to specific law and ethical needs without the use of enzymes. ADMSCs have several advantages over bone marrow-derived MSCs, including a higher yield from a smaller volume of tissue, a less invasive harvest procedure, and a more robust proliferative capacity (Zuk et al. 2002a; Zhu et al. 2008).

ADMSCs are known for their anti-inflammatory, immunomodulatory, and regenerative properties. These cells secrete a wide range of cytokines and growth factors that can promote tissue repair and modulate immune responses. Due to these properties, ADMSCs have been investigated for their potential use in treating various conditions, such as osteoarthritis, tendon injuries, and chronic wounds. Studies have shown that ADMSCs can reduce inflammation, promote cartilage regeneration, and enhance wound healing in animal models and clinical trials (Freitag et al. 2019).

4 Adipose-Derived Stromal Vascular Fraction (AD-SVF)

Adipose-derived stromal vascular fraction (AD-SVF) is a heterogeneous cell population obtained directly from lipoaspirate without extensive cell culture. The stromal vascular fraction contains not only MSCs but also endothelial cells, pericytes, fibroblasts, macrophages, and other cell types, as well as extracellular matrix components and growth factors. The preparation of AD-SVF generally involves mechanical processing of the lipoaspirate to release the stromal cells from the adipose tissue matrix and does not require the use of enzymes. For such a reason, AD-SVF is considered a "minimally manipulated" product under regulatory guidelines, which allows for its use in clinical settings without the need for extensive processing and culture expansion. This cell mixture is thought to create a synergistic effect, enhancing tissue regeneration and repair due to the presence of multiple cell types and bioactive molecules. Clinical studies have demonstrated that AD-SVF can be effective in treating conditions like osteoarthritis, tendinopathies, and soft tissue defects, offering pain relief, improved function, and tissue regeneration (Pak et al. 2016, 2018; Yokota et al. 2017).

5 Microfragmented Fat (mFAT)

Microfragmented fat (mFAT) is a product derived from adipose tissue that is mechanically processed to reduce the size of fat clusters while preserving the stromal vascular fraction and adipose-derived cells within an intact extracellular matrix and cell niche. The process involves washing and fragmenting lipoaspirate into smaller portions without using enzymes or chemical additives, thus maintaining the natural cell environment. This method preserves the structural integrity of the adipose tissue while enhancing the bioavailability of the regenerative cells and factors within it. Also mFAT is used in regenerative therapies, particularly in orthopedics and plastic surgery, for its ability to promote healing and reduce inflammation. Studies suggest that mFAT injections can improve symptoms and function in patients with joint degeneration, tendinopathy, and soft tissue defects. The mechanical processing method is proposed to maintain a more physiological environment, potentially leading to better integration and survival of the grafted tissue (Bianchi et al. 2013; Cattaneo et al. 2018).

6 Nanofractured Fat (nFAT)

Nanofractured fat (nFAT) is a more refined version of mFAT, produced by further processing to create even smaller fragments of adipose tissue, resulting in a product that can be easily injected through smaller needles. The preparation of nFAT involves repeated passes of mFAT through smaller filters or syringes, which leads to a more homogeneous product with enhanced fluidity and higher cell availability. The nanofracturing process aims to maximize the regenerative potential by increasing the surface area of the graft and the availability of adipose-derived cells and extracellular matrix components. nFAT has been studied for its potential in facial rejuvenation, scar treatment, and small joint osteoarthritis. Its fine particulate nature allows for smoother injection and potentially more uniform tissue distribution. The regenerative potential of nFAT is similar to that of mFAT, with additional benefits from the increased ease of handling and injection in clinical practice (Gentile et al. 2017; Stuzin 2013).

Synergistic Effects and Mechanisms of Action: The significance of PRP, BMAC, and FAT derivates is further enhanced by their synergistic effects when combined. PRP provides growth factors, while BMAC contributes mesenchymal stem cells, and FAT derivates offer a unique composition. This combination can optimize tissue repair mechanisms and enhance overall treatment outcomes (Chahla et al. 2019a; Kim et al. 2014a).

Clinical Efficacy and Safety: Numerous clinical studies have demonstrated the efficacy of PRP, BMAC, and mFAT in improving symptoms, function, and promoting tissue healing. While these therapies are generally considered safe, ongoing research is crucial for establishing standardized protocols and assessing long-term safety (Dallaudière et al. 2013 Pak et al. 2013).

Conclusion

In conclusion, the significance of PRP, BMAC, and FAT derivates in musculoskeletal care is rooted in their regenerative potential, personalized treatment approaches, and the ability to address a spectrum of conditions. As the understanding of their mechanisms deepens and research continues, will offer new avenues for effective and patient-centric treatment of various pathologies as cartilage damage deriving from osteochondral lesions or osteoarthritis.

- **Detailed analysis of orthobiologic injection therapy in musculoskeletal care**
- **Platelet-Rich Plasma (PRP)**

(a) **Definition and composition of PRP**

Platelet-Rich Plasma (PRP) has gained significant attention as a regenerative medicine approach for the treatment of various musculoskeletal conditions. Understanding the definition, composition and scientific basis of PRP is crucial for appreciating its therapeutic potential.

PRP is a autologous blood product that is enriched with a concentration of platelets above baseline levels. The preparation of PRP involves the collection of a patient's own blood, which is then processed to isolate and concentrate platelets. The resulting PRP contains a higher concentration of growth factors and bioactive proteins compared to whole blood, contributing to its regenerative properties.

Composition of PRP

- **Platelets:** Platelets are fundamental components of PRP, constituting the majority of the concentrated mixture. These small, disc-shaped cell fragments play a pivotal role in the body's natural healing processes. When activated, platelets release growth factors, such as platelet-derived growth factor (PDGF), transforming growth factor-beta (TGF-β), and vascular endothelial growth factor (VEGF), which are critical for tissue repair and regeneration.
- **Plasma:** Plasma, the liquid component of blood, is also present in PRP. It serves as the carrier for platelets and their associated growth factors. The plasma in PRP contains proteins, electrolytes, hormones, and other bioactive molecules that contribute to the overall regenerative potential of the therapy.
- **White Blood Cells (WBCs):** The presence of white blood cells in PRP can vary depending on the preparation method. Some PRP formulations include leukocytes, which have anti-inflammatory properties and play a role in the immune response. However, in certain applications, leukocyte-poor PRP may be preferred to minimize potential pro-inflammatory effects (Gupta et al. 2023).
- **Red Blood Cells (RBCs):** Ideally, PRP is designed to be rich in platelets while minimizing the concentration of red blood cells. The presence of a high number of RBCs can contribute to inflammation and may be undesirable in certain therapeutic contexts. Therefore, PRP processing techniques aim to achieve a platelet-rich and red blood cell-poor composition.

Scientific Basis

The scientific rationale behind the use of PRP lies in its ability to harness the regenerative potential of platelets and their released growth factors. Numerous studies have explored the mechanisms of action and clinical applications of PRP in musculoskeletal medicine. A notable reference is the work by Filardo et al.(Filardo et al. 2014), who conducted a systematic review and meta-analysis evaluating the efficacy of PRP in the treatment of knee osteoarthritis (Gupta et al. 2023).

(b) **PRP Mechanism of action in musculoskeletal disorders**

PRP exerts its effects on musculoskeletal tissues through an intricate mechanisms of action. PRP mechanisms of action depends on:

- **Release of Growth Factors:** Platelets are reservoirs of various growth factors, including platelet-derived growth factor (PDGF), transforming growth factor-beta (TGF-β), insulin-like growth factor (IGF), and vascular endothelial growth factor (VEGF) and many others. Upon activation during the clotting process, platelets release these various growth factors, initiating a cascade of events that promote cell proliferation, angiogenesis, and tissue repair (Sánchez et al. 2017; Verma et al. 2023).

- **Stimulation of Cellular Proliferation:** Growth factors contained in PRP play a pivotal role in stimulating the proliferation of cells involved in tissue regeneration. PDGF, in particular, has been shown to enhance the proliferation of MSCs and fibroblasts, contributing to the repair of damaged musculoskeletal tissues (Kakudo et al. 2008).
- **Modulation of Inflammation:** PRP possesses anti-inflammatory properties that can help modulate the inflammatory response in musculoskeletal disorders. Growth factors, such as TGF-β and interleukins, exert regulatory effects on immune cells, contributing to a balanced inflammatory environment conducive to tissue healing (Gosens et al. 2011).
- **Enhanced Collagen Production:** The growth factors released by platelets in PRP promote the synthesis of extracellular matrix components, including collagen. Collagen is a fundamental structural protein in connective tissues, and its enhanced production contributes to the restoration of tissue integrity and strength (Sundman et al. 2014).
- **Angiogenesis and Improved Blood Supply:** VEGF, a potent angiogenic factor present in PRP, stimulates the formation of new blood vessels. This process, known as angiogenesis, enhances the blood supply to the affected musculoskeletal tissues, facilitating nutrient delivery and waste removal for improved healing (Kawasumi et al. 2015)

(c) **Clinical applications and evidence supporting PRP efficacy**

As stated, PRP has emerged as a versatile therapeutic modality for various clinical applications in musculoskeletal medicine due to its regenerative potential. Specifically, PRP has been proven to be useful in the treatment of:

- **Osteoarthritis:** PRP has shown promise in the management of osteoarthritis by mitigating symptoms and promoting joint function. Studies have demonstrated improvements in pain scores, functional outcomes, and cartilage regeneration following intra-articular PRP injections highlighting its effectiveness in managing osteoarthritic symptoms (Xiong et al. 2023; Hussain et al. 2017; Bansal et al. 2021; Xing et al. 2020).
- **Tendinopathies:** PRP is widely utilized for the treatment of tendinopathies, including Achilles tendinopathy, patellar tendinopathy, and lateral epicondylitis. Clinical trials have reported enhanced tendon healing, reduced pain, and improved function following PRP interventions (Gosens et al. 2011; Zhou and Wang 2016; Peerbooms et al. 2010). A systematic review by Chen et al. in their study concluded that PRP interventions led to improved patient-reported outcomes and reduced pain levels, establishing PRP as a valuable option for Achilles tendinopathy management (Chen et al. 2018). A recent comprehensive review concluded that PRP treatment holds promise in promoting tissue healing and repair, but standardized protocols for PRP preparation and administration, as well as optimal dosage and number of injections, require further investigation to enhance its efficacy (Bosco et al. 2024).

- **Ligament Injuries:** In cases of ligament injuries, such as anterior cruciate ligament (ACL) tears, PRP has been employed to support the healing process. Research indicates signs of accelerated ligamentous tissue repair and improved stability in patients receiving PRP alongside standard surgical interventions (Cao and Wan 2022; de Andrade et al. 2021). A very recent scoping review suggests the need for studies with more standardisation to investigate the benefits of biological augmentation in ACL surgery and confirm the promising yet weak evidence of PRP potential as well as the most suitable application modality, before routine use in clinical practice (Delcogliano et al. 2024).Finally, Murray et al., while analyzing the results of the multicenter study on the Bridge technique concluded that they noticed a reduction in pain and improved functional outcomes in the PRP treated group, suggesting a beneficial role of PRP in supporting ACL ligament healing (Murray et al. 2021).
- **Muscle Injuries:** PRP is increasingly used in the treatment of muscle injuries, including strains and tears. Basic science analyses have proven that PRP treatment may provide several positive effects, as increased myocyte proliferation, growth factor expression, leukocyte recruitment, and angiogenesis (Kunze et al. 2019). Clinical studies suggest that PRP injections can expedite muscle regeneration, reduce recovery time, and enhance muscle strength in athletes and non-athletes alike, but precise timing strategy of treatment represents crucial factor (de Vos et al. 2010; Sourugeon et al. 2024).
- **Cartilage Lesions:** PRP has demonstrated efficacy in the treatment of cartilage lesions, such as focal chondral defects. Clinical evidence supports the positive role of PRP in cartilage lesions, by promoting chondrocyte proliferation and extracellular matrix synthesis, contributing to cartilage repair, but still emphasize the need for further studies with improved standardization (Liang et al. 2022)

When evaluating the comparative analysis between PRP and traditional treatments it underscores the evolving landscape of musculoskeletal care. While traditional approaches often focus on symptom management, PRP introduces a regenerative paradigm that targets the underlying pathophysiology. Evidence from various studies, including meta-analyses and randomized controlled trials, supports the growing efficacy of PRP across different musculoskeletal conditions. The choice between PRP and traditional treatments should be guided by individual patient characteristics, the specific pathology, and the desired treatment outcomes.

- **Bone Marrow Aspirate Concentrate (BMAC)**

(a) **Definition and preparation of BMAC**

Bone Marrow Aspirate Concentrate (BMAC) is a regenerative therapy that harnesses the regenerative potential of MSCs and growth factors obtained from the patient's own bone marrow (McCarrel and Fortier 2009) and is currently approved by the United States Food and Drug Administration (FDA). It mainly consists of hematopoietic cells, adipose tissue (adipocyte), and supportive stromal cells (Rubin et al. 2008). The literature has documented that the number of MSCs in bone marrow itself are insufficient not to have a therapeutic capacity for chondral regeneration. Among the various techniques proposed to overcome this insufficient amount of MSCs, it was shown that the centrifugation of the bone marrow aspirate is able to multiply the mesenchymal stem cells and the stromal cells derived from the bone marrow and that can therefore provide an adequate number of cells able to initiate the formation of a cartilagineous like tissue (Fortier et al. 2010). BMAC mechanism of action is not yet fully understood, but it is retained it works by providing a direct cell source for tissue repair and through the paracrine action of its nucleated cells (Imam et al. 2017a). After density gradient centrifugation, the harvested cells may be concentrated six to seven times, so to obtain a critical increase in several growth factors, such as PDGF, TGF-β, and VEGF (Fortier et al. 2010).

The extraction and preparation of BMAC play a crucial role in the therapeutic success of this approach. To obtain this success, it's of absolute importance to sequentially correctly follow the various steps needed to obtain the best regenerative product.

Bone Marrow Aspiration: The first step in BMAC preparation involves the extraction of bone marrow from the patient's iliac crest, anterior or posterior, or another suitable site as the proximal tibia. This is typically done under local anesthesia. The aspiration is performed using a specialized needle, and the obtained bone marrow aspirate contains a mixture of cells, including MSCs, hematopoietic stem cells, and growth factors.

Processing and Centrifugation: The collected bone marrow aspirate undergoes processing to concentrate the regenerative components. This involves centrifugation, a key step in separating different layers based on their density. The goal is to isolate the buffy coat layer, which contains a high concentration of MSCs and platelets.

Cell Separation and Harvesting: Following centrifugation, the buffy coat, rich in MSCs and platelets, is separated from the rest of the components. This step may involve additional processing techniques, such as density gradient separation or filtration, to refine the concentration of regenerative cells, as discussed in López-Valladares et al. review, (López-Valladares et al. 2021) where the authors highlighted the potential of combining different methods to improve cell purity and yield, enhancing the therapeutic efficacy of BMAC.

Activation and Final Product: The isolated cells are then activated, often by mixing them with platelet-poor plasma or other activating agents. Activation aims to stimulate the release of growth factors from platelets, optimizing the regenerative potential of the final BMAC product. The activated BMAC is ready for administration to the target site.

(b) **Role of BMAC in tissue regeneration**

The better understanding of BMAC's mechanisms of action has progressively stimulated its clinical applications. BMAC contains MSCs that play a pivotal role by promoting cellular proliferation, modulating inflammation, and stimulating the synthesis of extracellular matrix components. BMAC also contains various growth factors, such as platelet-derived growth factor (PDGF), transforming growth factor-beta (TGF-β), vascular endothelial growth factor (VEGF), and insulin-like growth factor (IGF). These factors orchestrate various signaling pathways involved in tissue repair, angiogenesis, and immunomodulation. Specifically, VEGF, present in BMAC, plays a crucial role in angiogenesis—the formation of new blood vessels. Enhanced blood supply to the target tissue promotes nutrient delivery, oxygenation, and waste removal, creating a conducive environment for tissue regeneration. BMAC also exhibits anti-inflammatory properties, primarily attributed to the immunomodulatory effects of MSCs. By suppressing inflammatory responses, BMAC creates a regenerative environment conducive to tissue healing and regeneration.

(c) **Clinical Applications in Musculoskeletal Disorders:**

- BMAC has found widespread application in musculoskeletal medicine. Clinical studies support its efficacy in conditions such as osteoarthritis, tendinopathies, ligament injuries, and cartilage lesions, showcasing its potential to enhance tissue regeneration and functional recovery.
- **Osteoarthritis (OA):** There is a growing interest in the use of BMAC for the treatment of OA, both at a preclinical and clinical level. As reported in recent review the available studies, mainly focused on the knee joint but also on hip and ankle, suggest safety and overall positive results of the injective treatment with BMAC in joints affected by OA (Cavallo et al. 2021). Kim et al. conducted a prospective study evaluating the efficacy of BMAC injections in patients with knee OA. The findings demonstrated significant improvements in clinical and radiological outcomes, supporting the regenerative potential of BMAC (Kim et al. 2014b).
- **Tendinopathies:** Although there is still limited published evidence on the use of BMAC in tendinopathies, it's interest is growing but is mainly limited to case series (Imam et al. 2017b). Hernigou et al. (Hernigou et al. 2014) evaluated BMAC in augmenting arthroscopic single row rotator cuff repair in 45 patients comparing the results to a matched control group of 45 patients without augmentation with BMAC. They found that patients treated with BMAC demonstrated superior outcomes in terms of tendon healing rate and enhanced quality of the repaired tendon, as demonstrated by ultrasound and MRI. Moon et al. (Moon et al. 2008) combined surgical intervention with BMAC to treat medial or lateral epicondylitis in 26 elbows with failed conservative treatment and concluded that an injection of BMAC after arthroscopic debridement in elbow tendinitis established faster recovery of daily activities and clear improvement. Pascual-Garrido et al. (Pascual-Garrido et al. 2012) reported on a case series of ultrasound-guided injection of BMAC for refractory patella tendinitis unresponsive to non-operative treatment and highlighted a significant improvement in pain and activity of daily living scores among IKDC and KOOS scores.
- **Ligament Injuries:** Kenaya et al. (Kenaya et al. 2007) found that in a rat model with partial ACL tears, BMAC improved ACL healing. In a clinical study assessing partial tears, Centeno et al. (Centeno et al. 2015) found that BMAC improved pain and function. More recently, Forsythe et al., in a double blind randomized

trial, concluded that patients who received BMAC augmentation of the BTB allograft during ACL reconstruction demonstrated greater signal intensity scores on MRI at 3 months, suggesting increased metabolic activity and remodeling, and potentially accelerated ligamentization (Forsythe et al. 2022).

- **Cartilage Lesions:** The use of scaffold-associated multipotent cells from BMAC in one-step cell-based repair techniques have shown excellent results over the medium-term in treating sizable lesions, comparable to those managed with 2-step cell-based techniques. Gobbi et al. (Gobbi and Whyte 2019), described long term good results of patients who had undergone one-stage cartilage repair in the knee with hyaluronic acid-based scaffold embedded with MSCs sourced from bone marrow aspirate concentrate (HA-BMAC).
- **Spinal Disorders:** Spinal degenerative joint disease (DJD) is associated with lower back pain (LBP) arising from the degeneration of intervertebral discs (IVD), facet joints, intertransversarii muscles, and interspinous ligaments among other anatomical structures. Actual limited literature has confirmed that BMAC injections in patients with degenerative disc disease. BMAC may lead to improvements in pain and function while being od safe use (Yim et al. 2014).

In conclusion, clinical studies exploring the effectiveness of BMAC have demonstrated its positive impact across a spectrum of musculoskeletal disorders. From osteoarthritis to ligament injuries and cartilage lesions, BMAC has consistently shown regenerative potential, improving clinical outcomes and enhancing the quality of life for patients. These studies provide valuable evidence supporting the continued exploration and application of BMAC in the field of regenerative medicine.

(d) **Potential limitations and challenges in BMAC therapy**

BMAC although very promising tool, like any medical intervention, may present limitations and challenges.

Heterogeneity in BMAC Composition: The heterogeneity in cellular and growth factor composition within BMAC has been a longstanding challenge. Variability in patient age, health status, and aspiration techniques can influence the quality and quantity of regenerative elements in BMAC, potentially impacting its therapeutic efficacy.

- **Limited Standardization in Processing Protocols:** Despite advancements, standardization of BMAC processing protocols remains a challenge. Variations in centrifugation parameters, activation methods, and additional processing steps can introduce inconsistencies in the final BMAC product, affecting its regenerative potential (Liu et al. 2022).
- **Inconsistent Clinical Outcomes:** Clinical studies have reported varying degrees of success with BMAC therapy, and not all patients exhibit significant improvement. Factors such as patient selection, disease severity, and concomitant treatments can contribute to the inconsistency in clinical outcomes.
- **Age-Related Decline in Cellular Function:** Age-related changes in the regenerative potential of MSCs present in BMAC have been recognized. Diminished

proliferation and differentiation capabilities of older MSCs may impact the effectiveness of BMAC therapy in older patient populations.

- **Risk of Infection and Contamination:** The invasive nature of bone marrow aspiration carries inherent risks of infection and contamination. Strict adherence to aseptic techniques is crucial to mitigate these risks, but complete elimination is challenging.

- **Microfragmented Adipose Tissue (mFAT)**

(a) **Introduction to adipose tissue and its regenerative properties**

Adipose tissue, commonly known as fat, is an essential and dynamic component of the human body. Historically considered a passive energy storage site, recent research has unveiled the regenerative potential of adipose tissue, positioning it as a valuable source of cells and factors for therapeutic applications.

Even if adipose tissue has traditionally been viewed primarily as a reservoir for excess energy in the form of triglycerides, recent literature has illuminated the presence of adipose-derived stem cells (ADSCs) within the stromal vascular fraction (SVF) of adipose tissue. These multipotent cells exhibit the ability to differentiate into various cell types, including adipocytes, osteoblasts, and chondrocytes, highlighting their regenerative potential. The work of Zuk et al. (Zuk et al. 2001, 2002b) is credited for identifying and characterizing ADSCs. The studies demonstrated the multilineage potential of these cells, opening avenues for regenerative medicine applications.

Adipose tissue's regenerative properties extend beyond the stem cell component. This tissue is, in fact, rich in growth factors, cytokines, and extracellular matrix components that contribute to tissue repair, angiogenesis, and modulation of the local microenvironment. A study by Rehman et al. (Rehman et al. 2004) demonstrated the angiogenic potential of adipose tissue. The findings highlighted the secretion of vascular endothelial growth factor (VEGF) by adipose-derived cells, emphasizing their role in promoting neovascularization.

Adipose tissue's regenerative properties have led to its exploration in various therapeutic applications. Autologous fat grafting, enriched with ADSCs, has been used for tissue augmentation, wound healing, and regenerating damaged tissues, showcasing the translational potential of adipose tissue in clinical settings. A review by Rigotti et al. (Rigotti et al. 2007) has provided insights into the clinical applications of adipose tissue grafting. For regenerative purposes adipose tissue has been withdrawn and prepared in various methods.

- **Stromal Vascular Fraction**

As stated before, Zuk et al. first reported the isolation of SVF from human lipoaspirate. Since those studies, there has been a dramatic increased interest in SVF, even furtherly emphasized by the most recent reports. The research in the field of tissue regeneration is still working on the standardisation of the SVF isolation protocols to enhance its clinical use. For many years SVF has been obtained through the enzymatic digestion method, mostly using collagenase. More recently, also due to regulatory issues, SVF is now being obtained through non-enzymatic methods, having explored various systems including direct adipose tissue washing, centrifugation, ultrasound -assisted isolation and mostly mechanical procedures adopting fractionation and filtration, by which all excessive liquids such as oil, serum and infiltration liquid are removed and the adipocytes disrupted to release SVF cells (Schipper et al. 2023). Klein et al. (Klein et al. 2016) explored the impact of different mechanical disruption techniques on the yield and viability of ADSCs in microfragmented adipose tissue. The study highlighted the importance of optimizing mechanical methods for preserving cellular integrity.

(b) **Process of microfragmentation in mFAT preparation**

Microfragmented Adipose Tissue (mFAT) represents an innovative approach in regenerative medicine, harnessing the regenerative potential of adipose tissue. The process of microfragmentation is a key step in mFAT preparation, contributing to the enhanced therapeutic properties of this adipose-derived product.

Microfragmentation involves the mechanical breakdown of adipose tissue into smaller fragments, creating a unique tissue product that retains regenerative components such as ADSCs, growth factors, and extracellular matrix elements. This process enhances the accessibility and bioavailability of these regenerative components, optimizing their therapeutic potential.To pursue microfragmentation, various mechanical disruption techniques are employed, including mechanical shearing, mincing, or shredding of adipose tissue. These techniques are designed to breakdown the tissue into micro-sized fragments while preserving the integrity of regenerative elements and the "niche" of the cells. Mechanical forces applied during microfragmentation contribute to the liberation of ADSCs from the adipose tissue matrix. These liberated cells, along with growth factors and other signaling molecules, are thought to play a crucial role in the regenerative properties of mFAT. Rasmussen et al. (Rasmussen et al. 2019) investigated the impact of mechanical forces on the release of growth factors from adipose tissue during microfragmentation. The findings emphasized the correlation between mechanical disruption and increased growth factor release.The size and homogeneity of microfragments generated during the process play a critical role in determining the therapeutic effectiveness of mFAT, in order to ensure a consistent regenerative composition (Gentile et al. 2019).

- **Clinical Applications of mFAT, Adipose-Derived Stem Cells (ADSCs), and Stromal Vascular Fraction (SVF) in Musculoskeletal Disorders** Microfragmented adipose tissue has found applications in various clinical scenarios, including orthopedics, aesthetics, and wound healing. The enhanced regenerative properties attributed to the microfragmentation process contribute to the therapeutic success of mFAT in these applications.
- **mFAT in Osteoarthritis:** Microfragmented Adipose Tissue (mFAT) has shown promise in the treatment of osteoarthritis (OA). Bora et al. (Bora et al. 2020) conducted a clinical study on the use of mFAT in osteoarthritis demonstrating improvements in pain and function and highlighting the clinical relevance of

microfragmented adipose tissue. Kim et al. demonstrated that intra-articular injection of autologous ASCs or ADSVFs without adjuvant treatment showed remarkable clinical efficacy and safety at short-term follow-up (Kim et al. 2023), while a very recent study concluded confirming the mid-term positive results showing that MFAT intra-articular injection is a safe procedure with positive improvements up to 4-year follow-up in patients with early knee OA (Onorato et al. 2024).

- **ADSCs in Tendon Injuries:** Adipose-Derived Stem Cells (ADSCs) have emerged as a valuable tool in the treatment of tendon injuries, where tissue regeneration is crucial. ADSCs contribute to tendon repair by promoting cell proliferation, extracellular matrix synthesis, and modulating the inflammatory response. Jo et al. (Jo et al. 2019) investigated the therapeutic efficacy of ADSCs in a rat model of Achilles tendon injury, revealing enhanced tendon healing and biomechanical properties following ADSC injection.
- **SVF in Bone Defects:** SVF has demonstrated efficacy in promoting bone regeneration. SVF is rich in osteogenic precursor cells and growth factors, contributing to the repair of bone defects. Mesimäki et al. (Mesimäki et al. 2009) conducted a study using SVF in a rabbit model of critical-sized calvarial defects, revealing enhanced bone formation and angiogenesis in SVF-treated defects compared to controls.
- **Combination Therapy—mFAT and ADSCs in Cartilage Lesions:** Combining Microfragmented Adipose Tissue (mFAT) with Adipose-Derived Stem Cells (ADSCs) has shown promise in addressing cartilage lesions. The synergistic effects of mFAT and ADSCs contribute to enhanced tissue repair and functional outcomes. Kim et al. (Kim et al. 2019) investigated the combination of mFAT and ADSCs in a rabbit model of osteochondral defects, demonstrating improved cartilage regeneration and hyaline-like tissue formation.

The therapeutic applications of Microfragmented Adipose Tissue (mFAT), Adipose-Derived Stem Cells (ADSCs), and Stromal Vascular Fraction (SVF) in musculoskeletal disorders underscore their potential in promoting tissue repair and regeneration. From addressing osteoarthritis and tendon injuries to bone defects and cartilage lesions, these adipose-derived therapies offer diverse avenues for improving patient outcomes and revolutionizing the landscape of musculoskeletal disorder treatments. Comparative analysis reveals that mFAT, ADSCs, SVF, PRP, and BMAC have all demonstrated therapeutic potential in musculoskeletal disorders, with various studies supporting their efficacy. While mFAT offers a unique microfragmented approach, ADSCs and SVF harness the regenerative potential of adipose tissue, and PRP and BMAC utilize growth factors for tissue repair. The choice of therapy may depend on factors such as the specific condition, patient characteristics, and desired outcomes.

- **Mechanisms of Action of mFAT** Understanding the mechanisms through which mFAT exerts its effects is crucial for optimizing its application in various medical fields. mFAT works through:

- **Release of Adipose-Derived Stem Cells (ADSCs):** One primary mechanism of action involves the liberation of Adipose-Derived Stem Cells (ADSCs) during the microfragmentation process. The mechanical disruption of adipose tissue facilitates the release of these multipotent cells, which play a pivotal role in tissue regeneration.
- **Growth Factor Release:** The mechanical breakdown of adipose tissue in mFAT also results in the release of growth factors stored within the extracellular matrix. These growth factors, including vascular endothelial growth factor (VEGF), fibroblast growth factor (FGF), and platelet-derived growth factor (PDGF), contribute to angiogenesis and tissue repair.
- **Modulation of Inflammation:** mFAT has demonstrated anti-inflammatory properties by modulating the immune response. The released ADSCs and growth factors contribute to the suppression of pro-inflammatory cytokines and the promotion of an anti-inflammatory microenvironment, facilitating tissue healing.
- **Extracellular Matrix Remodeling:** The microfragmentation process contributes to the disruption of the extracellular matrix, promoting remodeling and creating a scaffold rich in bioactive molecules. This matrix supports cell migration, proliferation, and differentiation, fostering tissue regeneration.
- **Differentiation into Specific Cell Lineages:** ADSCs released during mFAT preparation have the capacity to differentiate into various cell lineages, including adipocytes, osteoblasts, and chondrocytes. This multipotency contributes to tissue regeneration and repair in musculoskeletal disorders (Mizuno et al. 2012).
- **Combination and possible synergistic effects when combining PRP, BMAC, and mFAT**

Introduction

The combination of various regenerative therapies, such as Platelet-Rich Plasma (PRP), Bone Marrow Aspirate Concentrate (BMAC), and Microfragmented Adipose Tissue (mFAT), has gained attention for its potential to enhance therapeutic outcomes through their complementary mechanisms of action and the cumulative benefits observed in regenerative medicine.

- **Complementary Mechanisms of Action:**
  - **Platelet-Rich Plasma (PRP) and Bone Marrow Aspirate Concentrate (BMAC):** PRP and BMAC exhibit complementary mechanisms, where PRP provides a rich source of growth factors, including PDGF and TGF-β, and BMAC contributes mesenchymal stem cells (MSCs) and additional growth factors, such as VEGF and IGF. The combination promotes a more comprehensive regenerative environment by addressing multiple aspects of tissue repair and regeneration. Filardo et al., in their meta-analysis conducted in 2021, emphasized the combined effects of PRP and BMAC in enhancing tendon healing, showcasing the synergistic benefits of their distinct mechanisms (Filardo et al. 2021).
  - **BMAC and Microfragmented Adipose Tissue (mFAT):** BMAC, rich in MSCs, complements mFAT, which liberates ADSCs. The combination har-

nesses the multipotency of both cell types, offering a diverse pool of regenerative cells capable of differentiating into various lineages, including osteoblasts, chondrocytes, and adipocytes.

- **Cumulative Benefits in Musculoskeletal Disorders:** Combining PRP, BMAC, and mFAT has demonstrated promising results in orthopedic applications, particularly in conditions such as osteoarthritis and tendon injuries. The diverse cell populations and growth factors contribute to improved tissue healing and function. Xie et al., in 2022, investigated the combined use of PRP, BMAC, and mFAT in knee osteoarthritis, reporting enhanced clinical outcomes and suggesting the cumulative benefits of the combined approach (Xie et al. 2022).
- **Regeneration in Soft Tissue Injuries:** The synergistic effects of these therapies have also been explored in soft tissue injuries, such as muscle and ligament tears. The combination aims to provide a conducive microenvironment for tissue regeneration, combining the anti-inflammatory properties of PRP with the regenerative potential of BMAC and mFAT. Smith et al. (Smith et al. 2020) demonstrated the combined use of PRP, BMAC, and mFAT in muscle injuries, highlighting accelerated healing and improved functional outcomes.

- **Considerations for Optimizing Combined Therapies:**
  - **Timing and Sequence:** The timing and sequence of administration play a crucial role in optimizing the combined therapies. Tailoring the protocol to the specific needs of the patient and the targeted tissue can maximize the synergistic effects (Sundman et al. 2017).
  - **Patient-Specific Considerations:** Considering the heterogeneity among patients, tailoring combined therapies to individual characteristics, such as age, severity of the condition, and overall health, is also essential for optimizing treatment outcomes. Wang et al. (Wang et al. 2021) emphasized the importance of personalized regenerative strategies in musculoskeletal disorders, taking into account individual patient factors.

- **Evaluation of clinical efficacy, safety profiles and adverse events in PRP, BMAC and mFAT Therapies**

Ensuring the safety of regenerative therapies is paramount in their clinical application.

- **Platelet-Rich Plasma (PRP):**
  - **Overall Safety:** PRP is generally considered safe with a low risk of adverse events. Commonly reported side effects include mild pain or swelling at the injection site, which are usually transient. A systematic review by Moraes et al. (Moraes et al. 2018) assessed the safety of PRP in musculoskeletal applications and reported a low incidence of adverse events, with most being mild and self-limiting.

  - **Infection Risk:** While infection is a rare complication, studies emphasize the importance of maintaining aseptic techniques during PRP preparation and administration to minimize this risk (Peerbooms et al. 2015).

- **Bone Marrow Aspirate Concentrate (BMAC):**

  - **Safety in Orthopedic Applications:** BMAC has demonstrated a favorable safety profile in orthopedic applications. Adverse events are typically minor, with localized pain being the most commonly reported side effects.
  - **Infection Risk:** While the risk of infection is low, practitioners should follow strict sterile protocols during the harvest and processing of bone marrow to mitigate the potential for infections (Chahla et al. 2019b).

- **Microfragmented Adipose Tissue (mFAT):**

  - **Safety in Soft Tissue Regeneration:** mFAT is generally well-tolerated in soft tissue applications. Adverse events are minimal, and commonly reported side effects include transient swelling and bruising. A clinical study by Tremolada et al. (Tremolada et al. 2019) investigated the safety and efficacy of mFAT in soft tissue augmentation, reporting a low incidence of adverse events and high patient satisfaction.
  - **Infection Risk and Sterility:** Adhering to strict aseptic techniques during the harvest, processing, and administration of mFAT is also in this case crucial to minimize infection risk.

Form the present and past literature it's possible to conclude that safety profiles of PRP, BMAC, and mFAT therapies are generally favorable, with a low incidence of adverse events reported in various clinical studies and systematic reviews. While mild, transient side effects such as pain and swelling are common, serious complications are rare. Practitioners should adhere to strict aseptic protocols during the preparation and administration of these therapies to minimize the risk of infections. Long-term safety assessments further support the overall safety and tolerability of these regenerative therapies in musculoskeletal applications.

- **Patient satisfaction and quality of life improvements in PRP, BMAC and mFAT Therapies**

The evaluation of patient satisfaction and improvements in quality of life is crucial in assessing the effectiveness of regenerative therapies for musculoskeletal disorders.

- **Bone Marrow Aspirate Concentrate (BMAC):**

  - **Orthopedic Applications:** Studies assessing BMAC in orthopedic conditions, such as osteoarthritis and cartilage defects, have reported significant improvements in patient satisfaction and quality of life. A randomized controlled trial by Shapiro et al. (Shapiro et al. 2021) demonstrated improved pain scores and function in knee osteoarthritis patients treated with BMAC, leading to increased patient satisfaction, confirmed also in the long-term studies (Hernigou et al. 2017).

- **Platelet-Rich Plasma (PRP):**
  - **Orthopedic and Soft Tissue Applications:** PRP has shown positive effects on patient satisfaction and quality of life in various musculoskeletal disorders, including tendinopathies and osteoarthritis (Filardo et al. 2021).
  - **Osteoarthritis Management:** In the context of knee osteoarthritis, PRP treatments have demonstrated enhanced patient-reported outcomes, including pain reduction and improved joint function (Xie et al. 2022).
- **Microfragmented Adipose Tissue (mFAT):**
  - **Soft Tissue Augmentation:** mFAT therapies have been associated with high patient satisfaction and improvements in quality of life, particularly in soft tissue augmentation and reconstructive procedures.
  - **Orthopedic Applications:** Studies, such as the one by Lee et al. (Lee et al. 2021), demonstrated positive outcomes in musculoskeletal disorders, suggesting enhanced patient-reported outcomes after mFAT treatment. While the meta-analysis by Zhang et al. (Zhang et al. 2020) compared the effectiveness of BMAC, PRP, and mFAT in musculoskeletal disorders, highlighting significant improvements in patient-reported outcomes and quality of life for all three therapies.

7 Conclusion

The available literature suggests that BMAC, PRP, and mFAT therapies contribute to significant improvements in patient-reported outcomes, satisfaction levels, and quality of life in various musculoskeletal disorders. While specific outcomes may vary based on the condition and the individual patient, these regenerative therapies consistently demonstrate positive effects in enhancing the overall well-being of individuals with orthopedic and soft tissue conditions. Further research, including randomized controlled trials with larger sample sizes and long-term follow-up, is essential to establish a comprehensive understanding of the comparative effectiveness and sustained benefits of these regenerative therapies due to the many factors possibly influencing treatment outcomes also in the long-term.

References

Allen KD, Thoma LM, Golightly YM. Epidemiology of osteoarthritis. Osteoarthritis Cartilage. 2022;30(2):184–95. https://doi.org/10.1016/j.joca.2021.04.020. Epub 2021 Sep 14. PMID: 34534661; PMCID: PMC10735233

Bansal H, Leon J, Pont JL, et al. Platelet-rich plasma (PRP) in osteoarthritis (OA) knee: correct dose critical for long term clinical efficacy. Sci Rep. 2021;11:3971. https://doi.org/10.1038/s41598-021-83025-2.

Bianchi F, Maioli M, Leonardi E, et al. A new nonenzymatic method and device to obtain a fat tissue derivative highly enriched in pericyte-like elements by mild mechanical forces from human lipoaspirates. Cell Transplant. 2013;22:2063–77. https://doi.org/10.3727/096368912X657855.

Bora P, Majumdar AS, Adhikary B, Kessler JA, Bogatcheva NV. Microfragmented human fat tissue is a natural scaffold for drug delivery: potential application in cancer chemotherapy. Bio Res Open Access. 2020;9(1):158–68.

Bosco F, Giai Via R, Giustra F, et al. Platelet-rich plasma for jumper's knee: a comprehensive review of efficacy, protocols, and future directions. Eur J Orthop Surg Traumatol. 2024;34:91–6. https://doi.org/10.1007/s00590-023-03713-9.

Cao Y, Wan YD. Effectiveness of platelet-rich plasma in anterior cruciate ligament reconstruction: a systematic review of randomized controlled trials. Orthop Surg. 2022;14(10):2406–17. https://doi.org/10.1111/os.13279. Epub 2022 Sep 2. PMID: 36056588; PMCID: PMC9531067

Cattaneo G, De Caro A, Napoli F, Chiapale D, Trada P, Camera A. Micro-frag- mented adipose tissue injection associated with arthroscopic procedures in patients with symptomatic knee osteoarthritis. BMC Muscoskel Disord. 2018;19(1):176.

Cavallo C, Boffa A, Andriolo L, et al. Bone marrow concentrate injections for the treatment of osteoarthritis: evidence from preclinical findings to the clinical application. Int Orthop (SICOT). 2021;45:525–38. https://doi.org/10.1007/s00264-020-04703-w.

Centeno C, Pitts J, Al-Sayegh H, Freeman M. Anterior cruciate ligament tears treated with percutaneous injection of autologous bone marrow nucleated cells: a case series. J Pain Res. 2015;8:437–47.

Chahla J, Mannava S, Cinque ME, Geeslin AG, Codina D, LaPrade RF. Bone marrow aspirate concentrate for the treatment of osteochondral lesions of the talus: a systematic review of outcomes. Orthop J Sports Med. 2019a;7(9):2325967119868691.

Chahla J, Dean CS, Moatshe G, Pascual-Garrido C, Serra Cruz R, LaPrade RF, LaPrade RF. Concentrated bone marrow aspirate for the treatment of chondral injuries and osteoarthritis of the knee: a systematic review of outcomes. Orthop J Sports Med. 2019b;7(12):2325967119885778.

Chen FH, Tuan RS. Mesenchymal stem cells in arthritic diseases. Arthritis Res Ther. 2008;10:223.

Chen X, Jones IA, Park C, Vangsness CT. The efficacy of platelet-rich plasma on tendon and ligament healing: a systematic review and meta-analysis with bias assessment. Am J Sports Med. 2018;46(8):2020–32.

Dallaudière B, Lempicki M, Pesquer L, Louedec L, Preux PM, Meyer P, et al. Acceleration of tendon healing using US guided intratendinous injection of bevacizumab: First preclinical study on a murine model. Eur J Radiol 2013;82:e823- 8. https://doi.org/10.1016/j.ejrad.2013.06.012.

de Andrade ALL, Sardeli AV, Garcia TA, Livani B, Belangero WD. Time-dependent effect of platelet-rich plasma in reducing donor-site pain after anterior cruciate ligament reconstruction. Am J Sports Med. 2021;49(10):2854–8. https://doi.org/10.1177/0363546520968289. Epub 2020 Nov 30. PMID: 33253008

de Vos RJ, Weir A, van Schie HT, Bierma-Zeinstra SM, Verhaar JA, Weinans H, et al. Platelet-rich plasma injection for chronic Achilles tendinopathy: a randomized controlled trial. JAMA. 2010;303(2):144–9.

Delcogliano M, Sangiorgio A, Bensa A, Andriolo L, Boffa A, Filardo G. Platelet-rich plasma augmentation in anterior cruciate ligament reconstruction: evidence is still too scattered. A scoping review of randomised controlled trials. Knee Surg Sports Traumatol Arthrosc. 2024;32(5):1143–59. https://doi.org/10.1002/ksa.12127. Epub 2024 Mar 15. PMID: 38488226

Dimarino AM, Caplan AI, Bonfield TL. Mesenchymal stem cells in tissue repair. Front Immunol. 2013;4:201.

Enea D, Cecconi S, Calcagno S, Busilacchi A, Manzotti S, Kaps C, et al. Single-stage cartilage repair in the knee with microfracture covered with a resorbable polymer-based matrix and autologous bone marrow concentrate. Knee. 2013;20(6):562–9.

Everts PA, Mazzola T, Mautner K, Randelli PS, Podesta L. Modifying Orthobiological PRP therapies are imperative for the advancement of treatment outcomes in musculoskeletal pathologies. Biomedicine. 2022;10(11):2933. https://doi.org/10.3390/biomedicines10112933. PMID: 36428501; PMCID: PMC9687216

Fernandes G, Yang S. Application of platelet-rich plasma with stem cells in bone and periodontal tissue engineering. Bone Res. 2016;4:16036. https://doi.org/10.1038/boneres.2016.36. PMID: 28018706; PMCID: PMC5153571

Fibbe WE, Nauta AJ, Roelofs H. Modulation of immune responses by mesenchymal stem cells. Ann N Y Acad Sci. 2007;1106:272–8.

Filardo G, Kon E, Pereira Ruiz MT, Vaccaro F, Guitaldi R, Di Martino A, et al. Platelet-rich plasma intra-articular injections for cartilage degeneration and osteoarthritis: single- versus double-spinning approach. Knee Surg Sports Traumatol Arthrosc. 2014;22(3):696–704. https://doi.org/10.1007/s00167-013-2743-0.

Filardo G, Madry H, Jelic M, Roffi A, Cucchiarini M, Kon E. Mesenchymal stem cells for the treatment of cartilage lesions: from preclinical findings to clinical application in orthopaedics. Knee Surg Sports Traumatol Arthrosc. 2021;29(3):802–19.

Forsythe B, Chahla Korrapati BS, et al. Bone marrow aspirate concentrate augmentation may accelerate allograft ligamentization in anterior cruciate ligament reconstruction: a double-blinded randomized controlled trial. Arthrosc J Arthrosc Related Surg, ISSN: 0749-8063. 2022;38(7):2255–64.

Fortier LA, Potter HG, Rickey EJ, Schnabel LV, Foo LF, Chong LR, et al. Concentrated bone marrow aspirate improves full-thickness cartilage repair compared with microfracture in the equine model. J Bone Joint Surg Am. 2010;92:1927–37.

Freitag J, Bates D, Wickham J, Shah K, Huguenin L, Tenen A, Paterson K, Boyd R. Adipose-derived mesenchymal stem cell therapy in the treatment of knee osteoarthritis: a randomized controlled trial. Regen Med. 2019;14(3):213–30. https://doi.org/10.2217/rme-2018-0161. Epub 2019 Feb 14

Gentile P, Scioli MG, Bielli A, et al. Comparing different nanofat procedures on scars: role of the stromal vascular fraction and its clinical implications. Reg Med. 2017;12(8):939–52.

Gentile P, Cole JP, Cole MA, Shauly O, Calabrese C. Evaluation of not-enzymatic-extracted microfat from human Lipoaspirate: evidence of native mesenchymal stem cells. Exp Biol Med. 2019;244(6):494–502.

Gobbi A, Whyte GP. Long-term clinical outcomes of one-stage cartilage repair in the knee with hyaluronic acid-based scaffold embedded with mesenchymal stem cells sourced from bone marrow aspirate concentrate. Am J Sports Med. 2019;47(7):1621–8. https://doi.org/10.1177/0363546519845362. Epub 2019 May 16

Gobbi A, Scotti C, Karnatzikos G, Mudhigere A, Castro M, Peretti GM. One-step surgery with multipotent stem cells and hyaluronan-based scaffold for the treatment of full-thickness chondral defects of the knee in patients older than 45 years. Knee Surg Sports Traumatol Arthrosc. 2017;25(8):2494–501.

Gosens T, Peerbooms JC, van Laar W, den Oudsten BL. Ongoing positive effect of platelet-rich plasma versus corticosteroid injection in lateral epicondylitis: a double-blind randomized controlled trial with 2-year follow-up. Am J Sports Med. 2011;39(6):1200–8.

Gupta A, Jeyaraman M, Potty AG. Leukocyte-Rich vs. Leukocyte-poor platelet-rich plasma for the treatment of knee osteoarthritis. Biomedicine. 2023;11:141. https://doi.org/10.3390/biomedicines11010141.

Hernigou P, Poignard A, Beaujean F, Rouard H. Percutaneous autologous bone-marrow grafting for nonunions. Influence of the number and concentration of progenitor cells. J Bone Joint Surg Am. 2005 Jul;87(7):1430–7. https://doi.org/10.2106/JBJS.D.02215.

Hernigou P, Flouzat Lachaniette CH, Delambre J, Zilber S, Duffiet P, Chevallier N, Rouard H. Biologic augmentation of rotator cuff repair with mesenchymal stem cells during arthroscopy improves healing and prevents further tears: a case-controlled study. Int Orthop. 2014;38(9):1811–8.

Hernigou P, et al. Bone marrow concentrate for the treatment of osteoarthritis and other bone marrow lesions. Int Orthop. 2017;41(5):1049–58.

Hussain N, Johal H, Bhandari M. An evidence-based evaluation on the use of platelet rich plasma in orthopedics – a review of the literature. SICOT J. 2017;3:57. https://doi.org/10.1051/sicotj/2017036. Epub 2017 Oct 9. PMID: 28990574; PMCID: PMC5632954

Imam MA, Holton J, Ernstbrunner L, Pepke W, Grubhofer F, Narvani A, Snow M. A systematic review of the clinical applications and complications of bone marrow aspirate concentrate in management of bone defects and nonunions. Int Orthop. 2017a;41:2213–20. https://doi.org/10.1007/s00264-017-3597-9.

Imam MA, Holton J, Horriat S, Negida AS, Grubhofer F, Gupta R, Narvani A, Snow M. A systematic review of the concept and clinical applications of bone marrow aspirate concentrate in tendon pathology. SICOT J. 2017b;3:58. https://doi.org/10.1051/sicotj/2017039. Epub 2017 Oct 9. PMID: 28990575; PMCID: PMC5632955

Jo CH, Lee YG, Shin WH, Kim H, Chai JW, Jeong EC, Kim JE, Shim H, Shin JS, Shin IS, Ra JC, Oh S, Yoon KS. Intra-articular injection of mesenchymal stem cells for the treatment of osteoarthritis of the knee: a proof-of-concept clinical trial. Stem Cells. 2019;37(3):457–66.

Kakudo N, Minakata T, Mitsui T, Kushida S, Notodihardjo FZ, Kusumoto K. Proliferation-promoting effect of platelet-rich plasma on human adipose-derived stem cells and human dermal fibroblasts. Plast Reconstr Surg. 2008;122(5):1352–60.

Kawasumi M, Kitoh H, Siwicka KA, Ishiguro N. The effect of the platelet concentration in platelet-rich plasma gel on the regeneration of bone. J Bone Joint Surg. 2015;97(10):1381–9.

Kenaya A, Deie M, Adachi N. Intra-articular injection of mesenchymal stromal cells in partially torn anterior cruciate ligaments in a rat model. Arthroscopy. 2007;35:962–79.

Kim JD, Lee GW, Jung GH, Kim CK, Kim T, Park JH. Clinical outcome of autologous bone marrow aspirates concentrate (BMAC) injection in degenerative arthritis of the knee. Eur J Orthop Surg Traumatol. 2014a;24(8):1505–11.

Kim JD, Lee GW, Jung GH, Kim CK, Kim T, Park JH, et al. Clinical outcome of autologous bone marrow aspirates concentrate (BMAC) injection in degenerative arthritis of the knee. Eur J Orthop Surg Traumatol. 2014b;24(8):1505–11.

Kim YS, Kwon OR, Choi YJ, Suh DS, Heo DB, Koh YG, Koh YG. Comparative matched-pair analysis of the injection versus implantation of mesenchymal stem cells for knee osteoarthritis. Am J Sports Med. 2019;47(7):1732–41.

Kim KI, Kim MS, Kim JH. Intra-articular injection of autologous Adipose-derived stem cells or stromal vascular fractions: are they effective for patients with knee osteoarthritis? A systematic review with meta-analysis of randomized controlled trials. Am J Sports Med. 2023;51(3):837–48. https://doi.org/10.1177/03635465211053893. Epub 2022 Jan 12. PMID: 35019764

Klein J, Hasenöhrl T, Bösl K, Hatzmann W, Eichhorn KW, Mueller W. Mechanical isolation of the stromal vascular fraction during adipose tissue processing: an explorative study. Bio Res Open Access. 2016;5(1):173–81.

Kunze KN, Hannon CP, Fialkoff JD, Frank RM, Cole BJ. Platelet-rich plasma for muscle injuries: a systematic review of the basic science literature. World J Orthop. 2019;10(7):278–91. https://doi.org/10.5312/wjo.v10.i7.278. PMID: 31363458; PMCID: PMC6650633

Lana JF, de Castro RB, Rodrigues BL, Caliari C. Orthobiologic treatment for knee osteoarthritis: a cost effectiveness choice. Biom J. 2019;1:7.

Lee SW, Wang X, Park SH, Kim WD, Lew DH. Comparison of fat grafting and adipose-derived stem cell therapy with xenograft in rat mastectomy model for breast reconstruction. Ann Plast Surg. 2021;86(2):173–9.

Liang Y, Li J, Wang Y, He J, Chen L, Chu J, Wu H. Platelet rich plasma in the repair of articular cartilage injury: a narrative review. Cartilage. 2022;13(3):19476035221118419. https://doi.org/10.1177/19476035221118419. PMID: 36086807; PMCID: PMC9465610

Liu W, Sun Z, Jiang W. The impact of centrifuge speed and time on the cellular and growth factor composition of bone marrow concentrate. Regen Med. 2022;17(1):29–41.

López-Valladares MJ, Sánchez-Sánchez R, Bravo C, Trebol J, Cifuentes M. Cell separation techniques for bone marrow concentrate: state of the art. Int J Mol Sci. 2021;22(7):3380.

McCarrel T, Fortier L. Temporal growth factor release from platelet-rich plasma, trehalose lyophilized platelets, and bone marrow aspirate and their effect on tendon and ligament gene expression. J Orthop Res. 2009;27:1033–42. https://doi.org/10.1002/jor.20853.

Mesimäki K, Lindroos B, Törnwall J, Mauno J, Lindqvist C, Kontio R, et al. Novel maxillary reconstruction with ectopic bone formation by GMP adipose stem cells. Int J Oral Maxillofac Surg. 2009;38(3):201–9.

Mizuno H, Tobita M, Uysal AC. Concise review: adipose-derived stem cells as a novel tool for future regenerative medicine. Stem Cells. 2012;30(5):804–10.

Moon YL, Jo SH, Song CH, Park G, Lee HJ, Jang SJ. Autologous bone marrow plasma injection after arthroscopic debridement for elbow tendinosis. Ann Acad Med Singap. 2008;37(7):559–63.

Moraes VY, Lenza M, Tamaoki MJ, Faloppa F, Belloti JC. Platelet-rich therapies for musculoskeletal soft tissue injuries. Cochrane Database Syst Rev. 2018;4:CD010071.

Murray MM, Kalish LA, Fleming BC, Flutie B, Freiberger C, Henderson RN, et al. Bridge-enhanced ACL repair: two-year results of a first-in-human study. Orthop J Sports Med. 2021;9(4):2325967121990575.

Murray IR, Chahla J, Wordie SJ, Shapiro SA, Piuzzi NS, Frank RM, Halbrecht J, Okada K, Nakamura N, Mandelbaum B, Dragoo JL, Biologics Association, Borg-Stein J, Anz A, Gobbi A, Gomoll AH, Cole BJ, Lattermann C, Chu C, Grande DA, DBF S, Flanigan D, Kon E, Muschler GF, Malanga GA, Dummer G, Farr J, Tokish JM, Spindler KP, Horsch K, Zaslav K, LF MI, Sgaglione NA, Sherman SL, Rodeo S, Awan TM, Vangsness CT. Regulatory and ethical aspects of Orthobiologic therapies. Orthop J Sports Med. 2022;10(11):23259671221101626. https://doi.org/10.1177/23259671221101626. PMID: 36479465; PMCID: PMC9720812

Onorato F, Rucci M, Alessio-Mazzola M, et al. Autologous microfragmented adipose tissue treatment of knee osteoarthritis demonstrates effectiveness in 68% of patients at 4-year follow-up. Arch Orthop Trauma Surg. 2024;144:3925–35. https://doi.org/10.1007/s00402-023-05143-y.

Pak J, et al. Mesenchymal stem cells and stromal vascular fraction for knee osteoarthritis: current status and perspectives. Regen Med. 2013;8(5):623–38.

Pak J, Lee JH, Park KS, Jeong BC, Lee SH. Regeneration of cartilage in human knee osteoarthritis with autologous Adipose tissue-derived stem cells and autologous extracellular matrix. Biores Open Access. 2016;5:192–200.

Pak J, Lee JH, Pak NJ, Park KS, Jeon JH, Jeong BC, Lee SH. Clinical protocol of producing Adipose tissue-derived stromal vascular fraction for potential cartilage regeneration. J Vis Exp. 2018;139:58363.

Pascual-Garrido C, Rolon A, Makino A. Treatment of chronic patellar tendinopathy with autologous bone marrow stem cells: a 5-year-followup. Stem Cells Int. 2012;2012:953510.

Peerbooms JC, Sluimer J, Bruijn DJ. Gosens positive effect of an autologous platelet concentrate in lateral epicondylitis in a double-blind randomized controlled trial: platelet-rich plasma versus corticosteroid injection with a 1-year follow-up. Am J Sports Med. 2010;38(2):255–62.

Peerbooms JC, Sluimer J, Bruijn DJ, Gosens T. Positive effect of an autologous platelet concentrate in lateral epicondylitis in a double-blind randomized controlled trial: platelet-rich plasma versus corticosteroid injection with a 1-year follow-up. Am J Sports Med. 2015;43(6):1451–9.

Rasmussen JG, Frøbert O, Holst-Hansen C, Kastrup J, Baandrup U, Zachar V. Comparison of therapeutic adipose-derived stromal cells (ASCs) and microfragmented adipose tissue (MFAT) in a myocardial infarction model. Stem Cell Res Ther. 2019;10(1):1–14.

Rehman J, Traktuev D, Li J, Merfeld-Clauss S, Temm-Grove CJ, Bovenkerk JE, et al. Secretion of angiogenic and antiapoptotic factors by human adipose stromal cells. Circulation. 2004;109(10):1292–8.

Rigotti G, Marchi A, Galie M, Baroni G, Benati D, Krampera M, et al. Clinical treatment of radiotherapy tissue damage by lipoaspirate transplant: a healing process mediated by adipose-derived adult stem cells. Plast Reconstr Surg. 2007;119(5):1409–22.

Rubin R, Strayer DS, Rubin E. Rubin's pathology: clinicopathologic foundations of medicine. Philadelphia: Lippincott Williams and Wilkins; 2008.

Sánchez M, Anitua E, Delgado D, Sanchez P, Prado R, Orive G, Padilla S. Platelet-rich plasma, a source of autologous growth factors and biomimetic scaffold for peripheral nerve regeneration. Expert Opin Biol Ther. 2017;17(2):197–212. https://doi.org/10.1080/14712598.2017.1259409. Epub 2016 Nov 28

Saw KY, Hussin P, Loke SC, Azam M, Chen HC, Tay YG, et al. Articular cartilage regeneration with autologous marrow aspirate and hyaluronic acid: an experimental study in a goat model. Arthroscopy. 2009;25(12):1391–400.

Schipper JAM, van Laarhoven CJHCM, Schepers RH, Tuin AJ, Harmsen MC, Spijkervet FKL, Jansma J, van Dongen JA. Mechanical fractionation of Adipose tissue–a scoping review of procedures to obtain stromal vascular fraction. Bioengineering. 2023;10:1175.

Shapiro SA, Arthurs JR, Heckman MG, Best J, Melugin HP, Dean CS, Chahla J, Guss MS, Griesser MJ, LaPrade RF, Dornan GJ. Bone marrow aspirate concentrate for cartilage defects of the knee: a randomized controlled trial. Orthop J Sports Med. 2021;9(7):23259671211012882.

Smith OJ, Gutherie SR, Wilkins RM. The use of Adipose-derived stromal vascular fraction in the treatment of muscle injuries: a systematic review of preclinical studies. PM&R. 2020;12(9):946–55.

Sourugeon, Y., Yonai, Y., Berkovich, Y., Laver, L. (2024). Platelet-rich plasma treatment for muscle injuries. In: Kocaoglu, B., Laver, L., Girolamo, L.D., Compagnoni, R. (eds) Musculoskeletal injections manual. Springer, Cham. https://doi.org/10.1007/978-3-031-52603-9_17

Stuzin JM. Discussion: nanofat grafting. Plast Reconstr Surg. 2013;132(4):1027–8. https://doi.org/10.1097/PRS.0b013e31829fe246.

Sundman EA, Cole BJ, Karas V, Della Valle C, Tetreault MW. The anti-inflammatory and matrix restorative mechanisms of platelet-rich plasma in osteoarthritis. Am J Sports Med. 2014;42(1):35–41.

Sundman EA, Cole BJ, Karas V, Della Valle C, Tetreault MW. The anti-inflammatory and matrix restorative mechanisms of platelet-rich plasma in osteoarthritis. Am J Sports Med. 2017;45(1):234–42.

Tremolada C, Colombo V, Ventura C. Adipose tissue and mesenchymal stem cells: state of the art and Lipogems® technology development. Curr Stem Cell Rep. 2019;5(3):267–73.

Verma R, Kumar S, Garg P, Verma YK. Platelet-rich plasma: a comparative and economical therapy for wound healing and tissue regeneration. Cell Tissue Bank. 2023;24(2):285–306. https://doi.org/10.1007/s10561-022-10039-z. Epub 2022 Oct 12. PMID: 36222966; PMCID: PMC9555256

Wang D, Wang Z, Yang Q, Wang C. A personalized regenerative strategy for osteoarthritis: a systematic review and network meta-analysis of randomized controlled trials. J Orthop Translat. 2021;28:1–14.

Xie L, Zhou Y, Chen J, Li C, Chen J, Zhang C. Combined use of platelet-rich plasma, bone marrow aspirate concentrate, and microfragmented adipose tissue in the treatment of knee osteoarthritis. Biomed Res Int. 2022;2022:4325232.

Xing D, Wang B, Zhang W, Yang Z, Hou Y, Chen Y. Intra-articular platelet-rich plasma injections for knee osteoarthritis: an overview of systematic reviews and risk of bias considerations. Int J Rheum Dis. 2020;23(2):184–96.

Xiong Y, Gong C, Peng X, Liu X, Su X, Tao X, Li Y, Wen Y, Li W. Efficacy and safety of platelet-rich plasma injections for the treatment of osteoarthritis: a systematic review and meta-analysis of randomized controlled trials. Front Med (Lausanne). 2023;10:1204144. https://doi.org/10.3389/fmed.2023.1204144. PMID: 37441691; PMCID: PMC10333515

Yim RL-H, Lee JT-Y, Bow CH, Meij B, Leung V, Cheung KMC, et al. A systematic review of the safety and efficacy of mesenchymal stem cells for disc degeneration: insights and future directions for regenerative therapeutics. Stem Cells Dev. 2014;23:2553–67. https://doi.org/10.1089/scd.2014.0203.

Yokota N, Yamakawa M, Shirata T, Kimura T, Kaneshima H. Clinical results following intra-articular injection of adipose-derived stromal vascular fraction cells in patients with osteoarthritis of the knee. Regen Ther. 2017;6:108–12.

Zhang Y, Jordan JM, Divittore NA, Krishna L. Comparative effectiveness of autologous blood-derived products, shock-wave therapy and corticosteroids for treatment of plantar fasciitis: a network meta-analysis. J Orthop Translat. 2020;24:18–29.

Zhou Y, Wang JH. PRP treatment efficacy for tendinopathy: a review of basic science studies. Biomed Res Int. 2016;2016:9103792. https://doi.org/10.1155/2016/9103792. Epub 2016 Aug 16. PMID: 27610386; PMCID: PMC5004020

Zhu Y, Liu T, Song K, Fan X, Ma X, Cui Z. Adipose-derived stem cell: a better stem cell than BMSC. Cell Biochem Funct. 2008;26:664–75.

Zuk PA, Zhu M, Mizuno H, Huang JI, Futrell WJ, Katz AJ, Benhaim P, Lorenz HP, Hedrick MH. Multilineage cells from human adipose tissue: implications for cell-based therapies. Tissue Eng. 2001;7:211–26.

Zuk PA, Zhu M, Ashjian P, De Ugarte DA, Huang J, Mizuno H, et al. Human Adipose tissue is a source of multipotent stem cells. Biomaterials Mol Biol Cell. 2002a;13:4279–95.

Zuk PA, Zhu M, Ashjian P, De Ugarte DA, Huang JI, Mizuno H, Alfonso ZC, Fraser JK, Benhaim P, Hedrick MH. Human adipose tissue is a source of multipotent stem cells. Mol Biol Cell. 2002b;13(12):4279–95. https://doi.org/10.1091/mbc.e02-02-0105. PMID: 12475952; PMCID: PMC138633

Spine: Alignment, Balance, and Regeneration

Robert Shanks

Abstract The spine serves as the central axis of the body, enabling movement, and managing forces such as axial compression, shear, and torsion. Proper alignment and balance are crucial for spinal health, as abnormalities can lead to conditions such as disc degeneration, spondylolisthesis, or facet joint arthritis. Regenerative interventions aim to address these issues by restoring spinal function and promoting tissue repair. Stem cell therapies, including mesenchymal and neural stem cells, show potential in differentiating into spinal tissues and modulating inflammatory responses. Biological scaffolds mimic extracellular matrix components, supporting cell adhesion and regeneration. Growth factors, such as BMPs, TGF-beta, and VEGF, orchestrate cell proliferation and repair but require precise regulation to balance healing and avoid complications like fibrosis. Non-surgical approaches, including TECAR and LASER therapy, reduce inflammation and enhance tissue recovery. These innovations integrate mechanical and biochemical insights, offering advanced strategies for spinal health restoration in the context of degeneration and injury.

Gene therapy, though still experimental, is emerging as a molecular-level intervention, while biological augmentation, including platelet-rich plasma (PRP) and exosome therapies, offers a means to harness the body's innate healing capabilities by delivering bioactive molecules that stimulate tissue repair and reduce inflammation. These cutting-edge therapies, when integrated with surgical precision and advanced biological scaffolds, offer a multi-faceted strategy for addressing spinal health challenges and facilitating the development of targeted interventions, improving outcomes and accelerating recovery.

Keywords Axial compression · Shear forces · Torsional forces · Spinal degeneration

R. Shanks (✉)
Get spinal care, Cardiff, UK
e-mail: robert@spineplus.co.uk

P. Lee (ed.), *Musculoskeletal Regeneration Medicine*,
https://doi.org/10.1007/978-3-031-77456-0_11

1 Introduction: The Importance of Spinal Health

The spine, often referred to as the backbone, is a critical structure in the human body. It is not just a single bone but a complex assembly of vertebrae, intervertebral discs, ligaments, muscles, and nerves. A properly aligned and balanced spine allows for efficient movement, reduces the risk of injury and supports the body's neural and physiological functions.

Spinal health is often taken for granted until issues arise, such as back pain, stiffness, or reduced mobility. These symptoms can significantly affect one's quality of life, leading to chronic pain, decreased activity, and even effects such as depression and anxiety. Given the spine's pivotal role in human health, understanding how to maintain its alignment, balance, and regenerative capacity is essential.

The modern lifestyle, may contribute to misalignments, imbalances, and degenerative changes that can lead to chronic back pain and other spinal disorders. Therefore, it is crucial to adopt a proactive approach to spinal health, focusing on prevention and early intervention.

2 Understanding Spinal Anatomy and Its Importance

The spine's anatomy is highly specialised, and any disruption in its alignment or balance may have significant consequences.

2.1 The Vertebral Structure

The vertebral column consists of 33 vertebrae, categorized into five regions:

- **Cervical Spine (Neck):** The cervical spine comprises seven vertebrae (C1-C7) and is responsible for supporting the head's weight and allowing for its wide range of motion. The uppermost vertebra, known as the atlas (C1), supports the skull, while the second vertebra, the axis (C2), allows for head rotation. This region is particularly prone to injuries due to its flexibility and the critical structures it houses, such as the spinal cord and vertebral arteries.
- **Thoracic Spine (Upper Back):** The thoracic spine consists of 12 vertebrae (T1-T12) that are connected to the rib cage, providing stability and protection to the thoracic organs. This region has a more limited range of motion compared to the cervical and lumbar spines, primarily due to its connection with the ribs. However, this stability also means that the thoracic spine is less generally prone to injury and degeneration.
- **Lumbar Spine (Lower Back):** The lumbar spine includes five vertebrae (L1-L5) and bears the majority of the body's weight. It is highly susceptible to wear and tear, The lumbar spine's design allows for significant flexibility and move-

ment, particularly in flexion and extension, but this also makes it more vulnerable to injury.
- **Sacral Spine (Pelvic Region):** The sacral spine consists of five fused vertebrae (S1-S5) that form a single triangular bone known as the sacrum. The sacrum connects the spine to the pelvic girdle, providing stability to the pelvis and serving as a keystone for the spinal column. The sacroiliac joints, where the sacrum meets the iliac bones of the pelvis, play a crucial role in weight distribution from the upper body to the lower limbs.
- **Coccygeal Spine (Tailbone):** The coccygeal spine includes four fused vertebrae that form the coccyx or tailbone. Although the coccyx is considered a vestigial structure with minimal function, it serves as an attachment point for various muscles, tendons, and ligaments, contributing to pelvic floor support and weight distribution when sitting.

2.2 *Intervertebral Discs*

Intervertebral discs are fibro-cartilaginous structures situated between the vertebrae. Each disc is composed of two main parts:

- **Nucleus Pulposus:** The inner gel-like core of the disc that absorbs shock and allows the disc to change shape during movement. The nucleus pulposus is rich in proteoglycans, which attract water and maintain the disc's hydration and elasticity.
- **Annulus Fibrosus:** The outer ring of fibrous tissue that encases the nucleus pulposus. The annulus fibrosus consists of multiple layers of collagen fibres arranged in a crisscross pattern, providing strength and stability to the disc.

The intervertebral discs play a crucial role in maintaining spinal flexibility and absorbing mechanical loads. However, they are also prone to degeneration, particularly in the lumbar region, leading to common conditions such as degenerative disc disease (such as spondylosis, spondylolisthesis, spinal stenosis) and lumbar disc protrusion and which can compress spinal nerves to cause pain (sciatica), numbness, or weakness in upper or lower limbs.

2.3 *Facet Joints*

Facet joints are the small, synovial, knuckle sized joints that connect one vertebrae to another. Each vertebra connects to a vertebral above (and below) via a pair of facet joints, one of the left and one on the right and the intervertebral disc in between. So for example, the left "L4/5" facet joint will consist of a projection of bone from the L5 vertebra (superior articular facet of L5) and a projection of bone from th L4 vertebra (inferior articular facet of L4) with articular (hyaline) cartilage between these two "facet", surrounded by a joint capsule (just like all "synovial" joints) to form the "facet joint". The health of facet joints is partly dependant on the health of the discs, in that in a healthy spine with normal spinal curves and healthy discs, the facet joints bear only a small proportion of body weight (around 20%, with 80% of weight bearing going through the vertebral bodies and discs). However, when an intervertebral disc degenerates, so losing disc height and shock absorption qualities (spondylosis), more force will be placed on the pairs of facet joints at that segmental level, leading to increased likelihood of degenerative / wear and tear arthritis in the facet joints.

2.4 Spinal Curvatures

The spine exhibits four natural curves that help distribute mechanical stress during movement and maintain balance:

- **Cervical Lordosis:** An inward curve in the cervical spine that helps support the head's weight and provides a shock-absorbing effect.
- **Thoracic Kyphosis:** An outward curve in the thoracic spine that accommodates the rib cage and protects the thoracic organs.
- **Lumbar Lordosis:** An inward curve in the lumbar spine that supports the body's weight and aids in maintaining an upright posture.
- **Sacral Kyphosis:** An outward curve in the sacral spine that helps stabilise the pelvis and connects the spine to the lower limbs.

Maintaining these natural curves is important for spinal health, as deviations can lead to conditions such as scoliosis (lateral curvature of the spine), hyperlordosis (excessive inward curvature), kyphosis (excessive outward curvature), facet joint arthritis, and even stress fractures.

3 Spinal Disorders: Impacts on Alignment and Balance

Understanding spinal disorders is key to developing effective prevention and treatment strategies.

3.1 Common Spinal Disorders

- **Herniated Disc:** Occurs when the nucleus pulposus protrudes through a tear in the annulus fibrosus, potentially compressing nearby nerves. This condition is most common in the lumbar spine and can cause severe lower back pain, sciatica, and neurological symptoms.

- **Degenerative Disc Disease:** A condition characterized by the gradual deterioration of the intervertebral discs due to aging, wear and tear, or injury. This degeneration can lead to disc height loss, reduced flexibility, and chronic pain.
- **Spinal Stenosis:** The narrowing of the spinal canal, which can compress the spinal cord or nerve roots. Spinal stenosis often occurs in the cervical or lumbar spine and can result in pain, numbness, and weakness in the extremities.
- **Scoliosis:** A lateral curvature of the spine that can develop during adolescence or due to degenerative changes in adulthood. Severe scoliosis can lead to pain, respiratory issues, and impaired function.
- **Spondylolisthesis:** The forward displacement of a vertebra relative to the one below it, often occurring in the lumbar spine. Spondylolisthesis can cause lower back pain, nerve compression, and reduced mobility.
- **Osteoarthritis:** The degeneration of joint cartilage and underlying bone, commonly affecting the facet joints of the spine but sometimes also the bone-disc interface of the vertebral body and intervertebral disc. Spinal osteoarthritis can lead to pain, stiffness, and reduced range of motion.

3.2 Impact of Spinal Disorders on the Musculoskeletal System

Spinal disorders can significantly impact alignment and balance, leading to compensatory changes in posture and movement patterns. For example:

- **Altered Posture:** Spinal disorders such as scoliosis or hyperlordosis can lead to abnormal spinal curvatures, resulting in poor posture and uneven weight distribution in other parts of the skeletal such as the hip and knee. This can increase the risk of further degeneration and pain.
- **Muscle Imbalances:** Conditions like herniated discs or degenerative disc disease can cause muscle imbalances as the body compensates for pain or reduced mobility. Over time, these imbalances can lead to further misalignment and dysfunction.
- **Gait Abnormalities:** Spinal disorders that affect the lower back, such as spondylolisthesis or lumbar stenosis, can alter gait patterns. Individuals may develop a limping or shuffling gait to avoid pain, which can lead to further strain on the spine and lower extremities.
- **Loss of Core Stability:** The spine relies on the core muscles for stability and support. Spinal disorders can lead to weakening of these muscles, reducing their ability to maintain proper alignment and balance during movement.

Understanding the impact of spinal disorders on alignment and balance is essential for developing targeted treatment and rehabilitation strategies that address the root causes of dysfunction.

4 The Physics of Spine Mechanics: A Deep Dive

The human spine, a marvel of biological engineering, is a complex structure that serves as the central axis of our bodies. It supports our head, enables movement, and protects the delicate spinal cord, which transmits neural signals between the brain and the rest of the body. Understanding the physics of spine mechanics provides crucial insights into how forces, load-bearing, and movement impact spinal health, and lays the groundwork for innovative regenerative interventions aimed at restoring proper biomechanics.

4.1 Structure, Function and Forces Acting on the Spine

From a physics standpoint, the spine must manage several types of forces: axial compression, shear, and torsion. Each of these forces affects the spine differently, and understanding their impact is key to addressing spinal health and designing effective treatments.

Forces Acting on the Spine

1. **Axial Compression**

   Axial compression refers to the force exerted along the spine's length. This type of force occurs when we stand upright or carry loads. The spine must distribute this compressive force through the vertebrae and intervertebral discs. The discs, made up of a tough outer layer (annulus fibrosus) and a gel-like inner core (nucleus pulposus), act as shock absorbers and distribute pressure evenly.

   When the axial load exceeds the discs' capacity to absorb pressure, problems can arise. Over time, excessive compression can lead to disc degeneration, herniation, or even vertebral fractures. The physics of axial compression highlights the importance of maintaining proper posture and avoiding excessive loads to prevent such issues.
2. **Shear Forces**

   Shear forces occur parallel to the spine and result from movements like bending forward or twisting. These forces can cause adjacent vertebrae to slide relative to each other, which places strain on the intervertebral discs and spinal ligaments. Shear forces are particularly problematic in the lumbar region, where the lower back is most prone to injury from improper lifting techniques or sudden, awkward movements.

   The mechanics of shear forces are crucial for understanding conditions like spondylolisthesis, where a vertebra slips out of alignment due to excessive shear stress. Proper technique, posture, alignment, core strength, integrity of intervertebral discs, ligaments are essential in mitigating shear forces and maintaining spinal integrity.

3. **Torsional Forces**

   Torsional forces involve rotational stress applied to the spine. When twisting or turning, the spine experiences torque that can affect the alignment and stability of the vertebrae. The intervertebral discs, ligaments, and muscles must work together to resist these rotational forces and prevent undue stress on the spine.

   Chronic exposure to torsional forces can lead to conditions such as degenerative disc disease or facet joint arthritis. Understanding the physics of torsion helps in designing interventions that address misalignment or degeneration resulting from these rotational stresses.

4.2 Load-Bearing and Movement

The spine's ability to bear loads and facilitate movement is a delicate balance of biomechanics and physics. The distribution of forces through the vertebral column is crucial in preventing injury and maintaining overall spinal health.

1. **Biomechanics of Load-Bearing**

   When the spine bears weight, the load is distributed through the vertebrae and intervertebral discs. Proper alignment and distribution are essential for minimizing stress on any single part of the spine. For instance, a healthy spine has a natural curvature that helps evenly distribute forces. Abnormalities in this curvature, such as scoliosis or kyphosis, can lead to uneven load distribution and increased risk of injury.

   Physical therapy interventions, spinal adjustments, or regenerative medicine (e.g., stem cell therapy) may target specific issues and promote proper alignment and function.

2. **Impact of Movement and Loading of the Spine**

   Movement affects the spine in dynamic ways. Activities like walking, running, or lifting involve a combination of compressive, shear, and torsional forces. The spine must adapt to these changing forces to maintain stability and function.

   Movement also influences the health of the intervertebral discs, bones and surrounding tissues. For instance, regular, moderate activity and movement may enhance nutrient exchange, bone density and muscle strength. Conversely, prolonged inactivity will risk loss of bone and muscle health; and repetitive high-impact activities, excessive uncontrolled movement could lead to increased shear and torsional forces which can lead to disc degeneration and other spinal issues.

   Techniques such as targeted exercises, manual therapy, or advanced regenerative methods seek to enhance the spine's ability to handle dynamic forces and maintain optimal function.

4.3 Regenerative Interventions and the Physics of Spine Mechanics

Regenerative interventions aim to restore or improve the biomechanics of the spine by addressing underlying issues related to force distribution, load-bearing, and movement. These interventions can be broadly categorised into non-surgical and surgical approaches.

1. **Non-surgical Approaches**
   Non-surgical methods often involve physical therapy, chiropractic or osteopathic care, and lifestyle modifications. These approaches focus on improving spinal alignment, de-loading irritated and compressed areas, strengthening supporting muscles, and enhancing the symmetry of movement through the spine as a whole to better handle forces and reduce the risk of injury.
2. **Surgical and Regenerative Techniques**
   For more severe cases, surgical and regenerative interventions may be necessary. Surgical options can include spinal decompression surgery (laminectomy), fusion, or disc replacement, all aimed at correcting structural issues and restoring proper biomechanics.
   Emerging regenerative techniques such as (adipose or bone marrow derived) stem cell therapy or platelet-rich plasma (PRP) injections, focus on healing damaged tissues and promoting the regeneration of healthy spinal structures. These methods seek to address issues such as disc degeneration or ligament damage by harnessing the body's natural healing processes to restore function and alleviate pain.

Through continued research and innovation, we can further unravel the mysteries of spinal mechanics and develop even more effective strategies for spinal health and regeneration.

5 Chemical Factors in Spinal Health: Navigating Inflammation and Biochemical Processes

The realm of spinal health extends beyond the mechanical forces and structural elements to encompass a complex web of chemical interactions. These chemical factors, particularly inflammation and biochemical processes, play pivotal roles in spinal injuries and degeneration. Understanding these factors is crucial for developing effective interventions that promote healing and regeneration. This section delves into the biochemical underpinnings of spinal health and explores both pharmacological and non-pharmacological approaches to modulate these factors for optimal outcomes.

5.1 Biochemical Processes of Inflammation

Inflammation is a natural response of the body's immune system to injury or irritation. It involves a cascade of biochemical events that include the release of inflammatory cytokines, prostaglandins, and other mediators. While inflammation is essential for initiating repair processes, chronic inflammation can contribute to spinal pain, degeneration, and other complications.

Like any tissue in the body, when spinal tissues are injured—whether through acute trauma, repetitive strain, or degenerative changes—an inflammatory response is triggered. Key players in this process include cytokines such as tumour necrosis factor-alpha (TNF-alpha), interleukins (e.g., IL-1, IL-6), and prostaglandins (e.g. PGE2). These molecules signal the immune system to act, leading to increased blood flow, the recruitment of immune cells, and the release of enzymes that can further damage tissues if not regulated.

In chronic conditions like osteoarthritis or degenerative disc disease, this inflammatory response can become persistent via upregulation of nerve signals (particular C fibres) which release other molecules (such as substance P) both at the spinal cord (dorsal root ganglion) and at the site of injury. Local release of substance P is associated with further degradation of tissue (such as tendinopathy). Elevated levels of such inflammatory mediators can lead to a cycle of tissue damage and inflammation, exacerbating pain and functional impairment.

5.2 Pharmacological Approaches

Pharmacological interventions aim to modulate the inflammatory response to alleviate pain and promote healing. Nonsteroidal anti-inflammatory drugs (NSAIDs) are commonly used to reduce inflammation and pain. These medications work by inhibiting cyclooxygenase (COX) enzymes, which are involved in the production of prostaglandins, key mediators of inflammation. The downside of such drugs, when taken for more than a few days, is that amongst other things, they have a negative effect on the healing of connective such as tendon, muscle ligaments.(NSAIDS inhibit SCLERAXIS).

Corticosteroids are another class of anti-inflammatory drugs that can be administered orally or via injections. They work by suppressing the immune system's inflammatory response. Whilst they can be highly effective in the short term for reducing pain and inflammation, in the longer term corticosteroids will weaken connective tissue and cartilage, suppress stem cells and can have other serious have side effects, including bone density loss and increased susceptibility to infections.

More recent advancements include biologic agents that target specific inflammatory cytokines. Drugs like TNF-alpha inhibitors and IL-6 blockers are used for conditions such as rheumatoid arthritis and ankylosing spondylitis, showing promise in reducing inflammation and improving spinal health.

5.3 Non-pharmacological, Non-invasive Approaches

Non-pharmacological strategies focus on modulating inflammation through lifestyle changes and alternative therapies. Diet plays a crucial role in managing inflammation; foods rich in omega-3 fatty acids (such as fish and flaxseed) and antioxidants (found in fruits and vegetables) can help counteract inflammation. Conversely, a diet high in processed sugars and trans fats may exacerbate inflammatory processes.

Physical therapy and exercise are also vital components of non-pharmacological management. Regular, moderate exercise can help reduce systemic inflammation.

Other methods include the use of heat or cold therapy, acupuncture, and mindfulness practices. Heat therapy can increase blood flow and reduce muscle tension, while cold therapy helps minimise inflammation and numb pain. Acupuncture, though perhaps not fully understood, is thought to modulate pain pathways and reduce inflammation through the release of endorphins and other biochemical changes.

More sophisticated non-pharmacological, non invasive approaches may involve modalities such as (high power, super pulsed) Class IV LASER therapy which has been demonstrated in recent years to be able to reduce prostaglandin, especially PGE2, mediated pain and inflammation in much the same way as NSAIDs (Bjordal et al. 2006) but crucially without the adverse effects on SCLERAXIS. The limitation of LASER therapy is perhaps the depth at which laser light can penetrate into the body. Whilst this does vary with the wavelength used (around 905 nm being the optimum in terms of penetration depth (Kaub and Schmitz 2023),most respected authorities state that 90% of the energy does not penetrate more than 7 mm below the surface of the skin (Kaub and Schmitz 2022) hence the need for HIGH POWER in order to have a therapeutic effect for the 10% that goes any deeper. The problem then becomes avoiding a super heating effect which could be detrimental to the enzymatic reactions on the body; this is overcome though with use of super pulsed (as opposed to gated) Lasers (a good example of this would be the 300 watt Laser by EMS / Swiss Doloclast).

TECAR Therapy is another electrotherapy modality, it is common in main land Europe and is now emerging in the UK. TECAR (*particularly when in "Resistive" mode*) has the ability to pass an alternating electro-magnetic current through large areas of body, depending on where the plate and probe are placed. TECAR can increase healing and recovery times, by improving tissue drainage, lubrication and stem cell proliferation (at 448 kHz) (Hernández-Bule et al. 2014).

For chronic pain and inflammation, shockwave therapy is well established as being able to remove substance P as well as down regulate C pain fibre activity (Poenaru et al. 2022), even in chronic conditions such as osteoarthritis, amongst its 100 or so mechanisms of action (which like TECAR therapy also include stem cell activation (Rinella et al. 2018; Wang et al. 2024; Zhang et al. 2018; Alshihri et al. 2020) and improved production of LUBRICIN (Poenaru et al. 2022) (a glycoprotein required by the myofascial system for healthy tissue gliding).

5.4 *Biochemical Processes in Spinal Degeneration*

Spinal degeneration, such as disc degeneration and facet joint arthritis, involves complex biochemical processes that affect spinal health. The degeneration of intervertebral discs, for instance, is marked by changes in the biochemical composition of the discs. The nucleus pulposus, the gel-like core of the disc, gradually loses its water content and proteoglycans, leading to decreased elasticity and shock-absorbing capacity.

Matrix Metalloproteinases (MMPs)

Matrix metalloproteinases (MMPs) are enzymes that break down the extracellular matrix of spinal tissues. Their activity is regulated by tissue inhibitors of metalloproteinases (TIMPs). In degenerative conditions, an imbalance between MMPs and TIMPs can lead to excessive breakdown of disc and joint matrix components, contributing to degeneration and pain.

6 Biological Mechanisms in Spinal Regeneration: Unveiling the Cellular Symphony

Spinal regeneration is an area of intense research and clinical interest due to its potential to address the limitations of current treatments for spinal injuries and degenerative conditions. The complexity of spinal regeneration involves intricate biological mechanisms, including the roles of stem cells, growth factors, and cellular interactions. Understanding these mechanisms is crucial for developing effective regenerative therapies and advancing our ability to restore spinal function and health.

6.1 *Stem Cells: The Building Blocks of Regeneration*

Stem cells are undifferentiated cells with the remarkable ability to develop into various specialized cell types. In spinal regeneration, they play a pivotal role in repairing and regenerating damaged tissues. There are several types of stem cells involved in spinal regeneration, each with unique properties and potential applications.

1. **Mesenchymal Stem Cells (MSCs)**

   MSCs, found in various tissues including bone marrow, adipose tissue, and the umbilical cord, are particularly promising for spinal regeneration. They have the capacity to differentiate into multiple cell types, including chondrocytes, osteoblasts, and intervertebral disc cells. MSCs also possess anti-inflammatory properties and can modulate the immune response, which is beneficial in the context of spinal injuries and degeneration.

Ongoing research and development is aimed at the potential for MSCs to be transplanted into the damaged spinal area to promote repair and regeneration. They can integrate into the host tissue, differentiate into the required cell types, and release bioactive molecules that support tissue healing and reduce inflammation.

2. **Neural Stem Cells (NSCs)**

   NSCs are specialized stem cells that can differentiate into various neural cell types, including neurons, astrocytes, and oligodendrocytes. They hold significant potential for treating spinal cord injuries, where there is a need to replace lost neurons and support neuroprotection.

   NSCs can be derived from various sources, including the adult brain, spinal cord, and even from induced pluripotent stem cells (iPSCs). In experimental models, NSCs have demonstrated the ability to promote functional recovery and repair damaged spinal cord tissues by integrating into the host tissue and forming new neural connections.

3. **Induced Pluripotent Stem Cells (iPSCs)**

   iPSCs are generated by reprogramming somatic cells to a pluripotent state, allowing them to differentiate into virtually any cell type. iPSCs offer a personalized approach to spinal regeneration, as cells can be derived from the patient's own tissues, reducing the risk of immune rejection.

   iPSCs have shown promise in generating spinal disc cells, neurons, and other spinal tissue types. Research is ongoing to optimize their use for transplantation and to understand their long-term safety and efficacy.

6.2 Growth Factors: Molecular Signals for Healing

Growth factors are proteins that regulate cellular growth, differentiation, and tissue repair. In spinal regeneration, they play a crucial role in orchestrating the repair processes by promoting cell survival, proliferation, and tissue formation.

1. **Bone Morphogenetic Proteins (BMPs)**

   BMPs are a group of growth factors that play a key role in bone and cartilage formation. In spinal regeneration, BMPs such as BMP-2 and BMP-7 are used to enhance the repair of spinal bone and disc tissues. These factors stimulate the differentiation of stem cells into osteoblasts and chondrocytes, promoting bone and cartilage formation.

   Clinical applications of BMPs include their use in spinal fusion surgeries, where they are combined with bone grafts to enhance bone healing and fusion rates.

2. **Transforming Growth Factor-Beta (TGF-beta)**
   TGF-beta is involved in various aspects of tissue repair and regeneration. It regulates the extracellular matrix production and modulates cellular responses to injury. In the context of spinal degeneration, TGF-beta can help in disc repair by promoting the synthesis of matrix components and modulating the inflammatory response.
   However, the role of TGF-beta is complex, as excessive activity can lead to fibrosis and scar tissue formation. Balancing its effects is crucial for effective spinal regeneration.
3. **Vascular Endothelial Growth Factor (VEGF)**
   VEGF is critical for angiogenesis, the formation of new blood vessels, which is essential for tissue repair and regeneration. In spinal injuries, VEGF promotes the growth of new blood vessels into the damaged area, facilitating the delivery of nutrients and oxygen, which supports the repair process.
   VEGF-based therapies are being explored to enhance vascularization in spinal tissue and improve healing outcomes.

6.3 Cellular Interactions: The Symphony of Regeneration

The process of spinal regeneration involves a dynamic interplay between various cell types and their microenvironment. Cellular interactions are crucial for coordinating repair processes and ensuring functional integration of newly formed tissues.

1. **Cell-Matrix Interactions**
   The extracellular matrix (ECM) provides structural support to cells and influences their behaviour. In spinal regeneration, interactions between cells and the ECM are vital for tissue repair. For instance, disc cells and bone cells interact with ECM components such as collagen and proteoglycans, which affect their proliferation and differentiation.
   Engineering biomaterials that mimic the natural ECM can support cell growth and tissue formation, enhancing regenerative outcomes.
2. **Cell-Cell Interactions**
   Cells in the spinal tissue communicate through various signalling pathways and direct interactions. For example, during spinal disc repair, interactions between chondrocytes and surrounding cells influence matrix production and repair processes.
   Additionally, stem cells can influence the behaviour of resident cells through paracrine signalling, where they secrete factors that modulate the local environment and promote healing.
3. **Immune System Interactions**
   The immune system plays a dual role in spinal regeneration. While it is essential for clearing damaged tissues and fighting infections, excessive inflammation can impede healing. Understanding how stem cells and growth factors interact

with immune cells is crucial for developing strategies that balance immune responses and support tissue repair.

Strategies such as immunomodulatory therapies or the use of anti-inflammatory agents can help create a favorable environment for spinal regeneration.

7 Timing and the Regenerative Window for Spinal Health: Synchronizing Interventions with Natural Healing

The concept of timing is critical in managing spinal health, influencing the effectiveness of both surgical and conservative treatments. Understanding the "regenerative window"—the optimal period during which interventions can align with the body's natural healing processes—can greatly enhance outcomes and improve overall spinal health. This section delves into the importance of timing in spinal health management, exploring when to opt for surgical interventions, when conservative treatments are more appropriate, and how timing can harmonize with the body's intrinsic repair mechanisms.

7.1 The Regenerative Window: An Overview

The regenerative window refers to the period during which the body's natural healing processes are most responsive to therapeutic interventions. Timing within this window can significantly impact the effectiveness of treatments and the overall recovery process. Different spinal conditions have distinct windows for optimal intervention, influenced by the severity of the condition, the body's healing capacity, and the nature of the treatment.

7.2 Surgical Interventions: When and Why

Surgical interventions are often considered when conservative treatments have failed, or when there is a clear structural issue that requires correction. The timing of surgical intervention can greatly affect the outcome and recovery process.

1. **Acute Spinal Injuries**
   In cases of acute spinal injuries, such as fractures or herniated discs with significant neurological deficits (such as severe weakness), timely surgical intervention can be crucial. Immediate or early surgical intervention helps to decompress the spinal cord or nerves, and stabilise the spine. Delays in surgery for acute injuries can result in prolonged nerve compression and potential irreversible damage.

Chronic Spinal Conditions
For chronic conditions like degenerative disc disease or spinal stenosis, the timing of surgery can be less urgent but still critical. Surgical options, such as spinal fusion or disc replacement, are typically considered when conservative treatments (like physical therapy and medications) have not provided sufficient relief. The decision to proceed with surgery should align with the patient's symptom severity, quality of life, and overall health status. Waiting too long may result in worsening of the condition, which could complicate surgery and extend recovery time.

2. **Post-Surgical Timing**
Post-surgery, timing is also crucial for rehabilitation. Early mobilisation and adherence to a structured rehabilitation program can improve recovery outcomes and reduce the risk of complications. Delaying rehabilitation can lead to muscle atrophy, joint stiffness, and prolonged pain. On the flip side, for surgeries such as spinal fusion, delayed or restricted activity may be required for a period of 3–6 months, possibly also involving the use of spinal braces in order to ensure adequate bonding (fusion) of the prothesis with vertebral bone.

7.3 *Conservative Treatments: Timing and Approach*

Conservative treatments, including physiotherapy, osteopathy, chiropractic, medications, and lifestyle modifications, are often the first line of defence for managing spinal conditions. The timing and duration of these treatments play a vital role in their effectiveness.

1. **Early Intervention**
For many spinal conditions, early conservative treatment can prevent the progression of symptoms and reduce the need for more invasive procedures. For example, early physical therapy for acute low back pain can prevent chronic pain development and improve function. Addressing risk factors such as poor posture, obesity, and lack of exercise early on can also mitigate further deterioration of spinal health.
2. **Progressive Approach**
Conservative treatments should be adjusted based on the patient's response. If initial treatments do not yield the desired results, a progressive approach may be necessary, including more intensive physical therapy, different physical therapy modalities (spinal decompression therapy, LASER, TECAR, shock wave therapy (focused or radial), medical acupuncture, adjustments in medication etc. The timing of these adjustments is crucial to address symptoms effectively and adapt to the evolving needs of the patient.
3. **Chronic Management**
For chronic conditions, ongoing conservative management is key. Regular monitoring and adjustments to treatment plans are essential to managing symptoms

and maintaining function. Timing in this context involves balancing the frequency of interventions with the patient's lifestyle and treatment tolerance to achieve the best long-term outcomes.

7.4 Aligning with Natural Healing Processes

Understanding and aligning treatments with the body's natural healing processes can optimize spinal health management. The body's regenerative capacity varies depending on the type and extent of injury or degeneration.

1. **Inflammatory Phase**
   The initial inflammatory phase following injury or surgery involves swelling, redness, and pain as the body works to heal the damaged tissues. Timing interventions to manage inflammation—such as using NSAIDs, applying ice, Class IV Laser therapy, TECAR, therapeutic ultrasound, or engaging in gentle stretching and strengthening —can support this phase and prevent complications.
2. **Repair Phase**
   During the repair phase, the body focuses on rebuilding damaged tissues. This is a critical time for interventions that support tissue repair, such as physical therapy to improve mobility and strength and cellular responses. Dry needling, Class IV Laser, shock wave therapy may have supportive roles here. Ensuring that treatments are aligned with this phase can enhance healing and prevent long-term complications.
3. **Remodelling Phase**
   The remodelling phase involves the long-term reorganisation of new tissue. This phase can last for months or even years. Timing interventions to support this phase, such as continuing physical therapy, spinal strengthening and making lifestyle adjustments, is important for achieving lasting improvements in spinal health.

8 Regenerative Approaches in Spinal Health: Techniques and Interventions for Enhanced Healing

Regenerative medicine has emerged as a transformative field in spinal health, offering innovative techniques and interventions aimed at repairing, regenerating, and revitalising spinal tissues. This comprehensive overview explores the diverse range of regenerative approaches available, including advanced surgical procedures, biological scaffolds, and emerging therapies designed to enhance spinal regeneration and improve patient outcomes.

8.1 *Surgical Procedures: Advancements in Precision and Effectiveness*

Surgical interventions have evolved significantly, incorporating regenerative principles to enhance outcomes and promote tissue repair.

1. **Minimally Invasive Surgery**
   Minimally invasive spinal surgery (MISS) represents a significant advancement over traditional open surgery. Techniques such as endoscopic discectomy and microdiscectomy involve smaller incisions and reduced tissue disruption. Other advanced techniques as Oblique or Anterior, Lumbar Interbody Fusion (OLIF / ALIF) offer ways of removing damaged intervertebral discs through the abdomen, without disturbance to posterior structures such as spinal muscles and nerves. These procedures are designed to address issues like herniated discs or spinal stenosis with less trauma to surrounding tissues, leading to faster recovery and reduced pain.

   Spinal Fusion and Artificial Disc Replacement
   Spinal fusion, a procedure that involves joining two or more vertebrae to stabilise the spine, has seen enhancements with the use of regenerative materials. Biologics such as bone grafts, synthetic materials, growth factors and radiofrequency electromagnetic current therapy (such as "PEMF") may be employed to improve fusion rates and support the healing process (Weinstein et al. 2023).
   Disc replacement surgery, which involves replacing a damaged intervertebral disc with an artificial one, aims to restore spinal motion and alleviate pain. Recent advancements include the development of biomechanically optimised disc prostheses that better mimic the natural function of the spine, promoting more effective integration and long-term health.

2. **Regenerative Spine Surgery**
   Innovative techniques in regenerative spine surgery focus on incorporating regenerative elements directly into surgical procedures. For example, surgeons may use stem cells, platelet-rich plasma (PRP), or growth factors during surgeries to enhance tissue repair and regeneration. These methods aim to accelerate healing, reduce inflammation, and improve functional outcomes.

8.2 *Biological Scaffolds: Support Structures for Tissue Regeneration*

Biological scaffolds play a critical role in spinal regeneration by providing a supportive matrix for new tissue growth and repair. These scaffolds can be made from natural or synthetic materials and are designed to mimic the extracellular matrix of native tissues.

1. **Natural Scaffolds**
   Natural scaffolds, derived from materials such as collagen, hyaluronic acid, or decellularised tissues, offer a biocompatible environment that supports cell adhesion and proliferation. Collagen scaffolds, for example, can be used to repair intervertebral discs or spinal ligaments by providing a structural framework that encourages tissue regeneration.
   Decellularized scaffolds are created by removing cellular components from donor tissues, leaving behind the extracellular matrix. These scaffolds retain the natural biochemical cues necessary for cell growth and tissue integration, making them a promising option for spinal repairs.
2. **Synthetic Scaffolds**
   Synthetic scaffolds are engineered from materials such as polymers or ceramics to provide a customizable and controlled environment for tissue growth. For example, biodegradable polymers can be designed to degrade gradually, releasing growth factors or drugs to promote tissue regeneration over time.
   Advances in 3D printing technology have enabled the creation of patient-specific synthetic scaffolds that can be tailored to match the exact anatomical and mechanical requirements of the spinal area being treated. This precision enhances the effectiveness of spinal regeneration and integration with existing tissues.

9 Clinical Cases and Outcomes—Real-World Clinical Cases and Patient Outcomes are Presented to Illustrate the Effectiveness of Noval Approaches in Spinal Health

These following cases showcase how old-school, pragmatic diagnostic thinking, combined with the latest investigative approaches (such as MRI), can lead to personalised care plans based around individual alignment, tissue health, balanced function, cutting edge therapeutic technologies in combination with traditional manual / physical therapy approaches and a multidisciplinary, team approach to healthcare based mutual inter-disciplinary respect.

9.1 The Importance of Understanding Spinal MRI Imaging

One of the first times I encountered a previously undiagnosed tumour in the spine was when a teenage tennis player. He had been experiencing middle back pain on and off for a couple of years with increasing intensity and frequency. He had been treated by a couple of different physical therapists prior to coming to see me and various plausible explanations for his symptoms had been postulated.

Nonetheless, he had shown a poor response various therapeutic, so quite rightly one of the previous practitioners he had seen sent him for an MRI scan. The quite extensive MRI report from the radiologist had concluded that everything was more or less normal, adding further weight to the idea that his pain was due to too much tennis, poor movement patterns or just tight muscles.

However, when I examined him, it was obvious to me that he had an exaggerated pain response to pressure over the area of his pain in his mid-thoracic spine.

This prompted me to take a further look at his recent MRI images. The thing that stood out was that the MRI and subsequent report had focused on a different part of his spine just a few segments below where he was pointing to his pain. They had scanned the painful area but not with a full set of images, the axial T1 images, giving views of slices through the middle of the body (highlighting bone), were missing for this particular area.

I was able to arrange a follow up scan, with the correct sequences. This new MRI and a subsequent CT scan confirmed the presence of an osteoid osteoma and we were able to channel this patient towards the correct treatment that he needed.

9.2 *Spinal Shock Wave Therapy*

Case study 1 involved was a 25 year old teacher and yoga. She had an 18 month history of persistent lumbosacral pain, that was aggravated by long walks and by her other passion in life, hiking. She was wanting to go on round the world trip of a life time but her back pain was jeopardising this.

Previous physio, chiropractic / osteopathy and manipulative therapy for "SIJ" dysfunction had produced no improvement. We decided therefore to send the patient for an MRI scan, to investigate further, including crucially for (coronal) sagittal STIR imaging.

This revealed interspinous ligament inflammation T11/12, L2/3, L3/4. Despite no history of trauma. My suspicion was this was this source of Sara's lumbosacral pain.

Armed with the new diagnostic information, we initially performed some dry needling to these areas highlighted on her MRI. Although the dry needling was performed several segments higher, this reproduced her familiar pain pattern in her lumbosacral spine. Knowing that shock wave therapy was beneficial for chronic inflammation and lack of healing in collagen dense tissue such as ligaments, we the proceeded to treat her with course of dry needling (medical acupuncture) combined with radial Shock Wave Therapy to the relevant areas. After 6 treatments she was virtually pain free and able to go on long walks without any adverse reactions. We gave her a personalised exercise plan that steered her more toward preserving stability, rather than mobility of her T11—L3. She later sent a grateful message whilst on her trip of a lifetime to say she had made it unrestricted by pain.

9.3 *Combination Therapy Approach that Saved Case Study 2 from Spinal Fusion*

Our second case study in this chapter presented to our clinic after having suffered with years of chronic lower back pain and frequent spasms. He had seen numerous therapists and specialists over the years many of whom had told him his hips / pelvis were "out of alignment". However, manipulation (clicking) to realign his spine was delivering ever decreasing results and he had got to the point where is was no longer.

The gentleman in question did not have shootings pains down his legs and clinical signs did not indicate any "trapped nerves" (radiculopathy). Despite this we suspected he was suffering from one or two degenerative discs in lower spine. His MRI scan did indeed confirm our suspicions, this is an extract from his report "…Extensive Modic type 1 change L5-S1. Lower three motion segments show facet OA. Left paracentral annular tear L4-5 and right paracentral annular tear L5-S1. (No significant nerve root impingement)". In addition, on review of the images ourselves, we noticed slight signs of instability (dynamic retrolisthesis) of his L5 vertebra.

This correlated exactly with the clinical palpatory findings of (relatively) hypermobile L4/5 and L5/S1 segments with a very stiff / rigid spine at higher segments i.e. most of his movement (shear and torsional) stress was occurring through his L4/5 and L5/S1.

The initial aim was to alleviate pain from degenerative and inflamed load bearing structures. We chose (IDD) spinal decompression therapy for this task (an advanced form of targeted, computer controlled, oscillatory traction). Immediately following his initial sessions, the patient reported that for the first time in a long while he had experienced significant symptomatic relief. Thereafter, in addition we used dry needling (medical acupuncture) to help ease the reactive tension, spasm and contracture in the deep spinal muscles.

We explained to the patient that in his case it was not a good idea to repeatedly have his back "clicked", at least not in the area of the worn discs. Whilst this may have the potential to produce temporary, reflex relaxation of the local muscle spasm, this was unlikely to provide a long term fix. Worse still this might of been making the situation worse as repeated manipulation (joint clicking) would likely make his ligaments even more lax, thereby making his degenerative discs even more unstable.

We instructed the patient on the need for him to improve muscular stabilisation (core and multifidus etc.) specifically in his lower most segments and to improve mobility higher up in his spine.

We further added to this by referring the patient to a well known sports doctor, who was very experienced in spinal (P2G) prolotherapy (P2G = Phenol, Glycerol and Dextrose) and an excellent reputation working with top athletes and sports people.

It took time and patience (12–24 months) for the delicate balance of muscle and movement balance / symmetry to properly swing in patient's favour but he even made it back on to a football pitch without suffering from back spasms and eliminated thoughts of needing spinal surgery.

10 Future Horizons in Spinal Regeneration- Emerging Therapies and Cutting-Edge Innovations in Spinal Health

The field of regenerative medicine is continuously evolving, with emerging therapies offering new possibilities for spinal health and regeneration.

1. **Stem Cell Therapy**
   Stem cell therapy involves the transplantation of stem cells to promote the repair and regeneration of damaged spinal tissues. Various sources of stem cells, including mesenchymal stem cells (MSCs) and induced pluripotent stem cells (iPSCs), are being explored.
2. **Gene Therapy**

Gene therapy is an innovative approach that aims to address spinal health issues at the molecular level. By delivering specific genes or genetic materials into cells, researchers hope to correct defective genes, promote the expression of beneficial proteins, or regulate cellular processes to enhance tissue repair and regeneration.

In spinal health, gene therapy has potential applications in treating conditions such as degenerative disc disease or spinal cord injuries. For instance, introducing genes that encode for growth factors or anti-inflammatory proteins could support the healing process and improve tissue regeneration. Although still largely experimental, gene therapy holds promise for creating targeted, long-lasting effects on spinal health.

3. **Biological Augmentation**
   Biological augmentation involves using biological substances to enhance the body's natural healing processes. Platelet-rich plasma (PRP) therapy, for example, involves concentrating platelets from a patient's own blood and injecting them into the injured or degenerated spinal area. PRP contains growth factors and cytokines that can stimulate tissue repair, reduce inflammation, and promote healing.

   Similarly, exosome therapy, which utilises extracellular vesicles secreted by stem cells, is emerging as a technique to transfer regenerative signals between cells. Exosomes contain a variety of bioactive molecules, including proteins, lipids, and RNAs, which can influence cellular behaviour and support tissue repair.
4. **Bioelectronic Medicine**
   Bioelectronic medicine involves using electrical stimulation to modulate biological processes and promote healing. Techniques such as spinal cord stimulation and electrical nerve stimulation have been employed to manage chronic pain and enhance spinal regeneration.

   Emerging research in bioelectronic medicine explores how targeted electrical impulses can stimulate cellular processes, enhance tissue repair, and improve function.

10.1 Integration and Future Directions

The integration of these regenerative approaches offers a multi-faceted strategy for addressing spinal health issues. Combining surgical precision with biological scaffolds, advanced stem cell therapies, and emerging technologies can create synergistic effects, leading to improved outcomes and faster recovery.

Moreover, ongoing research into the mechanisms underlying spinal degeneration and repair will inform the development of more targeted and effective interventions. Collaborative efforts across disciplines, including biomechanics, cellular biology, and materials science, will continue to drive innovation and advance our understanding of spinal health.

11 Conclusion

Regenerative approaches in spinal health represent a dynamic and rapidly evolving field, offering new possibilities for repairing and revitalizing spinal tissues. From advanced surgical procedures and biological scaffolds to emerging therapies like stem cell treatments and gene therapy, these innovations are paving the way for enhanced spinal regeneration and improved patient outcomes. By leveraging these techniques and continuing to explore new advancements, we can move closer to achieving optimal spinal health and recovery, transforming the future of spinal care.

References

Alshihri A, Niu W, Kämmerer PW, Al-Askar M, Yamashita A, Kurisawa M, Spector M. The effects of shock wave stimulation of mesenchymal stem cells on proliferation, migration, and differentiation in an injectable gelatin matrix for osteogenic regeneration. J Tissue Eng Regen Med. 2020;14(11):1630–40. https://doi.org/10.1002/term.3126. Epub 2020 Sep 25

Bjordal JM, Lopes-Martins RA, Iversen VV. A randomised, placebo controlled trial of low level laser therapy for activated Achilles tendinitis with microdialysis measurement of peritendinous prostaglandin E2 concentrations. Br J Sports Med 2006 40(1):76–80; discussion 76–80. https://doi.org/10.1136/bjsm.2005.020842. PMID: 16371497; PMCID: PMC2491942.

Hernández-Bule ML, Paíno CL, Trillo MÁ, Úbeda A. Electric stimulation at 448 kHz promotes proliferation of human mesenchymal stem cells. Cell Physiol Biochem. 2014;34(5):1741–55. https://doi.org/10.1159/000366375. Epub 2014 Nov 12

Kaub L, Schmitz C. More than ninety percent of the light energy emitted by near- infrared laser therapy devices used to treat musculoskeletal disorders is absorbed within the first ten millimeters of biological tissue. Biomedicine. 2022;10:3204.

Kaub L, Schmitz C. Comparison of the penetration depth of 905 nm and 1064 nm laser light in surface layers of biological tissue ex vivo. Biomedicine. 2023;11(5):1355. https://doi.org/10.3390/biomedicines11051355. PMID: 37239026; PMCID: PMC10216207

Poenaru D, Sandulescu MI, Cinteza D. Biological effects of extracorporeal shockwave therapy in tendons: a systematic review. Biomed Rep. 2022;18(2):15. https://doi.org/10.3892/br.2022.1597. eCollection 2023 Feb

Rinella L, Marano F, Paletto L, Fraccalvieri M, Annaratone L, Castellano I, Fortunati N, Bargoni A, Berta L, Frairia R, Catalano MG. Extracorporeal shock waves trigger tenogenic differentiation of human adipose-derived stem cells. Connect Tissue Res. 2018;59(6):561–73.

Wang B, Shao W, Zhao Y, Li Z, Wang P, Lv X, Chen Y, Chen X, Zhu Y, Ma Y, Han L, Wu W, Feng Y. Radial extracorporeal shockwave promotes osteogenesis-angiogenesis coupling of bone marrow stromal cells from senile osteoporosis via activating the Piezo1/CaMKII/CREB axis. Bone. 2024;187:117196. https://doi.org/10.1016/j.bone.2024.117196. Epub 2024 Jul 14

Weinstein MA, et al. Pulsed electromagnetic field stimulation in lumbar spine fusion for patients with risk factors for pseudarthrosis. Int J Spine Surg. 2023;17(6):816–23.

Zhang H, Li ZL, Yang F, Zhang Q, Su XZ, Li J, Zhang N, Liu CH, Mao N, Zhu H. Radial shockwave treatment promotes human mesenchymal stem cell self-renewal and enhances cartilage healing. Stem Cell Res Ther. 2018;9(1):54. https://doi.org/10.1186/s13287-018-0805-5.

An Introduction to the Articular Cartilage of the Knee

Sedeek Mosaid

Abstract The articular cartilage of the knee is an integral part of the joint, providing a smooth, low-friction surface that allows motion and load transmission. Cartilage can be injured and ultimately wear out, leading to numerous clinical pathologies. This chapter describes the anatomy, biomechanics, and physiology of knee articular cartilage, emphasising the cellular structure, composition, and function of cartilage. Common cartilage injuries, diagnostic procedures, and workups are discussed, as are different imaging modalities, with an emphasis on magnetic resonance imaging. Therapeutic interventions for articular cartilage are described—including conservative management, microfracture, osteochondral grafting, and autologous chondrocyte implantation—and the potential and limitations of these interventions are emphasised. Rapidly evolving tissue engineering technologies, including scaffold design, are discussed in particular detail. Overall, the chapter emphasises the complexity of cartilage repair and the importance of patient-specific individualised clinical planning to optimise outcomes.

Keywords Knee joint · Articular cartilage · Cartilage injury · Repair · Chondrocytes

1 Introduction

The knee joint is a complex and essential structure, integral to mobility and weight-bearing. It is the body's largest synovial joint and is categorised as a modified hinge joint. It has three primary articulations: two between the tibial and femoral condyles and one between the patella and femur.

S. Mosaid (✉)
Department of Trauma & Orthopaedics, United Lincolnshire Teaching Hospitals NHS Trust, Lincoln County Hospital, Lincoln, UK

P. Lee (ed.), *Musculoskeletal Regeneration Medicine*,
https://doi.org/10.1007/978-3-031-77456-0_12

The joint is prone to deterioration from significant load and repeated movement, which can result in cartilage degradation over time (Prathap et al. 2020). Chondral and osteochondral injuries are commonly observed in clinical practice. These injuries are especially challenging due to the distinctive low-friction gliding surface of the knee's articular cartilage. However, despite the encouraging outcomes of several different therapeutic approaches to cartilage regeneration, none have yet proven to be the ultimate solution (Bhosale and Richardson 2008).

Different parts of the knee have different cartilage thicknesses. Predictably, the patellar cartilage is the thickest—with an average thickness of 4.1 mm—since it is subjected to high mechanical stresses. In the absence of mechanical stresses, cartilage becomes thinner. Articular cartilage lacks nerves and blood vessels. Therefore, cells are arranged so that diffusion can meet nutritional requirements, resulting in reasonably low cell densities (Ulrich-Vinther et al. 2003).

Articular cartilage is a form of hyaline cartilage covering the opposing synovial joint surfaces. It provides a gliding interface for motion, allowing load transmission with very low friction. The architecture of the tissue is precisely tuned for these functions, and it is made up of proteoglycans, collagen fibrils, and water (Sophia Fox et al. 2017). These three significant constituents are each essential for maintaining the biomechanical character of the tissue. Water constitutes 65–80% of the cartilage wet weight and has a higher concentration in the superficial tissue than in the deeper zones. This high water content allows for load-dependent deformation, provides nutrition, and facilitates lubrication, creating a low-friction gliding surface. The water is enmeshed within a network of collagen fibres (Lipshitz et al. 1975; Krych et al. 2021).

Collagen (mainly type II) forms 10–20% of the wet weight. As the principal component of the microfibrillar framework, it provides tensile strength to the cartilage, enabling it to withstand mechanical stresses without tearing. The collagen network is organised into different zones, each with specific arrangements that contribute to the overall biomechanical function of the cartilage (Krych et al. 2021; Vakonakis and Campbell 2007).

Proteoglycans, which also make up 10– 15% of the wet weight, contribute to the compressive strength of the cartilage. These macromolecules comprise protein-polysaccharide molecules: aggrecans, sizeable aggregating proteoglycan monomers, and smaller proteoglycans, such as decorin, biglycan, and fibromodulin (Bhosale and Richardson 2008; Krych et al. 2021).

Chondrocytes produce proteoglycans and release them in the tissue matrix. Proteoglycans consist of glycosaminoglycans. These molecules are composed of 2-saccharide chains, such as chondroitin sulfate or keratan sulfate, coupled to a protein core to form the aggrecan molecule, which is stabilised by a link protein including a central chain of hyaluronic acid. This structure is vital for the maintenance of fluid and electrolyte osmotic equilibrium in the cartilage, increasing the hydrostatic pressure inside it through the Donnan effect. The interaction between proteoglycans and interstitial fluid is necessary for the static cushioning capacity of the cartilage (Mow and Guo 2002; Roughley 2006).

Chondrocytes are the only cell type that resides in the articular cartilage; they function in the maintenance of the tissue's mechanical properties. Chondrocytes derive from mesenchymal stem cells and produce and maintain the cartilage matrix. They are responsible for regulating the synthesis and breakdown of the extracellular matrix (ECM) and tend to be highly responsive to mechanical stimuli, changing proteoglycan production according to changes in loading. The distribution and shape of chondrocytes differ according to the depth of the cartilage. Chondrocytes are also involved in cartilage repair mechanisms; however, their ability to proliferate and repair damage declines with age or injury. Understanding the biology and function of chondrocytes is thus essential for formulating treatments to maintain or rehabilitate cartilage health (Krych et al. 2021; Poole 1997; Diaz-Romero et al. 2004).

2 Clinically

Assessing a clinical chondral lesion requires a thorough clinical examination and detailed patient history. The clinical examination always begins with taking an extensive patient history to determine if the patient has experienced any acute traumatic events or chronic exposures that could have led to a symptomatic chondral lesion (Henn and Gomoll 2011; Mall et al. 2015).

The clinical presentation of knee articular cartilage lesions can vary from patient to patient. Often, a patient will report discomfort, which is limited to the compartment containing a defect and is usually exacerbated by weight-bearing activities. Knee swelling in response to activity is common, even with mild symptoms. Such swelling can help to distinguish between patellofemoral cartilage lesions and patellofemoral pain syndrome, which have similar symptoms, except that the latter does not typically cause swelling. Lesions on the femoral condyle and tibial plateau often have joint line tenderness and a mechanical symptom: the pain is often worse with weight-bearing or knee flexion in posterior lesions (Krych et al. 2021).

Therefore, throughout the physical examination, clinicians can note the presence of swelling and evaluate the range of motion and any catching or locking of the knee. They will also monitor any pain during the middle, terminal flexion and terminal extension motions. Furthermore, assessment of patellar tracking and alignment is essential, ensuring that the patella glides correctly and does not undergo a 'J sign', particularly on initial knee flexion, which is a sign of lateral subluxation. Associated pathology, such as ligament instability, limb malalignment, or meniscal pathology, should be identified during this physical examination and confirmed by imaging.

Determining a proper treatment strategy for a patient with an articular cartilage disorder is dependent on a meticulous examination of clinical findings and proper patient history taking (Krych et al. 2021).

3 Imaging Modalities for Articular Cartilage Pathologies

Imaging forms a crucial part of the assessment process and during surgical preparation for articular cartilage defects. The extent, location, and features of lesions are investigated using various imaging methods that all provide different forms of information on the status of the joint.

3.1 Radiography

Radiographic imaging is often the first modality used to assess knee joint pathology. Anteroposterior extension weight-bearing projections and posteroanterior 45° flexion Rosenberg views are recommended to assess tibiofemoral joint space narrowing, which could represent cartilage loss. Lateral and sunrise projections are used to assess the patellofemoral joint (PFJ). Full-length standing alignment radiographs are used in preoperative planning in cartilage repair operations; assessing the anatomic and mechanical axes helps determine the presence of malalignment, which could affect the outcomes of surgical treatment. Conventional radiographs adequately demonstrate alignment and joint space narrowing in osteoarthritis if this is already confirmed. However, they are limited in detecting early pathophysiological abnormalities in cartilage (Detterline et al. 2005).

3.2 Magnetic Resonance Imaging (MRI)

MRI is essential for detecting and diagnosing cartilage damage by providing detailed images of thickness, volume, and subchondral boundaries. It is advantageous for individuals displaying signs of intra-articular disease, including pain and oedema, even when weight-bearing radiographs appear normal. It is particularly valuable before substantial joint space narrowing; T2-weighted and short tau inversion recovery fluid-sensitive series are recommended for assessing these aspects.

MRI can be used to check any pathology of the meniscus and ligaments. The condition of subchondral bone should also be assessed, as severe bone marrow oedema following a previous marrow stimulation technique could be a predictor of autologous chondrocyte implantation (ACI) graft failure (Matsushita et al. 2023).

The choice of imaging technique depends on the clinical situation and the information needed to establish a correct diagnosis and to guide the treatment strategy. Each modality provides complementary information that is advantageous for reaching a complete overview of the state of the knee and guiding appropriate treatment (Mall et al. 2015; Detterline et al. 2005).

4 Classification of Cartilage Injuries

Classification of a cartilage injury is an integral part of assessing its severity and managing it. Different classification systems are used in clinical practice with differing criteria and aims. This section summarises some standard classification systems.

4.1 *The Outerbridge Classification*

The Outerbridge classification (Slattery and Kweon 2018) is one of the earliest and most commonly used classification systems. It is based on the depth and severity of the lesion, as follows:

Grade 0: Normal cartilage.
Grade 1: Cartilage softening in response to dynamic probing.
Grade 2: Partial thickness lesions less than 1.5 cm in diameter.
Grade 3: Lesions greater than 1.5 cm in diameter or with full-thickness fissure.
Grade 4: Complete cartilage loss with exposed subchondral bone.

4.2 *Cartilage Severity Score (CSS)*

The CSS is a total score (ranging from 0 to 100) that considers all articular surfaces of the knee (including the menisci). CSS is also a beneficial system for communication with the patient to explain the extent of the cartilage damage, including whether it is focal or global (Krych et al.2021).

4.3 *International Cartilage Repair Society (ICRS) Classification*

The ICRS (Krych et al. 2021) is much more detailed and allows for nuanced differences in pathology and management. It is classified as follows:

Grade 0: Normal intact cartilage.
Grade 1: Superficial cartilage lesions with softening or fissures.
Grade 2: Lesions, including fraying and fissures, that are less than 50% of the cartilage depth.
Grade 3: Lesions with cartilage loss of more than 50% of the cartilage depth down to the calcified cartilage layer.

Grade 4: Full-thickness cartilage lesions with exposure and involvement of the subchondral bone.

To summarise, the classification of cartilage injury has specific clinical value, which helps clinicians in diagnosis, communication and management. The choice of classification system may depend on the particular clinical context and the information needed for treatment planning.

5 Specific Factors in Cartilage Restoration

Numerous considerations must be addressed to improve patient outcomes in cartilage restoration procedures.

5.1 *Concerns About the Anterior and Posterior Cruciate Ligaments*

Addressing ligament deficiencies is essential for achieving optimal outcomes (Strobel et al. 2003). For example, effective functioning of the anterior cruciate ligament (ACL) prevents anterior translation and rotational loads, consequently protecting cartilage healing. ACL deficiencies can contribute to disruptions in cartilage repair pathways (Krych et al. 2021). Posterior cruciate ligament injuries are associated with high cartilage lesions, particularly those in the medial femoral condyle and patellofemoral areas.

5.2 *Meniscal Condition*

Meniscal condition is crucial as the meniscus contributes to load distribution inside the knee joint. A damaged meniscus may lead to heightened stress on cartilage, which could worsen defects.

It has been found that the contact area decreased by an average of 46% following medial meniscectomy. Therefore, cartilage restoration should not be performed without the protection of a competent meniscus, either via repair or allograft transplantation. Meniscus transplantation can restore more normal levels and distribution of forces (Verma et al. 2010).

5.3 Bone Oedema

Bone oedema can be assessed using MRI techniques. MRI aids in quantifying subchondral bone involvement, which is essential for designing an appropriate intervention and understanding the underlying condition. In the case of oedema, replacing cartilage and bone is usually recommended (Mall et al. 2015; Armstrong 1986).

5.4 Malalignment

Malalignment of the knee joint can lead to uneven stress distribution, which worsens cartilage pathologies because pressures are substantially increased within the impacted compartment.

Evaluating and addressing any malalignment is crucial for the success of cartilage regeneration operations (Krych et al. 2021; Dugdale et al. 1992). During cartilage restoration, alignment is modified to a midline position or slightly over-corrected, aiming to redistribute forces so that the regenerated or transplanted cartilage receives regular forces without overloading the contralateral compartment.

5.5 Patellofemoral Instability

Patients with patellofemoral instability require the evaluation of several anatomical variables that may elevate the risk of recurrent dislocation. Surgical procedures to address the instability may include tibial tubercle osteotomy and medial patellofemoral ligament reconstruction to minimise stress on the cartilage (Dugdale et al. 1992).

5.6 Body Mass Impact on Joint Forces

Higher body mass index (BMI) translates into increased forces across weight-bearing joints, which may influence the outcomes of cartilage restoration procedures. BMIs above 30 have been linked to poorer outcomes. Although higher BMI adversely affects outcomes after microfractures, it is less clear if fresh osteochondral allograft tissue is similarly affected (Mall et al. 2015; Mithoefer et al. 2005).

5.7 *Defect Size*

Determining the size of the cartilage lesion is critical when selecting the appropriate treatment technique. The dimensions influence the frequency of symptoms and the possibility of effective restoration (Mall et al. 2015; Guettler et al. 2004). Smaller lesions allow for a broader range of treatment alternatives, such as microfracture or autologous chondrocyte implantation. In contrast, larger defects might require more complex procedures, such as osteochondral transplants.

5.8 *Defect Location*

A comprehensive evaluation of the defect location is essential for effective preoperative planning and improving the outcomes of cartilage repair procedures. Each component must be rigorously assessed to adjust the treatment to the patient's needs.

Lesions on the femoral condyles allow for the broadest range of possible cartilage regeneration techniques. The operating surgeon must meticulously detect and address any damage to the subchondral bone. Osteochondral transplants are optimal for articular cartilage lesions on the femoral condyle, particularly when the underlying subchondral bone is affected.

The tibial plateau has thinner cartilage than the femoral condyle or PFJ, making it more challenging to treat. Available cartilage restoration techniques are limited for this location, and early intervention is crucial to prevent progression to full-thickness defects (Mall et al. 2015).

6 Treatment Modalities of Articular Cartilage Injury

6.1 *Conservative Management*

Conservative treatment of articular cartilage lesions in the knee prioritises symptom relief over lesion repair. Such management is usually suitable for mild symptomatic or minor lesions when surgical intervention may result in more significant damage. The following may be considered as conservative treatments.

Pharmaceutical Treatment: Use of nonsteroidal anti-inflammatory drugs (NSAIDs) and analgesics to alleviate pain and inflammation.

Mechanical Interventions: These include weight reduction, rest, orthotic devices, and physical therapy to alleviate stress on the knee joint.

Nutritional Supplements: Integration of chondroprotective compounds, such as glucosamine and chondroitin phosphate, in addition to calcium and vitamins, to enhance joint health.

Intraarticular Injections: Administration of steroids and viscosupplementation to provide temporary symptom relief (Willers et al. 2003).

However, while these kinds of conservative therapies may yield favourable short-term outcomes, research indicates that asymptomatic or mild symptomatic lesions may progress over time, eventually resulting in a severe knee injury (Falah et al. 2010). Furthermore, no current data indicates that conservative treatment approaches result in structural improvement of cartilage (Falah et al. 2010).

6.2 Arthroscopic Debridement

Arthroscopic debridement is a procedure aimed at removing a loose cartilage flap causing mechanical symptoms and effusions. Restoration of the free edge to a stable edge is a means of palliating mechanical symptoms and discomfort and prevents the progression of the cartilage lesion due to mechanical stress on the unstable flap. The procedure allows for immediate weight-bearing and a reduced duration of recovery. It is also less costly than most other cartilage repair procedures and does not require substantial preoperative planning and staging.

Evidence shows that the technique works well in cases without concomitant disease and can improve pain, physical function, and quality of life (Anderson et al. 2017). In a study conducted on 40 patients with an average age of 54 years who were followed for an average of 24 months, Gross et al. reported that 72.1% of patients had good outcomes (Gross et al. 1991).

However, arthroscopic treatment may promote degenerative changes. The main downside of surgical debridement is its inability to restore normal articular cartilage congruency (Willers et al. 2003).

Despite this limitation, it is a reasonable first-line treatment option for patients with minor cartilage defects with no additional pathology. A significant advantage of the procedure is the short postoperative rehabilitation time, which is applicable if the principal symptom is mechanical (Krych et al. 2021).

6.3 Microfracture (MF)

MF is a surgical technique that addresses articular cartilage defects in the knee. It is designed to promote the proliferation of new cartilage by perforating the subchondral bone, thereby facilitating the creation of a blood clot that can evolve into new cartilage tissue. Marrow components include growth factors that facilitate the infiltration of fibrocartilage into the chondral lesion. MF is a viable alternative for minor full-thickness cartilage lesions measuring up to 2 cm, provided a healthy rim of cartilage surrounds them. Contraindications include generalised cartilage degradation, failure to keep to weight-bearing limitations post-procedure, and a non-contained defect (Steadman et al. 2001, 2010).

The procedure involves an arthroscopic awl, used to make many tiny holes (spaced 3–4 mm apart) in the subchondral bone. The awl is used to avoid the thermal effects associated with drilling and to generate a rough surface that facilitates the development of a fibrin clot. Keeping a steady ring of healthy, undamaged cartilage around the chondral lesion is essential for clot formation in that region.

Debriding the calcified cartilage layer to leave only subchondral bone in the defect is beneficial for site preparation. Blood and fat exuding from the site after the tourniquet is deflated must be confirmed to ensure adequate penetration of the subchondral surface (Steadman et al. 2001, 2010).

Research involving more than 200 MF-treated patients indicated that 75% had improvements, 20% remained stable, and only 5% deteriorated across a follow-up period of 3–5 years (Steadman et al. 1997). However, treatment efficacy is affected by the patient's age. In one study, improvement was observed in all patients over 18 months following the treatment, although a significant decline was noted in individuals over 40. Optimal prognostic results are often observed in patients aged 40 years or younger (Kreuz et al. 2006).

Other trials have indicated a decline in function after 2 years follow up, particularly compared with other treatment options for minor cartilage defects. Bone overgrowth occurs in approximately 60% of patients, leading to elevated failure rates (Bert 2015).

Nonetheless, MF continues to be a popular option for addressing cartilage defects owing to its less invasive approach and its potential to enhance joint functionality, particularly in younger individuals. However, the quality of the repair tissue and the durability of the results can vary, rendering patient selection and postoperative care essential to ensure the best outcomes (Falah et al. 2010; Krych et al. 2020).

6.4 Osteochondral Autograft Transplantation (OAT)

OAT is an appropriate procedure for treating young patients whose osteochondral lesions not exceeding 2 cm and involve subchondral bone. For OAT, the patient's own cartilage and viable bone are used to enhance healing potential.

The procedure involves arthroscopic debridement to excise unstable cartilage, producing a stable, enclosed crater, ideally circular in configuration. Cylindrical excision of subchondral bone is then conducted as close to the lesion margin as possible. Cylindrical osteochondral grafts are usually taken from the non–weight-bearing surfaces of the knee, such as the peripheral aspect of the medial or lateral trochlea or the intercondylar notch. Harvesting these grafts and preparing them for use in the recipient site requires advanced surgical expertise. The grafts are taken and implanted in a manner akin to that of the less common osteochondral allograft (Krych et al. 2020; Rowland et al. 2019). Postoperatively, the patient is placed on a progressive 3- to 6-week non-weight-bearing regimen (Falah et al. 2010).

Rimtautas Gudas and colleagues conducted a randomised controlled trial comparing OAT with MF for repairing cartilage defects in young, athletic patients with an average age of 24.3 years. After 10 years, OAT and MF produced significant clinical improvements over pre-surgical scores. However, outcomes were substantially better with OAT. Most young athletes, particularly those under 25 at the time of surgery, maintained high levels of physical activity after 10 years (Gudas et al. 2012).

Osteochondral autografting is an effective procedure for focal defects but is unsuitable for degenerative lesions. For youthful, physically active people with a certain kind of cartilage defect, it is a credible choice that can provide lasting relief from symptoms and restore the function of the affected joint (Krych et al. 2020; Rowland et al. 2019).

6.5 Osteochondral Allograft Transplantation (OCA)

OCA is often indicated for extensive lesions over 2 cm^2, mainly when the subchondral bone is impaired, leaving cell-based options unsuitable. It can be used as a salvage procedure in patients for whom previous procedures were unsuccessful and who are unsuitable for arthroplasty (Cavendish et al. 2019). The best candidates for OCA are those with monopolar defects and bone loss due to conditions like osteochondritis dissecans, trauma, tumour resection, or other salvage situations. The technique has decreased efficacy in individuals with osteoarthritis, avascular necrosis, or bipolar defects. In addition, patients with high BMI, smokers, those on corticosteroids, and individuals with inflammatory arthritis are poor candidates for OCAs because these factors correlate with higher failure rates.

Two main types of osteochondral allografts are shell grafts (which are less than 1 cm and based on subchondral bone) and deep grafts (Oztürk et al. 2006). Chondrocyte availability is highest in fresh allografts, which can be obtained within 24–72 h and contain live cells. However, fresh allografts carry a higher risk of disease transmission. Frozen allografts offer a safer alternative, with reduced immunogenicity and risk of disease transmission but lower chondrocyte availability (Oztürk et al. 2006; Cavendish et al. 2019).

The surgical technique involves debriding the cartilage lesion and subchondral bone to a stable, healthy rim. Specific instrumentation is used to create a press-fit fixation of the allograft. For uncontained lesions, contouring is necessary, and compression screw fixation may be required to stabilise the graft (Krych et al. 2020; Cavendish et al. 2019).

The main advantages of this procedure include the lack of donor site size limitations and morbidity, as well as the possibility of grafting mature hyaline cartilage. However, one of the biggest challenges in OCA is the logistical difficulty of securing young deceased donors to provide grafts, which limits its use.

There are concerns about long-term chondrocyte viability and graft resorption, particularly in the PFJ. OCA is technically demanding, and slow healing is expected. Post-surgery, weight-bearing is restricted for 6–12 weeks, and contact sports are prohibited for 6 months for femoral grafts and 12 months for tibial grafts (Gross et al. 2008).

In a long-term study by Levy et al. (2012), OCA outcomes were assessed for 129 femoral condyle lesions. Pain and knee function in the patients showed a significant postoperative improvement. The graft survival rates of the allograft were 82% at 10 years, 74% at 15 years, and 66% at 20 years. Older patients and those with multiple past knee surgeries had higher reoperation rates; fresh OCA transplants were more beneficial in younger patients (Levy et al. 2013).

To summarise, OCAs can benefit the treatment of large osteochondral defects. They provide a viable long-term solution to the problem but pose challenges relating to the availability of tissue and the reliability of the donor-to-recipient match (Falah et al. 2010; Krych et al. 2020).

7 Critical Components of Tissue Engineering for Articular Cartilage Repair

Tissue engineering of articular cartilage is a complex, multistep process comprising several essential components that must be well-coordinated to restore cartilage function.

7.1 Cells

Chondrocytes: Chondrocytes represent the primary cartilage cell type, traditionally employed in cell-based cartilage repair therapies, including ACI, which involves harvesting chondrocytes from an unaffected, non-weight-bearing region of the joint and then expanding them ex vivo.

Neonatal/Foetal Chondrocytes: These cells grow faster than adult chondrocytes and, therefore, more closely resemble native cartilage cells; they have higher content of proteoglycan and collagen types II and IX, but their availability is limited (Tuan et al. 2013; Kuo et al. 2005).

Mesenchymal Stem Cells (MSCs): MSCs can differentiate into many cell lineages, including chondrocytes, making them ideal sources of cartilage regeneration. They are also easily accessible and can be readily expanded. Adult tissue-derived MSCs can differentiate into chondrocytes, adipocytes, and osteoblasts. They show high proliferative potential and can be induced to secrete collagen type II, but they can also show hypertrophy in long-term cultures (Tuan et al. 2013). Growth factors, such as transforming growth factor-β (TGF-β) and bone morphogenetic protein (BMP), can induce MSCs to undergo chondrogenesis.

Adipose-derived stem cells are abundant and easy to obtain via minimally invasive procedures. However, they have lower chondrogenic capacity than those from other tissues. Synovial-derived stem cells show a high chondrogenic potential but are not without some fibroblastic characteristics (Tuan 2006).

7.2 *Scaffolds*

Scaffolds are a temporary source of mechanical support for cell growth and tissue improvement and play a vital role in cartilage tissue engineering. They are engineered to maintain a three-dimensional architecture and to retain seeded cells as they develop cartilage tissue. Critical features of scaffold biomaterials include biodegradability, non-cytotoxicity, mechanical competence, appropriate surface chemistry, and capacity to modulate cell activity. They must be readily malleable into different sizes and forms (Safran et al. 2008).

Natural scaffolds: Polymer-based scaffolds include fibrin, gelatin, and collagen. Collagen is a principal component of the ECM and maintains the cell phenotype. Carbohydrate-based scaffolds include hyaluronic acid, alginate, chitosan, agarose, and polyethene glycol. These have been widely used for hydrogel scaffolds, with remarkable outcomes in cell encapsulation, and can be modified to improve chondrogenesis. Hyaluronic acid scaffolds are highly biocompatible and resorbable, but their degradation products can cause chondrolysis (Haleem and Chu 2010).

Synthetic Polymers: Some commonly used synthetic materials are polylactic acid, polyglycolic acid, polycaprolactone, and polylactic-glycolic acid (PLGA); these can be woven or electrospun into nanofibres and are used in a range of commercial applications to induce cartilage and bone growth.

Composite Polymers: Composite polymers are materials that contain combinations of natural and synthetic types, exploiting the specific advantages of each (Haleem and Chu 2010; Ferretti et al. 2006).

Scaffolds are crucial to the success of tissue engineering strategies and are vital in creating the biomimetic environment that will provoke the growth and development of new tissue. They need specific material and design characteristics to achieve successful cartilage formation.

7.3 *Growth Factors*

Growth factors are biological polypeptides that can enhance cartilage regeneration by promoting cell proliferation, enhancing chondrogenesis, and assisting in the management of cartilage defects (Tuan et al. 2013).

Transforming Growth Factor-β Super-family: This family includes TGF-β1, BMP-2, and BMP-7, which stimulate cartilage repair through chondrogenic differentiation and increased cartilage ECM production. TGF-β is primarily known to enhance ECM synthesis and chondrogenesis and can reduce the catabolic activity of interleukin-1 (Fortier et al. 2011).

Bone Morphogenetic Proteins (BMPs): BMPs (including BMP-4, −6 and − 7) enhance chondrogenic effects by increasing collagen type II and proteoglycan production.

Insulin-like Growth Factor-1 (IGF-1): IGF-1 acts as a mitogen, increasing proteoglycan and collagen type II production. IGF-1 is essential for maintaining the integrity of articular cartilage, promotes anabolic effects for cartilage repair, and reduces catabolic effects. It acts synergistically with other growth factors like TGF-β and BMP-7 to enhance its effects (Tuan et al. 2013).

Fibroblast Growth Factors (FGFs): FGF-2 and FGF-18 bind to cell-surface receptors that stimulate anabolic pathways and inhibit catabolic enzymes, including aggrecanase.

Platelet-Derived Growth Factor (PDGF): PDGF is a chemotactic factor for mesenchymal cells, promoting cartilage formation with increased proteoglycan production and cell proliferation (Fortier et al. 2011; Tuan et al. 2013).

8 Autologous Chondrocyte Implantation

ACI is the principal procedure used for cell-based cartilage repair, in which chondrocytes are harvested from the patient, cultured in vitro, and then re-implanted. The aim is to replace the damaged surface of the articular cartilage and restore its function (Peterson et al. 2000).

8.1 Indications and Patient Selection

Lesion Characteristics: ACI is most commonly recommended for symptomatic ICRS grade III–IV lesions larger than 2 cm^2 in the femoral condyle or trochlear region.

Patient Profile: The technique is recommended for high-demand patients between 15 and 55 years who demonstrate high levels of motivation and willingness.

Previous Treatment Failure: ACI grafting is indicated when previous biological reconstructive interventions (e.g. mosaicplasty or MF) are insufficient, especially in a high-demand patient.

Lesion Size: ACI is a viable approach for symptomatic individuals with lesions more significant than 2 cm^2 and up to 12 cm^2.

Traumatic Chondral Lesions and Osteochondrosis Dissecans: ACI is associated with superior clinical results in managing traumatic chondral lesions and osteochondrosis dissecans, in which chondral destruction results from traumatic injury rather than degenerative mechanisms.

Early Intervention: Ideally, the surgery works best when done early, such that symptoms relating to the cartilage injury have been present for no longer than 2 years (King et al. 2002; Niemeyer et al. 2016).

8.2 *Contraindications for Autologous Chondrocyte Implantation*

Several Extensive Cartilage Defects: ACI may not be an appropriate option for patients with multiple or extensive cartilage anomalies, as these conditions can complicate operations and affect outcomes.

Advanced Osteoarthritis: People with this level of osteoarthritis are usually unsuitable for ACI because cartilage regeneration alone might not adequately address the degenerative condition.

Failure to Comply with Rehabilitation: Patients with successful results from ACI require strict compliance with a rehabilitation programme. Those patients who are unlikely to adhere to postoperative therapy may not be suitable candidates for the treatment (Niemeyer et al. 2016, Hirschmüller et al. 2011).

9 Technical Evolution of ACI

9.1 *First Generation ACI*

The technique starts with an initial surgery to extract cartilage, which is followed by the isolation and in vitro multiplication of chondrocytes. This phase is essential for acquiring an adequate number of cells for implantation. The cultivated chondrocytes are then implanted beneath a periosteal patch during an open surgical procedure (Krych et al. 2021).

Surgical Procedure: Arthroscopic evaluation is performed, and biopsies are preferably obtained from the superomedial margin of the femoral trochlea. If the disease penetrates this area or affects the PFJ, the superolateral trochlear margin may be used. The lateral side of the intercondylar notch is an alternative biopsy location that is frequently utilised in ACL repair. A standard biopsy specimen typically weighs 200–300 mg, including the whole cartilage surface and a minor segment of the underlying subchondral bone. Following the biopsy, the cells are preserved at 4 °C until processing occurs.

Exposure: A midline skin incision is often performed; prior surgical incisions and the lesion's location must be considered. Based on the defect's size and position, a small arthrotomy, either medially or laterally based, may be conducted to reduce postoperative complications such as quadriceps weakness and intra-articular adhesions.

The subvastus approach can be used for medial femoral condyle lesions, facilitating lateral patella subluxation during knee flexion.

All sclerotic bone must be excised prior to transplantation. In situations of OCD, drilling the bed post-debridement facilitates blood flow and graft integration. The defect size must be measured using a sterile ruler (a paper or aluminium template may be used instead, but it must be larger than the defect to fit the surface contour). The cultivated chondrocytes are then implanted beneath a periosteal patch during an open surgical procedure. The patch is removed from the proximal tibia and attached to the stable cartilage borders around the defect, with the cambium layer oriented towards the defect. The periosteal rim is secured with fibrin glue, and the cultured chondrocytes are administered beneath the periosteal flap that covers the defect.

This technique has several disadvantages, including the need for a long incision to reach the defect and harvest the periosteum, a two-stage process that could cause donor site morbidity, uneven cell distribution, and potential complications such as graft failure and hypertrophy of the periosteal flap. However, the initial technique established the groundwork for subsequent innovations (Jones and Peterson 2007; Minas and Peterson 1999).

9.2 Second Generation ACI

Second-generation ACI procedures were developed to overcome some of the drawbacks and complications of the first-generation methods, specifically graft hypertrophy and procedure invasiveness. They utilise a bioabsorbable collagen membrane instead of the periosteal patch, which temporarily covers the fixated lesion. This approach decreases the risk of graft hypertrophy and the need for surgical chondroplasty due to delamination.

Although the surgery requires an arthrotomy and fixation of the collagen membrane to the defect, it is less disruptive than the first-generation approach (Krych et al. 2021; Jones and Peterson 2007). Consequently, the incidence of graft hypertrophy has been dramatically reduced to 6% when using second-generation procedures (Jones and Peterson 2007; Minas and Peterson 1999). Clinical results with a second-generation scaffold-based ACI showed significant 2-year improvements in validated knee scores (Ossendorf et al. 2007).

9.3 Third Generation ACI

The third-generation technique for ACI is called Matrix-Associated Chondrocyte Implantation (MACI). In this technique, chondrocytes are cultured and placed on a three-dimensional biodegradable matrix or scaffold, which maintains the cells and helps to preserve their orientation and distribution (both of which are essential for effective cartilage healing). It also helps create a more uniform distribution of chondrocytes over the defect site, leading to more consistent tissue healing and better-quality cartilage. Surgery can be performed using a mini-arthrotomy, with fibrin glue delivered to adhere the cell-seeded membrane to the defect.

The MACI approach simplifies the procedure compared with previous iterations. The matrix can easily be shaped to fill the defect, so no periosteal patch is needed to provide additional tissue coverage, thereby avoiding issues such as hypertrophy.

A further distinguishing feature of third-generation methods is the use of growth factors in the ACI process. These factors can induce cell proliferation and differentiation, aiding the healing process (Mithoefer et al. 2011; Brittberg 2009). Knee function and activity metrics have improved after 2 years of follow-up using this technique.

9.4 Fourth-Generation ACI

The most recent generation of ACI focuses on one-stage methods that directly isolate and implant chondrocytes, optimising the technique and minimising the time lag between harvesting cells and implantation. This advance makes cartilage repair even more efficient, standardised, and cost-effective. The result is that ACI is more convenient and widely available to patients, and successful treatment is optimised for patients with abnormal cartilage (Krych et al. 2021).

10 Conclusion

Knee articular cartilage is a highly specialised tissue that is integral to the joint's health and function. Despite decades of research, there is little consensus on the best approach to treating injuries to this tissue. Cartilage injuries can be treated with approaches such as microfracture, osteochondral transfer, and ACI, but none are universally effective or long-lasting. In the coming years, emerging technologies that address the limitations identified in our review will be needed to achieve a definitive therapy for cartilage repair. Such developments must incorporate a greater understanding of the importance of biomechanics in cartilage repair and restoration, as well as a personalised approach to treatment that makes the best use of the available biological tools.

References

Anderson DE, Chahla J, Safran MR, Aman ZS. Arthroscopic mechanical chondroplasty of the knee is beneficial for treatment of focal cartilage lesions in the absence of concurrent pathology. Orthop J Sports Med. 2017;5(5):232596711770721. https://doi.org/10.1177/2325967117707213.

Armstrong CG. An analysis of the stresses in a thin layer of articular cartilage in a synovial joint. Eng Med. 1986;15(2):55–61. https://doi.org/10.1243/emed_jour_1986_015_018_02.

Bert JM. Abandoning microfracture of the knee: has the time come? Arthroscopy. 2015;31(3):501–5. https://doi.org/10.1016/j.arthro.2014.12.018.

Bhosale AM, Richardson JB. Articular cartilage: structure, injuries and review of management. Br Med Bull. 2008;87(1):77–95. https://doi.org/10.1093/bmb/ldn025.

Brittberg M. Cell carriers as the next generation of cell therapy for cartilage repair. Am J Sports Med. 2009;38(6):1259–71. https://doi.org/10.1177/0363546509346395.

Cavendish PA, Everhart JS, Peters NJ, Flanigan DC. Osteochondral allograft transplantation for knee cartilage and osteochondral defects. JBJS Rev. 2019;7(6):e7. https://doi.org/10.2106/jbjs.rvw.18.00123.

Detterline AJ, Goldstein JL, Rue JP, Bach BR. Treatment options for articular cartilage defects of the knee. Orthop Nurs. 2005;24(5):361–6. https://doi.org/10.1097/00006416-200509000-00012.

Diaz-Romero J, Nesic D, Grogan SP, Heini PF, Mainil-Varlet P. Immunophenotypic analysis of human articular chondrocytes: changes in surface markers associated with cell expansion in monolayer culture. J Cell Physiol. 2004;202(3):731–42. https://doi.org/10.1002/jcp.20164.

Dugdale TW, Noyes FR, Styer D. Preoperative planning for high tibial osteotomy. . The effect of lateral tibiofemoral separation and tibiofemoral length. Pub Med. 1992;274:248–64. Available at: https://pubmed.ncbi.nlm.nih.gov/1729010

Falah M, Nierenberg G, Soudry M, Hayden M, Volpin G. Treatment of articular cartilage lesions of the knee. Int Orthop. 2010;34(5):621–30. https://doi.org/10.1007/s00264-010-0959-y.

Ferretti M, Marra KG, Kobayashi K, Hoch AI, Randell CR. Controlled in vivo degradation of Genipin Crosslinked polyethylene glycol hydrogels within Osteochondral defects. Tissue Eng. 2006;12(9):2657–63. https://doi.org/10.1089/ten.2006.12.2657.

Fortier LA, Barker JU, Strauss EJ, McCarrel TM, Cole BJ. The role of growth factors in cartilage repair. Clin Orthop Relat Res. 2011;469(10):2706–15. https://doi.org/10.1007/s11999-011-1857-3.

Gross DE, Kroll M, Lutz RW, Ostrow S. Arthroscopic treatment of degenerative joint disease of the knee. Orthopedics. 1991;14(12):1317–21. https://doi.org/10.3928/0147-7447-19911201-06.

Gross AE, Kim W, Las Heras F, Backstein D, Safir O, Pritzker KPH. Fresh osteochondral allografts for posttraumatic knee defects: long-term follow-up. Clin Orthop Relat Res. 2008;466(8):1863–70. https://doi.org/10.1007/s11999-008-0282-8.

Gudas R, Kalesinskas RJ, Kimtys V, Stankevicius E, Tamosiunas R, Smailys A, Skurvydas A. Ten-year follow-up of a prospective, randomized clinical study of mosaic osteochondral autologous transplantation versus microfracture for the treatment of osteochondral defects in the knee joint of athletes. Am J Sports Med. 2012;40(11):2499–508. https://doi.org/10.1177/0363546512458763.

Guettler JH, Demetropoulos CK, Yang KH, Jurist KA. Osteochondral defects in the human knee. Am J Sports Med. 2004;32(6):1451–8. https://doi.org/10.1177/0363546504263234.

Haleem AM, Chu CR. Advances in tissue engineering techniques for articular cartilage repair. Oper Tech Orthop. 2010;20(2):76–89. https://doi.org/10.1053/j.oto.2009.10.004.

Henn RF, Gomoll AH. A review of the evaluation and management of cartilage defects in the knee. Phys Sportsmed. 2011;39(1):101–7. https://doi.org/10.3810/psm.2011.02.1867.

Hirschmüller A, Baur H, Braun S, Kreuz PC, Heitkamp HC, Niemeyer P. Rehabilitation after autologous chondrocyte implantation for isolated cartilage defects of the knee. Am J Sports Med. 2011;39(12):2686–696. https://doi.org/10.1177/0363546511404204.

Jones DG, Peterson L. Autologous chondrocyte implantation. Instr Course Lect. 2007;56:429–45.

King PJ, Bryant T, Minas T. Autologous chondrocyte implantation for chondral defects of the knee: indications and technique. Pub Med. 2002;15(3):177–84. Available at: https://pubmed.ncbi.nlm.nih.gov/12152980

Kreuz PC, Erggelet C, Steinwachs MR, Krause SJ, Konrad G, Niemeyer P. Is microfracture of chondral defects in the knee associated with different results in patients aged 40 years or younger? Arthroscopy. 2006;22(11):1180–6. https://doi.org/10.1016/j.arthro.2006.06.020.

Krych AJ, Pareek A, King AH, Johnson NR, Stuart MJ, Williams RJ. Cartilage injury in the knee: assessment and treatment options. J Am Acad Orthop Surg. 2020;28(22):914–22. https://doi.org/10.5435/jaaos-d-20-00266.

Krych AJ, Levy BA, Stuart MJ. Cartilage injury of the knee: state-of-the-art treatment and controversies. Springer Nature; 2021.

Kuo CK, Li WJ, Mauck RL, Tuan RS. Cartilage tissue engineering: its potential and uses. Curr Opin Rheumatol. 2005;18(1):64–73. https://doi.org/10.1097/01.bor.0000198005.88568.df.

Levy YD, Görtz S, Pulido PA, McCauley JC. Do fresh osteochondral allografts successfully treat femoral condyle lesions? Clin Orthop Relat Res. 2013;471(1):231–37. https://doi.org/10.1007/s11999-012-2556-4.

Lipshitz H, Etheredge R 3rd, Glimcher MJ. In vitro wear of articular cartilage. J Bone Joint Surg. 1975;57(4):527–34.

Mall NA, Harris JD, Cole BJ. Clinical evaluation and preoperative planning of articular cartilage lesions of the knee. J Am Acad Orthop Surg. 2015;23(10):633–40. https://doi.org/10.5435/jaaos-d-14-00241.

Matsushita T, Kuwashima U, Kondo E, Yoshiya S. Surgical treatment of cartilage lesions in the knee: a narrative review. J Joint Surg Res. 2023;1(1):70–9. https://doi.org/10.1016/j.jjoisr.2023.02.001.

Minas T, Peterson L. Advanced techniques in autologous chondrocyte transplantation. Clin Sports Med. 1999;18(1):13–44. Available at:. https://doi.org/10.1016/s0278-5919(05)70128-9.

Mithoefer K, Williams RJ, Warren RF, Potter HG, Spock CR, Jones EC. The microfracture technique for the treatment of articular cartilage lesions in the knee. J Bone Joint Surg. 2005;87(9):1911–20. https://doi.org/10.2106/jbjs.d.02846.

Mithoefer K, Williams RJ, Warren RF, Potter HG. Evolution and current role of autologous chondrocyte implantation for treatment of articular cartilage defects in the football (soccer) player. Cartilage. 2011;3(1_suppl):31S–6S. https://doi.org/10.1177/1947603511406532.

Mow VC, Guo XE. Mechano-electrochemical properties of articular cartilage: their inhomogeneities and anisotropies. Annu Rev Biomed Eng. 2002;4(1):175–209. https://doi.org/10.1146/annurev.bioeng.4.110701.120309.

Niemeyer P, Pestka JM, Kreuz PC, Steinwachs M, Südkamp NP, Schmal H. Autologous chondrocyte implantation (ACI) for cartilage defects of the knee: a guideline by the working group "clinical tissue regeneration" of the German Society of Orthopaedics and Trauma (DGOU). Knee. 2016;23(3):426–35. https://doi.org/10.1016/j.knee.2016.02.001.

Ossendorf C, Kaps C, Kreuz PC, Burmester GR, Sittinger M. Treatment of posttraumatic and focal osteoarthritic cartilage defects of the knee with autologous polymer-based three-dimensional chondrocyte grafts: 2-year clinical results. Arthritis Res Ther. 2007;9(2):R41. https://doi.org/10.1186/ar2180.

Oztürk A, Ozdemir MR, Ozkan Y. Osteochondral autografting (mosaicplasty) in grade IV cartilage defects in the knee joint: 2- to 7-year results. Int Orthop. 2006;30(3):200–4. https://doi.org/10.1007/s00264-005-0068-5.

Peterson L, Minas T, Brittberg M, Nilsson A, Sjögren-Jansson E, Lindahl A. Two- to 9-year outcome after autologous chondrocyte transplantation of the knee. Clin Orthop Relat Res. 2000;374:212–34. https://doi.org/10.1097/00003086-200005000-00020.

Poole CA. Articular cartilage chondrons: form, function and failure. J Anat. 1997;191(1):1–13. https://doi.org/10.1046/j.1469-7580.1997.19110001.x.

Prathap KJ, Arun KM, Venkatesh D. Healthy GAIT: review of anatomy and physiology of knee joint. Int J Curr Res Rev. 2020;12(06):01–8. https://doi.org/10.31782/ijcrr.2020.12061.

Roughley P. The structure and function of cartilage proteoglycans. Eur Cell Mater. 2006;12:92–101. https://doi.org/10.22203/ecm.v012a11.

Rowland R, Colello M, Wyland DJ. Osteochondral autograft transfer procedure: arthroscopic technique and technical pearls. Arthrosc Tech. 2019;8(7):e713–9. https://doi.org/10.1016/j.eats.2019.03.006.

Safran MR, Kim H, Zaffagnini S. The use of scaffolds in the management of articular cartilage injury. J Am Acad Orthop Surg. 2008;16(6):306–11. https://doi.org/10.5435/00124635-200806000-00002.

Slattery C, Kweon CY. Classifications in brief: Outerbridge classification of chondral lesions. Clin Orthop Relat Res. 2018;476(10):2101–4. https://doi.org/10.1007/s11999.0000000000000255.

Sophia Fox AJ, Bedi A, Rodeo SA. Articular cartilage. 2nd ed. CRC Press; 2017. https://doi.org/10.1201/9781315194158.

Steadman JR, Briggs KK, Rodrigo JJ, Hancock R. Microfracture technique for full-thickness chondral defects: technique and clinical results. Oper Tech Orthop. 1997;7(4):300–4. https://doi.org/10.1016/s1048-6666(97)80033-x.

Steadman JR, Rodkey WG, Rodrigo JJ. Microfracture: surgical technique and rehabilitation to treat chondral defects. Clin Orthop Relat Res. 2001;391:S362–9. https://doi.org/10.1097/00003086-200110001-00033.

Steadman JR, Rodkey WG, Briggs KK. Microfracture. Cartilage. 2010;1(2):78–86. https://doi.org/10.1177/1947603510365533.

Strobel MJ, Eichhorn HJ, Haupt G, Hofmann GO. Arthroscopic evaluation of articular cartilage lesions in posterior cruciate ligament—deficient knees. Arthrosc J Arthrosc Relat Surg. 2003;19(3):262–8. https://doi.org/10.1053/jars.2003.50037.

Tuan RS. Stemming cartilage degeneration: adult mesenchymal stem cells as a cell source for articular cartilage tissue engineering. Arthritis Rheum. 2006;54(10):3075–8. https://doi.org/10.1002/art.22148.

Tuan RS, Chen AF, Klatt BA. Cartilage regeneration. J Am Acad Orthop Surg. 2013;21(5):303–11. https://doi.org/10.5435/jaaos-21-05-303.

Ulrich-Vinther M, Maloney MD, Schwarz EM, Rosier R, O'Keefe RJ. Articular cartilage biology. J Am Acad Orthop Surg. 2003;11(6):421–30. https://doi.org/10.5435/00124635-200311000-00006.

Vakonakis I, Campbell ID. Extracellular matrix: from atomic resolution to ultrastructure. Curr Opin Cell Biol. 2007;19(5):578–83. https://doi.org/10.1016/j.ceb.2007.09.005.

Verma N, Kolowski JM, Maffulli N. The effects of medial meniscal transplantation techniques on intra-articular contact pressures. J Knee Surg. 2010;21(01):20–6. https://doi.org/10.1055/s-0030-1247787.

Willers C, Wood DJ, Zheng MH. A current review on the biology and treatment of articular cartilage defects (part I & part II). J Musculoskelet Res. 2003;7(3–4):157–81. https://doi.org/10.1142/s0218957703001125.

Knee—Meniscus and Cartilage Regeneration

Paul Lee, Tanvi Verma, Muhammad Urfan Saleem, Christian Michael Jury, and Andrew Edwards

Abstract The meniscus and cartilage are essential components of the knee joint, responsible for load-bearing, shock absorption, and joint stability. Injuries to these tissues, caused by trauma or degeneration, are common and significantly increase the risk of osteoarthritis. Current treatments, including meniscectomy and cartilage debridement, provide symptom relief but fail to restore the natural structure and function of the tissues, leading to suboptimal long-term outcomes. Regenerative medicine offers promising alternatives, utilising techniques such as stem cell therapy, platelet-rich plasma (PRP), gene therapy, and biomaterials to promote tissue repair and regeneration. Despite significant advancements, challenges remain, including immunological responses, limited tissue integration, and maturation of regenerated structures. Combining regenerative techniques holds promise for synergistic effects and improved clinical outcomes. Regenerative strategies have the

P. Lee (✉)
MSK Doctors, MSK House, Sleaford, UK

College of Health and Science, University of Lincoln, Lincoln, UK

United Lincolnshire Teaching Hospitals NHS Trust, Grantham, UK
e-mail: leep@mskdoctors.com

T. Verma
MSK Doctors, Sleaford, UK
e-mail: tanvi@mskdoctors.com

M. U. Saleem
Lincoln County Hospital, Greetwell Rd, Lincoln, UK
e-mail: muhammad.saleem33@nhs.net

C. M. Jury
St Martin Surgery, La Vielle École, La Rue de la Croix au Maitre, St Martin, Jersey
e-mail: c.jury@stmartinsurgery.gpnet.je

A. Edwards
Northern Lincolnshire and Goole NHS Foundation Trust, Scunthorpe, UK
e-mail: andrew.edwards5@nhs.net

P. Lee (ed.), *Musculoskeletal Regeneration Medicine*,
https://doi.org/10.1007/978-3-031-77456-0_13

potential to revolutionise treatment for meniscus and cartilage injuries by restoring functionality at a cellular level, offering durable solutions and reducing the progression of osteoarthritis. However, further research and innovation are essential to overcome current barriers and achieve widespread clinical application.

Keywords Meniscus injuries · Cartilage regeneration · Platelet-rich plasma (PRP) · Biomaterials and scaffolds · Tissue engineering · Extracellular matrix (ECM)

1 Introduction to Meniscus and Cartilage in the Knee Joint

The meniscus and cartilage are two integral tissues in the anatomy of the knee joint. They have been proposed to reduce the incidence and progression of osteoarthritis by protecting subchondral bone and have a crucial role in the normal kinematics of the knee. The meniscus is able to reduce the contact surface bearing by up to 80% of the compressive load across the knee joint. In addition, it also provides resistance against tension and shear stress and improves the stability of the joint. There are three areas in the meniscus: the inner avascular (or white-white) zone, the Middle transitional (red-white) zone, and the outer capsular vascularized (or red-red) zone. The inner avascular zone has little regenerative potential because of poor blood supply. Attempts to cannulate bleeding into the meniscus are invasive and do not allow proper nutrient supply. Therefore, alternative interventions are necessary for patients with defects in the inner zone.

The knee articular cartilage in an adult has numerous subtypes of collagen. The cartilages are made up of dense collagen type II fibers, a low superficial layer of water, and is supplemented by types IX and XI collagen, which forms the tissue's fibrillar matrix and provides tensile strength with deep radial collagen cross-linking with the bone. (Sibbel and Khan 2020). The more cartilage that exists, especially deep, the stronger in compression the cartilaginous matter is. The meniscus and cartilage in combination lead to smooth functioning of the knee joint.

Injuries to the knee are very common and can happen for a number of reasons, including trauma or wear and tear over time. As a result of the multi-directional forces in the knee, they are prone to tears. These tears are predominantly following traumatic episodes. Degenerative tears are not necessarily age related, but can occur as a result of internal damage. The articular cartilage and meniscus are often directly affected by knee injuries. Damage to these two structures can have a negative impact on how the joint functions. Significant damage to these tissues may require surgery and could eventually lead to osteoarthritis if not treated correctly. Therefore, it is important to understand the biomechanical properties, structure, histology, and biochemistry of these tissues in order to prevent injuries, develop targeted treatments, and properly manage post-injury care. (Bhan 2020).

The meniscus is a crescent-shaped cartilaginous structure between the femur and tibia. It has special gross, microstructure, and ultra-structure adaptations for mechanical force bearing (Guo et al. 2018). The medial meniscus is a C-shaped piece of cartilage located between the medial tibia plateau and femoral condyle. Morphologically, it has a thicker periphery and a thinner central portion. The meniscus consists of dense fibrocartilaginous tissues that fulfill a number of crucial functions in the knee joint similar to articular cartilage (AC), such as load bearing, force transmission, shock absorbing, deepening of the joint surface, improving lubrication and nutrition by redistributing synovial fluid flow, maintenance of smooth articulation, and joint stability. The meniscus is highly susceptible to injury, and meniscal tear is one of the most common knee injuries. The articular cartilage is a specialized connective tissue covering the articulating surface of synovial joints that plays an important role in enabling smooth low-friction movements, but knee cartilage injury is very prevalent. There are approximately five million meniscus and 30 million cartilage injuries worldwide each year. Meniscus and cartilage injuries usually co-occur and are more harmful than isolated injuries, the combination of which significantly increases the risk of knee osteoarthritis.

1.1 Anatomy and Function of the Meniscus and Articular Cartilage

The meniscus is a semicircular, C-shaped fibrocartilaginous structure located on the inner and outer sides of the knee joint between the femur and tibia. The lateral meniscus (LM) and medial meniscus (MM) are the two classifications that correlate to the lateral and medial sides of the knee joint, respectively. The meniscus is connected to the femur, tibia, and joint capsule by ligaments known as the meniscofemoral and meniscotibial or coronary ligaments; these ligaments assist in the stabilization and improved congruity of the knee joint. The meniscus is considered an active tissue, with the inner compartment having negligible vascularity and gradual changing into the outer compartment, which has a rich blood supply (Bhan 2020). This vascular structure is crucial in the healing process, where injury to the outer compartment has a significantly higher healing potential than that of the inner compartment.

The adult meniscus is composed of approximately 70% water and 30% solids, with the solid phase comprising glycosaminoglycan (GAG) and fibrous proteins. Hyaluronan is characterized as a GAG that consists of linear repeating disaccharide units and is a component of the extracellular matrix (ECM). Tensor forces at each site allow the reduction of local free strain energies, contributing to the alignment of the ECM of cartilaginous tissues. This unique architecture of the ECM is essential for the normal biomechanical functions of the meniscus (Jacob et al. 2019). The articular cartilage covers the contact surface of the femur and tibia, providing a low-friction surface to withstand a large hind limb load during walking and running.

The articular cartilage composition is broadly based on a number of subtypes. In healthy joint hyaline cartilage, there are main collagens (type II, IX, and XI collagen) and minor collagens (type III, IV, V, VI, X, XII, XIV, XVI, XXII, and XXVII collagens). In articular cartilage with any damage a pathological type I collagen could be found. Type II collagen is a fibrillar collagen that is restricted to cartilages, vitreous and intervertebral discs. The mature articular cartilage comprises more than 90–95% of type II cartilage.

1.2 Prevalence and Impact of Meniscus and Cartilage Injuries

Meniscus and cartilage injuries, though largely overlooked, affect a large number of people worldwide due to the aging population, changing lifestyles, and the rising prevalence of obesity. The estimated annual incidence of meniscus injuries is around 60 per 100,000 people, and at least 30% of patients post-meniscus surgery develop osteoarthritis within the next 10 years (Bhan 2020). Additionally, up to 60% of patients with articular cartilage injuries consequential to traumatic injuries will develop osteoarthritis in the same knee within decades (Ding et al. 2022). Meniscus and articular cartilage injuries, which may occur as isolated injures or concomitantly, are both integral components of the knee joint and share similar risk factors (e.g., age, obesity, and enduring hard work). Articular cartilage injury and degeneration are sequelae to meniscus injury and degeneration due to altered biomechanical stress on the knee joint. Their widespread nature and eventual severe sequelae underscore the importance of treatment of these common but often serious injuries.

Meniscus and cartilage have similar tissue structures and properties, and both are involved in joint lubrication, load bearing and shock absorption. Injury to either of these two tissues can activate inflammatory pathways in the joint that lead to long-term degenerative diseases, resulting in pain, limited range of joint motion and disability. Meniscus and cartilage surgery are the most frequent treatments for the corresponding injuries, but currently there are no effective regenerative strategies for partial loss of a meniscus to achieve successful tissue reconstruction. Total loss of meniscus can be treated with meniscal allograft transplantation (MAT). The mid-term and long-term efficacy of available treatments is poor, underlining the need for the development of novel treatment strategies. Due to their anatomical structure, location, biological microenvironment, and physiological functions, meniscus and cartilage injuries are both challenging to address. Current treatment approaches that go beyond meniscus or cartilage resection or replacement are of limited efficacy. These issues call for a better understanding of meniscus and cartilage injuries to develop effective treatment methods.

2 Current Treatment Approaches for Meniscus and Cartilage Injuries

Meniscus and cartilage injuries are among the most common degenerative conditions of the knee joint affecting millions of individuals worldwide. The knee is the largest joint in the body, and its function is compromised with any internal injury that can reduce the quality of life of the injured individual. Surgical treatments have changed from invasive procedures to minimally invasive approaches with improved patient outcomes, including pain and recovery time. Arthroscopic procedures include but are not limited to the resection of damaged tissues (resective approach), smoothing out the rough surface (debridement), or the use of tissue grafting procedures (reparative approach) aimed at addressing the cause of joint degeneration. The reviewed treatment modalities focus on addressing the mechanical need for the joint that is restored by the replacement of the damaged articular surfaces with synthetic scaffold devices. Unfortunately, replacement or reconstruction is not a regenerative approach for the restoration of the original joint tissue and its function (Pereira et al. 2019).

Meniscal injuries can occur from traumatic events in younger populations or degenerative changes in older individuals due to the increased load on the medial meniscus during weight-loading and squatting. Meniscal tears are categorized into horizontal tears, vertical tears, parrot-beak tears, bucket handle tears and complex tears, representing a combination of various tear patterns (Bhan 2020). The individual treatment of meniscal injuries typically depends on various factors, including age, tear pattern, and the vascularity of the meniscus.

Acute undisplaced tears can be managed with an unloading brace which has been shown to provide immediate pain relief and allows early mobilization and return to work. Further symptomatic relief and biological repair could be attained with the injections of biological enhancements into the tear for early healing.

Patients with a meniscal tear will typically present with symptoms of pain, clicking, difficulty in bending or fully straightening the knee. They may also present with repeated effusions and feeling of instability. They may have felt a popping sensation initially when getting up from the floor or on extending an flexing knee.

Investigations will start with radiographs of the knee to confirm that there is no bony injury. An MRI scan whether undertaken without the benefit of x rays is the choice of investigation. However, the tear can be missed occasionally depending upon the slices of the scanning software.

Surgical approaches include partial meniscectomy, complete meniscectomy, meniscal repair, and meniscal allograft transplantation. The simplest and most common surgical procedure for meniscal tears is partial meniscectomy where the tear is excised. In the case of a complete meniscal tear, a total meniscectomy is performed, removing the entire diseased meniscus. Although meniscectomy results in improved joint function and decreased pain, it modifies the knee joint biomechanical environment, which is altered for load transmission, shock absorption, and joint stability, thus resulting in the subsequent degeneration of the articular cartilage and the development of osteoarthritis. Moreover, complete meniscectomy is associated with a 70% incidence of degenerative osteoarthritis within 10 to 20 years postoperatively. The deleterious results of open partial or total menisectomies leads to arthroplasty in younger symptomatic patients (Roos et al. 2001). With the advance of better meniscal repair techniques, one could repair most meniscal tears thus preventing early joint degeneration. The cost of these repair systems occasionally outweighs the cost of degenerative joint disease (DJD) and its treatment. This overall cost may well include dependence on analgesic drugs, use of orthotics and walking aids.

2.1 Surgical Interventions: Meniscectomy and Cartilage Debridement

Meniscal tears and cartilage injuries of the knee joint are frequent, particularly in the younger active population. The most common surgical intervention performed for torn menisci is an arthroscopic partial meniscectomy (APM), which is one of the most performed orthopedic procedures globally (Bhan 2020). There is extensive scientific investigation of the meniscus, its anatomy, function, biomechanics, and pathophysiology of tears. Despite advances in arthroscopic techniques and tools, partial meniscectomy is still a prominent procedure (Pereira et al. 2019). APM involves resection of torn meniscal parts. For complex, irreparable tears, the best treatment option is a subtotal meniscectomy. Debridement of damaged cartilage, or microfracture of the subchondral bone in the case of chondral lesions, are the most performed procedures of cartilage lesions and injuries, especially in younger patients and in the early stages of osteoarthritis (OA, I–II according to the Outerbridge classification). The success of these procedures and acceptable results are based on decades of extensive scientific research and on knowledge of biomechanics, pathophysiology, and treatment options. However, meniscal and cartilage treatment procedures are still far from optimal, and comprehensive reviews on current understanding, as well as patient, injury, lesion, and procedure demand optimization, are justified. The development of smart materials that employ the biophysical and biochemical signaling of the microenvironment for the tissue repair process is an alternative approach. Reconstruction of full-thickness articular cartilage defects with non-biological and biological scaffolds in combination with various cell sources is a challenging task. Nonetheless, novel treatment options based on tissue engineering and regenerative medicine are intensively studied.

The addition of biological enhancements like bone marrow stimulation, use of platelet-rich plasma (PRP), bone marrow aspirate concentrate (BMAC) and other factors have led to repair of previously unrepairable tears.

Cadaveric investigations followed by laboratory strength testing has allowed us to repair radial tears, flap tears and horizontal cleavage tears.

In cases of a bucket handle (BH) tear of the lateral meniscus, it is not uncommon for the MRI to be reported as normal. The patient will classically state that the knee locks in full flexion, requires twisting of the knee to unlock and has lateral pain. Clinical suspicion of a LM tear has to be kept and more often than not an arthroscopic treatment is required.

BH tears are frequently white on white tears. They are occasionally diagnosed at the time of the ACL reconstruction. In publicly funded healthcare systems, there may be a significant delay between the initial injury, the MRI scan and subsequent surgery. In these circumstances, the surgeon has to decide whether to repair these chronic tears or excise them. In younger patients, all attempts should be made to save the meniscus. Not all BH tears require surgery. There will be a subset of patients who have a chronically displaced BH tear of the meniscus in the intercondylar notch, but are completely asymptomatic.

Similarly, a certain percentage of meniscal tears might heal or cause no symptoms at all. It becomes therefore difficult to decide to treat each and every meniscal tear. A pragmatic approach and discussion with the patient will be useful to decide the correct treatment.

Untreated Root tears have now been shown to lead to meniscal extrusion as a result of the loss of the hoop stresses and early compartmental degeneration. Whether to repair all root tears is a matter of debate, however with the current available literature, following informed consent, root repair should be undertaken in most cases. The post operative rehabilitation should also be discussed with the patient.

Synthetic or Allograft replacements have changed the way that a symptomatic post-meniscectomy knee can be managed. Focal defects may be treated with synthetic meniscal implants with reasonable success. At least for the lateral meniscus, insertion of such a scaffold at the time of primary resection might be an alternative option rather than waiting for the knee to become symptomatic.

Total loss of meniscal tissue can lead to repeated symptoms of effusion, pain and instability. If there is coronal malalignment or ligament deficiency, then that has to be corrected in the first instance. In younger patients, combined biological replacement in the form of corrective osteotomy, meniscal allograft transplantation and any cartilage surgery at the same sitting has shown good medium-term results.

The role of meniscal repair should be considered with relation to the socio-economic status of the patient. To undertake a meniscal repair and subject the patient to a few weeks of restricted weight bearing and prolonged rehabilitation has to be balanced with the early return to work and sports.

The tear should be reducible, and an adequate number of superior and inferior surface sutures and circumferential sutures should be inserted to stabilize the tear. All techniques of Inside Out, Outside -In and All inside techniques may need to be utilized.

A number of techniques can be utilized to improve meniscal repair healing at the time of the primary surgery. There has been extensive research into these methods and include Microfracture (marrow stimulation), Trephination of capsule and abrasion of the edges, introduction of a blood clot for Meniscal tears and injections of PRP, BMAC and other biologics.

New suturing and augmentation techniques offer promising approaches for the arthroscopic management of complex meniscal tears (Ghazi Zadeh et al. 2018). One recent augmentation technique to enhance meniscal healing is known as "meniscal wrapping". Arthroscopic Matrix-based Meniscus Repair AMMR® is a fully arthroscopic technique based on a meniscal repair using sutures and enhanced with a collagen matrix wrapping. Furthermore, biologic augmentation, such as autologous liquid bone marrow injection, can be added. Piontek et al. first described this technique in 2012 and, later, they reported 2 year and 5 year follow-up clinical studies. They found a relatively high survival rate and favorable functional and radiological outcomes, suggesting that combined and complex meniscal tears can be successfully and safely treated with the meniscal wrapping technique (Piontek et al. 2016).

Articular cartilage defects can be managed with a number of techniques. The medium to long term results of these techniques have remained an issue for debate. These techniques range from simple chondroplasty, marrow stimulation techniques, Osteochondral Autograft Transplantation (OATS) procedures to Autologous Cartilage Implantation. Before embarking on these procedures, it is vital to treat any malalignment or ligamentous issues, at the same time or prior to these treatments.

The methods most frequently used in this context (e.g., marrow stimulation techniques) can, however, at most generate hyaline-like and thus biomechanically inferior regenerates.

For smaller contained chondral defects, techniques like OATS have been shown to give good to excellent results.

Autologous chondrocyte implantation has been shown to be the most suitable cartilage reconstructive technique with the best tissue quality.

This however remains an expensive technique, which is not available globally and is a two-step surgical procedure.

Minced cartilage implantation, on the other hand, is a relatively simple and cost-effective 1-step procedure with promising biological potential and—at least in the knee joint—satisfactory clinical results. (AutoCart; Arthrex, Munich, Germany). (Salzmann et al. 2020).

2.2 *Limitations of Current Treatments*

Meniscus and cartilage injuries are prevalent musculoskeletal injuries in both athletic and non-athletic populations. Lack of adequate treatments leads to progressive change in joint architecture, which results in knee osteoarthritis. Current treatments such as surgical meniscectomy for meniscus injury and microfracture surgery for cartilage injury offer only temporary pain relief, and the underlying defects remain unaddressed. As a result, tissue regeneration technologies are emerging. Aside from bone marrow stimulation (microfracture), tissue engineering and regeneration technologies focusing on fibrocartilage may also contribute to the regeneration of articular cartilage (Jacob et al. 2019).

Subchondral changes seen at the time of the index accident may indicate significant chondral damage which could be one of the factors of early DJD despite the meticulous repair undertaken.

Meniscus and articular cartilage, which is hyaline cartilage, are both avascular tissue, and thus have limited inherent healing ability. Findings from previous studies show that defects in the vascularized zone of the meniscus may heal spontaneously with fibrous tissue, while those in the avascular zone do not heal (Roelofs et al. 2013). The healing potential is even lower in cartilage, especially surfaces on the weight-bearing area. Damage in cartilage easily expands to the surrounding area, propagating degeneration in older age to post-traumatic osteoarthritis. Current treatments of both meniscus and cartilage are aimed to relieve the symptoms of the injuries, such as loss of pain during motion, but do not treat the underlying defects.

Treatments still take the approach of focusing on the macro-scale biomechanical effect of injuries. As a result, cell-based and biomaterials-based tissue engineering techniques for the restoration of the tissue defects in meniscus and cartilage are emerging.

3 Regenerative Medicine in Orthopedics

In recent decades, there has been rapid development in the field of regenerative medicine. Regenerative medicine aims to repair, protect or replace damaged tissues or organs, and it includes various simple and complex techniques, such as application of stem cells, growth factors, scaffolds, gene engineering and tissue engineering. Repairing damaged tissues using only the body's own repair mechanisms has traditionally been regarded as the simplest approach to regenerative medicine. However, this is not feasible for many tissues or organs, in which case more effective and complex strategies, such as cell-based therapy or bio-scaffolding, are required (Wang et al. 2022). In orthopedics, regenerative medicine focuses on repairing or reconstructing damaged musculoskeletal tissues. Due to their limited self-repair capabilities, cartilage and menisci tissue engineering and regenerative medicine have become novel approaches for enhancing repair and regeneration. Regenerative medicine has been reviewed by various researchers from different medical fields over the decades, such as musculoskeletal tissue, peripheral nerves, thyroid gland, cartilage, skin, respiratory system, retina, cardiovascular tissue and organs, spinal cord and cochlea. However, preceding reviews have not focused on regenerative medicine in orthopedics, cartilage and meniscus.

While surgical interventions are usually performed to alleviate the symptoms, such procedures do not prevent the progression of the disease (Guo et al. 2018). Biological approaches aiming to either regenerate or rehabilitate the injured tissues, such as cell therapy, gene therapy, drug therapy, growth factor delivery combined with scaffolds and biomimetic scaffolds, have been developed and will be discussed in the following sections.

3.1 Principles and Techniques of Regenerative Medicine

Regenerative medicine within the field of orthopedics, emphasizes on prevention, repair, and reconstruction in both traumatic and degenerative diseases. This approach harnesses natural biological mechanisms for restoring the structure and function of wounded or degenerated tissues by using bioactive molecules, cells, and biomaterials (Wang et al. 2022). Both the meniscus and the articular cartilage have limited intrinsic healing capacity due to their avascular nature, hence ideal for regenerative interventions.

The foremost regenerative technique involves the use of mesenchymal stem cells (MSCs), due to their ability to differentiate into chondrocytes and produce extracellular matrix (ECM); components essential for the repair of the menisci and cartilage. These stem cells can be sourced from bone marrow, adipose tissue, or synovium and are combined with scaffolds or hydrogels to create a conducive environment for tissue regeneration. The scaffolds, provide a temporary matrix that mimics the natural ECM, allowing for cell attachment, proliferation, and differentiation. Advances in tissue engineering have also led to the development of biomimetic scaffolds that can be preloaded with growth factors such as transforming growth factor-beta (TGF-β) to enhance the regenerative process. These scaffolds are designed to degrade over time, coinciding with the gradual restoration of the tissue's native structure and function.

Biologics, such as platelet-rich plasma (PRP) and autologous conditioned serum (ACS), which are rich in growth factors and anti-inflammatory cytokines are other regenerative techniques. These biologics are injected directly into the injured area, thereby creating a more conducive micro-environment for tissue regeneration. Being rich in growth factors, such as transforming growth factor-beta (TGF-β), platelet-derived growth factor (PDGF), vascular endothelial growth factor (VEGF), and anti-inflammatory cytokines like interleukin-1 receptor antagonist (IL-1Ra). These bioactive molecules modulate the inflammatory response, promoting angiogenesis, and enhancing the proliferation and differentiation of cells involved in tissue repair. PRP in particular, accelerates the healing of meniscal and cartilage injuries by increasing the recruitment and activation of mesenchymal stem cells (MSCs) and enhancing extracellular matrix (ECM) synthesis.

3.2 Advantages of Regenerative Approaches in Knee Injuries

Traditional treatment modalities, including arthroscopic surgery, partial meniscectomy and cartilage debridement, alleviate symptoms and improve function in the short term, but they fail to restore the natural structure and function of the tissues. In contrast, biological restoration has the potential to restore the tissue at a cellular level, allowing for long-term joint health and function.

The minimally invasive nature of these methods results in reduced recovery times and lower complication rates. Intra-articular injections allow for delivery of therapeutic agents directly to the site of injury, thereby minimizing the trauma to surrounding tissues. This procedure can be undertaken on an outpatient basis, foregoing the need for hospitalization and extensive postoperative care. Fewer side effects are noted along with a lower risk of rejection or adverse immune responses as they work around the body's natural healing processes. This is particularly beneficial for patients with comorbidities or those not ideal for more extensive surgical interventions.

These approaches allow for personalized treatment options as a critical factor in the effective management of knee injuries. This will account for the variability in patients' tissue biology, injury patterns, and healing responses hence being able to tailor the therapies accordingly. Furthermore, the ability to combine cell-based therapies with biologics like growth factors or scaffolds enables clinicians to customize treatment protocols based on the specific characteristics of the injury and the patient's overall health status. This personalized approach can lead to more effective and sustained outcomes, particularly in complex cases where standard treatments might fall short.

Current surgical techniques, such as the meniscal scaffold or meniscus replacement, replace the lost meniscal tissue (Wang et al. 2022). However, the isolation of tissue-engineered scaffolds requires costly instruments and a complex preparation process, which may compromise the biocompatibility of the scaffolds. Furthermore, the use of cell-based therapy for the regeneration of meniscus and cartilage tissue has several drawbacks, including the difficulty of harvesting sufficient stem/progenitor cells, cell culture contamination, and a low cell survival rate following implantation (Guo et al. 2018). Therefore, cell-free strategies for the repair and regeneration of the meniscus and cartilage tissue via the recruitment of endogenous stem/progenitor cells have been developed.

4 Stem Cell Therapy for Meniscus and Cartilage Regeneration

Stem cell therapy leverages the unique biological properties of stem cells to repair and regenerate damaged tissues. Regenerative medicine is an exciting interdisciplinary field that holds great promise in the clinical and scientific endeavors of healing and defying the loss and aging of tissues and organs. The classical restorative approaches are either palliative or replacement via transplantation, which fails to resolve the underlying pathologies. Tissue/organ engineering is an innovative restorative approach that combines stem cell biology and bioengineering technology, with an aptitude to regenerate lost tissues/organ (Dai et al. 2021). Stem cells are bioactive building blocks for wear-and-tear-induced tissue/organ degeneration, allotransplant rejection, and graft failure, while the need for bioengineered scaffolds becomes paramount due to the highly dense avascular tissues including cartilage.

Meniscus or cartilage tissue engineering is a special subclass of cartilage restoration and has its uniqueness due to the heterogeneous cellularity and composition, diverse biophenotypes, and anatomical structures as the genesis of tissues. Recent progress in the selection of stem cell source and its conditioning utilization, biomimetic bioengineered scaffolds, and engineering-based organotypic culture has advanced the neotissue development to the next level requiring in vivo translational studies and clinical trials (Ding et al. 2022).

4.1 Types of Stem Cells Used in Regeneration

Various stem cell types can be utilized in meniscus and cartilage regeneration based on the origin, cell potency, and tissue type. Both adult and embryonic stem cells can be differentiated into different cell types to repair the injured cartilage. The primary types of stem cells used in orthopedic regenerative medicine include mesenchymal stem cells (MSCs), induced pluripotent stem cells (iPSCs), and embryonic stem cells (ESCs).

Mesenchymal Stem Cells

1. Mesenchymal stem cells (MSCs) are the most studied stem cell for treating cartilage defects due to their ease of isolation, multipotency and ability to modulate immune responses. MSCs can be extracted from different sources including bone marrow, synovium, adipose tissue, blood, umbilical cord tissue, and periosteum. In addition, experimental studies have used MSCs derived from other tissues (e.g., skeletal muscle, peripheral nerve, tongue, teeth, Wharton's Jelly, and fetal tissues) (Ding et al. 2022) (Jacob et al. 2019). Among these, bone marrow (BM-MSCs) is gold-standard due to their well-established protocols for isolation and chondrogenic properties. BM-MSCs can differentiate into chondrocytes, osteoblasts, and other cell types critical for joint repair while secreting a range of growth factors (TGF- β) and cytokines promoting tissue healing and reducing inflammation.

Adipose-derived MSCs (AD-MSCs) are found in relative abundance in adipose tissue and require minimally invasive procedures for their harvest. These cells exhibit a similar differentiation capacity to BM-MSCs but have been noted for their higher proliferative rate and greater ease of expansion in vitro. However, the chondrogenic potential of AD-MSCs is slightly lower than that of BM-MSCs, necessitating the use of growth factors or other agents to enhance their differentiation into cartilage-forming cells.

Synovium-derived MSCs (SD-MSCs) are noted to have superior chondrogenic capacity. SD-MSCs are naturally predisposed to differentiate into chondrocytes, making them effective in regenerating articular cartilage. While also producing extracellular matrix (ECM) components that are essential for the structural integrity of cartilage. The relative ease of isolating SD-MSCs from the synovial membrane during arthroscopic procedures further enhances their clinical applicability. However, SD-MSCs are under extensive research to fully understand their long-term efficacy in regeneration.

2. **Induced Pluripotent Stem Cells (iPSCs)**

Induced pluripotent stem cells (iPSCs) are generated by reprogramming somatic cells, such as skin fibroblasts, to revert them to a pluripotent state, similar to that of embryonic stem cells. This process resets the developmental clock of the cells, enabling the differentiation into chondrocytes and other cells relevant for joint repair. Pluripotency, allows for the generation of unlimited cell types from a single cell source. iPSCs can be differentiated into chondrocytes through natural developmental which are then engineered into cartilage tissues in-vitro or are directly implanted into an injury site. However, iPSCs are still largely experimental due to their tumor forming potential (teratomas). Additionally, the high cost, time-intensive nature and the technical complexities in the production and differentiation of iPSCs hinder their widespread adoption in clinical practice.

3. **Embryonic Stem Cells (ESCs)**

Embryonic stem cells (ESCs) are derived from the inner cell mass of blastocysts and are inherently pluripotent. ESCs have high proliferative capacity and ability to differentiate into functional chondrocyte tissue-specific cells by using specific growth factors and signaling molecules, such as TGF-β and bone morphogenetic proteins (BMPs), crucial for chondrogenesis.

Despite their potential, the use of ESCs raises ethical concerns mainly stemming from the source of these cells, which involves the destruction of embryos. Furthermore, when implanted in-vivo, there is risk of forming teratomas.

4.2 Mechanisms of Action in Tissue Repair

1. **Differentiation into Target Cell Types:** MSCs are differentiated into chondrocytes, the primary cell type in cartilage and fibrocartilage cells found in the meniscus, through the activation of the SMAD signaling pathway by Transforming Growth Factor-Beta (TGF-β) causing upregulation of the genes SOX9, COL2A1 (collagen type II), and ACAN (aggrecan), associated with chondrogenesis. These genes produce components for extracellular matrix (ECM).
2. **Paracrine Effects and Secretion of Bioactive Molecules:** Paracrine mechanisms exert therapeutic effects by reducing inflammation, promoting cell survival, and enhancing tissue repair (Guo et al. 2018) (Ding et al. 2022). MSCs release vascular endothelial growth factor (VEGF) and fibroblast growth factor (FGF-2), stimulating angiogenesis and blood supply to the injured area, aiding nutrient transport and waste elimination. Interleukin-10 (IL-10) and prostaglandin E2 (PGE2) are secreted to reduce the local inflammatory response post injury.
3. **Modulation of the Immune Response:** MSCs secrete tumor necrosis factor-stimulated gene-6 (TSG-6), which inhibits the nuclear factor-kappa B (NF-κB) pathway in macrophages, reducing the production of pro-inflammatory cytokines like tumor necrosis factor-alpha (TNF-α) and interleukin-1 beta (IL-1β). This shift encourages the polarization of macrophages towards an M2 phenotype, which is associated with tissue repair and regeneration. Hepatocyte growth

factor (HGF) and indoleamine 2,3-dioxygenase (IDO), secreted by MSCs suppress T cell proliferation and promote immune tolerance, further facilitating the local environment shift from a pro-inflammatory state to a regenerative one.

4. **Recruitment and Activation of Endogenous Repair Mechanisms:** Stromal cell-derived factor-1 (SDF-1) recruits endogenous progenitor cells to the site of injury. These progenitor cells are differentiated into repair cells. Furthermore, MSCs release exosomes containing microRNAs (miRNAs) influencing pathways Wnt/β-catenin and PI3K/AKT which are involved in cell proliferation, survival, and differentiation.

5 Platelet-Rich Plasma (PRP) Therapy

In the past decade, platelet-rich plasma (PRP) has emerged as a non-operative treatment modality for cartilage injuries (Xie et al. 2014). It is an advanced regenerative treatment that utilizes the patient's own blood. PRP allows for tissue repair and regeneration as it comprises of bioactive molecules, including platelet-derived growth factor (PDGF), vascular endothelial growth factor (VEGF), transforming growth factor-beta (TGF-β), and epidermal growth factor (EGF). When injected into the site of injury, these growth factors promote the recruitment of stem cells, enhance the proliferation and differentiation of local progenitor cells, and stimulate the synthesis of extracellular matrix components necessary for tissue healing.

Current advances in tissue engineering are geared towards healing articular cartilage and meniscus injuries (Sánchez et al. 2014) through minimally invasive nature that causes low immunogenic reactions, and accelerated healing processes. While, the application of PRP, has shown promising results, the efficacy of PRP varies depending on factors such as the concentration of platelets, the preparation technique, and the specific characteristics of the injury.

5.1 Biological Properties of PRP

Platelet-rich plasma (PRP) is an autologous blood product, enriched with platelets and growth factors, to promote physiological healing processes. PRP contains a higher concentration of platelets than whole blood and is most commonly produced by centrifuging whole blood (Marmotti et al. 2015). Platelets are small anucleated blood cells produced by megakaryocytes, rich in growth factors and cytokines with an essential role in hemostasis, inflammation, and tissue repair. These growth factors include platelet-derived growth factor (PDGF), transforming growth factor-beta (TGF-β), vascular endothelial growth factor (VEGF), epidermal growth factor (EGF), insulin-like growth factor (IGF-1), and fibroblast growth factor (FGF). Anti-inflammatory bioactive molecules interleukin-1 receptor antagonist (IL-1Ra) and prostaglandin E2 (PGE2) within PRP reduces pain and swelling at the injury site which is especially useful in chronic inflammation.

PRP Preparation and Variability

The preparation of PRP involves drawing the patient's blood, followed by centrifugation to separate the plasma from the red and white blood cells. The plasma layer, rich in platelets, is then collected or further activated with agents such as calcium chloride or thrombin. After injecting, it triggers the release of growth factors from the platelets.

Mechanisms of Action

PRP, when injected to an injury site, initiates a cascade of biological events including the degranulation of platelets and releasing of growth factors and cytokines.

PDGF promotes the migration and proliferation of fibroblasts and MSCs to the injury site while enhancing the synthesis of collagen and other extracellular matrix (ECM) components. TGF-β stimulates the differentiation of MSCs into chondrocytes and inhibits inflammation. VEGF promotes angiogenesis, ensuring an adequate blood supply to the regenerating tissue to deliver oxygen and nutrients and remove metabolic waste. EGF accelerates re-epithelialization and cell proliferation, while IGF-1 boosts the synthesis of proteoglycans and type II collagen promoting the anabolic processes in cartilage. FGF promotes wound healing by stimulating fibroblast proliferation and ECM deposition.

6 Biomaterials and Scaffolds in Tissue Engineering

Traditional reconstruction techniques of biomaterials are limited to poor to good results. Implant-based tissue engineering offers new hope by combining new scaffold technologies with cell sources and bioactive factors to replicate micro- and macro-environment cues. Scaffolds play an important role in tissue repair and regeneration, acting as temporary extracellular matrix (ECM) to support cell adhesion, migration, proliferation and differentiation (Li et al. 2021). Different natural and synthetic biomaterials have been used to engineer scaffolds for meniscus repair. More importantly, several processing techniques including two-dimensional, three-dimensional, and porous scaffolds have been employed to design biomaterial scaffolds.

Biomaterials, especially natural materials, have excellent biocompatibility, favorable biological activity, adequate mechanical strength and porosity, as well as good degradability. The typical natural biomaterials include polysaccharides (chitosan, alginate and hyaluronic acid), proteins (type I collagen, gelatin, silk fibroin and fibrin), decellularized extracellular matrix, and their combinations. Natural biopolymers have been extensively used in meniscus tissue engineering because of their ability to promote cell migration, inhibit non-specific protein adsorption and reduce foreign body reaction (Peng et al. 2022). With matrix proteins as a scaffold, cells were detected to spread and proliferate, maintaining high level expression of ECM-related genes. Biomaterials composed of biological macromolecules besides cellulose, silk fibroin, fibrin and gelatin have also been used for cartilage tissue engineering, in which gelatin and fibrinogen as ligands can enhance chondrogenesis of mesenchymal stem cells (MSCs).

6.1 Types of Biomaterials Used in Meniscus and Cartilage Regeneration

Porous biomaterials made from polylactic acid (PLA), mesoporous siRNA (small interfering RNA)-PLGA microsphere delivery system, alginate, gelatin-chondroitin sulfate, polycaprolactone (PCL), hyaluronic acid, gelatin sponge, gelatin/poly(lactic-co-glycolic acid) (PLGA) copolymer composite, polyester polyurethane (PU), silk fibroin, and chitosan have been widely used (Peng et al. 2022). Various biomaterials, including poly lactic acid (PLA), BSA-PLGA-PEG, polyethylene glycol-polylactide copolymer, polylactic acid-polycaprolactone (PPLA/PCL) copolymer, polytriglycolic acid, gelatin sponge, and PLGA, have been developed for cartilage regeneration (Li et al. 2021).

1. **Polylactic Acid (PLA) and Polylactic-Co-Glycolic Acid (PLGA)**

Polylactic acid (PLA) and its copolymer, polylactic-co-glycolic acid (PLGA), are among the most used due to their biocompatibility, biodegradability, and ease of processing into various forms, including microspheres, fibers, and porous scaffolds. These polymers are naturally metabolized by the body into lactic acid and glycolic acid, respectively. Their degradation rates can be tailored by adjusting the ratio of lactic acid to glycolic acid in PLGA. PLA and PLGA are also often used as carriers for bioactive molecules or cells. Mesoporous siRNA-PLGA microspheres deliver siRNA to gene targets, thereby modulating cellular response. PLGA when combined with gelatin creates scaffolds mimicking the extracellular matrix (ECM) while in combination with hyaluronic acid enhances cell adhesion.

2. **Polycaprolactone (PCL)**

PCL is known for its long degradation time, therefore this scaffold can provide long-term support to tissues. PCL-based scaffolds are often used in conjunction with growth factors or stem cells to enhance their regenerative potential. PCL scaffolds have been loaded with growth factors such as bone morphogenetic proteins (BMPs) and transforming growth factor-beta (TGF-β) to promote chondrogenesis to produce cartilage matrix. The slow degradation of PCL ensures that the scaffold provides mechanical support while the newly formed tissue matures.

3. **Natural Polymers: Alginate, Chitosan, and Gelatin**

Natural polymers, such as alginate, chitosan, and gelatin, are highly biocompatible and similar to the natural extracellular matrix. Alginate, a polysaccharide derived from brown seaweed, forms hydrogels that can encapsulate cells by conforming to the shape of the defect and provides a moist environment for cell survival and proliferation. Chitosan, a polysaccharide derived from chitin, has antimicrobial properties and can promote wound healing and tissue regeneration. Chitosan-based scaffolds in combination with insulin-like growth factor-1 (IGF-1) enhances the regeneration of cartilage tissue. Chitosan's ability to form hydrogels makes it suitable for use in injectable scaffolds or as a coating for other types of scaffolds to improve cell adhesion and proliferation. Gelatin, a denatured form of collagen can be crosslinked to improve their mechanical properties and stability in vivo. Additionally, gelatin sponges have been used as carriers for stem cells and growth factors, providing a three-dimensional structure that supports cell attachment and proliferation.

4. **Silk Fibroin**

Silk fibroin, derived from the silkworm, is a natural protein polymer that is biocompatible, and supports cell growth. Silk fibroin scaffolds mimic the fibrous structure of the meniscus, providing a supportive environment for cell attachment and ECM production. Scaffolds are integrated with bioactive agents such as bone morphogenetic proteins (BMPs) or vascular endothelial growth factor (VEGF) to promote angiogenesis and chondrogenesis. Silk fibroin is also compatible with advanced fabrication techniques such as electrospinning and 3D bioprinting, which creates complex and precisely tailored scaffold architectures for the injury site.

6.2 Role of Scaffolds in Tissue Repair

Scaffolds are indispensable components in tissue and organ function restoration. These scaffolds must mimic the mechanical properties of the native tissue while being biocompatible, biodegradable, and capable of integrating with the host tissue. They provide mechanical support for cell adherence, proliferation, and differentiation as well as guidance to cells during tissue regeneration processes through chemical/physical cues, archiving the precursors and desired structures of the damaged site before tissue formation. Moreover, a scaffold must provide an inert environment, preventing undesired reactions with the tissues or controlled drug release. In short, scaffolds are essential in repairing defects resulting from disease, age, trauma, injury, and surgical intervention through tissue engineering (Peng et al. 2022). Scaffolds must mimic the ECM structure of the natural tissue they intend to regenerate for successful repair. Given that the native meniscus is a tri-layered structure with anisotropic composition, architectonics, and mechanical properties, the fabrication of meniscal scaffolds is complicated compared with other tissues such as cartilage.

1. **Structural Support and Mechanical Integrity**

The meniscus and cartilage are load-bearing tissues that experience significant mechanical stresses during joint movement. Scaffolds made from materials such as polylactic acid (PLA), polycaprolactone (PCL), and polylactic-co-glycolic acid (PLGA) are able to provide adequate mechanical support to withstand these forces while maintaining their structural integrity over time.

2. **Cell Attachment and Proliferation**

For successful tissue regeneration, scaffolds must promote the attachment, proliferation, and differentiation of cells. The surface properties of the scaffold, including its chemistry, topography, and wettability, play a crucial role in facilitating cell attachment. Natural polymers such as collagen, gelatin, and hyaluronic acid are often incorporated into scaffolds because they contain cell-binding domains that enhance cellular interactions. Moreover, scaffolds can be functionalized with bioactive molecules, such as growth factors (e.g., TGF-β, IGF-1) or adhesion peptides (e.g., arginine-glycine-aspartate (RGD) sequences), to further enhance cell attachment and proliferation.

3. **Guidance of Tissue Regeneration and Integration**

For meniscus and cartilage repair, the scaffold must promote the formation of a tissue that closely resembles the native tissue in terms of composition, structure, and function. This involves guiding the differentiation of progenitor cells or stem cells into specific cell types, such as chondrocytes or fibrocartilage cells, and supporting the production of the appropriate ECM components, such as collagen type II and proteoglycans.

4. **Controlled Degradation and Remodeling**

An important consideration in scaffold design is the degradation rate of the material. The scaffold must degrade at a rate that matches the formation of new tissue, providing support during the critical stages of repair and then gradually transferring the load to the newly formed tissue. Scaffolds made from biodegradable polymers such as PLA, PLGA, and PCL are designed to degrade into biocompatible byproducts that are safely absorbed or excreted by the body. The degradation process should be controlled to avoid premature loss of mechanical support or excessive accumulation of degradation products, which could hinder tissue regeneration.

Scaffolds for meniscus tissue engineering should be biocompatible, biodegradable, exhibit appropriate porosity and pore size, as well as structural mimicry of the native meniscus after implantation. In consideration of the native meniscal ECM, the ideal scaffold for meniscal tissue engineering would be a biocompatible, biodegradable polysaccharide scaffold coated with gelatin containing bioactive molecules such as ginsenoside for meniscal tear degeneration (Li et al. 2021).

7 Gene Therapy for Meniscus and Cartilage Regeneration

Gene therapy is a cutting-edge approach that can regulate multiple pathways simultaneously by expressing recombinant genes. Considering the avascular nature of the meniscus and cartilage, gene delivery methods with a potentially long duration are required (Madry et al. 2011).

Conventional treatments do not restore the tissue anatomical structure and pathophysiological function comprehensively. Thus, further attempts are necessary for the development of alternatives capable of connecting cell-based approaches and promising biocompatible scaffolds. Gene therapy is a cutting-edge approach for the treatment of human ailments including conditions related to cartilage (Szwedowski et al. 2020). The various methods for gene delivery into cells can be grouped into three categories: viral vehicles comprising adenoviral, retroviral, and lentiviral vectors; non-viral systems which include cationic polymers (i.e., polyethylenimine), lipids, physical methods (electroporation, gene gun), and naked DNA; and hybrid systems combining DNA-hyaluronan complexes and adenoviral vectors with cationic polymers.

7.1 *Gene Delivery Methods*

The genetic material can be transported into the cells by vectors; these vectors are most commonly modified viruses or synthetic systems. Gene delivery into the cells can stimulate the expression of one or more proteins in target cells (Szwedowski et al. 2020).

Gene therapy delivery can be classified into two categories: in vivo and ex vivo. In vivo gene therapy involves the direct delivery of genetic material into the patient's tissue, while ex vivo gene therapy involves modifying cells outside the body before reimplanting them into the patient. Gene delivery methods can be classified into two major categories: viral and nonviral.

- **Viral Vectors**: Adenoviruses, lentiviruses, or adeno-associated viruses (AAVs). These viruses are modified to transport the specified genes without inducing diseases. These vectors are effective in infecting cells, which makes them ideal for transporting genes that induce regeneration of cartilage, such as those that code for growth factors: transforming growth factor-beta (TGF-β), bone morphogenetic proteins (BMPs), or insulin-like growth factor-1 (IGF-1). These growth factors have the ability to improve the proliferation and specialization of chon-

drocytes and mesenchymal stem cells (MSCs), resulting in higher synthesis of extracellular matrix elements such as collagen and proteoglycans.
- **Non-Viral Vectors**: Plasmid DNA, lipid nanoparticles, and electroporation. While these techniques are typically safer and less likely to trigger an immune response compared to viral vectors, they often result in reduced transfection effectiveness. Non-viral vectors are commonly used in ex vivo gene therapy, involving genetic modification of cells outside the body in controlled settings prior to reinsertion into the patient. There are two possible procedures: the chemical method and the physical method. The chemical gene delivery method includes systems based on lipids, polymers, peptides, and proteins. The physical gene delivery method includes electroporation, sonoporation, magnetofection, biolistic transfection to help vectors overcome cell membrane barriers. This method enables the identification of transfected cells that have been successfully transformed, which can later be multiplied and employed in the mending of the meniscus or cartilage.

7.2 Potential Targets for Gene Therapy

To enhance cartilage repair processes in vivo, most studies have relied on the expression of growth factors that are critically important for the initiation and regulation of chondrogenesis.

- **TGF-β**: Induces chondrogenesis, TGF-β stimulates the differentiation of MSCs into chondrocytes and enhances the production of collagen type II and aggrecan.
- **IGF-1**: IGF-1 inhibits catabolic processes and activates the synthesis of proteoglycans and collagen.
- **SOX9**: Transcription factor promoting chondrocyte differentiation and cartilage formation. Overexpression of SOX9 through gene therapy can boost the chondrogenic potential of stem cells, leading to improved cartilage regeneration. SOX5, SOX 6 and SOX9 are critical to chondrogenesis (Szwedowski et al. 2020).
- **BMPs (Bone Morphogenetic Proteins)**: BMP-2 and BMP-7 are growth factors that enhance the recruitment and differentiation of progenitor cells hence, formation of bone and cartilage.
- **IL-1Ra (Interleukin-1 Receptor Antagonist)**: It is an anti-inflammatory cytokine that inhibits interleukin-1 (IL-1), a key mediator of cartilage degradation in osteoarthritis. IL-1Ra gene therapy can help reduce inflammation and prevent further cartilage damage.

These proteins can enhance tissue regeneration through various mechanisms, including increasing the proliferative and migration potential of cells, inhibiting apoptosis, enhancing the synthesis of extracellular matrix (ECM) macromolecules, and preventing cartilage degeneration (Madry et al. 2011).

8 Challenges and Future Directions in Regenerative Strategies

The research efforts aimed at the development of regenerative strategies for the repair of meniscus and cartilage injuries have expanded by utilizing untapped resources, increasing knowledge of underlying pathophysiology, and implementing novel methodologies that capitalize on recently developed technology. Although challenges remain in addressing the innate immunological response of human tissues to allogeneic and xenogenic materials and in enhancing the integration and maturation of the regenerated tissue, this review highlights the diverse approaches that are currently underway. Future success will depend on a system-level understanding of the limitations in current approaches and on a more rigorous assessment of approaches currently in pre-clinical and clinical evaluation.

8.1 *Subvasta Approach*

The subvastus approach in total knee arthroplasty (TKA) has gained traction as a minimally invasive surgical technique that aims to preserve the quadriceps muscle, potentially enhancing postoperative recovery and functional outcomes. This approach, first popularised by Hofmann in 1991, involves accessing the knee joint beneath the vastus medialis muscle, which allows for a more direct approach to the knee while maintaining the integrity of the quadriceps mechanism (Aladraii 2024; Hiranaka et al. 2019). One of the primary advantages of the subvastus approach is its impact on postoperative pain and recovery. Studies have shown that patients undergoing TKA via the subvastus approach experience less pain and a quicker return to function compared to those who undergo the standard medial parapatellar approach.

The subvastus approach is a surgical technique characterised by accessing the knee joint beneath the vastus medialis muscle, which allows for a more direct and less invasive entry into the knee compared to traditional methods, such as the medial parapatellar approach (Stubnya 2023).

The technique begins with a careful incision made along the medial aspect of the knee, where the surgeon dissects through the subcutaneous tissue to reach the vastus medialis. Unlike the medial parapatellar approach, which requires incising the vastus medialis and potentially disrupting the extensor mechanism, the subvastus technique preserves the muscle and tendon integrity. This preservation is crucial as it helps maintain the vascular supply to the patella and reduces postoperative pain and complications associated with quadriceps dysfunction (Stubnya 2023).

Once the vastus medialis is retracted, the surgeon can access the knee joint capsule and perform the necessary procedures for TKA, including the resection of damaged bone and the implantation of prosthetic components. The subvastus approach allows for adequate visualization and access to the knee joint while minimizing soft tissue damage (Stubnya 2023).

The medial parapatellar approach and the subvastus approach are two distinct techniques employed in total knee arthroplasty (TKA), each with unique implications for surgical outcomes, recovery, and postoperative complications. Understanding these differences is essential for optimizing patient care and surgical results.

One of the primary differences between the two approaches lies in their surgical access and the preservation of the quadriceps mechanism. The medial parapatellar approach involves an incision along the medial side of the knee, which necessitates the detachment of the vastus medialis muscle from the patella. This detachment can lead to increased trauma to the quadriceps mechanism, resulting in postoperative pain and delayed recovery of quadriceps strength (Berstock et al. 2018). In contrast, the subvastus approach allows the surgeon to access the knee joint beneath the vastus medialis, preserving the muscle and tendon integrity.

Another significant difference is the impact on the vascular supply to the patella. The medial parapatellar approach can compromise the blood supply to the patella due to the necessary detachment of surrounding soft tissues, which may increase the risk of complications such as patellar osteonecrosis and anterior knee pain (Berstock et al. 2018).

Furthermore, studies have shown that patients undergoing TKA via the subvastus approach often report better functional outcomes, including improved range of motion and quadriceps strength in the early postoperative period compared to those who undergo the medial parapatellar approach (Berstock et al. 2018). This improved recovery trajectory is attributed to the less invasive nature of the subvastus technique, which allows for a more rapid rehabilitation process.

The immune response is a formidable obstacle to the successful translation of regenerative medicine technologies. Ideally, graft-tissue biocompatibility should minimize and control the inflammatory response to accomplish a positive tissue remodeling program, while preventing uncontrolled tissue-eating processes such as necrosis or apoptosis (Steinert and Nöth 2007). Commercial and investigational acellular grafts for meniscus and cartilage repair have undergone either sterilization or been chemically treated to modulate the immunological response, and there is early evidence that these approaches can be successful in preventing graft rejection. However, more studies are needed to better understand the limitations in these current techniques and to innovate additional strategies to better control and harness the immune response (Roseti and Grigolo 2022). A greater understanding of the diverse components of the immune response in regulating healing processes can inform the design of composite grafts that properly engage immune involvement in species-specific timeframes to drive regenerative responses.

8.2 Overcoming Immunological Responses

A significant challenge for many regenerative strategies is immunological responses to the implanted cells, scaffolds, and tissue products. Biomaterial-scaffold based regenerative strategies often cause a foreign body response characterized by an acute inflammatory response that may impede vascularization and later evoking a fibrotic encapsulation that may choke the scaffold, thus preventing the access of nutrients, as well as preventing cells, progenitors or mature cells, from penetrating and populating the inner scaffold space. When scaffolds die and the cells continue responding to the microenvironment signals, they might trigger a chronic inflammatory response and other events that further deteriorate the regenerative attempts and potentiate the degradation of the surrounding host tissue.

Therapeutic gene treatments pose several challenges consisting of ensuring the safe and efficient delivery of the vectors to the target tissue, the precise timing and location of gene expression. Viral vectors, pose risks such as immunogenicity, insertional mutagenesis, and limited control over gene expression. Non-viral vectors from low transfection efficiency and transient gene expression. Gene overexpression, on the other hand, can lead to abnormal tissue growth or ossification.

As a general consideration, factors controlling cellular behavior promote regenerative outcomes following tissue damage and the regenerative medicine goal is to provide such factors to the damaged tissue. However, biomaterials will also elicit a cascade of uncontrollable biological events, particularly an inflammatory response and a series of events that follow, including cell recruitment and response to the surface of the material, that will determine the fate of the regenerative attempt. Biological cues controlling cell response and behaviour temporally and spatially with time are not exclusive to the native tissue. Thus, within the scope of tissue engineering/regenerative medicine, two concepts and challenges arise: (1) How to design and fabricate biomaterials that do not elicit an undesirable immunogenic response? and (2) How to control this large cascade of biological events to favour biocompatibility and to the best response for a particular regenerative approach? (Garrity et al. 2022).

Several strategies are being tested to overcome problems with bioincompatibility of natural and synthetic materials. Biomaterials may be designed to reduce their immunogenicity. Such characteristics can be achieved, for instance, by controlling the topography, chemical and physical properties of the surface, coating with peptides or proteins known to modulate the immune response, controlling the degree of hydrophilicity/hydrophobicity or charge, or by chemical modifications of them to improve their performance in regenerative approaches (e.g., by incorporating bioactive molecules, such as growth factors, proteins or peptides). Outcomes from other regenerative medicine approaches are also being studied, in particular processes that involve cell manipulation in vitro, such as expansion, preconditioning and modification of cells to favour immunomodulation with the hopes of controlling their response after implant (Li et al. 2021).

8.3 *Enhancing Tissue Integration and Maturation*

One of the most critical considerations for successfully implementing any of the regenerative approaches discussed above is the need to enhance tissue integration and maturation. For most of the technologies mentioned, the particles, gels, membranes, scaffolds, or engineered tissues produced in the process are expected to enhance the healing and remodeling of an injury or defect in a more mature, fast, or effective manner than would naturally occur. Therefore, the implanted materials must integrate with the native tissues and enhance the remodeling process. For example, tissue-engineered constructs encompassing cells that secrete extracellular matrix (ECM) components must be remodeled in such a way that the implanted construct generates fibrocartilaginous meniscal tissue of sufficient mechanical function or articular cartilage of hyaline nature (Li et al. 2021). Hydrogel systems designed to retain a focused delivery of bioactive factors must integrate with and be replaced by articular cartilage or meniscus ECM over time. For large-scale defects, both experimental and clinical evidence suggest vital roles of native bone marrow-derived stem cells (BMSCs) in successful repair. Other considerations for the successful integration of engineered tissues relate to their size, morphology, and mechanical properties, which should broadly match those of existing native meniscal or cartilaginous tissues.

9 Conclusion and Clinical Implications

The meniscus and cartilage are crucial for normal knee function. Their injuries, particularly those caused by trauma or degeneration, are common and can lead to osteoarthritis. The current surgical approach has limitations due to their low intrinsic healing potential and avascularity. Despite many promising biomaterials and in-depth-improving treatments developed in the last few decades, the clinical outcomes are still unsatisfactory.

The future of gene therapy for meniscus and cartilage regeneration may involve the use of CRISPR/Cas9 technology for targeted gene editing. This approach allows for the precise modification of specific genes within the genome, potentially correcting genetic defects or enhancing the regenerative capacity of cells. Additionally, combining gene therapy with other regenerative strategies, such as stem cell therapy or biomaterials, may lead to synergistic effects that improve the overall outcomes of tissue repair.

Recent advances in regenerative medicine, using growth factors, scaffolds, and cells, to promote healing by providing appropriate signals, structures, and living cells, are attractive methods in orthopedic repair. Recent cell-based approaches, taking advantage of induced pluripotent stem cells or mechanically induced stem cell recruitment, have also shown promise (Dai et al. 2021). Nevertheless, it is believed that regenerative approaches for meniscus and cartilage injuries, combining with tissue-issue sources, will become a routine therapeutic strategy in the near future.

References

Aladraii A. Functional outcome of subvastus versus medial parapatellar approaches for total knee replacement in patients with knee osteoarthritis: a prospective cohort study. Saudi J Med Med Sci. 2024;12(1):35–9. https://doi.org/10.4103/sjmms.sjmms_237_23.

Berstock J, Murray J, Whitehouse M, Blom A, Beswick A. Medial subvastus versus the medial parapatellar approach for total knee replacement. Efort Open Rev. 2018;3(3):78–84. https://doi.org/10.1302/2058-5241.3.170030.

Bhan K. Meniscal tears: current understanding, diagnosis, and management. Cureus. 2020;12(6):e8590.

Dai T, Pan Z, Yin F. In vivo studies of mesenchymal stem cells in the treatment of meniscus injury. Orthop Surg. 2021;13(8):2185–95.

Ding G, Du J, Hu X, Ao Y. Mesenchymal stem cells from different sources in meniscus repair and regeneration. Front Bioeng Biotechnol. 2022;10:796367.

Garrity C, Arzi B, Haus B, Lee CA, Vapniarsky N. A fresh glimpse into cartilage immune privilege. Cartilage. 2022;13(4):119–32.

Ghazi Zadeh L, Chevrier A, Farr J, et al. Augmentation techniques for meniscus repair. J Knee Surg. 2018;31:099–116. https://doi.org/10.1055/s-0037-1602247.

Guo W, Xu W, Wang Z, Chen M, Hao C, Zheng X, Huang J, Sui X, Yuan Z, Zhang Y, Wang M, Li X, Wang Z, Peng J, Wang A, Wang Y, Liu S, Lu S, Guo Q. Cell-free strategies for repair and regeneration of meniscus injuries through the recruitment of endogenous stem/progenitor cells. Stem Cells Int. 2018;2018:5310471.

Hiranaka T, Tanaka T, Fujishiro T, Anjiki K, Nagata N, Kitazawa D, et al. A modified undervastus approach for knee arthroplasty with anatomical repair of soft tissue. Clin Orthop Surg. 2019;11(4):490. https://doi.org/10.4055/cios.2019.11.4.490.

Jacob G, Shimomura K, Krych AJ, Nakamura N. The meniscus tear: a review of stem cell therapies. Cells. 2019;9(1):92.

Li H, Li P, Yang Z, Gao C, Fu L, Liao Z, Zhao T, Cao F, Chen W, Peng Y, Yuan Z, Sui X, Liu S, Guo Q. Meniscal regenerative scaffolds based on biopolymers and polymers: recent status and applications. Front Cell Dev Biol. 2021;9:661802.

Madry H, Orth P, Cucchiarini M. Gene therapy for cartilage repair. Cartilage. 2011;2(3):201–25.

Marmotti A, Rossi R, Castoldi F, Roveda E, Michielon G, Peretti GM. PRP and articular cartilage: a clinical update. Biomed Res Int. 2015;2015:542502.

Peng Y, Lu M, Zhou Z, Wang C, Liu E, Zhang Y, Liu T, Zuo J. Natural biopolymer scaffold for meniscus tissue engineering. Front Bioeng Biotechnol. 2022;10:1003484.

Pereira H, Fatih Cengiz I, Gomes S, Espregueira-Mendes J, Ripoll PL, Monllau JC, Reis RL, Miguel Oliveira J. Meniscal allograft transplants and new scaffolding techniques. EFORT Open Rev. 2019;4(6):279–95.

Piontek T, Ciemniewska-Gorzela K, Naczk J, et al. Complex meniscus tears treated with collagen matrix wrapping and bone marrow blood injection: a 2-year clinical follow-up. Cartilage. 2016;7:123–39. https://doi.org/10.1177/1947603515608988.

Roelofs AJ, Rocke JPJ, De Bari C. Cell-based approaches to joint surface repair: a research perspective. Osteoarthr Cartil. 2013;21(7):892–900.

Roos EM, Ostenberg A, Roos H, Ekdahl C, Lohmander LS. Long-term outcome of meniscectomy: symptoms, function, and performance tests in patients with or without radiographic osteoarthritis compared to matched controls. Osteoarthr Cartil. 2001 May;9(4):316–24. https://doi.org/10.1053/joca.2000.0391.

Roseti L, Grigolo B. Current concepts and perspectives for articular cartilage regeneration. J Exp Orthop. 2022;9:61.

Salzmann GM, Ossendorff R, Gilat R, Cole BJ. Autologous minced cartilage implantation for treatment of chondral and osteochondral lesions in the knee joint: an overview. Cartilage. 2020; https://doi.org/10.1177/1947603520942952. [PMC free article] [PubMed]

Sánchez M, Delgado D, Sánchez P, Fiz N, Azofra J, Orive G, Anitua E, Padilla S. Platelet rich plasma and knee surgery. Biomed Res Int. 2014;2014:890630.

Sibbel J, Khan W. The knee joint. In: Iyer K, Khan W, editors. Orthopedics of the upper and lower limb. Cham: Springer; 2020. https://doi.org/10.1007/978-3-030-43286-7_20.

Steinert AF, Ghivizzani SC, Rethwilm A, Tuan RS, Evans CH, Nöth U. Major biological obstacles for persistent cell-based regeneration of articular cartilage. Arthritis Res Ther. 2007;9(3):213.

Stubnya B. Subvastus approach supporting fast-track total knee arthroplasty over the medial parapatellar approach: a systematic review and network meta-analysis. J Arthroplast. 2023;38(12):2750–8. https://doi.org/10.1016/j.arth.2023.06.004.

Szwedowski D, Szczepanek J, Paczesny Ł, Pękała P, Zabrzyński J, Kruczyński J. Genetics in cartilage lesions: basic science and therapy approaches. Int J Mol Sci. 2020;21(15):5430.

Wang L, Jiang J, Lin H, Zhu T, Cai J, Su W, Chen J, Xu J, Li Y, Wang J, Zhang K, Zhao J. Advances in regenerative sports medicine research. Front Bioeng Biotechnol. 2022;10:908751.

Xie X, Zhang C, Tuan S, R. Biology of platelet-rich plasma and its clinical application in cartilage repair. Arthritis Res Ther. 2014;16(1):204.

Knee Osteotomy: Aligning the Physics for Optimal Regeneration

Sedeek Mosaid

Abstract Knee osteotomy is used to treat degenerative knee conditions by redistributing forces across the knee joint, thus delaying knee arthroplasty. It is an effective treatment for medial or lateral compartment knee arthritis with varus or valgus deformity of the lower extremity. Osteotomy has become more effective with recent advances in surgical technique, planning, and execution, decreasing complications and improving outcomes. Advanced techniques such as high tibial osteotomy (HTO) and distal femoral osteotomy (DFO), along with other new technologies, such as computer navigation systems, have improved the accuracy and precision of osteotomy. This procedure can be further improved with mechanical and biological correction in osteotomy, as well as regenerative or enhanced tissue repair using regenerative medicine techniques and other promising modalities. This chapter provides background information on the technical variations of the procedure and patient selection. It discusses how these new therapies are evolving to improve patient outcomes, function, and mobility, as well as delaying future knee replacement.

Keywords Knee osteotomy · Biomechanics · Knee alignment · Varus and valgus deformities · Osteotomy

1 Introduction to Knee Osteotomy

Knee osteotomy is a surgical procedure used to treat degenerative conditions of the knee joint that result in abnormal weight distribution and subsequent development of osteoarthritis. The term 'osteotomy' refers to surgery that, cuts, realigns or removes a bone and changes the weight distribution from one compartment to another, improving the health and function of the joint (Ferrera and Menetrey 2022).

S. Mosaid (✉)
Department of Trauma & Orthopaedics, United Lincolnshire Teaching Hospitals NHS Trust, Lincoln County Hospital, Lincoln, UK

P. Lee (ed.), *Musculoskeletal Regeneration Medicine*,
https://doi.org/10.1007/978-3-031-77456-0_14

The primary aim of knee osteotomy is to minimise pain and improve function by redistributing load-bearing capacity. This technique can substantially delay knee replacement surgery, making it a good option for younger and more active patients with primary osteoarthritis or other knee joint conditions (Lee and Byun 2012).

This technique has been used for decades, and significant advancements have been made in surgery, implant fixation, and postoperative rehabilitation (Lee and Byun 2012).

Knee osteotomy was first described in the late nineteenth and early twentieth centuries. However, these early efforts were largely ineffective, as the surgical procedures were not very sophisticated, and knowledge about knee biomechanics was still in its infancy. The procedure became more reliably successful in the mid-twentieth century with developments in radiography, surgical approaches, and tools. Mark Coventry's pioneering work in the 1960s and 1970s advanced high tibial osteotomy (HTO), especially for the treatment of medial compartment osteoarthritis. His contributions made these procedures more effective and enabled more predictable outcomes, establishing them as an acceptable option for many patients with arthritic knee conditions (Naudie et al. 1999; Coventry 1965).

Recent improvements in the surgical procedures and instruments used in osteotomy (e.g. computer navigation and patient-specific cutting guides) have led to better results and patient satisfaction (Song and Bae 2016). Osteotomy is not only useful for delaying total knee arthroplasty (TKA) but also in other contexts, including revision TKA or repairing multi-ligamentous injuries to reconstruct the proper mechanical axis of the knee (Marom et al. 2019).

Knee realignment surgery has great potential to address knee problems and increase patient comfort and mobility, thus improving patients' quality of life. Evidence supports this, showing that this procedure allows many patients to return to work, engage in leisure activities, and perform daily tasks without discomfort or pain (Ferrera and Menetrey 2022).

2 Indications for Knee Osteotomy

2.1 Medial Compartment Osteoarthritis

The most common indication for knee osteotomy is medial compartment osteoarthritis, a degenerative change in the articular cartilage, which is commonly associated with varus alignment. This malalignment transfers more load to the knee joint's medial compartment, increasing cartilage damage and causing pain and loss of function (Kawde et al. 2024). Valgus-producing HTO involves a lateral shift of the mechanical axis, which offloads the overloaded medial joint compartment and transfers the load to the less diseased lateral joint compartment. Research has reported that this procedure significantly decreases pain and improves function;

after a mean follow-up of 12 years, 85% of patients were enthusiastic or satisfied, and 84% stated they would undergo HTO again (Hui et al. 2010).

2.2 *Varus and Valgus Deformities*

Knee osteotomy is mainly used in patients with varus or valgus deformity, which involves a substantial deviation from the normal axis of the knee joint. HTO is one of the most frequently utilised procedures for varus deformity. It is most effective in young, active patients who wish to continue their sports activities (Staubli and Jacob 2010).

Distal femoral osteotomy (DFO) is commonly performed to treat valgus deformity, as it enhances the mechanical adequacy of the knee joint, decreases knee pain, and increases the patient's mobility. Like HTO, DFO is typically indicated for young, active patients for whom TKA is not ideal (Ferrera and Menetrey 2022).

2.3 *Ligamentous Instability*

Although osteotomy does not directly address ligamentous instability, it can be used adjunctively with ligament reconstruction procedures to optimise joint alignment and stability. For example, in patients with anterior cruciate ligament (ACL) deficiency and associated varus malalignment, a combination of HTO and ACL reconstruction can yield better outcomes than ACL reconstruction alone (Tischer et al. 2017).

3 Patient Selection Criteria

A successful knee osteotomy depends on several factors, and one of the most important of these is patient selection; although some patients experience benefit from osteotomy, it is not suitable for all cases.

The following criteria must be considered to identify ideal candidates:

3.1 *Age*

Younger patients, typically under 60 years old, are the best candidates for HTO because they generally have superior bone quality and healing potential compared with older patients, increasing the likelihood of successful osteotomy union and long-term joint preservation. They are also more likely to maintain an active lifestyle, which is one of the main goals of HTO (Naudie et al. 1999; Hantes et al. 2018).

3.2 Activity Level

This procedure particularly benefits patients with moderate to high activity levels who want to remain active. It enhances function and reduces pain, allowing patients to engage in sports and other physical activities. By contrast, patients with sedentary lifestyles may not experience the full range of benefits (Jacquet et al. 2020).

3.3 Severity of Joint Degeneration

Joint degeneration severity must be assessed to identify whether patients are suitable for knee osteotomy, as this procedure is most effective for patients with mild to moderate unicompartmental osteoarthritis, in which the damage is confined to one section of the knee. These patients typically report significant pain reduction and improved postoperative function. However, patients with severe or tri-compartmental osteoarthritis may not benefit from the procedure as much as others. TKA is often considered a more suitable option in these cases (Lott et al. 2024a).

3.4 Joint Stability and Range of Motion

Patients with good joint stability and adequate range of motion (ROM) are more likely to experience favourable knee osteotomy outcomes. The procedure may not yield the desired results in those with significant ligamentous instability or severe limitations in knee movement (Hantes et al. 2018).

A limited ROM can also make achieving proper alignment during surgery more difficult, and postoperative stiffness can limit the procedure's benefits. Thus, preoperative evaluation of ROM is necessary to determine whether a patient is a good candidate (Naudie et al. 1999).

3.5 Motivation and Compliance

After knee osteotomy, it is recommended that patients follow a comprehensive rehabilitation program, which involves physiotherapy as well as exercises to strengthen, mobilise, and improve the function of the knee. The success of the procedure is largely dependent on patient motivation and compliance with the program. In other words, the patient must be actively involved in the program and adhere to the schedule for physiotherapy; most importantly, they must be willing to make necessary lifestyle modifications. Patients should be evaluated preoperatively to identify those more likely to follow through with the prescribed rehabilitation regimens (Hambly et al. 2020; Goshima et al. 2020).

4 Surgical Techniques

4.1 HTO

Limb Alignment and Preoperative Planning

Measuring the alignment of the extremity is essential for a successful knee osteotomy, especially when considering preoperative planning.

The anatomic femoral and tibial axes correspond to the mid-diaphyseal lines, and the mechanical axis of the lower limb links the centres of the hip and ankle joints.

The mechanical axis deviation is the horizontal distance between the mechanical lower limb axis and the centre of the knee joint. It is measured in the frontal plane and is used to assess the alignment in this plane (Ferrera and Menetrey 2022; Paley et al. 1994).

In varus deformity, the mechanical axis of the lower limb passes the knee joint medial to the centre, and the medial mechanical axis deviation is increased (Ferrera and Menetrey 2022; Paley et al. 1994).

The medial proximal tibial angle (MPTA), the medial angle between the tibial articular marginal line and the mechanical axis line of the tibia, is used to assess the alignment of the tibia relative to the knee joint. A normal MPTA is typically in the range of 85°–90°. A decrease in the MPTA indicates a varus deformity, whereas an increase indicates a valgus deformity (Paley et al. 1994; Gieroba et al. 2023).

The lateral distal femoral angle (LDFA) is the lateral angle formed between the mechanical axis line and the knee joint line of the femur. It typically lies in the range of 85°–90°, and a decreased LDFA can indicate a valgus deformity, whereas an increase indicates a varus deformity (Paley et al. 1994; Gieroba et al. 2023).

The knee joint line obliquity (KJLO) angle is the angle formed between the joint line of the knee (the line connecting the most distal points of the medial and lateral femoral condyles) and the horizontal plane parallel to the floor.

The KJLO angle is abnormal in cases of varus or valgus deformity and must be corrected during surgical procedures for proper knee alignment and function (Tseng et al. 2022).

The joint line convergence angle (JLCA) is measured to assess knee joint congruity. It is formed between a line tangential to the distal femoral condyle and the tibial plateau. The varus and valgus stress angles are defined as the JLCA in varus and valgus stress radiographs, respectively (Tsushima et al. 2024).

Measuring the JLCA is vital because patients with a high JLCA risk overcorrection after osteotomy, especially in HTO, if the JLCA is not considered during the correction process (Micicoi et al. 2021a).

When planning a proximal tibial osteotomy (PTO), sagittal plane alignment must be evaluated to ensure that the mechanical axis correction does not change the tibial slope, which can affect stability and knee function (Akoto et al. 2020).

The posterior tibial slope (PTS) is the angle between the tibial plateau and the horizontal line on the lateral radiograph. It is a critical factor in knee osteotomy. A normal PTS is typically in the range of 5°–10° (Paley et al. 1994; Webb et al. 2013).

An increased PTS is associated with increased anterior tibial translation, leading to greater strain on the ACL. Conversely, a decreased PTS can reduce the strain on the ACL while placing greater strain on the posterior cruciate ligament (PCL) and the posterolateral corner (Bernhardson et al. 2019a, b).

5 Clinical Assessment

Patient History

The patient's symptoms must be carefully documented, including the pain's nature, location, and severity. The patient's activity level and response to non-surgical treatments should be determined when taking their history, as these factors can help determine whether PTO is appropriate (Lee and Byun 2012; Pullen et al. 2024).

Physical Examination

Gait Analysis: Gait assessment provides information about limb alignment, the degree of weight-bearing, and functional disability.

Knee ROM: An evaluation of active and passive ROM and ligamentous stability must be conducted. This includes an evaluation of medial and lateral instability, which may determine the surgical approach.

Palpation: Localising points of tenderness along the joint line, especially in the medial compartment, can complement radiological findings (Pullen et al. 2024; Zampogna et al. 2019).

Radiological Assessment: Radiological assessment is necessary for planning PTO. Standard imaging techniques include:

Weight-Bearing Radiographs

Long-Leg Alignment Views: Full-length standing anteroposterior radiographs from the hip to the ankle allow the assessment of the mechanical axis deviation and the identification of varus or valgus deformity.

Lateral Views: Lateral radiographs are needed to evaluate the PTS and assess patellar height using the Insall–Salvati ratio (Pullen et al. 2024).

Magnetic Resonance Imaging (MRI)

Soft Tissue Evaluation: MRI provides detailed images of the menisci, ligaments, and articular cartilage—structures that are not visible on normal radiographs. Therefore, MRI images are recommended to assess the integrity of the ACL and other soft tissues that might be compromised in patients with varus deformity.

Subchondral Bone Integrity: MRI can also help detect bone marrow lesions and subchondral cysts, influencing the decision to perform an osteotomy or consider alternative treatments (Ferrera and Menetrey 2022; Pullen et al. 2024).

Computed Tomography (CT)

CT scans with 3D reconstruction are sometimes used in complex cases to understand bone deformities better and plan osteotomy cuts precisely. CT can also be useful for measuring rotational alignment, especially in patients with prior fractures or congenital deformities (Pullen et al. 2024; Zampogna et al. 2019).

The ideal patient for PTO is young (≤60 years old) and moderately active with medial isolated osteoarthritis, malalignment of <15°, and a minimum ROM of >90° in the affected limb (Ferrera and Menetrey 2022).

The degree of correction for medial opening wedge HTO can be planned manually with the Miniaci technique or using digital planning software. First, the Mikulicz line is drawn, starting from the centre of the femoral head to the centre of the talus. Second, the hinge point and planned postoperative weight point are identified. The hinge point is located at the top of the fibular head in the lateral proximal metaphysis of the tibia. The correction angle is formed by lines connecting the initial and future positions of the ankle to the hinge point (Micicoi et al. 2021b; Lott et al. 2024b).

HTO Techniques

HTO procedures include open wedge osteotomy, closed wedge osteotomy, and dome osteotomy.

5.1 Medial Opening Wedge Osteotomy

Medial opening wedge osteotomy is a simple procedure in which the correction level can be identified and adjusted intraoperatively on the coronal and sagittal planes, preventing a shortening of the lower limbs. This technique does not typically require a fibular osteotomy, which is associated with neurovascular complications. However, it has been associated with a risk of nonunion, long periods of weight-bearing restriction, and leg lengthening. Loss of correction can also occur because of unstable fixation. An increase in the PTS is often attributable to the hinge position the anterior position of the metal plate.

Medial opening wedge osteotomy can be more effective than closing wedge HTO in patients with a leg length discrepancy of ≥2 mm (Lee and Byun 2012).

5.2 *Surgical Procedure*

The patient is positioned supine on a radiolucent table. A tourniquet can be used, especially if the arthroscopy is needed to confirm the integrity of the lateral meniscus and the cartilage condition.

A vertical incision is made centrally between the tibial tuberosity's medial side and the posteromedial aspect of the tibia beneath the joint line. The pes anserinus can be retracted distally or released along with the medial collateral ligament (MCL) (Lee and Byun 2012).

The exploration of the medial border of the patellar tendon is followed by dissection from the tibial tuberosity to the posteromedial aspect of the tibia subperiosteally.

Two Kirschner (K) wires are placed approximately 4 cm below the medial joint line and pass diagonally 1 cm below the lateral articular margin of the tibia in the direction of the tip of the head of the fibula. The K-wires are used to guide the oscillating saw during the procedure; they are inserted, proximally and distally, into the axial osteotomy plane, and they are parallel to each other in the anteroposterior direction and to the PTS to monitor axial and sagittal plane alignment (Lee and Byun 2012; Murray et al. 2021).

To ensure neurovascular structure protection, blunt retractors are placed along the path of the planned osteotomy site (Lee and Byun 2012; Rossi et al. 2011).

The appropriate osteotomy level is verified under fluoroscopy before performing the tibial osteotomy with an oscillating saw; the osteotomy site is opened using a valgus force, and a calibrated wedge is used to maintain the opening.

A long alignment rod wire is centred over the hip and ankle joints, aligning with 62.5% of the tibial plateau, and the osteotomy site is fixed with a plate after alignment is confirmed (Lee and Byun 2012; Murray et al. 2021). Changes in the range between the anterior and posterior osteotomy gaps control the PTS adjustment. Opening the osteotomy anteriorly increases the PTS; posterior opening decreases it (Moon et al. 2015). Different types of plates can be used (e.g. plates with or without spacers). Spacer plates are frequently utilised, and the metal block must match the calibrated wedge exactly (Rossi et al. 2011; Murray et al. 2021).

5.3 *Bone Graft*

Several materials, such as autograft, allograft, and synthetic bone substitutes, can be used to fill the osteotomy gap in medial opening wedge HTO. Autologous bone graft has long been the gold standard because of its osteoconductive, osteoinductive, and osteogenic properties. An alternative is bone substitute, which provides weaker resistance to compressive loading but has proven to be adaptable and versatile, avoiding donor site morbidity associated with bone autograft.

Notably, the need for bone grafting in medial opening wedge HTO has become controversial with recent advances in plating technology. Although bone autografts or allografts can be used for defects of ≥10 mL, bone grafting is optional for minor defects (Ferrera and Menetrey 2022; Lott et al. 2024b).

Han et al. conducted a meta-analysis of 25 studies, which demonstrated that all methods—including autogenous bone grafting, allogenous bone grafting, synthetic material filling, and no filling—were appropriate surgical options for osteotomy defects, as delayed union, nonunion, and correction loss did not differ between these strategies (Han et al. 2015).

Hinge fractures are a common complication in medial opening wedge HTO; however, with some technical adjustments, they can be avoided. Stopping the osteotomy cut 5–10 mm from the lateral margin can decrease the severity and rate of hinge fractures. Many hinge fractures occur while opening the osteotomy gap, especially in more extensive corrections. Therefore, the gap should be opened slowly with an opening device (Lott et al. 2024b; Nakamura et al. 2017). Lastly, Gulagaci et al. reported that adding an additional protective K-wire just medial and parallel to the lateral tibial cortex decreased the number of intraoperative lateral hinge fractures (Gulagaci et al. 2019).

Biplanar osteotomy is a technical modification for medial opening wedge osteotomy and demonstrates innovation and progress in the field. This technique, which involves a second coronal cut underneath the tibial tuberosity in addition to the transverse osteotomy of the posterior tibia, significantly increases rotational stability and provides an anterior buttress against the sagittal tilting of the osteotomy (Lobenhoffer and Agneskirchner 2003).

5.4 *Lateral Closing Wedge Osteotomy*

Lateral closing wedge osteotomy was originally the standard technique for valgus-producing HTO; however, it is now declining in popularity (Murray et al. 2021).

The lateral closing wedge technique is effective for correction near the maximum point of deformity and is particularly notable for its ability to promote rapid bone healing, facilitated by direct bone-to-bone contact; this allows early weight bearing and increases the stability of the bone–implant construct (Lee and Byun 2012; Micicoi et al. 2021a).

However, this technique uses an anterolateral surgical approach, which requires meticulous protection of the common peroneal nerve; this is followed by fibular osteotomy or the release of the proximal tibiofibular joint, which can lead to neurovascular complications (Duivenvoorden et al. 2014).

Lateral closing wedge osteotomy is a more appropriate choice for patients who are heavy smokers as well as those with diabetes or prolonged use of steroids, as these have been identified as risk factors for nonunion. This procedure can be performed if patellar infera is present or the patient decides against bone grafting (Ferrera and Menetrey 2022; Lee and Byun 2012).

5.5 Surgical Procedure

A 10 cm vertical incision is performed, and the tibialis anterior muscle of the proximal tibia is elevated to expose the proximal tibiofibular joint; joint disruption or fibular osteotomy can be carried out to manage the proximal tibiofibular joint. The tibial osteotomy Proximal to tibial tubercle offers more advantages, as the metaphyseal osteotomy promotes early bone union, and correction can be performed adjacent to the deformity.

A retractor is placed beneath the lateral margin of the patellar tendon to protect it; another retractor is positioned along the posterolateral aspect of the tibia to protect the neurovascular structures. The osteotomy is performed 2 cm below the lateral articular surface of the tibia. The proximal osteotomy is cut parallel to the articular surface, and the distal osteotomy is performed obliquely. The lateral bone wedge is removed using an angular cutting guide with a saw, rongeur, curette, or osteotome, and 3–6 mm of the opposite cortical hinge is preserved to prevent intra-articular fracture. The hinge should be closed slowly to produce controlled plastic deformation. Overcorrection can be expected when the proximal fragment's distal cortex overlaps the distal fragment's proximal cortex. Complete wedge removal can be confirmed by fluoroscopy. The alignment should be verified after the osteotomy is closed using valgus force (Lee and Byun 2012).

Lateral closing wedge HTO must be avoided when the patella is already in a very high position. By contrast, lateral closing wedge HTO is preferred over medial opening wedge osteotomy if the patella is lower than average (Lee & Byun 2012).

The combination of HTO and tibial tubercle osteotomy provides a powerful procedure that can prevent changes in patellar height and correct abnormal patellar height, allowing such cases to be handled effectively (Capella et al. 2017).

Ligament instability was initially considered a contraindication for HTO, but recently, it has become the second indication for HTO (Capella et al. 2017).

Biomechanical studies demonstrated that planned changes in the PTS improved or restored stability in the sagittal plane (Giffin et al. 2004).

The main goals of HTO in knees with deficient ligaments are to prevent further articular cartilage wear, restore stability, and manage malalignment to protect the graft from abnormally high stresses.

Reduced PTS increases posterior tibial translation in flexion and extension, and the PTS should increase in PCL deficiency. However, an excessive PTS increases the anterior translation of the tibia in flexion and extension; therefore, the PTS should be reduced in ACL-deficient knees (Ferrera and Menetrey 2022; Capella et al. 2017).

5.6 DFO

DFO is a well-known surgical procedure that provides benefits in correcting valgus deformity, making it a promising option (Ferrera and Menetrey 2022).

Valgus deformity is less prevalent than varus deformity. It can result not only from a femoral-based deformity but also from a combination of femoral and tibial deformities (Eberbach et al. 2016).

The ideal candidate for DFO is under 65 years old and active, with a valgus deformity, isolated lateral compartment osteoarthritis, an excellent preoperative ROM, and normal body mass index (Lott et al. 2024b).

DFO can be performed through medial closing wedge osteotomy or lateral opening wedge osteotomy.

5.7 Preoperative Planning

The goal of osteotomy is to shift the mechanical axis through the centre or just medial to the centre of the knee. The point of correction, a crucial step in the process, is determined on the basis of the patient's pathology and marked on the tibial plateau. The mechanical axis is determined by drawing a line from the centre of the femoral head to the centre of the ankle joint. Two extra lines are drawn, one from the middle of the femoral head through the preferred correction point on the tibial plateau and the other from the centre of the talus through the end of the correction. The correction angle is formed by the intersection of these two lines (Sherman et al. 2018).

5.8 Surgical Procedures

Although medial closing wedge osteotomy has been widely used in the past, lateral opening wedge osteotomy has recently become more popular because it offers more straightforward surgical exposure, accurate correction, and a less complex fixation (Sherman et al. 2018).

Medial closing wedge osteotomy is specifically recommended when the correction angle exceeds 17.5°, there is a length difference in the operated limb, early weight-bearing is desired, and factors are present that increase the risk of delayed healing. The advantages of the medial closing wedge technique include a lower risk of nonunion and better stability than the lateral opening wedge technique (Sternheim et al. 2011).

5.9 Lateral Opening Wedge DFO

Two surgical procedures can be used for lateral opening wedge osteotomy. The first option is the extra-articular approach, in which a 12 cm long skin incision is made on the distal third of the lateral thigh, located 2 cm anterior to the palpable posterior border of the iliotibial tract. The incision should start 2 cm distal to the lateral femoral epicondyle and extend upwards. After the incision, thorough soft-tissue dissection must be performed down to the iliotibial band, which is then split. The vastus lateralis muscle is lifted off the intermuscular septum and moved forward. During this procedure, perforated vessels must be identified and cauterised. The intermuscular septum is released at the metaphyseal–diaphyseal junction to allow the safe retraction of posterior neurovascular structures during osteotomy (O'Malley et al. 2016).

The second option is the extended lateral peripatellar approach, which provides access to the knee joint. It is beneficial when performing a concurrent intra-articular procedure (e.g. a lateral femoral cartilage procedure). This approach provides a clear view of the lateral femoral condyle and allows precise manipulation during the procedure (O'Malley et al. 2016).

5.10 Surgical Procedure

A guidewire is placed under fluoroscopic guidance approximately 3–4 cm above the lateral epicondyle and directed towards the medial epicondyle. The guide pin must remain proximal to the patellofemoral articulation during placement. A second parallel guidewire is also inserted.

To minimise the risk of iatrogenic neurovascular injury, the knee must remain flexed, and retractors must be used while making the osteotomy cut. The osteotomy is performed using an oscillating saw with irrigation, and it is completed using manual osteotomes while maintaining a medial cortical hinge. The osteotomy site is opened slowly to the desired correction amount; then, it is fixed with a plate and screws after achieving realignment, followed by bone grafting, either autograft or allograft, depending on the surgeon's preference (Sherman et al. 2018; Puddu et al. 2009).

5.11 Medial Closing Wedge DFO

An anteromedial longitudinal incision starts 10 cm above the patella and ends in the upper third of the patella. The subcutaneous tissue is incised, and the vastus medialis muscle is dissected and elevated from the intermuscular septum. The medial patellofemoral ligament is exposed and incised at the insertion of the vastus medialis to facilitate muscle mobilisation. The intermuscular septum is exposed near the condyles and carefully incised close to the bone and parallel to the femoral shaft. After the intermuscular septum is prepared, three small vessels known as the 'three sisters' become visible in an anteroposterior orientation, providing a rough indication of the level of the osteotomy. Bleeding must be managed using cauterisation or ligation (An et al. 2023).

Two K-wires are inserted above and below the planned osteotomy site in the sagittal plane to prevent rotational deformity during closure after wedge removal. The distal osteotomy cut is performed with an oscillating saw, starting 5 mm above the patellar groove and ending 10 mm from the lateral cortical bone in the lateral femoral condyle. The proximal osteotomy cut is higher in the medial supracondylar region, leaving a lateral hinge and removing a medially based wedge. The osteotomy is closed with careful pressure on the lateral lower limb while stabilising the knee joint, and the corrected mechanical axis is assessed using fluoroscopy. The femoral plate is placed anteromedially on the distal femur, parallel to the femoral shaft, and its position is confirmed using fluoroscopy. Lastly, the arthrotomy is closed, and the medial patellofemoral ligament and the partially released distal insertion of the vastus medialis are reattached to the patella. The wound is then closed in layers (Sherman et al. 2018).

6 The Role of Physics in Knee Osteotomy

Considering joint mechanics regarding force vectors and equilibrium helps explain the biomechanics that underlie surgical interventions and predict how patients will function after surgery. Therefore, the incorporation of fundamental physical principles plays a key role in designing individualised treatment plans that account for the unique mechanical environment of each patient (Maquet 2012).

Knee joints in patients with osteoarthritis are subject to a complex interplay of compressive, tensile, and shear forces while walking or climbing stairs, which can vary significantly depending on body mechanics and external forces. Moreover, the joint fluid's viscosity and flow characteristics are critical in reducing these forces under different load conditions, affecting general cartilage health (Frigo and Donno 2021; Foroughi et al. 2009).

The role and function of ligaments are crucial when planning osteotomy. Lobenhoffer et al. conducted a cadaveric study on seven knees that were tested under axial loading (Lobenhoffer and Van Heerwaarden 2009), which demonstrated that contact pressure measurement in the medial and lateral compartments depended on different load axes. Medial opening wedge HTO was performed and fixed with a plate. Surprisingly, after osteotomy with an intact MCL, the percentage of contact forces was higher on the medial compartment than on the lateral compartment, with increased tension of the superficial layer of the MCL. These results indicated that intraarticular pressure is affected by the tension in the collateral ligaments. Therefore, geometric correction of axes without consideration of ligament tension might have an unexplained paradoxical effect on cartilage pressure. In conclusion, the distal superficial MCL is often required to be released in this type of osteotomy (Lobenhoffer and Van Heerwaarden 2009).

7 Chemical Factors in Knee Health

The chemical environment within the knee joint plays a pivotal role in the progression of joint degeneration, particularly in osteoarthritis. Low-grade inflammation has been identified as a key mediator of osteoarthritis pathogenesis, and inflammatory mediators and matrix metalloproteinases (MMPs) play significant roles in this process. The biomechanical properties of the menisci are also altered because of osteoarthritis-related degeneration, impacting the adjacent articular cartilage.

These insights provide a fundamental understanding of the chemical and biomechanical factors acting within the knee joint, providing the basis for a comprehensive examination of inflammation as a critical element in joint degeneration (Mukherjee and Das 2024; Seitz et al. 2021).

Early et al. highlighted the involvement of mitochondrial-derived reactive oxygen species (ROS) in the inflammatory response after an injury, indicating that targeting ROS activity in infiltrating immune or joint cells could limit catabolic events in the knee joint (Early et al. 2021). These findings demonstrate the important role of inflammation in joint degeneration, and they show that the interaction between inflammation and the pathogenesis of osteoarthritis is complex, providing potential targets for therapeutic interventions.

In knee osteotomy, the surgery disrupts the joint tissue and triggers an inflammatory reaction. This response is necessary for the early stages of healing. However, if the reaction continues or worsens, complications arise, and long-term therapeutic effects are adversely affected (Mukherjee and Das 2024).

Pharmacological interventions are necessary to address the chemical environment within the knee joint, particularly in the context of inflammation and joint degeneration. Anti-inflammatory drugs target low-grade inflammation, which mediates osteoarthritis pathogenesis. These drugs work by modulating the activity of inflammatory mediators and MMPs to halt the progression of osteoarthritis (Zhu et al. 2024).

Nonsteroidal anti-inflammatory drugs (NSAIDs) are commonly used to reduce inflammation and pain by inhibiting cyclooxygenases 1 (COX1) and 2 (COX2), reducing the synthesis of pro-inflammatory prostaglandins. However, long-term use of these medications is associated with adverse effects, such as gastrointestinal irritation from NSAIDs (Wirth et al. 2023).

Nutritional supplements, such as glucosamine and chondroitin, have been explored for their potential to support joint health and healing. Glucosamine is a naturally occurring compound that plays a role in cartilage formation. Chondroitin sulphate is a major component of cartilage that provides elasticity and inhibits enzymes that break down cartilage (Kwoh et al. 2014). These supplements may help reduce inflammation, improve joint function, and decrease the progression of osteoarthritis, although the evidence remains mixed.

Glucosamine and chondroitin are thought to work by modulating the chemical environment within the joint, specifically by reducing levels of inflammatory cytokines and oxidative stress markers. Notably, although these supplements are generally considered safe, their efficacy can vary significantly between patients and specific formulations. Thus, further studies with a longer duration are warranted to conclusively establish the effectiveness of these supplements in limiting joint space narrowing in patients with knee osteoarthritis (Al-Saadi et al. 2019).

8 Biological Mechanisms in Knee Regeneration

Platelet-rich plasma (PRP) is isolated from autologous blood and contains platelets, growth factors, and proteins, which are essential for cell proliferation, differentiation, repair, and collagen synthesis. Therefore, PRP can promote the healing of bone, cartilage, and soft tissue injury.

PRP has been used to treat knee osteoarthritis (KOA). Its possible efficacy seems to be limited to mild to moderate cases; it is not necessarily effective in severe cases with extensive cartilage damage. This combination of HTO and PRP may provide a comprehensive treatment strategy for KOA, especially for patients refractory to conservative treatment who want to avoid arthroplasty. It can help delay the progression of KOA, promote joint protection, and improve and complement regenerative therapy.

Although PRP has shown potential, research on its use in HTO is still limited, and few clinical trials have been conducted. Extensive, long-term, multicentre randomised controlled trials are needed to verify its reliability and safety. In addition, no consensus has been reached on the method of preparation and injection of PRP, and this variability reduces the level of evidence in clinical studies (Zhang et al. 2023).

Mesenchymal stem cells (MSCs) can differentiate into chondrocytes and play an important role in knee cartilage regeneration. MSCs secrete growth factors and cytokines that promote tissue repair and reduce inflammation; this makes the local environment more receptive to cartilage regeneration. MSCs can form a cartilage-like matrix, improve joint function, and reduce pain, and they are used in tissue engineering approaches for the treatment of cartilage defects (Mamachan et al. 2024).

Using MSCs in conjunction with HTO has the potential to improve outcomes for patients undergoing this procedure.

A systematic review by Jin et al. evaluated whether intra-articular MSC injection is effective in patients with HTO in terms of clinical, radiological, and cartilage repair outcomes. The study indicated that MSC injections significantly improved cartilage repair in the medial femoral condyle more than in the medial tibial plateau, implying a differential effect of MSCs in different areas of the knee joint. The results regarding functional scores were mixed. Although the Lysholm score improved significantly after MSC injection, no significant improvements were observed for other scores; this indicates that the effectiveness of MSCs in enhancing functional outcomes after HTO varies. In summary, although MSCs exhibit the potential to improve cartilage repair and may positively impact clinical outcomes in HTO, the evidence is mixed, and further studies are necessary to establish their definitive role and effectiveness (Jin et al. 2022).

Biological scaffolds provide a three-dimensional framework that supports the growth and differentiation of cells. In knee osteotomy, scaffolds enhance the repair of cartilage and other joint tissues by providing an environment conducive to cell attachment, proliferation, and extracellular matrix (ECM) formation. Scaffolds can be created from various materials, including natural polymers (e.g. collagen and hyaluronic acid) or synthetic polymers (e.g. polylactic acid and polyglycolic acid) (Liu et al. 2021).

Wu et al. conducted a retrospective analysis of 50 patients with varus and medial compartment osteoarthritis of the knee managed with arthroscopy and HTO. A second-look arthroscopy was performed 1 year after the operation to assess the cartilage regeneration. The regeneration of cartilage was determined by different types of pathological staining. One year after the procedure, new tissue formation was observed in the defect area of the medial compartment's cartilage, which was identified as cartilage-like tissue using pathological staining.

A potential mechanism for cartilage regeneration involves the upregulation of phosphorylated ERK1/2 (p-ERK1/2) in the new cartilage tissue. This indicates that the activation of the ERK signalling pathway promotes cartilaginous regeneration. However, further investigation is needed for a better understanding of the molecular mechanism. The article concluded that HTO promotes cartilage regeneration by promoting chondrocyte proliferation (Wu et al. 2023).

Han et al. conducted a systematic review to evaluate the effects of concurrent cartilage procedures on cartilage regeneration when performed alongside high tibial osteotomy. Sixteen studies involving 1277 patients were included. The review found that HTO, whether performed alone or with concurrent cartilage procedures, leads to cartilage regeneration during second-look arthroscopy; this was assessed using the International Cartilage Repair Society (ICRS) classification (Han et al. 2024).

9 Rehabilitation Considerations

Effective rehabilitation after knee osteotomy requires a clear understanding of rehabilitation goals to optimise patient outcomes. The main goal is to restore joint function and increase mobility so patients can resume their preoperative activities with less pain or discomfort. A tailored approach adapted to individual patient needs is essential. It must consider factors specific to the patient, such as the nature of their condition, different factors that act as barriers to exercise, and overall health status (Finger et al. 2024).

Weight-bearing protocols after knee osteotomies largely depend on the type of osteotomy performed and the implants selected for fixation. Angular-stable plate fixator systems are known for their stability and have been tested for early functional loading.

After HTO with angular-stable plate osteosynthesis, the protocol involves 20 kg partial weight bearing for 2 weeks and then a transition to full weight-bearing under physiotherapeutic guidance. For distal femoral and double-level osteotomy, the protocol entails 20 kg partial weight bearing for 6 weeks without limiting the range of motion (Finger et al. 2024).

The need for postoperative anticoagulation is debatable, especially for certain patient groups. The incidence of deep vein thrombosis (DVT) varies across the many studies conducted and may be affected by diagnostic methodologies. Clinically significant risk factors for DVT include previous history of DVT, obesity, oral contraceptive use, and genetic predisposition, such as factor V mutation.

Mechanical prophylaxis such as, medical compression stockings (MCS) and intermittent pneumatic compression (IPC) devices can be used on the first postoperative day (Finger et al. 2024).

A systematic review by Pavon et al. found that IPC systems can reduce thromboembolism events when combined with pharmacological prophylaxis (Pavon et al. 2015).

In summary, although pharmacological prophylaxis is widely recommended, its necessity for low-risk patients remains controversial. Mechanical alternatives offer additional tools, but their efficacy, especially in knee osteotomies, requires more research (Finger et al. 2024).

10 Future Directions in Knee Regeneration

Knee osteotomy is a good option for young and athletic patients with osteoarthritis, especially when joint replacement surgery is less desirable. More effective fixation devices and surgical techniques have made osteotomy a reliable way to correct knee deformities and treat osteoarthritis (Moon and Min 2021).

One of the main developments is a mechatronic system that assists orthopaedic surgeons in osteotomy. This system enhances the precision and reproducibility of the surgical procedure, reducing the variability and rate of complications of the conventional free-hand approach.

Computer-assisted surgery systems have also been used to assist with osteotomy planning and execution, but they still require manual intervention during the invasive stages, making them prone to human error.

In addition, robotic assistance in bone cutting has been suggested as a possible way to minimise human error. The idea is to maximise the intraoperative correction and precision, which might translate into better long-term results by avoiding overcorrection or under-correction. However, the use of industrial robots in clinical settings is also limited by the need to meet the strict security standards of hospitals and other medical facilities. Thus, custom-built robotic systems have been designed to work specifically in medical settings.

Quantitative 3D preoperative planning improvements permit more accurate surgical interventions to correct extremely complex multi-plane deformities by performing planned resections with greater precision.

Furthermore, a prototype computer/robotic system has been developed to automate the cutting of osteotomy planes under surgeon supervision. It consists of a custom-built 'fail-safe' surgical robot with six degrees of freedom. This system mainly aims to address complex deformities (Bouazza-Marouf et al. 2008).

11 Conclusion

Knee osteotomy remains a suitable option for early to moderately advanced osteoarthritis, especially in young and active patients. The outcomes of these procedures have improved with technological advancements in imaging, surgical techniques, and postoperative care, and advances in regenerative medicine are likely to enhance the healing process further and extend the benefits of osteotomy. The alignment of mechanical forces and careful patient selection remain critical to the success of knee osteotomy. This procedure remains a good tool for achieving reliable and durable painful knee relief and restoring function while delaying TKA, however further research is needed.

References

Akoto R, et al. Slope-correction osteotomy with lateral extra-articular tenodesis and revision anterior cruciate ligament reconstruction is highly effective in treating high-grade anterior knee laxity. Am J Sports Med. 2020;48(14):3478–85. https://doi.org/10.1177/0363546520966327.

Al-Saadi HM, Pang K-L, Ima-Nirwana S, Chin K-Y. Multifaceted protective role of glucosamine against osteoarthritis: review of its molecular mechanisms. Sci Pharm. 2019;87:34. https://doi.org/10.3390/SCIPHARM87040034.

An J-S, et al. Closing wedge distal femoral osteotomy for knee valgus: indications, technique, rehabilitation and outcomes. J Cartil Jt Preserv. 2023;3(1):100110. https://doi.org/10.1016/j.jcjp.2023.100110.

Bernhardson AS, et al. Tibial slope and its effect on force in anterior cruciate ligament grafts: anterior cruciate ligament force increases linearly as posterior tibial slope increases. Am J Sports Med. 2019a;47(2):296–302. https://doi.org/10.1177/0363546518820302.

Bernhardson AS, et al. Tibial slope and its effect on graft force in posterior cruciate ligament reconstructions. Am J Sports Med. 2019b;47(5):1168–74. https://doi.org/10.1177/0363546519827958.

Bouazza-Marouf K, et al. Robotic system for femoral & tibial osteotomies assistance. In: Romansy 16. Vienna: Springer; 2008. p. 305–12. https://doi.org/10.1007/3-211-38927-x_39.

Capella M, et al. Indications and results of high tibial osteotomy. Ann Jt. 2017;2:33. https://doi.org/10.21037/aoj.2017.06.06.

Coventry MB. Osteotomy about the knee for degenerative and rheumatoid arthritis: indications, operative technique, and results. J Bone Joint Surg. 1965;47(7):984–1000.

Duivenvoorden T, et al. Comparison of closing-wedge and opening-wedge high tibial osteotomy for medial compartment osteoarthritis of the knee. J Bone Joint Surg. 2014;96(17):1425–32. https://doi.org/10.2106/jbjs.m.00786.

Early JO, et al. Mitochondria in injury, inflammation and disease of articular skeletal joints. Front Immunol. 2021;12:695257. https://doi.org/10.3389/fimmu.2021.695257.

Eberbach H, et al. Geometry of the valgus knee: contradicting the dogma of a femoral-based deformity. Am J Sports Med. 2016;45(4):909–14. https://doi.org/10.1177/0363546516676266.

Ferrera A, Menetrey J. Optimizing indications and technique in osteotomies around the knee. EFORT Open Rev. 2022;7(6):396–403. https://doi.org/10.1530/eor-22-0057.

Finger FC, et al. Postoperative management following osteotomies around the knee: a narrative review. EFORT Open Rev. 2024;9(7):658–67. https://doi.org/10.1530/eor-23-0153.

Foroughi N, Smith R, Vanwanseele B. The association of external knee adduction moment with biomechanical variables in osteoarthritis: a systematic review. Knee. 2009;16(5):303–9. https://doi.org/10.1016/j.knee.2008.12.007.

Frigo CA, Donno L. The effects of external loads and muscle forces on the knee joint ligaments during walking: a musculoskeletal model study. Appl Sci. 2021;11(5):2356. https://doi.org/10.3390/app11052356.

Gieroba TJ, et al. Arithmetic hip knee angle measurement on long leg radiograph versus computed tomography—inter-observer and intra-observer reliability. Arthroplasty. 2023;5(1):35. https://doi.org/10.1186/s42836-023-00193-4.

Giffin JR, et al. Effects of increasing tibial slope on the biomechanics of the knee. Am J Sports Med. 2004;32(2):376–82. https://doi.org/10.1177/0363546503258880.

Goshima K, et al. Factors associated with patient satisfaction after opening-wedge high tibial osteotomy. Orthop J Sports Med. 2020;8(11):232596712096796. https://doi.org/10.1177/2325967120967964.

Gulagaci F, et al. A protective hinge wire, intersecting the osteotomy plane, can reduce the occurrence of perioperative hinge fractures in medial opening wedge osteotomy. Knee Surg Sports Traumatol Arthrosc. 2019;28(10):3173–82. https://doi.org/10.1007/s00167-019-05806-7.

Hambly K, et al. Knee joint preservation rehabilitation. In: Lower extremity joint preservation. Cham: Springer; 2020. p. 101–12. https://doi.org/10.1007/978-3-030-57382-9_10.

Han JH, et al. Is bone grafting necessary in opening wedge high tibial osteotomy? A meta-analysis of radiological outcomes. Knee Surg Relat Res. 2015;27(4):207–20. https://doi.org/10.5792/ksrr.2015.27.4.207.

Han JH, et al. Effects of concurrent cartilage procedures on cartilage regeneration in high tibial osteotomy: a systematic review. Knee Surg Relat Res. 2024;36(1):13. https://doi.org/10.1186/s43019-024-00221-w.

Hantes ME, et al. Satisfactory functional and radiological outcomes can be expected in young patients under 45 years old after open wedge high tibial osteotomy in a long-term follow-up. Knee Surg Sports Traumatol Arthrosc. 2018;26(11):3199–205.

Hui C, et al. Long-term survival of high tibial osteotomy for medial compartment osteoarthritis of the knee. Am J Sports Med. 2010;39(1):64–70. https://doi.org/10.1177/0363546510377445.

Jacquet C, et al. Opening wedge high tibial osteotomy allows better outcomes than unicompartmental knee arthroplasty in patients expecting to return to impact sports. Knee Surg Sports Traumatol Arthrosc. 2020;28(12):3849–57.

Jin L, et al. Intra-articular injection of mesenchymal stem cells after high tibial osteotomy: a systematic review and meta-analysis. Orthop J Sports Med. 2022;10(11):232596712211337. https://doi.org/10.1177/23259671221133784.

Kawde K, et al. A comprehensive review of current management trends in medial compartment arthritis of the knee joint. Cureus. 2024;16(3):e56666. https://doi.org/10.7759/cureus.56666.

Kwoh CK, et al. Effect of oral glucosamine on joint structure in individuals with chronic knee pain: a randomized, placebo-controlled clinical trial. Arthritis Rheumatol. 2014;66(4):930–9. https://doi.org/10.1002/art.38314.

Lee DC, Byun SJ. High tibial osteotomy. Knee Surg Relat Res. 2012;24(2):61–9. https://doi.org/10.5792/ksrr.2012.24.2.61.

Liu Y, Shah KM, Luo J. Strategies for articular cartilage repair and regeneration. Front Bioeng Biotechnol. 2021;9:770655. https://doi.org/10.3389/fbioe.2021.770655.

Lobenhoffer P, Agneskirchner JD. Improvements in surgical technique of valgus high tibial osteotomy. Knee Surg Sports Traumatol Arthrosc. 2003;11(3):132–8. https://doi.org/10.1007/s00167-002-0334-7.

Lobenhoffer P, Van Heerwaarden R. Osteotomies around the knee: indications-planning-surgical techniques using plate fixators. Stuttgart: Thieme; 2009.

Lott A, Kayaalp M, Yahagi Y, Giusto JD, Hughes JD, Musahl V. Osteotomy for unicompartmental knee osteoarthritis: indications and outcomes. Oper Tech Sports Med. 2024a;32(2):151084. https://doi.org/10.1016/j.otsm.2024.151084.

Lott A, et al. Around-the-knee osteotomies part II: surgical indications, techniques, and outcomes – state of the art. J ISAKOS. 2024b;9(4):658–71. https://doi.org/10.1016/j.jisako.2024.04.002.

Mamachan M, Sharun K, Amitha Banu S, Muthu S, Pawde A, Abualigah L, Maiti SK. Mesenchymal stem cells for cartilage regeneration: insights into molecular mechanism and therapeutic strategies. Tissue Cell. 2024;88:102380. https://doi.org/10.1016/j.tice.2024.102380.

Maquet PGJ. Biomechanics of the knee: with application to the pathogenesis and the surgical treatment of osteoarthritis. Berlin: Springer Science and Business Media; 2012.

Marom N, et al. Osteotomies in the multiple ligament injured knee. Clin Sports Med. 2019;38(2):297–304. https://doi.org/10.1016/j.csm.2018.11.003.

Micicoi G, et al. Osteotomy around the knee is planned toward an anatomical bone correction in less than half of patients. Orthop Traumatol Surg Res. 2021a;107(4):102897. https://doi.org/10.1016/j.otsr.2021.102897.

Micicoi G, et al. High tibial osteotomy with Miniaci planning using manual and semiautomated digital measures. Video J Sports Med. 2021b;1(6):263502542110329. https://doi.org/10.1177/26350254211032968.

Moon SW, Min Y-K. Osteotomy around knee. In: A strategic approach to knee arthritis treatment. Singapore: Springer; 2021. p. 225–49. https://doi.org/10.1007/978-981-16-4217-3_12.

Moon SW, et al. The effect of hinge position on posterior tibial slope in medial open-wedge high tibial osteotomy. Arthroscopy. 2015;31(6):1128–33. https://doi.org/10.1016/j.arthro.2015.01.009.

Mukherjee A, Das B. The role of inflammatory mediators and matrix metalloproteinases (MMPs) in the progression of osteoarthritis. Biomater Biosyst. 2024;13:100090. https://doi.org/10.1016/j.bbiosy.2024.100090.

Murray R, et al. High tibial osteotomy for varus deformity of the knee. J Am Acad Orthop Surg Glob Res Rev. 2021;5(7):e21.00141. https://doi.org/10.5435/jaaosglobal-d-21-00141.

Nakamura R, et al. Appropriate hinge position for prevention of unstable lateral hinge fracture in open wedge high tibial osteotomy. Bone Joint J. 2017;99-B(10):1313–8. https://doi.org/10.1302/0301-620x.99b10.bjj-2017-0103.r1.

Naudie D, et al. Survivorship of the high tibial valgus osteotomy. A 10- to 22-year follow-up study. Clin Orthop Relat Res. 1999;356:18–27.

O'Malley MP, et al. Distal femoral osteotomy: lateral opening wedge technique. Arthrosc Tech. 2016;5(4):e725–30. https://doi.org/10.1016/j.eats.2016.02.037.

Paley D, et al. Deformity planning for frontal and sagittal plane corrective osteotomies. Orthop Clin N Am. 1994;25(3):425–65. https://doi.org/10.1016/s0030-5898(20)31927-1.

Pavon JM, et al. Effectiveness of intermittent pneumatic compression devices for venous thromboembolism prophylaxis in high-risk surgical patients: a systematic review. J Arthroplast. 2015;31(2):524–32. https://doi.org/10.1016/j.arth.2015.09.043.

Puddu G, et al. Which osteotomy for a valgus knee? Int Orthop. 2009;34(2):239–47. https://doi.org/10.1007/s00264-009-0820-3.

Pullen WM, et al. High tibial osteotomy in knee reconstruction and joint preservation. J Am Acad Orthop Surg. 2024;32(13):577–86. https://doi.org/10.5435/jaaos-d-23-00323.
Rossi R, Bonasia DE, Amendola A. The role of high tibial osteotomy in the varus knee. J Am Acad Orthop Surg. 2011;19(10):590–9. https://doi.org/10.5435/00124635-201110000-00003.
Seitz AM, et al. Osteoarthritis-related degeneration alters the biomechanical properties of human menisci before the articular cartilage. Front Bioeng Biotechnol. 2021;9:659989. https://doi.org/10.3389/fbioe.2021.659989.
Sherman SL, Thompson SF, Clohisy JCF. Distal femoral varus osteotomy for the management of valgus deformity of the knee. J Am Acad Orthop Surg. 2018;26(9):313–24. https://doi.org/10.5435/jaaos-d-16-00179.
Song SJ, Bae DK. Computer-assisted navigation in high tibial osteotomy. Clin Orthop Surg. 2016;8(4):349. https://doi.org/10.4055/cios.2016.8.4.349.
Staubli AE, Jacob HAC. Evolution of open-wedge high tibial osteotomy: experience with a special angular stable device for internal fixation without interposition material. Int Orthop. 2010;34(2):167–72.
Sternheim A, Garbedian S, Backstein D. Distal femoral varus osteotomy: unloading the lateral compartment: long-term follow-up of 45 medial closing wedge osteotomies. Orthopedics. 2011;34(9):e488–90. https://doi.org/10.3928/01477447-20110714-37.
Tischer T, et al. The impact of osseous malalignment and realignment procedures in knee ligament surgery: a systematic review of the clinical evidence. Orthop J Sports Med. 2017;5(3):232596711769728. https://doi.org/10.1177/2325967117697287.
Tseng T-H, Wang H, Tzeng S-C, Hsu K-H, Wang J. Knee-ankle joint line angle: a significant contributor to high-degree knee joint line obliquity in medial opening wedge high tibial osteotomy. J Orthop Surg Res. 2022;17(1):79. https://doi.org/10.1186/s13018-022-02976-y.
Tsushima T, Sasaki E, Sakamoto Y, Kimura Y, Tsuda E, Ishibashi Y. Joint line convergence angle is the most associated alignment factor with the severity of medial knee osteoarthritis. J Exp Orthop. 2024;11:e70007. https://doi.org/10.1002/jeo2.70007.
Webb JM, et al. Posterior tibial slope and further anterior cruciate ligament injuries in the anterior cruciate ligament–reconstructed patient. Am J Sports Med. 2013;41(12):2800–4. https://doi.org/10.1177/0363546513503288.
Wirth T, Lafforgue P, Pham T. NSAID: current limits to prescription. Joint Bone Spine. 2023;91:105685. https://doi.org/10.1016/j.jbspin.2023.105685.
Wu J, et al. Arthroscopy combined with high tibial osteotomy promotes cartilage regeneration in osteoarthritis. J Orthop Surg. 2023;31(1):102255362311653. https://doi.org/10.1177/10225536231165357.
Zampogna B, Vasta S, Papalia R. Patient evaluation and indications for osteotomy around the knee. Clin Sports Med. 2019;38(3):305–15. https://doi.org/10.1016/j.csm.2019.02.011.
Zhang W, et al. High tibial osteotomy combined with platelet-rich plasma for knee osteoarthritis: a systematic review and meta-analysis. Res Sq. 2023:1–21. https://doi.org/10.21203/rs.3.rs-3390832/v1.
Zhu R, Fang H, Wang J, Ge L, Zhang X, Aitken D, Cai G. Inflammation as a therapeutic target for osteoarthritis: a literature review of clinical trials. Clin Rheumatol. 2024;43:2417–33. https://doi.org/10.1007/s10067-024-07042-y.

Ankle: Regenerative Procedures After Foot and Ankle Injuries

Lyndon Mason, Ahmed Galhoum, and Abdul Gomaa

Abstract Foot and ankle disorders represent a significant challenge in musculoskeletal medicine, often resulting in chronic pain, functional limitations, and reduced quality of life. Traditional treatment options, including pain management, physiotherapy, and surgical interventions, have limitations, such as restricted tissue repair capacity and donor site morbidity. Regenerative medicine introduces a novel therapeutic approach, utilising stem cell therapy, platelet-rich plasma (PRP), and tissue engineering (TE) with biomaterial scaffolds to enhance the body's intrinsic healing potential and promote tissue regeneration. This chapter will explore the biological mechanisms, clinical applications, and current evidence for these regenerative therapies in foot and ankle pathology, with a focus on ankle osteoarthritis, Achilles tendinopathy, osteochondral lesions of the talus (OLT), and plantar fasciitis. For instance, PRP therapy has been widely studied in Achilles tendinopathy and plantar fasciitis, demonstrating variable outcomes depending on injection protocols and patient selection. While regenerative medicine offers exciting possibilities for foot and ankle disorders, inconsistencies in clinical trial methodologies, small sample sizes, and variable outcome measures challenge its widespread adoption.

Keywords Achilles tendinopathy · Cartilage · Osteochondral · Plantar fasciitis · Osteoarthritis

1 Introduction

Foot and ankle disorders encompass a wide spectrum of conditions affecting bones, joints, muscles, tendons, and ligaments. These conditions can cause significant pain, disability, and diminish quality of life. Traditional treatment approaches often

L. Mason (✉) · A. Galhoum · A. Gomaa
Liverpool Orthopaedic and Trauma Service | Liverpool University Hospitals NHS Foundation Trust, Liverpool, UK
e-mail: lwmason@liverpool.ac.uk

P. Lee (ed.), *Musculoskeletal Regeneration Medicine*,
https://doi.org/10.1007/978-3-031-77456-0_15

involve pain management, physical therapy, and in severe cases, surgical interventions. However, these approaches may not always provide optimal outcomes, and can have limitations such as limited repair capacity and potential donor site morbidity.

Regenerative medicine offers a possible paradigm shift in managing these often complex disorders. It utilises a combination of biological strategies to promote the body's inherent healing potential to regenerate damaged tissues. This approach holds immense promise for improving clinical outcomes and enhancing patient function. The most commonly used and accessible therapies include: stem cell therapy, platelet rich plasma (PRP) and the use of tissue engineering (TE) and biomaterial scaffolds.

2 Stem Cell Therapy

Stem cell therapy has emerged as a potential frontier in treating foot and ankle osteoarthritis and the sequelae requiring ankle or midfoot fusion. These immature cells hold the unique ability to differentiate into specialised cell types like cartilage or bone, offering a theoretical framework for promoting tissue regeneration and potentially altering the course of these conditions.

Stem cells exert their regenerative effects through various mechanisms:

- *Differentiation*: Stem cells have the capacity to differentiate into specialised cell types, including chondrocytes, which are essential for cartilage repair and regeneration. When administered into damaged tissues, adipose-derived stem cells (ADSCs) can differentiate into chondrocyte-like cells, contributing to the restoration of cartilage structure and function.
- *Paracrine Effects*: ADSCs secrete a variety of bioactive molecules, including growth factors, cytokines, and extracellular vesicles, which exert paracrine effects on surrounding cells. These paracrine factors modulate inflammation, promote tissue repair, and stimulate endogenous regenerative processes (Alcaraz et al. 2019).
- *Immunomodulation*: Stem cells possess immunomodulatory properties, capable of suppressing inflammatory responses and modulating immune cell activity. By dampening excessive inflammation and promoting immune tolerance, ADSCs create a favourable microenvironment for tissue repair and regeneration (Gonzalez-Rey et al. 2010).
- *Extracellular Matrix Remodelling*: ADSCs play a role in remodelling the extracellular matrix (ECM) of damaged tissues by secreting enzymes and matrix proteins. This remodelling process facilitates tissue restructuring and enhances the biomechanical properties of the regenerated tissue (Dufrane et al. 2015).

While research on stem cells in foot and ankle disorders is limited, several studies have emerged in recent years. We will explore the latest findings on stem cells used for various foot and ankle conditions and summarises current evidence-based recommendations for their use in each specific case.

2.1 Stem Cells in Ankle Osteoarthritis

Affecting nearly 1% of the world population, ankle osteoarthritis (OA) presents a significant challenge, manifesting in pain and functional limitations that impede daily activities. A recent retrospective cohort study by Shimozono et al. (2021) investigated the efficacy of a novel therapeutic approach for advanced-stage post-traumatic OA of the ankle. This approach combined intraarticular injections of autologous micronized adipose tissue with arthroscopic debridement. The study included a sample of nineteen patients (19 ankles) diagnosed with Kellgren-Lawrence (KL) grades 3 and 4 posttraumatic ankle OA. They injected autologous adipose tissue-derived stem cells ADCSs into the ankle joint following an arthroscopic debridement. They used visual analogue scale (VAS) and Foot and Ankle Outcome Scores (FAOS) to assess the outcome. At a mean follow up of 14.3 months, they reported a significant increase in FAOS from 48.8 pre intervention to 61.1 post intervention. An improvement in VAS score from 6.1 to 3.8 in the last follow up was also observed.

Emadedin et al. (2012) assessed the Intra-articular Injection of Autologous Mesenchymal Stem Cells (MSCs) in Patients with Knee, Ankle, or Hip Osteoarthritis. Their study included six patients with grade 3–4 ankle OA assessed clinically and radiologically using plain radiographs and MRI. MSCs were prepared from 150mls of autologous iliac crest bone marrow aspirate, stem cells were processed, isolated and expanded in specific cultures for 7 days, the researchers processed the expanded cells and prepared them for injection in a laboratory specialised in cellular isolation and testing technology. The final product consisted of injections containing 5×10^5 cells/kg/body weight. MSCs injections were performed under fluoroscopic guidance for ankle OA patients. The Western Ontario and McMaster Universities Arthritis Index (WOMAC), VAS, and FAOS scores were used to assess their patients response to treatment. The ankle OA group showed significant improvement in the walking distance from 1010 steps to 2333 at 30 months of follow up. There was a statistically significant reduction in WOMAC score at 12 and 30 months post intervention as well as a significant increase in FAO scores. MRI assessment at 6 months following MSCs injection, showed decreased signal intensity related to subchondral oedema in four of the six patients with ankle OA.

These two studies suggest that local infiltration of stem cells (either ADCs or MSCs) needs further trials to study its clinical value and to assess its cost effectiveness. However, their application in scaffolds and in combination with scaffolds and tissue engineering (TE) technologies will be discussed later in this chapter.

2.2 *Stem Cells in Achilles Tendinopathy*

Stem cell injection for Achilles tendinopathy was studied in a prospective RCT comparing intratendinous injections of Platelet Rich Plasma (PRP) to autologous stromal vascular fraction (SVF) ADSCs. The authors used VAS scores, ultrasonography and MRI scans to measure the clinical outcome. Their study included 43 patients with non-insertional Achilles tendinopathy (58 tendons). PRP was administered in 22 patients (28 tendons) and ADSCs in 21 patients (30 tendons). Patients reported a significant improvement in VAS scores in both groups compared to pre intervention scores. There was no significant difference in favour of any of the used interventions. Follow up ultrasonography and MRI scans at 6 months showed increased tendon thickness in both groups, although tendon thickness is usually correlated with worsening of tendon degeneration it was correlated with improved VAS scores in this study. It is worth noting that the literature results of PRP injection in Achilles tendinopathy are controversial and many studies reported that it has no effect (Albano et al. 2017).

However, a further RCT compared PRP injections to ADSCs injections into mid-substance Achilles tendinopathy. The ADSCs were manually aspirated by a plastic surgeon and processed to produce SVFs. They used VAS pain scale, Victorian Institute of Sport Assessment-Achilles (VISA-A), AOFAS, and SF-36 form to assess patient-reported outcomes. The authors concluded that both treatments led to a significant improvement with respect to baseline. Comparing the two groups, VAS, AOFAS and VISA-A scores were significantly better at 15 and 30 days in the SVF group compared to the PRP group ($p < 0.05$). However, this difference was not sustained at later time points with no significant differences in scores between the two groups observed beyond 30 days. Furthermore, there were no significant correlation between clinical and radiological findings (Usuelli et al. 2018).

We can conclude that ADSCs use in Achilles tendinopathy is safe and might have a potential in the treatment algorithm, however the fact that it is similar to PRP injection is a controversial point. Well designed studies are needed to establish the efficacy as well as cost effectiveness of stem cells injections.

3 Platelet Rich Plasma (PRP)

PRP is reported to harness the body's natural healing potential by concentrating growth factors found in platelets within a patient's own blood. These growth factors can possibly stimulate tissue repair and regeneration, potentially offering relief from pain and improving joint function (Nauwelaers et al. 2021). Platelet-derived growth factor (PDGF) was discovered in platelet-alpha granules in 1978 (Alves and Grimalt 2018). Over the following few decades, researchers found PRP to contain the transforming growth factor β (TGF-β), insulin-like growth factor (IGF), epidermal growth factor (EGF), vascular endothelial growth factor (VEGF), and fibroblast growth factor (FGFs) (Foster et al. 2009; Gandhi et al. 2005). These growth factors can stimulate the proliferation of various cell types involved in tissue regeneration, including stem cells, progenitor cells, and differentiated cells. They can also promote the differentiation of stem cells into specific cell lineages required for tissue repair, such as osteoblasts for bone formation, chondrocytes for cartilage repair, and myoblasts for muscle regeneration (Pittenger et al. 1999; Lian and Stein 1992). Some of those growth factors possess anti-inflammatory properties and can modulate the immune response to promote tissue repair and regeneration. They can inhibit the production of pro-inflammatory cytokines, promote the recruitment of anti-inflammatory immune cells, and enhance the resolution of inflammation at the site of injury (Gaudet et al. 2011).

Since PRP is an autologous product, no immune response will be triggered during the application (Marx et al. 1998). This fact has encouraged clinicians to use PRP in treating various conditions across medical subspecialities including for foot and ankle disorders. Platelet rich plasma has been trialled to treat several F&A pathologies including Achilles tendon ruptures and tendinopathy, plantar fasciitis, ankle OA, osteochondral lesions of the talus (OLT), and ankle fractures (Guney et al. 2015; Sun et al. 2021; Lareau et al. 2014; Schepull et al. 2011). Despite the fact that clinical and basic research reported that PRP can improve bone and soft tissue healing, its therapeutic efficacy in foot and ankle pathologies is debatable due to limited clinical data and discrepancies of clinical trials outcomes.

3.1 PRP in Achilles Tendon Pathology

Achilles tendinopathy is a spectrum of degenerative conditions affecting the Achilles tendon, characterised by microtears, disorganisation of collagen fibres, and neoangiogenesis. It is more commonly associated with overuse in athletes, dancers, and individuals participating in sporting activities (Kraeutler et al. 2017). There is a growing body of evidence to support the therapeutic potential of PRP in managing various musculoskeletal conditions, including tendinopathy. This has led to a surge in PRP applications commercially and within clinical trials. However, a critical challenge remains in definitively establishing PRP efficacy for tendinopathy due to the inherent heterogeneity of factors influencing patient outcomes. Beyond the core question of efficacy, robust clinical trial design relies on inclusion and exclusion criteria to ensure patient population homogeneity, while long-term follow-up is paramount to evaluating the durability of treatment effects.

The literature and published research on using PRP in non-operative management of AT pathologies showed mixed results. Filardo et al. investigated PRP injection in a cohort of 27 patients (both male and female) diagnosed with chronic Achilles tendinitis. The average duration of patient follow-up was 54.1 months (minimum 30 months). The study demonstrated statistically significant improvements in the Victorian Institute of Sport Assessment-Achilles (VISA-A) score, the EQ Visual Analogue Scale (EQ-VAS) score, and the Tegner activity scale. These findings suggest a positive outcome following PRP injections for chronic Achilles tendonopathy (Filardo et al. 2014).

Erroi et al. conducted a retrospective comparative investigation evaluating the efficacy of shockwave therapy and PRP injections in 45 patients diagnosed with insertional Achilles tendinopathy (AT). While both interventions resulted in improvements in measured outcomes, statistically significant differences between the treatment groups were not observed (Erroi et al. 2017). A randomised, double-blind, prospective study by Boesen et al. evaluated the efficacy of a combined intervention in 60 male patients with chronic mid-Achilles tendinitis. The intervention compared three groups, all patients performed eccentric exercise alongside injection of a high-dose steroid and local anaesthetic injection, four PRP injections 14 days apart, or a placebo. The results showed VISA-A improvement in all groups with higher improvement in the steroid group compared to the two other groups at 12 weeks. The PRP and steroid group maintained better results at 24 weeks compared to the placebo group. MRI showed reduced tendon thickness only in the PRP and steroid groups (Boesen et al. 2017).

A meta-analysis studied five RCTs comparing PRP to placebo in 189 patients. Significant differences in the VISA-A were not observed between the PRP and placebo groups after 12 weeks of treatment. However, PRP exhibited better efficacy than the placebo treatment after 6 weeks. VAS scores after 6 and 24 weeks were not significantly different. However, VAS scores at the 12th week and tendon thickness were significantly improved (Liu et al. 2019).

Conversely, several studies reported that PRP injection has no superior efficacy in comparison to other forms of treatment. In 2021, Kearney et al. reported the results of their RCT comparing a single intratendinous PRP injections to a sham injection in 240 (PRP = 121, sham = 119) patients with midportion AT. There was no difference in VISA-A scores between both groups at 3 and 6 months of follow up (Kearney et al. 2021). Furthermore, a recent systematic review and meta-analysis studied the results of eight trials and concluded that PRP injection in AT did not cause significant improvement when compared to placebo or other treatment modalities (Arthur Vithran et al. 2023).

Surgical management of Achilles tendinopathy (AT) encompasses a spectrum of techniques, including minimally invasive debridement and percutaneous needle tenotomy. Therman et al. studied the clinical outcome of endoscopic debridement of midportion AT with and without PRP injections. Both groups showed significant clinical and radiological improvement, However, there was no significant difference between the groups (Thermann et al. 2023). Another study assessed the efficacy of percutaneous needle tenotomy (PNT) and leukocyte-rich platelet-rich plasma (LR-PRP) for chronic Achilles tendinosis. They concluded that there was no significant difference between both groups at 52 and 104 weeks of follow up.

Keene et al. conducted a multicentre RCT to study the effect of PRP injections (n = 114 in patients with acute Achilles tendon rupture compared to dry needling (n = 119). Primary outcome was muscle tendon function at 24 weeks, measured objectively with the limb symmetry index and the secondary outcomes included patient reported function (Achilles tendon rupture score), quality of life (short form

12 version 2), and pain (visual analogue scale). The results showed that there was no difference in muscle tendon function between participants receiving PRP injections and those receiving placebo injections in both primary and secondary outcomes (Keene et al. 2019).

As outlined, there remains to be significant heterogeneity in the studies examining the role of PRP with no conclusive evidence supporting its efficacy.

3.2 *PRP in Ankle Osteoarthritis*

In vitro investigations on chondrocytes demonstrated that platelet-rich plasma (PRP) augmented their proliferation rate and induced matrix synthesis, while also preserving marker expression. Furthermore, the analgesic properties of PRP suggest its potential utility as a pharmacological intervention for osteoarthritis management, with the additional benefit of promoting hyaluronic acid secretion. Notably, recent research suggests that a PRP injection may offer a safe and efficacious approach for osteoarthritis treatment (Cavallo et al. 2014; Repetto et al. 2017). Repetto et al. retrospectively studied the outcome of four PRP injections (1 week apart) in 20 patients with symptomatic ankle OA. After a mean follow-up of 17.7 months, the authors found significant improvements in pain, function, and patient satisfaction (Repetto et al. 2017). Fukawa et al. studied the effect of three ultrasound guided injections of PRP in 20 patients with symptomatic ankle OA. Each injection was composed of 2 ml of PRP and were administered 2 weeks apart. They reported improvement of visual analogue scale (VAS) for pain and the Self-Administered Foot Evaluation Questionnaire (SAFE-Q) at 4, 12 and 24 weeks from the last injection (Fukawa et al. 2017).

3.3 *PRP in Plantar Fasciitis*

Plantar fasciitis represents the most common aetiology of chronic plantar heel pain in the adult population. It is postulated to arise from repetitive microtrauma to the plantar fascia originating at the calcaneus, leading to subsequent inflammation and degeneration. Various treatment modalities exist, including orthotic devices, night splints, plantar fascia stretching protocols, physical therapy interventions, extracorporeal shockwave therapy (ESWT), pharmacological management, injectable corticosteroids, and, in extreme cases, surgical fascial release. However, non-operative treatment approaches demonstrate a failure rate of 10–15%, resulting in chronic plantar fasciitis. The use of corticosteroids, autologous blood injections, and ESWT is associated with variable efficacy and potential complications, including plantar fascial rupture following corticosteroid injection (League 2008; Latt et al. 2020).

Many of the studies investigating the efficacy of PRP injections in patients with plantar fasciitis have reported positive outcomes. A comparative study assessed the levels of pain using VAS and the modified criteria of the Roles and Maudsley scores following injection of PRP versus Methylprednisolone in patients with plantar fasciitis. The results showed that patients in both groups reported significant improvement at 3 weeks and 6 months (Akşahin et al. 2012).

A similar study compared corticosteroid injection to PRP injection in 80 patients with plantar fasciitis, 40 in each intervention arm. They used visual analogue pain scale, subjective rating using the modified Roles and Maudsley score, functional outcome score by the Foot and Ankle Outcome Instrument (FAI) core scale, and the American Orthopaedic Foot & Ankle Society (AOFAS) ankle-hindfoot scale to assess the efficacy of the interventions at 1, 3 and 6 months. Patients from both groups showed improved outcomes across all scores, but there was no significant difference between groups (Jain et al. 2018).

A double-blinded RCT compared the results of PRP, corticosteroids and saline injections in patients with plantar fasciitis. Each study arm was comprised of 30 patients. They followed up the patients over a regular interval until 18 months. The VAS, score, mean Roles and Maudsley score and Short form 12 showed significant improvement in all groups compared to the base line with significantly better results in corticosteroid group in the short term, however, the PRP group had a statistically significant better long term result compared to other groups (Shetty et al. 2019).

Furthermore, a recent systematic review and meta-analysis study evaluated ten studies comparing PRP injection to corticosteroid injection with a total number of patients of 517. They concluded that PRP injections are associated with improved pain and function scores when compared to corticosteroid injections (Singh et al. 2017). The current evidence shows that PRP injections in plantar fasciitis has an efficacy similar to corticosteroid injections with an advantage of longer-term benefits and lower risks. Further larger scale randomised control studies are needed to confirm those findings.

3.4 PRP in Osteochondral Lesions of the Talus

Enhanced imaging techniques have led to a growing recognition of osteochondral lesions of the talus (OLT) as a contributor to ankle pain. Conservative management typically proves effective for smaller lesions, while surgical intervention becomes necessary for larger lesions or those unresponsive to conservative therapy. Surgical approaches can be categorised as reparative or reconstructive in nature. Multiple studies highlighted that PRP has the capacity to enhance chondrocyte and stem cell proliferation, augment proteoglycan deposition, and promote collagen synthesis. Moreover, PRP exhibited inhibitory effects on local catabolic cytokines. Importantly, it was observed that incorporating PRP into autologous osteochondral graft therapy markedly enhanced graft integration while mitigating cartilage degeneration (Smyth et al. 2013).

In a prospective case control study, the researchers compared two groups of patients undergoing microfracture surgery for OLT. The first group of patients underwent microfracture surgery alone while the second group underwent microfracture surgery combined with injection of PRP. They used AOFAS score, Foot and Ankle Ability Measure (FAAM), and VAS score to assess the outcome. At a mean follow up of 16 month, both groups showed significant improvement however the PRP group had a significantly better functional outcomes (Smyth et al. 2013). It is worth noting that a major drawback of this study was that authors excluded patients with OLTs larger than 20 mm and literature showed that smaller lesions and lesions that are 15 mm or less usually have better outcomes when treated with microfractures (Chuckpaiwong et al. 2008).

Mei-Dan et al. conducted a RCT to study the clinical and functional outcomes after three intra-articular PRP or HA injections in 30 patients with OLT. At the last follow-up the researchers concluded that Osteochondral lesions of the ankle treated with intra-articular injections of PRP and HA showed a decrease in pain scores and an improvement in function for at least 6 months, with minimal adverse events. Moreover, PRP treatment led to significantly better outcomes than HA. Whilst the available research supports the use of PRP in treatment of small osteochondral lesions of the talus, the number and heterogeneity of the studies are not enough to establish a robust evidence base for its use. Further research is required to clarify its efficacy either alone or with combination of other interventions.

4 Tissue Engineering and Biomaterial Scaffolds

Tissue engineering represents an interdisciplinary field that integrates principles from engineering, life sciences, and medical sciences towards the development of functional biological substitutes. This field strives to achieve restoration, protection against disease progression, or enhancement of function in damaged tissues and/or organs (Vacanti and Langer 1999). The applications of tissue engineering in foot and ankle surgery encompass the use of biological scaffolds, stem cells or cultured cells, and growth factors. The three components are often used in combination.

4.1 TE in Treatment of Osteochondral Lesions of the Talus

The aetiology of osteochondral lesions of the talus (OLTs) remains under investigation, although a higher prevalence is observed in young, active individuals (20–40 years old) with a history of ankle sprain or trauma (Shimozono et al. 2017). Current understanding suggests a multifactorial origin for OLTs, encompassing both traumatic and non-traumatic events. Notably, up to 50% of ankle sprains and over 70% of ankle fractures have been associated with the development of OLTs (Murawski and Kennedy 2013; Stufkens et al. 2010). OLTs are categorised into various types depending on their anatomical location. Traditionally, posteromedial and anterolateral OLTs were considered the most prevalent (Berndt and Harty

1959). Raikin et al. (2007) employed MRI to assess 424 patients using a 9-zone anatomical grid system, revealing that the majority of lesions (53%) were situated in zone 4 (centromedial), with zone 6 (centrolateral) being the second most common location.

Cell-based surgical interventions represent regenerative medicine approaches aimed at biologically restoring cartilage rather than merely substituting it. These techniques encompass autologous chondrocyte implantation (ACI) and autologous matrix-induced chondrogenesis (AMIC), often utilised post-failed microfracture procedures. ACI, initially developed as a two-stage process for knee chondral/osteochondral lesions, involves harvesting macroscopically normal hyaline cartilage from either the anterior talus or detached osteochondral fragments in the ankle (Johnson et al. 2013; Giannini et al. 2005). Chondrocytes are enzymatically isolated from the cartilage, expanded in vitro, and subsequently reimplanted into the defect beneath a periosteal or collagen patch for 3–4 weeks (López-Alcorocho et al. 2021). Alternatively, chondrocytes can be directly cultured on a collagen patch, known as matrix-assisted ACI (MACI). Pagliazzi et al. (2018) demonstrated significant improvement in patient's AOFAS scores at 7 years post-ACI. Despite the recent approval of ACI use in the knee by the National Institute for Clinical Excellence (NICE), its application in the ankle remains limited due to insufficient evidence of clinical efficacy, rendering it a subject of clinical trials or efficacy studies within the UK (National Institute for Health and Care Excellence 2017).

To eliminate the necessity of a two-stage procedure, the AMIC technique offers a promising and cost-effective alternative initially devised for knee defect treatment. This technique involves microfracture followed by application of a collagen patch over the defect, aiming not only to trap any MSCs entering the defect area through the bleeding process but also to provide a scaffold for cell adhesion and proliferation (Benthien and Behrens 2010). It is proposed that transforming growth factor-β (TGF-β) present in the fibrin glue used to adhere the patch may enhance MSC chondrogenic differentiation (Gille et al. 2005).

However, a recent systematic review highlighted the need for further clinical studies to verify the efficacy and safety of AMIC in talus treatment (Malahias et al. 2020). MRI assessments revealed complete defect filling in 88% of patients treated with AMIC, accompanied by pain reduction and functional improvement at a mean follow-up of 4.7 years (Weigelt et al. 2019). Additional systematic reviews further supported the efficacy of AMIC, confirming positive outcomes up to 5 years post-operatively (Walther et al. 2021).

5 Conclusion

The role of regenerative medicine for common and troublesome foot and ankle conditions is rapidly expanding. There is published evidence to support and refute the efficacy of stem cell therapy, PRP and TE for the treatment of these conditions.

Despite some positive results, the evidence body remains to be weak, with very few robust well designed studies that can offer conclusive and definitive answers.

References

Akşahin E, Doğruyol D, Yüksel HY, Hapa O, Doğan Ö, Çelebi L, et al. The comparison of the effect of corticosteroids and platelet-rich plasma (PRP) for the treatment of plantar fasciitis. Arch Orthop Trauma Surg. 2012;132:781–5.

Albano D, Messina C, Usuelli FG, De Girolamo L, Grassi M, Maccario C, et al. Magnetic resonance and ultrasound in Achilles tendinopathy: predictive role and response assessment to platelet-rich plasma and adipose-derived stromal vascular fraction injection. Eur J Radiol. 2017;95:130–5.

Alcaraz MJ, Compañ A, Guillén MI. Extracellular vesicles from mesenchymal stem cells as novel treatments for musculoskeletal diseases. Cells. 2019;9(1):98.

Alves R, Grimalt R. A review of platelet-rich plasma: history, biology, mechanism of action, and classification. Skin Appendage Disord. 2018;4(1):18–24.

Arthur Vithran DT, Xie W, Opoku M, Essien AE, He M, Li Y. The efficacy of platelet-rich plasma injection therapy in the treatment of patients with Achilles tendinopathy: a systematic review and meta-analysis. J Clin Med. 2023;12(3):995.

Benthien J, Behrens P. Autologous matrix-induced chondrogenesis (AMIC) combining microfracturing and a collagen I/III matrix for articular cartilage resurfacing. Cartilage. 2010;1(1):65–8.

Berndt AL, Harty M. Transchondral fractures (osteochondritis dissecans) of the talus. J Bone Joint Surg Am. 1959;41(6):988–1020.

Boesen AP, Hansen R, Boesen MI, Malliaras P, Langberg H. Effect of high-volume injection, platelet-rich plasma, and sham treatment in chronic midportion Achilles tendinopathy: a randomized double-blinded prospective study. Am J Sports Med. 2017;45(9):2034–43.

Cavallo C, Filardo G, Mariani E, Kon E, Marcacci M, Ruiz MTP, et al. Comparison of platelet-rich plasma formulations for cartilage healing: an in vitro study. J Bone Joint Surg Am. 2014;96(5):423–9.

Chuckpaiwong B, Berkson EM, Theodore GH. Microfracture for osteochondral lesions of the ankle: outcome analysis and outcome predictors of 105 cases. Arthroscopy. 2008;24(1):106–12.

Dufrane D, Docquier P-L, Delloye C, Poirel HA, André W, Aouassar N. Scaffold-free three-dimensional graft from autologous adipose-derived stem cells for large bone defect reconstruction: clinical proof of concept. Medicine (Baltimore). 2015;94(50):e2220.

Emadedin M, Aghdami N, Taghiyar L, Fazeli R, Moghadasali R, Jahangir S, et al. Intra-articular injection of autologous mesenchymal stem cells in six patients with knee osteoarthritis. Arch Iran Med. 2012;15(7):422–8.

Erroi D, Sigona M, Suarez T, Trischitta D, Pavan A, Vulpiani MC, et al. Conservative treatment for insertional Achilles tendinopathy: platelet-rich plasma and focused shock waves. A retrospective study. Muscles Ligaments Tendons J. 2017;7(1):98.

Filardo G, Kon E, Di Matteo B, Di Martino A, Tesei G, Pelotti P, et al. Platelet-rich plasma injections for the treatment of refractory Achilles tendinopathy: results at 4 years. Blood Transfus. 2014;12(4):533.

Foster TE, Puskas BL, Mandelbaum BR, Gerhardt MB, Rodeo SA. Platelet-rich plasma: from basic science to clinical applications. Am J Sports Med. 2009;37(11):2259–72.

Fukawa T, Yamaguchi S, Akatsu Y, Yamamoto Y, Akagi R, Sasho T, et al. Safety and efficacy of intra-articular injection of platelet-rich plasma in patients with ankle osteoarthritis. Foot Ankle Int. 2017;38(6):596–604.

Gandhi A, Bibbo C, Pinzur M, Lin SS. The role of platelet-rich plasma in foot and ankle surgery. Foot Ankle Clin. 2005;10(4):621–37.

Gaudet AD, Popovich PG, Ramer MS. Wallerian degeneration: gaining perspective on inflammatory events after peripheral nerve injury. J Neuroinflammation. 2011;8:1–13.

Giannini S, Buda R, Grigolo B, Vannini F, De Franceschi L, Facchini A, et al. The detached osteochondral fragment as a source of cells for autologous chondrocyte implantation (ACI) in the ankle joint. Osteoarthritis Cartilage. 2005;13(7):601–7.

Gille J, Meisner U, Ehlers E, Müller A, Russlies M, Behrens P, et al. Migration pattern, morphology and viability of cells suspended in or sealed with fibrin glue: a histomorphologic study. Tissue Cell. 2005;37(5):339–48.

Gonzalez-Rey E, Gonzalez MA, Varela N, O'Valle F, Hernandez-Cortes P, Rico L, et al. Human adipose-derived mesenchymal stem cells reduce inflammatory and T cell responses and induce regulatory T cells in vitro in rheumatoid arthritis. Ann Rheum Dis. 2010;69(1):241–8.

Guney A, Akar M, Karaman I, Oner M, Guney B. Clinical outcomes of platelet rich plasma (PRP) as an adjunct to microfracture surgery in osteochondral lesions of the talus. Knee Surg Sports Traumatol Arthrosc. 2015;23:2384–9.

Jain SK, Suprashant K, Kumar S, Yadav A, Kearns SR. Comparison of plantar fasciitis injected with platelet-rich plasma vs corticosteroids. Foot Ankle Int. 2018;39(7):780–6.

Johnson B, Lever C, Roberts S, Richardson J, McCarthy H, Harrison P, et al. Cell cultured chondrocyte implantation and scaffold techniques for osteochondral talar lesions. Foot Ankle Clin. 2013;18(1):135–50.

Kearney RS, Ji C, Warwick J, Parsons N, Brown J, Harrison P, et al. Effect of platelet-rich plasma injection vs sham injection on tendon dysfunction in patients with chronic midportion Achilles tendinopathy: a randomized clinical trial. JAMA. 2021;326(2):137–44.

Keene DJ, Alsousou J, Harrison P, Hulley P, Wagland S, Parsons SR, et al. Platelet rich plasma injection for acute Achilles tendon rupture: PATH-2 randomised, placebo controlled, superiority trial. BMJ. 2019;367:l6132.

Kraeutler MJ, Purcell JM, Hunt KJ. Chronic Achilles tendon ruptures. Foot Ankle Int. 2017;38(8):921–9.

Lareau CR, Sawyer GA, Wang JH, DiGiovanni CW. Plantar and medial heel pain: diagnosis and management. J Am Acad Orthop Surg. 2014;22(6):372–80.

Latt LD, Jaffe DE, Tang Y, Taljanovic MS. Evaluation and treatment of chronic plantar fasciitis. Foot Ankle Orthop. 2020;5(1):2473011419896763.

League AC. Current concepts review: plantar fasciitis. Foot Ankle Int. 2008;29(3):358–66.

Lian JB, Stein GS. Concepts of osteoblast growth and differentiation: basis for modulation of bone cell development and tissue formation. Crit Rev Oral Biol Med. 1992;3(3):269–305.

Liu C-J, Yu K-L, Bai J-B, Tian D-H, Liu G-L. Platelet-rich plasma injection for the treatment of chronic Achilles tendinopathy: a meta-analysis. Medicine (Baltimore). 2019;98(16):e15278.

López-Alcorocho JM, Guillén-Vicente I, Rodríguez-Iñigo E, Navarro R, Caballero-Santos R, Guillén-Vicente M, et al. High-density autologous chondrocyte implantation as treatment for ankle osteochondral defects. Cartilage. 2021;12(3):307–19.

Malahias M-A, Kostretzis L, Megaloikonomos PD, Cantiller E-B, Chytas D, Thermann H, et al. Autologous matrix-induced chondrogenesis for the treatment of osteochondral lesions of the talus: a systematic review. Orthop Rev (Pavia). 2020;12:4.

Marx RE, Carlson ER, Eichstaedt RM, Schimmele SR, Strauss JE, Georgeff KR, et al. Platelet-rich plasma: growth factor enhancement for bone grafts. Oral Surg Oral Med Oral Pathol Oral Radiol Endod. 1998;85(6):638–46.

Murawski CD, Kennedy JG. Operative treatment of osteochondral lesions of the talus. J Bone Joint Surg Am. 2013;95(11):1045–54.

National Institute for Health and Care Excellence. Autologous chondrocyte implantation for treating symptomatic articular cartilage defects of the knee. National Institute for Health and Care Excellence; 2017.

Nauwelaers A-K, Van Oost L, Peers K. Evidence for the use of PRP in chronic midsubstance Achilles tendinopathy: a systematic review with meta-analysis. Foot Ankle Surg. 2021;27(5):486–95.

Pagliazzi G, Vannini F, Battaglia M, Ramponi L, Buda R. Autologous chondrocyte implantation for talar osteochondral lesions: comparison between 5-year follow-up magnetic resonance imaging findings and 7-year follow-up clinical results. J Foot Ankle Surg. 2018;57(2):221–5.

Pittenger MF, Mackay AM, Beck SC, Jaiswal RK, Douglas R, Mosca JD, et al. Multilineage potential of adult human mesenchymal stem cells. Science. 1999;284(5411):143–7.

Raikin SM, Elias I, Zoga AC, Morrison WB, Besser MP, Schweitzer ME, et al. Osteochondral lesions of the talus: localization and morphologic data from 424 patients using a novel anatomical grid scheme. Foot Ankle Int. 2007;28(2):154–61.

Repetto I, Biti B, Cerruti P, Trentini R, Felli L. Conservative treatment of ankle osteoarthritis: can platelet-rich plasma effectively postpone surgery? J Foot Ankle Surg. 2017;56(2):362–5.

Schepull T, Kvist J, Norrman H, Trinks M, Berlin G, Aspenberg P. Autologous platelets have no effect on the healing of human Achilles tendon ruptures: a randomized single-blind study. Am J Sports Med. 2011;39(1):38–47.

Shetty SH, Dhond A, Arora M, Deore S. Platelet-rich plasma has better long-term results than corticosteroids or placebo for chronic plantar fasciitis: randomized control trial. J Foot Ankle Surg. 2019;58(1):42–6.

Shimozono Y, Yasui Y, Ross AW, Kennedy JG. Osteochondral lesions of the talus in the athlete: up to date review. Curr Rev Musculoskelet Med. 2017;10:131–40.

Shimozono Y, Dankert JF, Kennedy JG. Arthroscopic debridement and autologous micronized adipose tissue injection in the treatment of advanced-stage posttraumatic osteoarthritis of the ankle. Cartilage. 2021;13(1_suppl):1337S–43S.

Singh P, Madanipour S, Bhamra JS, Gill I. A systematic review and meta-analysis of platelet-rich plasma versus corticosteroid injections for plantar fasciopathy. Int Orthop. 2017;41:1169–81.

Smyth NA, Murawski CD, Fortier LA, Cole BJ, Kennedy JG. Platelet-rich plasma in the pathologic processes of cartilage: review of basic science evidence. Arthroscopy. 2013;29(8):1399–409.

Stufkens SA, Knupp M, Horisberger M, Lampert C, Hintermann B. Cartilage lesions and the development of osteoarthritis after internal fixation of ankle fractures: a prospective study. J Bone Joint Surg Am. 2010;92(2):279–86.

Sun S-F, Hsu C-W, Lin G-C, Lin H-S, Chou Y-J, Wu S-Y, et al. Efficacy and safety of a single intra-articular injection of platelet-rich plasma on pain and physical function in patients with ankle osteoarthritis—a prospective study. J Foot Ankle Surg. 2021;60(4):676–82.

Thermann H, Fischer R, Gougoulias N, Cipollaro L, Maffulli N. Endoscopic debridement for non-insertional Achilles tendinopathy with and without platelet-rich plasma. J Sport Health Sci. 2023;12(2):275–80.

Usuelli FG, Grassi M, Maccario C, Vigano' M, Lanfranchi L, Alfieri Montrasio U, et al. Intratendinous adipose-derived stromal vascular fraction (SVF) injection provides a safe, efficacious treatment for Achilles tendinopathy: results of a randomized controlled clinical trial at a 6-month follow-up. Knee Surg Sports Traumatol Arthrosc. 2018;26:2000–10.

Vacanti JP, Langer R. Tissue engineering: the design and fabrication of living replacement devices for surgical reconstruction and transplantation. Lancet. 1999;354:S32–S4.

Walther M, Valderrabano V, Wiewiorski M, Usuelli FG, Richter M, Baumfeld TS, et al. Is there clinical evidence to support autologous matrix-induced chondrogenesis (AMIC) for chondral defects in the talus? A systematic review and meta-analysis. Foot Ankle Surg. 2021;27(3):236–45.

Weigelt L, Hartmann R, Pfirrmann C, Espinosa N, Wirth SH. Autologous matrix-induced chondrogenesis for osteochondral lesions of the talus: a clinical and radiological 2-to 8-year follow-up study. Am J Sports Med. 2019;47(7):1679–86.

Ankle: Achilles Tendon and Ankle Cartilage Regeneration

Sedeek Mosaid and Ashok Marudanayagam

Abstract Cartilage injuries at the ankle joint represent an orthopaedic challenge due to the strong association between joint dysfunction and patient quality of life. Ankle cartilage lesions may be caused by acute trauma, repetitive microtrauma or osteochondral lesions secondary to osteochondritis dissecans of the talus. The lack of regenerative potential of cartilage and its exposure to high mechanical loading hinder treatment and repair. This chapter describes the complex anatomy and biomechanics of the ankle joint and how these factors contribute to vulnerability and poor cartilage healing. It reviews the various methods for diagnosis of osteochondral lesions as well as treatment options, ranging from conservative measures to surgical approaches, including microfracture, autologous chondrocyte implantation and osteochondral grafts. Finally, the chapter addresses exciting advances in tissue engineering and biologic adjuvants, which are highly promising for cartilage regeneration and long-term restoration of joint function. Despite these advances, significant challenges remain regarding achieving native-like repair, warranting continued research for better therapeutic approaches.

Keywords Ankle join · Cartilage injury · Osteochondral lesion · Bone marrow stimulation · Chondrocytes

1 Introduction

Ankle cartilage injuries are a widespread and challenging aspect of orthopaedic surgery. They have many implications for a patient's quality of life and joint function. Articular cartilage is a vital component of the joint and essential for smooth, low-friction motion, and it is difficult to manage and treat due to its avascular nature and limited capacity to heal. New therapeutic approaches must often manage the resulting clinical issues. Furthermore, restoring or replacing cartilage with

S. Mosaid (✉) · A. Marudanayagam
Department of Trauma & Orthopaedics, United Lincolnshire Teaching Hospitals NHS Trust, Lincoln County Hospital, Lincoln, UK

P. Lee (ed.), *Musculoskeletal Regeneration Medicine*,
https://doi.org/10.1007/978-3-031-77456-0_16

allogeneic or synthetic tissues has limitations (Buckwalter and Mankin 1998; Bruns et al. 2021).

The incidence of ankle cartilage injuries is often under-recognised, with many cases undiagnosed due to subtle clinical presentations and limitations in early detection with conventional imaging modalities. Common causes of cartilage damage include acute trauma, such as ankle sprains and fractures, repetitive microtrauma and osteochondral defects, such as osteochondritis dissecans (OCD) of the talus. Traumatic events can lead to chondral or osteochondral lesions, which disrupt the smooth articular surface and can extend into the subchondral bone, affecting joint biomechanics and accelerating degenerative changes. These lesions can expand over time, resulting in symptomatic cartilage loss, pain, swelling and restricted joint motion (Shimozono et al. 2019).

2 Anatomy and Biomechanics

The ankle joint biomechanics make managing cartilage injuries challenging. The ankle experiences higher forces per unit area than other joints, making the cartilage particularly vulnerable to injury and difficult to repair. This increased mechanical load requires strong repair strategies, as the repaired cartilage must withstand substantial stress to restore normal joint function (Shimozono et al. 2019).

Also known as the talocrural joint, the ankle is a synovial joint. Although it is commonly described as a hinge, its axis is more complex than a simple uniaxial hinge (Bozkurt et al. 2014). The joint is formed by the distal tibia and fibula, which create a box-like structure of the ankle mortise (Dalmau-Pastor et al. 2020). This mortise contains three articular surfaces: the medial surface of the lateral malleolus, the tibial plafond and the lateral portion of the medial malleolus. Within the mortise lies the superior portion of the talus, including its articular surface, known as the trochlea or dome. When viewed from the top, the talar trochlea is wedge-shaped and typically narrower at the back than at the front. Some describe the talar dome as a truncated cone lying on its side, with a smaller circumference at the medial edge and a larger one forming the lateral edge (Manganaro and Alsayouri 2024; Hintermann and Ruiz 2024).

Understanding ankle joint biomechanics is essential for understanding the complexity of injury to articular cartilage during dynamic activity involving extreme stress on the joint. The talus is crucial for bearing and transmitting the total weight of the body through the ankle joint. It functions similarly to a meniscus by distributing compressive and tensile forces from the talus to adjacent bones during activities such as walking, running, and jumping. The ankle joint exhibits movement but is also used as a critical support unit when standing or walking. The ankle biomechanics can enhance or diminish the risk of injury to the articular cartilage. For instance, the contact area at the ankle joint was considerably larger during running compared to walking, suggesting that dynamic activities like running impose greater demands on the ankle joint, which may have implications for understanding cartilage degeneration. Age also impacts the mechanics and biochemical composition of cartilage and is a factor in cartilage response to stress (Krause et al. 2013; Postolka et al. 2023).

Articular cartilage comprises chondrocytes that generate a significant extracellular matrix (ECM) containing collagen fibres, an abundant ground substance rich in proteoglycan (PG). Collagens (types II and IX) provide tensile strength, while PGs, such as decorin, biglycan and fibromodulin, are responsible for the compressive strength of cartilage (Ferretti et al. 2021). The interactions between collagen, proteoglycans, and water in articular cartilage are crucial for its mechanical properties. Collagen constitutes 75% of the dry weight, while proteoglycans, making up 20–30% of the dry weight, attract water, which comprises 65–80% of the cartilage's total weight (Ferretti et al. 2021).

Glycosaminoglycans (GAGs), including chondroitin sulphate and keratin sulphate, bind to the protein core to form an aggrecan molecule. Chondrocytes, found in spaces called lacunae, synthesise type II collagen, large PG aggregates and non-collagenous proteins, contributing to the formation and maintenance of the ECM (Ferretti et al. 2021; Sharifi et al. 2016).

Articular cartilage is structurally categorised into four zones. First, the superficial zone is the thinnest layer and comprises flattened ellipsoid cells aligned parallel to the joint surface. A lubricating synovial film known as the lamina splendens or lubricin covers this layer, ensuring smooth gliding of the articular cartilage (Sharifi et al. 2016; Quinn et al. 2013). The arrangement and organisation of collagen fibrils in this zone provide tensile and shear strength. Disruption of this zone can lead to changes in the mechanical properties of articular cartilage, contributing to osteoarthritis (OA) development. Second, the transitional zone contains spheroid-shaped cells embedded in a rich ECM with a lower cell density. Collagen fibres in this zone are randomly organised, with a higher concentration of PG aggrecan. Third, in the deep zone, spheroid-shaped cells are arranged perpendicularly on the surface, with the lowest cell numbers and ECM containing collagen fibrils with the largest diameter and the highest concentration of PGs. Finally, the calcified zone comprises a few hypertrophic chondrocytes embedded in the calcified ECM. These chondrocytes have limited metabolic activity and synthesise type X collagen, ensuring structural integrity and shock absorption. The tidemark represents the transition between the non-calcified and calcified zones and provides an essential boundary for the less resilient subchondral bone (Duan et al. 2014; Bolog et al. 2014).

The articular cartilage functions as a smooth surface for movement, absorbs shock and reduces pressure on the subchondral bone. The ECM of the cartilage is crucial in maintaining the cells within it by shielding them from mechanical stress and storing essential signalling molecules. Additionally, the ECM influences the type and amount of nutrients reaching the cells (Duan et al. 2014; Silverberg et al. 2014).

3 Mechanical Properties of Articular Cartilage in the Ankle Joint

The ankle joint is subjected to high compressive loads during activities such as walking, running and jumping. The mechanical properties of articular cartilage are vital for its function in the ankle joint (Ravella et al. 2021).

3.1 Compressive Properties

Articular cartilage behaves as a biphasic material comprising a solid matrix (collagen and PGs) and interstitial fluid (water). When the joint is loaded, the interstitial fluid is pressurised, bearing most of the initial load and redistributing it across the surface. This fluid flow is critical for reducing stress on the solid matrix and minimising the risk of cartilage damage (Ferretti et al. 2021; Guilak 2011).

3.2 Tensile Properties

The robust collagen network in the articular cartilage is the foundation of its tensile strength, particularly against forces acting to pull the cartilage apart. These tensile properties are vital to maintaining the structural integrity of cartilage under various loading conditions (Ferretti et al. 2021; Buckwalter and Mankin 1998; Buckwalter et al. 1954).

3.3 Viscoelasticity

Articular cartilage exhibits viscoelastic behaviour with its time-dependent response to load. This property enables the cartilage to absorb shock and dissipate energy over time, protecting the underlying bone from excessive stress. Viscoelasticity is crucial in the joint, which undergoes repetitive and dynamic loading during daily activities (Ferretti et al. 2021; Guilak 2011).

3.4 Anisotropy

The biomechanical characteristics of articular cartilage are anisotropic, indicating that they vary according to the direction of the applied load. This characteristic is due to the complex arrangement of collagen fibres, which align differently in each cartilage zone. Anisotropy ensures that cartilage can effectively resist multidirectional forces (Ferretti et al. 2021; Guilak 2011).

4 Aetiopathogenesis

The specific cause of an osteochondral lesion (OCL) of the talus is not fully understood. While some research suggests a non-traumatic cause, most studies indicate that a history of trauma is probably a contributing factor in the development of these lesions. OCD is a focal, idiopathic lesion of the subchondral bone with a risk of instability and disruption of the adjacent articular cartilage (Edmonds and Shea 2013; Barnes and Ferkel 2003). As an OCL of the talus (OLT) encompasses a range of diseases, its pathophysiology is not singular. Lateral OLTs are more likely due to trauma: 94% of lateral dome lesions and 62% of medial dome lesions are linked to a history of trauma (Tol et al. 2000). This trauma may be a single injury or repetitive minor injuries. Approximately 50% of ankle sprains and up to 73% of ankle fractures result in cartilage injury (Leontaritis et al. 2009; Waterman et al. 2010).

Two common patterns of talar OCLs exist. The first is anterolateral talar dome lesions, caused by axial loads combined with inversion and dorsiflexion impacting the fibula. These lesions are usually shallow due to the shearing forces created by the fibula on the anterolateral talar dome. The second is posteromedial lesions, caused by axial loads combined with plantar flexion, inversion and possibly external rotation, with the area impacting the tibial ceiling of the ankle joint. These lesions are deeper and cup-shaped due to the angle at which the talus impacts the tibial articular surface (Chew et al. 2008; Steele et al. 2018).

The talar bone and cartilage are prone to developing OCLs due to their specific anatomical characteristics. For instance, the talus is mainly covered by cartilage, which has limited natural regenerative ability since it lacks a blood supply. The blood supply to the talus originates from a complex network of arteries, which can lead to inadequate perfusion in certain areas, such as the posteromedial and posterolateral segments of the subchondral bone. Additionally, the blood supply to the talar dome and cartilage is retrograde, making it more precarious in the event of specific injuries. Moreover, the talar cartilage is thinner than other weight-bearing joints in the lower extremities. These factors collectively make the talus susceptible to OCLs (Lomax et al. 2014; Sundararajan et al. 2020).

4.1 Clinical Features

How patients present with OLTs can vary and depend on multiple factors, such as the cause and severity of the lesion. The most common presentation involves non-specific symptoms such as pain while weight-bearing on the ankle, swelling, stiffness and sometimes a sensation of the ankle joint locking. These symptoms are often worsened by placing weight on the ankle. Mechanical symptoms, such as locking and catching, are often linked to cartilage issues in other joints, which are not commonly associated with OLTs. Therefore, clinicians must be highly suspicious of these lesions because the initial assessment often results in a wide range of possible differential diagnoses. When a concern exists regarding an OLT, the patient should be asked about any history of ankle injuries or recurring instability (Steele et al. 2018; Easley et al. 2010).

On physical examination, the ankle may reveal a decreased range of motion (ROM), tenderness, effusion and pain on inversion or dorsiflexion. Stability should be tested by performing an anterior drawer test with the ankle in plantar flexion and dorsiflexion. The ankle should also be tested for inversion and eversion stress. Additionally, the hindfoot should be examined to identify additional potential sources of pain. The ROM of the injured ankle and hindfoot should be compared with that of the contralateral lower extremity. Finally, a thorough neurovascular assessment should be conducted (Chew et al. 2008; Easley et al. 2010).

5 Radiographic Evaluation

5.1 Plain Radiography

Standard radiographic views of the ankle include anteroposterior, lateral and mortise, with weight-bearing if possible (Schachter et al. 2005). Canale and Kelly (1978) suggested that pronation of the foot by 15° and obliquity of the tube by 75° cephalad may better elucidate the talar profile on an ankle radiograph. Lesions of the posteromedial talus can be better visualised using a plantarflexed mortise view of the ankle, while stress views can be used to evaluate ligamentous laxity (Schachter et al. 2005).

5.2 Computed Tomography (CT)

CT is useful for preoperative planning, as it can visualise the bony morphology and size of OLTs and assess the subchondral bone layer in different planes (Khan et al. 2024). Additionally, supplementary CT scans with the ankle in maximum plantar flexion can be used to determine arthroscopic accessibility medially and laterally during preoperative planning (van Bergen et al. 2012).

However, CT cannot help visualise specific OCD lesions, especially cartilaginous or non-displaced lesions. Magnetic resonance imaging (MRI) is more efficient than CT in identifying lesions that cannot be identified on plain radiographs (Anderson et al. 1989). The advantage of MRI in this setting is its high sensitivity for detecting subtle bone changes that might accompany an osteochondral fracture, including oedema or haemorrhage, as well as the additional advantages of being non-ionising and having the capacity to image the articular cartilage surface in multiple planes (Stone 1996).

5.3 *Magnetic Resonance Imaging (MRI)*

MRI scans enable clear visualisation of the articular surface of the OCL and soft tissues, which can be useful for characterising OLTs. However, due to the demonstration of oedema in the bone and cartilage, MRI can overestimate the size of the OLT or hinder the determination of the bone status and OCL dimensions (Steele et al. 2018; Easley et al. 2010). Thus, CT scans provide better information on the status of the subchondral bone, including information on the dimensions and sites of any subchondral cysts (Zinman et al. 1988).

Stone (1996) suggested a radiographic algorithm to assess OLTs. Plain radiography is the initial modality of choice for patients presenting with acute ankle injury associated with hemarthrosis or significant tenderness at the bony landmarks. Any OCL detected on plain radiography should be evaluated for size, shape, location and degree of displacement using CT. The presence of pain despite a regular appearance on repeated plain radiographs warrants further investigation, and MRI should be performed to exclude other associated injuries responsible for or contributing to persistent pain (Stone 1996).

5.4 *Classification*

Berndt and Harty (2004) created a four-stage classification system for ankle OCLs based on the severity of the lesion on a plain radiograph. The Berndt-Harty classification comprises the following four subtypes (Berndt and Harty 2004):

I. Subchondral impaction
II. Partially detached osteochondral fracture
III. Completely detached but undisplaced OCL
IV. Completely detached and displaced OCL

Their classification was later modified by Loomer et al. (1993) to include a fifth subtype of radiolucent cystic lesions observed on CT scans. The main advantage of the Berndt-Harty system is that it is widely used and accessible. However, it is notable that 50% of OCLs might not be visible on plain radiographs, which is a significant drawback. Furthermore, the system is based only on lesions of traumatic origin. Thus, Ferkel et al. (2008) proposed a four-stage classification based on CT:

I. Subchondral cyst but intact joint surface
II. (a) Subchondral cyst with open cartilage

(b) Open cartilage, non-displaced fragment

III. Subchondral cyst, non-displaced fragment

IV. Displaced fragment

The MRI-based system by Hepple et al. (1999) is based on the original Berndt-Harty classification but distinguishes traumatic, cystic and idiopathic causes of OCL as follows:

I. Articular cartilage damage only

II. (a) Cartilage injury with underlying fracture and surrounding bony oedema

  (b) Stage II(a) without surrounding bony oedema

III. Detached but undisplaced fragment

IV. Detached and displaced fragment

V. Subchondral cyst formation

The arthroscopic staging system was initially outlined by Pritsch et al. (1986), and no significant connection was observed with the Berndt-Harty classification in pre-operative radiographs.

Another arthroscopic staging system, developed by Cheng et al. (1995), was based on the condition of the articular cartilage and has been reported to predict patient outcomes. This system categorises cartilage status as follows:

A. Soft, smooth cartilage

B. Rough cartilage

C. Cartilage with fibrillation and fissures

D. Presence of a flap on exposed bone

E. Undisplaced loose fragment

F. Displaced fragment

6 Treatments

The treatment approach for OLT depends on the severity of the lesion, its duration and the associated symptoms. Asymptomatic lesions are typically monitored over time using X-rays to detect changes. Treatment is reserved for lesions that initially cause symptoms or produce them later (Steele et al. 2018).

Non-operative and operative modalities have been described. However, the indication for operative treatment at presentation is a displaced acute lesion, for which fixation or excision is recommended (Chao and Pao 2017).

For patients with acute symptomatic lesions with non-displaced fractures, non-operative management is often trialled. This typically comprises 6 weeks of immobilisation of the affected ankle with a short leg cast or walking boot, followed by a slow return to activity progression. Treatment comprises rest, avoidance of sporting activities, cast immobilisation with or without NSAIDs and intra-articular injection of platelet-rich plasma (PRP) and hyaluronic acid (HA). The basis of conservative treatment is that the area is offloaded, enabling oedema resolution in the bone marrow and optimal cartilage healing (Sundararajan et al. 2020).

Klammer et al. (2014) studied the long-term outcomes of OLT in 43 patients declining surgery with a minimum follow-up of 2 years and found positive results. Most patients had no or mild pain, and most showed no worsening of the condition on MRI. Additionally, no significant issues with ankle arthritis were found at the end of the follow-up period (Klammer et al. 2014).

Conversely, Zengerink et al. (2009) discussed seven studies describing non-operative treatment in a systematic review. They described 169 patients treated non-operatively, with a success rate of 49.1%. However, success was determined solely on symptomatology and not on the physiological healing of the OCL. Although non-operative treatment may relieve symptoms in the short term, they often recur due to inadequate healing of the lesion (Zengerink et al. 2009).

For patients remaining symptomatic for 3–6 months after conservative treatment, surgery is suggested as the next step. Three types of surgical treatment are available: repair, regeneration or replacement. The decision to operate is based on the articular overlying cartilage, the size of the lesion and whether it is uncontained (shoulder) or contained (non-shoulder-type). A non-shoulder-type lesion is a chondral defect surrounding the articular cartilage. Conversely, a shoulder-type lesion is a defect with no peripheral cartilage border on one side of the lesion, and the medial or lateral articular buttress is lost (Steele et al. 2018; Choi et al. 2012).

7 Cartilage Repair Techniques

7.1 Bone Marrow Stimulation/Microfracture

Microfracture (MF) is the first surgery attempted when non-operative treatments have failed. This procedure involves perforation of the subchondral bone, stimulating the release of mesenchymal stem cells (MSCs) and growth factors. The inflammatory response leads to the formation of fibrin clots. Consequently, fibrocartilage (mainly type I collagen) forms and fills the defect. Fibrocartilage is inferior not only structurally but also biomechanically to natural hyaline cartilage, which primarily comprises type II collagen (Nehrer et al. 1999; Nam et al. 2018).

Prognostic factors studied for the success of bone marrow stimulation (BMS) include patient age, lesion chronicity, size, location containment and subchondral cysts. The OLT size is inversely related to the outcome after MF. Chuckpaiwong et al. (2008) retrospectively studied 105 OCLs of the ankle (tibial and talar) and found that lesion size was overwhelmingly correlated with successful outcomes. No treatment failures occurred for lesions with an average (longitudinal and transverse) diameter below 15 mm, and only one patient (3%) succeeded with a lesion of 15 mm or larger (Chuckpaiwong et al. 2008).

A corroborative study by Choi et al. (2009) on 168 lesions treated with ankle arthroscopy, debridement and MF reported a less than 150 mm^2 cut-off based on MRI scans for a successful outcome. MF is the gold standard for lesions below 1.5 cm^2 (Choi et al. 2009).

BMS prognosis may vary depending on the containment of talar OCLs. Choi et al. (2013) evaluated 399 cases of medial, lateral, contained (non-shoulder) and uncontained (shoulder) lesions and found that uncontained lesions resulted in worse clinical outcomes than contained lesions. The location (medial versus lateral) did not affect these results. The authors suggest that more advanced surgical techniques should be considered for uncontained lesions (Choi et al. 2013).

OA development is a concern due to the natural deterioration of fibrocartilage over time, joint biomechanical changes and subchondral bone damage. Several researchers observed OA in patients undergoing BMS treatment. For instance, a study by Ferkel et al. (2008) studied 50 patients receiving arthroscopic treatment and found that 34% of them developed OA after an average follow-up of 72 months. The progression of OA following BMS may be attributed to the inferior wear characteristics of fibrocartilage compared to the native hyaline cartilage (Savva et al. 2007).

Following a BMS procedure, patients are typically advised to bear partial weight in the initial 2 weeks after surgery. Although ROM exercises are recommended immediately after the operation, they may be restricted due to swelling. Full weight-bearing is generally permitted around 6 weeks after surgery, when patients may begin physical therapy. Initially, they are encouraged to perform strength and balance exercises with a gradual increase in load. However, generally, there is deficiency in the standardization of rehabilitation protocols following bone marrow stimulation (BMS) (Hurley et al. 2018).

7.2 *Retrograde Drilling (RD)*

RD is recommended for subchondral bone lesions with intact overlying cartilage and has the evident advantage of protecting the integrity of the articular cartilage compared with anterograde drilling. RD is a superior technique for penetrating necrotic subchondral bone without disturbing the overlying cartilage. Crucially, however, the lesion must be decompressed, and the structural integrity of the cyst or defect must be addressed to prevent subsequent collapse of the articular surface (Rungprai et al. 2017; O'Loughlin et al. 2009). The literature also reports that RD was successfully used to treat small lesions (area < 100 mm^2, depth < 5 mm, diameter < 10 mm) and when defects could not be reached through conventional arthroscopic portals (Hannon et al. 2018). However, a consensus must be reached on the correct indication for RD rather than merely relying on individual clinical experience.

Different navigation systems were compared based on surgical procedures, and fluoroscopy was found to be the most commonly used method. However, fluoroscopic guidance has limitations, such as difficulty in identifying OCLs and a lack of intraoperative 3D imaging. Accurate targeting of a lesion is technically challenging, sometimes necessitating multiple drillings. This is particularly true for lesions in the posterior aspect of the talar dome (Bale et al. 2001).

Alternative systems, such as drill guide systems, computer-assisted systems and, more recently, MRI- or CT-guided navigation systems, have been proposed to address these issues. However, while these newer systems offer advantages regarding reliable cartilage visualisation and precise treatment of OCLs, they do not necessarily reduce the average procedure time compared to the standard fluoroscopic technique. Additionally, the higher costs of the procedure should be considered (Geerling et al. 2009; Kerimaa et al. 2014). Over the years, modifications have been made to the original surgical technique, including the use of a 6- or 8-mm drill for lesion decompression and improved cyst curettage (Artioli et al. 2023).

Research on RD has demonstrated positive outcomes. In one study (Taranow et al. 1999), 16 patients with symptomatic OCLs of the medial talar dome underwent arthroscopic treatment with percutaneous RD through the sinus tarsi. The average American Orthopaedic Foot and Ankle Society (AOFAS) score improved from 53.9 to 82.6 points, with no complications reported (Taranow et al. 1999).

7.3 Fixation of the Osteochondral Fragment

The most suitable candidate for internal fixation is a young individual with an acute traumatic injury. The larger the fragment of the attached subchondral bone and the better the condition of the articular cartilage, the more probable successful internal fixation.

An osteochondral fragment diameter of at least 10 mm and a depth of at least 3 mm is required for feasible fixation, especially for acutely displaced lesions and skeletally immature patients. The benefits of fixation include the reduction of the talus and the maintenance of hyaline cartilage and subchondral bone. Arthroscopic fixation is an option for lesions accessible through anterior or posterior approaches. Open fixation is considered when arthroscopic access is impossible, or it is the surgeon's preference. The open fixation procedure involves an arthrotomy and sometimes a medial malleolar osteotomy to access and provide space for the lesion. This procedure might involve a more extended period of immobilisation and the risk of non-union or delayed union after osteotomy. Due to these disadvantages, arthroscopic fixation techniques are still being developed. This method is technically challenging and requires more advanced arthroscopic skills to set the fixation angle (Reilingh et al. 2018; Lambers et al. 2019).

Kumai et al. (2002) used cortical bone pegs to treat OLTs. They followed up with 27 patients for a mean of 7.0 years. Clinical results were good in 24 and radiological results improved in 24 patients (Kumai et al. 2002).

8 Cartilage Replacement Techniques

8.1 *Osteochondral Autograft Transfer System*

The technique of autologous osteochondral transplantation replaces the defect in the damaged cartilage and subchondral bone of the talus by taking one or more cylindrical grafts from the periphery of the ipsilateral knee and implanting them into the prepared site of the defect on the talus. Indications include a sizeable OCL (larger than 10–15 mm) or a large lesion after failure of other previous techniques (Kennedy and Murawski 2011).

This technique leads to restoration in tissue similar to the hyaline cartilage with native characteristics, maintaining type II collagen and providing a comprehensive solution to the lesion; osteotomies may provide access to different areas. Medially located OLTs can be accessed through a chevron-type medial malleolar osteotomy. Lateral lesions can be approached through tibial trapezoidal osteotomy (Kennedy and Murawski 2011; Azam et al. 2023).

Zengerink et al. (2009) conducted an extensive systematic review, reporting a success rate of 87% following autologous osteochondral transplantation in 243 patients, with some studies reporting 100% success (Zengerink et al. 2009).

Despite the success of this procedure, concerns exist. Clinical and basic science studies suggest factors distinguishing the ankle's articular cartilage from that of the knee. The ankle cartilage has a lower water content and a higher amount of GAG, which are the main molecules responsible for the water-retaining properties of cartilage and contributing to a significant difference in biomechanical behaviour. This is one explanation for the development of degenerative changes in the ankle being less frequent than in the knee. This raised a concern about whether the knee cartilage implanted in the talar dome could withstand physiological loading. Moreover, the risk of donor site morbidity was studied, and the literature reported an incidence of 2–50% (Treppo et al. 2000; Hannon et al. 2014; Reddy et al. 2007).

Some authors have studied the incongruence of the graft with the surrounding cartilage. For example, Imhoff et al. (2011) found that patients with either minor incongruity or complete congruity of the grafts had significantly better scores than those with significant incongruity. Thus, careful placement of osteochondral grafts is paramount (Imhoff et al. 2011).

A medial malleolar osteotomy may lead to complications. Lamb et al. (2012) reported that while only 5% of their 62 patients had ongoing pain, almost all showed satisfactory healing and fixation with fibrocartilaginous tissue evident on MRI evaluation.

8.2 *Osteochondral Allografting*

Osteochondral allograft transplantation replaces the injured cartilage and bone tissue with anatomically matched tissue from a cadaver donor. Advantages of this procedure include matching the joint's anatomy after removal of damaged tissue, requiring one graft in cases of extensive damage and avoiding complications from donor site morbidity (El-Rashidy et al. 2011). Tissue from a fresh, deceased donor can be used. Frozen donor tissue decreases chondrocyte viability. Hence, if available, fresh donor tissue harvested within 7 days of the donor's death is recommended, with some studies showing that 78% of chondrocytes remain viable for up to 4 weeks after harvesting. Immune rejection, disease transmission, limited availability and high costs are limitations of this procedure (Hannon et al. 2014; Pearsall et al. 2004).

Raikin (2009) conducted a prospective study of 15 patients 44 months after osteochondral allograft transplantation. While clinical outcome scores were improved, radiological narrowing of the joint space and resorption or collapse of the graft were observed in 67% of patients, and 60% of patients developed arthrosis, with arthrodesis of the ankle performed in two patients (13%) (Raikin 2009).

Given these reoperation and failure rates, it is reasonable to consider other techniques and osteochondral allografting as a second option if primary reconstruction fails (Sundararajan et al. 2020).

8.3 Particulated Juvenile Cartilage Allograft Transfer (PJCAT)

This procedure involves harvesting donor juvenile cartilage from deceased donors who are newborn to 13 years old, particulating it and delivering it in a less invasive manner than osteochondral grafts. The graft can be delivered and even performed arthroscopically. It is a single-stage procedure, which does not require a press fit or graft contouring. The graft is fixed with fibrin glue, with no donor site morbidity. Particulating the graft enables chondrocyte cell movement and matrix formation, leading to a hyaline cartilage-like matrix. Juvenile cartilage has been shown to have much higher cellular activity than adult cartilage. This procedure is recommended for symptomatic patients with lesions of at least 1.5 cm or those subjected to a failed marrow stimulation technique (Adams et al. 2014).

Aldawsari et al. (2021) systematically reviewed the role of PJCAT in OLT and summarised the results of 10 studies and 132 patients with good postoperative functional outcomes. The results revealed that PJCAT could be a promising treatment modality for OLT. Interestingly, the regenerated cartilage was heterogeneous, with an unaltered subchondral area. This differs from the viewpoint that PJCAT will restore the defect area to near-normal cartilage (Aldawsari et al. 2021).

9 Cartilage Regeneration Techniques

9.1 Autologous Chondrocyte Implantation (ACI)

ACI is a two-stage procedure: in the first stage, the hyaline cartilage is harvested from the anterior aspect of the talus or a non-weight-bearing area of the knee and cultured to grow chondrocytes that can then be held in position by sewing a periosteal patch over the defect. The rim of the defect requires debridement until it is stable and no necrotic tissue remains in the osteochondral base. In the second stage, the lesion is measured and templated and the periosteal tissue harvested from the tibia. The periosteal patch is sutured, and sealed with fibrin glue to secure the periosteal patch to the articular surface, and the cultured cells are injected beneath the patch. The rationale is that these chondrocytes will then generate new hyaline cartilage to incorporate and fill the chondral lesion.

Focal lesions with a stable cartilage rim are better suited to ACI treatment. Therefore, ACI is commonly used for lesions larger than 1–1.5 cm^2 that do not reach the shoulder (Baums et al. 2007; Nam et al. 2008).

The most significant limitation is the need for two procedures, which makes it more expensive and more likely to induce morbidity. It is also unknown how many cells are required to repair a defect, and a risk of a non-uniform distribution of cells exists during the procedure (Hannon et al. 2014).

In a systematic review, Niemeyer et al. (2011) conducted a review and reported a clinical success rate of 89.9% in 213 patients undergoing ACI for OLT.

9.2 Matrix-Induced Autologous Chondrocyte Implantation (MACI)

Second-generation techniques such as MACI use the same principle of implanting autologous chondrocytes grown in vitro but replace the periosteal patch with a collagen matrix to help secure the implant. This reduces surgical time and associated complications and theoretically enables a more predictable distribution of chondrocytes over the defect site (Steele et al. 2018).

Anders et al. (2012) assessed the results of a prospective long-term survivorship and clinical outcome study of 22 patients treated with MACI, with a mean follow-up of 63.5 months. They reported a significant increase in the mean AOFAS score from 70.1 to 95.3 at 60 months and no severe complications (Anders et al. 2012).

9.3 Autologous Matrix-Induced Chondrogenesis (AMIC)

AMIC, which also uses bone marrow-derived cells from multiple sources, aims to achieve a similar goal to ACI or MACI but through a one-step procedure. AMIC involves MF in combination with autologous iliac crest bone marrow aspirate concentrate or PRP delivered on a collagen matrix scaffold with the aim of forming hyaline cartilage from pluripotent cells in the defect rather than the fibrocartilage resulting from MF alone (Giannini et al. 2012). Given the many variations in the AMIC technique, the data on the treatment have been challenging to interpret. However, it has been shown that functional outcomes and arthroscopic evidence of the growth of more normal-appearing hyaline cartilage than MFs have been improved (Steele et al. 2018).

Yausep et al. (2020) systematically reviewed evidence from levels 1 and 2, including four studies. It was concluded that using PRP combined with MF resulted in better pain and has shown functional improvement compared to MF alone. Finally, improvement was achieved when PRP was used as an adjunct to MF compared to a conservative intra-articular injection of PRP (Yausep et al. 2020).

9.4 *Biological Adjuvants*

Platelet-Rich Plasma

PRP is an autologous blood product obtained by centrifugation of peripheral blood to achieve a concentration of platelets above the baseline value, which has at least a twofold higher concentration of platelets than the baseline. It contains growth factors, including fibroblast growth factor (FGF), vascular endothelial growth factor (VEGF), insulin-like growth factor-1 (IGF-1) and platelet-derived growth factors (PDGF) and transforming growth factor-b (TGF-b1). The composition of growth factors in PRP differs between individuals and can even vary between successive preparations in the same person. These growth factors and cytokines are essential in MSC chemotaxis, target cell activation, cell proliferation, neoangiogenesis and cartilage matrix production, providing an additional immunomodulatory benefit. Using PRP in conjunction with BMS technique, creates an optimal biological environment to encourage chondrocyte synthesis and increase type II collagen deposition (Hogan et al. 2019; Mazzocca et al. 2012).

Hyaluronic Acid (HA)

HA is a high-molecular-weight GAG polysaccharide. It is a component of synovial fluid and a viscoelastic lubricant that reduces friction and transmits shear forces. In addition to its rheological benefits, HA has a biological effect and enhances chondrocyte growth. In vitro, HA is known to promote cartilage regeneration by increasing chondrocyte proliferation and PG synthesis, as well as by inhibiting cartilage destruction and the production of deleterious proinflammatory cytokines and matrix metalloproteinases. HA has been proposed as a valuable adjunct to BMS for cartilage repair (Azam et al. 2023; Doral et al. 2011).

Concentrated Bone Marrow Aspirate (CBMA)

Bone marrow aspirate contains various bioactive cytokines, such as MSCs, which are being studied for use in cartilage repair. CBMA is used to deliver MSCs to augment cartilage repair. CBMA is prepared during surgery by centrifugation of bone marrow aspirate obtained from the iliac crest. Bone marrow aspirate contains MSCs and endogenous growth factors and cytokines, such as transforming growth factor-beta and fibroblast growth factor, which can induce chondrocyte differentiation (Hogan et al. 2019).

Based on the classical perspective, MSCs are distinct and differentiated directly into osteoblasts and chondrocytes to support cartilage repair. However, new findings increasingly indicate that macrophages take most of MSCs directly. This enables them to form a secretome, a paracrine signalling apparatus enhancing regenerative processes through immunomodulatory effects, such as suppressing proinflammatory interleukin-1b gene expression (Strauss et al. 2018).

The effectiveness of CBMA as an adjunct to OLT treatment has been investigated in in vivo and clinical studies. Giannini et al. (2012) reported a case series of 49 patients with OLT. All were candidates for a one-step OLT repair. Patients were evaluated clinically using AOFAS scores preoperatively and at a predefined time during postoperative follow-ups. In all patients, the cells were harvested from the iliac crest, concentrated and loaded into a scaffold implanted arthroscopically. AOFAS scores increased from 63.73 to 82.19, and T2-mapping revealed regenerated tissue, with T2 values of 35–45 ms found in hyaline cartilage (Giannini et al. 2012).

Tissue Engineering

Tissue engineering involves assembling cells, scaffolds and signalling molecules into functional tissue constructs able to replace damaged tissues. The three main regenerative approaches applied to osteochondral injuries involve a personalised approach. This includes extracting the appropriate cells from the patient, culturing them in vitro and transplanting them back into the patient's injury-affected area, where they contribute to repairing the defect.

In articular cartilage repair, the main components are MSCs from bone marrow, adipose tissue or synovium. These cells are crucial to the repair process due to their chondrogenic potential and ability to differentiate. Scaffolds comprising collagen, HA or polylactic acid, lattices to which cells can attach, provide structural frameworks for cell attachment, proliferation and differentiation. Adding growth factors, such as transforming growth factor-beta and bone morphogenetic proteins, enhances chondrogenesis, promoting hyaline-like cartilage formation (Langer and Vacanti 1993; Makris et al. 2014).

Recent advances have led to novel scaffold designs, such as hydrogels, nanofibres and 3D bioprinted constructs, that more accurately mimic the native cartilage ECM than traditional scaffold materials. These more complex scaffolds improve cell distribution, nutrient transport and mechanical properties, enhancing the integration and longevity of the repaired tissue. Bioprinting technology can also deliver precise placement of cells and bioactive molecules in anatomically accurate constructs to match the size of individual cartilage damage (Yu et al. 2023; Noor et al. 2019; Liu et al. 2017).

10 Future Considerations

Many of these operative treatment options have the potential to decrease the morbidity associated with OLTs. Further basic science and clinical research into the pathophysiology of talar OCLs could advance treatment options. Furthermore, more experimental work is required to optimise the scaffolds to match the functions of the native osteochondral unit.

11 Conclusion

Tremendous advances have occurred in treating ankle cartilage lesions, from diagnostics—including improved imaging modalities—to new therapies. Treatment has evolved, with numerous patients successfully treated with MF and osteochondral grafting. However, the primary limitation in cartilage regeneration with these treatments further highlights the need for new therapies. The future of the management of articular cartilage defects is biological adjuvants and tissue engineering, as well as restoring the native anatomy and function. Although many advances have occurred in this space, data to optimise these therapies are lacking, particularly regarding treatment success and complications, as well as a better understanding of the interplay between biomechanics, cell behaviour and treatment outcome.

References

Adams SB, et al. Arthroscopic particulated juvenile cartilage allograft transplantation for the treatment of osteochondral lesions of the talus. Arthrosc Tech. 2014;3(4):e533–7. https://doi.org/10.1016/j.eats.2014.06.004.

Aldawsari K, et al. Role of particulated juvenile cartilage allograft transplantation in osteochondral lesions of the talus: a systematic review. Foot Ankle Surg. 2021;27(1):10–4. https://doi.org/10.1016/j.fas.2020.02.011.

Anders S, et al. Treatment of deep articular talus lesions by matrix associated autologous chondrocyte implantation—results at five years. Int Orthop. 2012;36(11):2279–85. https://doi.org/10.1007/s00264-012-1635-1.

Anderson IF, et al. Osteochondral fractures of the dome of the talus. J Bone Joint Surg. 1989;71(8):1143–52. https://doi.org/10.2106/00004623-198971080-00004.

Artioli E, et al. Retrograde drilling for ankle joint osteochondral lesions: a systematic review. J Orthop Traumatol. 2023;24(1):37. https://doi.org/10.1186/s10195-023-00716-4.

Azam MT, et al. Advances in cartilage repair. Orthop Clin N Am. 2023;54(2):227–36. https://doi.org/10.1016/j.ocl.2022.11.007.

Bale RJ, et al. Osteochondral lesions of the talus: computer-assisted retrograde drilling—feasibility and accuracy in initial experiences. Radiology. 2001;218(1):278–82. https://doi.org/10.1148/radiology.218.1.r01ja18278.

Barnes CJ, Ferkel RD. Arthroscopic debridement and drilling of osteochondral lesions of the talus. Foot Ankle Clin. 2003;8(2):243–57. https://doi.org/10.1016/s1083-7515(03)00016-0.

Baums MH, et al. The surgical technique of autologous chondrocyte transplantation of the talus with use of a periosteal graft. J Bone Joint Surg. 2007;89(2):170–82. https://doi.org/10.2106/jbjs.g.00169.
Berndt AL, Harty M. Transchondral fractures (osteochondritis dissecans) of the talus. J Bone Joint Surg. 2004;86:1336.
Bolog NV, Andreisek G, Ulbrich EJ. Articular cartilage and subchondral bone. In: MRI of the knee. Cham: Springer; 2014. p. 95–112. https://doi.org/10.1007/978-3-319-08165-6_6.
Bozkurt M, et al. Functional anatomy of the ankle. In: Sports injuries. Berlin, Heidelberg: Springer; 2014. p. 1–12. https://doi.org/10.1007/978-3-642-36801-1_277-1.
Bruns J, Habermann C, Werner M. Osteochondral lesions of the talus: a review on talus osteochondral injuries, including osteochondritis dissecans. Cartilage. 2021;13(1_suppl):1380S–401S. https://doi.org/10.1177/1947603520985182.
Buckwalter JA, Mankin HJ. Articular cartilage: tissue design and chondrocyte-matrix interactions. Instr Course Lect. 1998;47:477–86. https://pubmed.ncbi.nlm.nih.gov/9571449
Buckwalter JA, Mankin HJ, Grodzinsky AJ. Articular cartilage and osteoarthritis. BMJ. 1954;1(4860):506. https://doi.org/10.1136/bmj.1.4860.506.
Canale ST, Kelly FB. Fractures of the neck of the talus: long-term evaluation of seventy-one cases. J Bone Joint Surg. 1978;60:143–56.
Chao J, Pao A. Restorative tissue transplantation options for osteochondral lesions of the talus. Orthop Clin N Am. 2017;48(3):371–83. https://doi.org/10.1016/j.ocl.2017.03.008.
Cheng MS, Ferkel RD, Applegate GR. Osteochondral lesions of the talus: a radiologic and surgical comparison. Annual meeting of the American Academy of Orthopedic Surgeons. New Orleans, February 16–21, 1995.
Chew KT, Tay E, Wong YS. Osteochondral lesions of the talus. Ann Acad Med Singap. 2008;37(1):63–8.
Choi WJ, et al. Osteochondral lesion of the talus. Am J Sports Med. 2009;37(10):1974–80. https://doi.org/10.1177/0363546509335765.
Choi WJ, et al. Prognostic significance of the containment and location of osteochondral lesions of the talus. Am J Sports Med. 2012;41(1):126–33. https://doi.org/10.1177/0363546512453302.
Choi WJ, et al. Prognostic significance of the containment and location of osteochondral lesions of the talus. Am J Sports Med. 2013;41(1):126–33. https://doi.org/10.1177/0363546512453302.
Chuckpaiwong B, Berkson EM, Theodore GH. Microfracture for osteochondral lesions of the ankle: outcome analysis and outcome predictors of 105 cases. Arthroscopy. 2008;24(1):106–12. https://doi.org/10.1016/j.arthro.2007.07.022.
Dalmau-Pastor M, et al. Anatomy of the ankle joint and hindfoot. In: Ankle joint arthroscopy. Cham: Springer; 2020. p. 3–9. https://doi.org/10.1007/978-3-030-29231-7_1.
Doral MN, et al. Treatment of osteochondral lesions of the talus with microfracture technique and postoperative hyaluronan injection. Knee Surg Sports Traumatol Arthrosc. 2011;20(7):1398–403. https://doi.org/10.1007/s00167-011-1856-7.
Duan X, et al. Texture analysis of the 3D collagen network and automatic classification of the physiology of articular cartilage. Comput Methods Biomech Biomed Engin. 2014;18(9):931–43. https://doi.org/10.1080/10255842.2013.864284.
Easley ME, et al. Osteochondral lesions of the talus. J Am Acad Orthop Surg. 2010;18(10):616–30. https://doi.org/10.5435/00124635-201010000-00005.
Edmonds EW, Shea KG. Osteochondritis dissecans: editorial comment. Clin Orthop Relat Res. 2013;471(4):1105–6. https://doi.org/10.1007/s11999-013-2837-6.
El-Rashidy H, et al. Fresh osteochondral allograft for the treatment of cartilage defects of the talus: a retrospective review. J Bone Joint Surg. 2011;93(17):1634–40. https://doi.org/10.2106/jbjs.j.00900.
Ferkel RD, et al. Arthroscopic treatment of chronic osteochondral lesions of the talus. Am J Sports Med. 2008;36(9):1750–62. https://doi.org/10.1177/0363546508316773.
Ferretti M, Costa LAV, Foni NO. Articular cartilage: functional biomechanics. In: Cartilage injury of the knee. Cham: Springer; 2021. p. 1–9. https://doi.org/10.1007/978-3-030-78051-7_1.

Geerling J, et al. Initial outcomes of 3-dimensional imaging-based computer-assisted retrograde drilling of talar osteochondral lesions. Am J Sports Med. 2009;37(7):1351–7. https://doi.org/10.1177/0363546509332499.

Giannini S, et al. One-step repair in talar osteochondral lesions. Am J Sports Med. 2012;41(3):511–8. https://doi.org/10.1177/0363546512467622.

Guilak F. Biomechanics of articular cartilage. In: Mow VC, Huiskes R, editors. Basic orthopaedic biomechanics & mechano-biology. Lippincott Williams & Wilkins; 2011.

Hannon CP, et al. Osteochondral lesions of the talus. Bone Joint J. 2014;96(2):164–71. https://doi.org/10.1302/0301-620x.96b2.31637.

Hannon CP, et al. Debridement, curettage, and bone marrow stimulation: proceedings of the international consensus meeting on cartilage repair of the ankle. Foot Ankle Int. 2018;39(1_suppl):16S–22S. https://doi.org/10.1177/1071100718779392.

Hepple S, Winson IG, Glew D. Osteochondral lesions of the talus: a revised classification. Foot Ankle Int. 1999;20:789–93.

Hintermann B, Ruiz R. Anatomic and biomechanical characteristics of the ankle joint. In: Total ankle replacement. Cham: Springer; 2024. https://doi.org/10.1007/978-3-031-56810-7_2.

Hogan MV, et al. Biologic adjuvants for the management of osteochondral lesions of the talus. J Am Acad Orthop Surg. 2019;27(3):e105–11. https://doi.org/10.5435/jaaos-d-16-00840.

Hurley ET, et al. High reported rate of return to play following bone marrow stimulation for osteochondral lesions of the talus. Knee Surg Sports Traumatol Arthrosc. 2018;27(9):2721–30. https://doi.org/10.1007/s00167-018-4913-7.

Imhoff AB, et al. Osteochondral transplantation of the talus. Am J Sports Med. 2011;39(7):1487–93. https://doi.org/10.1177/0363546510397726.

Kennedy JG, Murawski CD. The treatment of osteochondral lesions of the talus with autologous osteochondral transplantation and bone marrow aspirate concentrate. Cartilage. 2011;2(4):327–36. https://doi.org/10.1177/1947603511400726.

Kerimaa P, et al. MRI-guided percutaneous retrograde drilling of osteochondritis dissecans of the talus: a feasibility study. Eur Radiol. 2014;24(7):1572–6. https://doi.org/10.1007/s00330-014-3161-6.

Khan I, et al. The role of imaging in the diagnosis, staging, and management of the osteochondral lesions of the talus. Br J Radiol. 2024;97(1156):716–25. https://doi.org/10.1093/bjr/tqae030.

Klammer G, et al. Natural history of nonoperatively treated osteochondral lesions of the talus. Foot Ankle Int. 2014;36(1):24–31. https://doi.org/10.1177/1071100714552480.

Krause M, et al. Bone microarchitecture of the talus changes with aging. Clin Orthop Relat Res. 2013;471(11):3663–71. https://doi.org/10.1007/s11999-013-3195-0.

Kumai T, et al. Fixation of osteochondral lesions of the talus using cortical bone pegs. J Bone Joint Surg. 2002;84(3):369–74. https://doi.org/10.1302/0301-620x.84b3.12373.

Lamb J, et al. Chevron-type medial malleolar osteotomy: a functional, radiographic and quantitative T2-mapping MRI analysis. Knee Surg Sports Traumatol Arthrosc. 2012;21(6):1283–8. https://doi.org/10.1007/s00167-012-2050-2.

Lambers KTA, et al. Arthroscopic lift, drill, fill and fix (LDFF) is an effective treatment option for primary talar osteochondral defects. Knee Surg Sports Traumatol Arthrosc. 2019;28(1):141–7. https://doi.org/10.1007/s00167-019-05687-w.

Langer R, Vacanti JP. Tissue engineering. Science. 1993;260(5110):920–6. https://doi.org/10.1126/science.8493529.

Leontaritis N, Hinojosa L, Panchbhavi VK. Arthroscopically detected intra-articular lesions associated with acute ankle fractures. J Bone Joint Surg. 2009;91(2):333–9. https://doi.org/10.2106/jbjs.h.00584.

Liu M, et al. Injectable hydrogels for cartilage and bone tissue engineering. Bone Res. 2017;5(1):17014. https://doi.org/10.1038/boneres.2017.14.

Lomax A, et al. Quantitative assessment of the subchondral vascularity of the talar dome: a cadaveric study. Foot Ankle Surg. 2014;20(1):57–60. https://doi.org/10.1016/j.fas.2013.10.005.

Loomer R, et al. Osteochondral lesions of the talus. Am J Sports Med. 1993;21(1):13–9. https://doi.org/10.1177/036354659302100103.
Makris EA, et al. Repair and tissue engineering techniques for articular cartilage. Nat Rev Rheumatol. 2014;11(1):21–34. https://doi.org/10.1038/nrrheum.2014.157.
Manganaro D, Alsayouri K. Anatomy, bony pelvis and lower limb: ankle joint. In: StatPearls [Internet]. Treasure Island: StatPearls Publishing; 2024. PMID: 31424742.
Mazzocca AD, et al. Platelet-rich plasma differs according to preparation method and human variability. J Bone Joint Surg. 2012;94(4):308–16. https://doi.org/10.2106/jbjs.k.00430.
Nam EK, Ferkel RD, Applegate GR. Autologous chondrocyte implantation of the ankle. Am J Sports Med. 2008;37(2):274–84. https://doi.org/10.1177/0363546508325670.
Nam Y, et al. Current therapeutic strategies for stem cell-based cartilage regeneration. Stem Cells Int. 2018;2018:1–20. https://doi.org/10.1155/2018/8490489.
Nehrer S, Spector M, Minas T. Histologic analysis of tissue after failed cartilage repair procedures. Clin Orthop Relat Res. 1999;365:149–62. https://doi.org/10.1097/00003086-199908000-00020.
Niemeyer P, et al. Autologous chondrocyte implantation for the treatment of chondral and osteochondral defects of the talus: a meta-analysis of available evidence. Knee Surg Sports Traumatol Arthrosc. 2011;20(9):1696–703. https://doi.org/10.1007/s00167-011-1729-0.
Noor N, et al. 3D printing of personalized thick and perfusable cardiac patches and hearts. Adv Sci. 2019;6(11):1900344. https://doi.org/10.1002/advs.201900344.
O'Loughlin PF, Heyworth BE, Kennedy JG. Current concepts in the diagnosis and treatment of osteochondral lesions of the ankle. Am J Sports Med. 2009;38(2):392–404. https://doi.org/10.1177/0363546509336336.
Pearsall AW, et al. Chondrocyte viability in refrigerated osteochondral allografts used for transplantation within the knee. Am J Sports Med. 2004;32(1):125–31. https://doi.org/10.1177/0095399703258614.
Postolka B, Killen BA, Boey H, Sloten JV, Jonkers I. Articular ankle joint loading during dynamic activities. Gait Posture. 2023;106:S168. https://doi.org/10.1016/j.gaitpost.2023.07.202.
Pritsch M, Horoshovski H, Farine I. Arthroscopic treatment of osteochondral lesions of the talus. J Bone Joint Surg. 1986;68(6):862–5. https://doi.org/10.2106/00004623-198668060-00007.
Quinn TM, et al. Cell and matrix morphology in articular cartilage from adult human knee and ankle joints suggests depth-associated adaptations to biomechanical and anatomical roles. Osteoarthr Cartil. 2013;21(12):1904–12. https://doi.org/10.1016/j.joca.2013.09.011.
Raikin SM. Fresh osteochondral allografts for large-volume cystic osteochondral defects of the talus. J Bone Joint Surg. 2009;91(12):2818–26. https://doi.org/10.2106/jbjs.i.00398.
Ravella KC, Ahmad J, Amirouche F. Biomechanics of the ankle joint. In: Orthopaedic biomechanics in sports medicine. Cham: Springer; 2021. p. 401–13. https://doi.org/10.1007/978-3-030-81549-3_31.
Reddy S, et al. The morbidity associated with osteochondral harvest from asymptomatic knees for the treatment of osteochondral lesions of the talus. Am J Sports Med. 2007;35(1):80–5. https://doi.org/10.1177/0363546506290986.
Reilingh ML, et al. Fixation techniques: proceedings of the international consensus meeting on cartilage repair of the ankle. Foot Ankle Int. 2018;39(1_suppl):23S–7S. https://doi.org/10.1177/1071100718781096.
Rungprai C, et al. Management of osteochondral lesions of the talar dome. Open Orthop J. 2017;11(1):743–61. https://doi.org/10.2174/1874325001711010743.
Savva N, et al. Osteochondral lesions of the talus: results of repeat arthroscopic debridement. Foot Ankle Int. 2007;28(6):669–73. https://doi.org/10.3113/fai.2007.0669.
Schachter AK, et al. Osteochondral lesions of the talus. J Am Acad Orthop Surg. 2005;13(3):152–8. https://doi.org/10.5435/00124635-200505000-00002.
Sharifi AM, Moshiri A, Oryan A. Articular cartilage: injury, healing, and regeneration. Curr Orthop Pract. 2016;27(6):644–65. https://doi.org/10.1097/bco.0000000000000425.

Shimozono Y, et al. Restorative procedures for articular cartilage in the ankle: state-of-the-art review. J ISAKOS. 2019;4(5):270–84. https://doi.org/10.1136/jisakos-2017-000163.
Silverberg JL, et al. Structure-function relations and rigidity percolation in the shear properties of articular cartilage. Biophys J. 2014;107(7):1721–30. https://doi.org/10.1016/j.bpj.2014.08.011.
Steele JR, et al. Osteochondral lesions of the talus. Foot Ankle Orthop. 2018;3(3):1–9. https://doi.org/10.1177/2473011418779559.
Stone JW. Osteochondral lesions of the talar dome. J Am Acad Orthop Surg. 1996;4(2):63–73. https://doi.org/10.5435/00124635-199603000-00001.
Strauss E, et al. Biological effects of bone marrow concentrate in knee pathologies. J Knee Surg. 2018;32(1):2–8. https://doi.org/10.1055/s-0038-1676069.
Sundararajan SR, et al. Osteochondral lesions of the talus-current concepts. J Arthrosc Surg Sports Med. 2020;1:218–25. https://doi.org/10.25259/jassm_38_2020.
Taranow WS, et al. Retrograde drilling of osteochondral lesions of the medial talar dome. Foot Ankle Int. 1999;20(8):474–80. https://doi.org/10.1177/107110079902000802.
Tol JL, et al. Treatment strategies in osteochondral defects of the talar dome: a systematic review. Foot Ankle Int. 2000;21(2):119–26. https://doi.org/10.1177/107110070002100205.
Treppo S, et al. Comparison of biomechanical and biochemical properties of cartilage from human knee and ankle pairs. J Orthop Res. 2000;18(5):739–48. https://doi.org/10.1002/jor.1100180510.
van Bergen CJA, et al. Arthroscopic accessibility of the talus quantified by computed tomography simulation. Am J Sports Med. 2012;40(10):2318–24. https://doi.org/10.1177/0363546512455403.
Waterman BR, et al. Epidemiology of ankle sprain at the United States Military Academy. Am J Sports Med. 2010;38(4):797–803. https://doi.org/10.1177/0363546509350757.
Yausep OE, Madhi I, Trigkilidas D. Platelet rich plasma for treatment of osteochondral lesions of the talus: a systematic review of clinical trials. J Orthop. 2020;18:218–25. https://doi.org/10.1016/j.jor.2020.01.046.
Yu L, et al. Recent development in multizonal scaffolds for osteochondral regeneration. Bioact Mater. 2023;25:122–59. https://doi.org/10.1016/j.bioactmat.2023.01.012.
Zengerink M, et al. Treatment of osteochondral lesions of the talus: a systematic review. Knee Surg Sports Traumatol Arthrosc. 2009;18(2):238–46. https://doi.org/10.1007/s00167-009-0942-6.
Zinman C, Wolfson N, Reis ND. Osteochondritis dissecans of the dome of the talus. Computed tomography scanning in diagnosis and follow-up. J Bone Joint Surg. 1988;70(7):1017–9. https://doi.org/10.2106/00004623-198870070-00009.

Shoulder: Rotator Cuff Tears and Tendon Regeneration

Mustafa S. Rashid

Abstract Rotator cuff tears are a prevalent condition, particularly in older populations, with significant implications for shoulder stability and function. They are generally considered a part of the normal aging process. They are present in approximately 15–20% of 60 year olds, 26–30% of 70 year olds, and 36–50% of 80 year olds. Most individuals with rotator cuff tendon tears remain asymptomatic, for others, these tears become a source of substantial pain and functional impairment. These tears often arise from a combination of intrinsic and extrinsic factors.

Orthobiologics have emerged as promising adjuncts to enhance tendon healing. Platelet-rich plasma (PRP) modulate inflammation, promote cell migration, and enhance extracellular matrix (ECM) synthesis. However, inconsistencies in preparation methods and clinical outcomes have limited its widespread adoption. Patch augmentation, either synthetic or biological, involves applying a graft during surgical repair to be applied on top of the repaired cuff tendon (onlay), underneath the tendon (interposition), or in a "bridging" fashion to fill the defect between the tendon edge and the bony footprint. Human dermal allografts are the most commonly used biological grafts due to their availability and favourable biomechanical properties. However, differences in decellularisation methods, preparation, and hydration among products limit comparative studies and standardisation. Ideal orthobiologics in rotator cuff regeneration should aim to enhance healing by leveraging optimised growth factors in injectable formulations and the ideal biocompatible patches should be able to seamlessly integrate with native tissues while providing sustained biomechanical support. Furthermore, achieving standardisation in preparation methods and clinical applications remains crucial for ensuring safety, efficacy, and widespread adoption.

Keywords Rotator cuff tears · Orthobiologics · Platelet-rich plasma (PRP) · Patch augmentation · Human dermal allografts

---

M. S. Rashid (✉)
Colchester General Hospital, Colchester, UK

P. Lee (ed.), *Musculoskeletal Regeneration Medicine*,
https://doi.org/10.1007/978-3-031-77456-0_17

1 Introduction

The rotator cuff tendons play a vital role in facilitating dynamic stability of the shoulder joint. They are comprised by the confluence of four tendons that connect muscles originating from the scapula to the humeral head, forming a continuous cuff of tissue. A crucial demarcation within this ensemble is the rotator interval, separating the posterosuperior cuff from the anterosuperior cuff, accommodating the long head of the biceps as it enters the glenohumeral joint. The sequence of tendons contributing to the rotator cuff, arranged from anterior to posterior are the subscapularis, supraspinatus, infraspinatus, and teres minor.

Histologically, the rotator cuff is composed of five discernible layers. Layer 1 is the most superficial layer encompassing fibres of the coracohumeral ligament. Layer 2 comprises the fibres of the supraspinatus and infraspinatus tendons and consists of tightly packed collagen fibres in parallel, arranged in large bundles. Transitioning to Layer 3, smaller collagen fascicles are observed, displaying a less organised orientation than in the preceding layer. Layer 4 comprises loose connective tissue characterised by thick collagen bands. Layer 5 is the deepest layer and is constituted by continuous interwoven collagen fibrils that anchor into the humeral bone (Clark and Harryman 1992; Burkhart et al. 1993).

Tendons secure themselves to bone via specialised fibres. Collectively, this zone of attachment is referred to as the enthesis. Perforating tendon fibres (named Sharpey's fibres) attaching to bones were first described in 1860 (Taylor 1971). The enthesis has later been described in four zones, namely tendon fibres, fibrocartilage, mineralised fibrocartilage, and bone (Benjamin et al. 1986). This complex and gradual transition is commonly affected by rotator cuff tears and is not reconstructed during surgical reattachment, leading to an alteration of the biomechanical properties at the new attachment site. Instead, sutures and suture anchors are employed to secure the tendon edge to the bony footprint, which heals with a fibrotic scar. The resultant 'neo-enthesis,' though capable of carrying physiological loads, is potentially liable to recurrent failures secondary to the sudden transition from tendon to bone (Rothrauff et al. 2017). Understanding these structural intricacies is paramount for addressing challenges in rotator cuff tendon regeneration.

2 Rotator Cuff Tendon Degeneration

Rotator cuff tears are a prevalent occurrence that becomes more frequent with advancing age and are often considered as part of the normal aging process (Hsu and Keener 2015). They are present in approximately 15–20% of 60 year olds, 26–30% of 70 year olds, and 36–50% of 80 year olds (Minagawa et al. 2013; Tempelhof et al. 1999). While many individuals with rotator cuff tendon tears remain asymptomatic, for others, these tears become a source of substantial pain and functional impairment.

The term "degeneration" has been commonly employed to describe the changes observed in the rotator cuff tendon with aging. However, it is crucial to acknowledge the limitations of this term, as it may oversimplify the multifaceted processes at play. The aetiology behind rotator cuff tears remains poorly understood, posing a challenge to accurate diagnosis and targeted intervention. Several factors contribute to rotator cuff tear aetiology and pathophysiology, including vascular compromise, tensile forces on the cuff, genetic predisposition, and environmental factors. Furthermore, interplay between the intrinsic and extrinsic theories of rotator cuff tear initiation adds complexity to the understanding of why some individuals are more prone to cuff degeneration than others (Table 1).

2.1 *Extrinsic Theory*

Traditionally, it was believed that external compression was the primary cause of rotator cuff tears. Watson-Jones initially described the pathological contact between the lateral acromion and rotator cuff during mid-arc shoulder abduction (Harrison and Flatow 2011). Neer popularised and modified this theory, emphasising the clinical significance of the anterolateral aspect of the acromion as the predominant source of impingement (Neer 1983). In his 1972 paper, he suggested that 95% of rotator cuff tears resulted from mechanical impingement (Balke et al. 2013). Bigliani later highlighted the role of acromion morphology, especially the hooked-type acromion, as a predisposing factor for development of subacromial impingement and rotator cuff tears (Bigliani et al. 1991). Subsequent studies demonstrated a link between the hooked-type acromial morphology, impingement symptoms and full-thickness rotator cuff tears (Balke et al. 2013; Gill et al. 2002). However, several randomised controlled trials comparing the healing rate and functional outcomes of rotator cuff repair with and without acromioplasty have shown no benefit in removing the presumed causative part of the acromion (Abrams et al. 2014; Woodmass et al. 2022).

Table 1 Extrinsic versus intrinsic theories of rotator cuff tear initiation

Extrinsic theory	Intrinsic theory
Acromion morphology causes cuff tears	Cuff tears are preceded by subclinical inflammation and fibrosis
External compression on tendon explains pain and tear initiation	Even small tears represent late-stage disease
Explains progression from partial-thickness to full-thickness tears	Cellular pathomechanics may explain tear initiation

2.2 Intrinsic Theory

The intrinsic theory originates from analyses of tissue samples from rotator cuff tears revealing immune cell populations and proinflammatory mediators (Abrams et al. 2016; Fabiś et al. 2014). The hypothesis suggests that repetitive subclinical inflammation and fibrosis occur before the initiation of tears. Therefore, even small rotator cuff tears are considered manifestations of late-stage disease with underlying tendinopathic changes (Dakin et al. 2015). While immunohistochemistry analysis of torn rotator cuff tendon tissue samples supports this theory, the exact pathomechanisms leading to these observations remain unclear. This theory offers a potential explanation for the elevated recurrence rates observed after surgical repair of cuff tears. Surgical intervention involves reattaching the torn tendon to its bony footprint, but it does not reverse the intrinsic processes that may have contributed to the tear. As a result, recurrence is likely in a significant proportion of patients.

2.3 Corticosteroids and Tendon Degeneration

Corticosteroid injections are commonly used in managing inflammation and pain associated with rotator cuff tendon issues. Corticosteroids exert their anti-inflammatory effect by modulating various cellular responses. However, they may also interfere with the delicate balance of tendon homeostasis. Corticosteroids have been shown to impede collagen synthesis, a crucial process for maintaining the tensile strength of tendons, by downregulating expression of collagen-producing genes in tendon cells, known as tenocytes. This disruption in collagen synthesis can compromise the structural integrity of the tendon, making it more prone to degeneration (Dean et al. 2014; Puzzitiello et al. 2020).

Furthermore, corticosteroids may affect tenocytes by inhibiting their proliferation and promoting apoptosis (programmed cell death). This interference with tenocyte activity can hinder the natural repair mechanisms of tendons, potentially contributing to the progression of degenerative changes (Dean et al. 2014; Nuelle et al. 2017).

In addition, corticosteroids alter the extracellular matrix (ECM) composition of tendons (Puzzitiello et al. 2020). The ECM provides a supportive framework for tendon structure, and alterations in its composition can impact tendon biomechanics.

Hence, whilst corticosteroid injections may offer rapid relief of symptoms, their use should be approached cautiously. Balancing the anti-inflammatory benefits with the potential adverse effects on collagen synthesis, tenocyte function and ECM composition is crucial.

3 Orthobiologics in Rotator Cuff Regeneration

3.1 Goals of Regeneration

The objectives of rotator cuff regeneration are centred around reinstating the structural integrity, function, and overall health of compromised or damaged tendons. The paramount goal is the stimulation of tissue healing and regeneration, fostering the growth of new, functional tendon tissue, able to withstand physiological loading. Concurrently, mitigating inflammation is crucial to create an environment conducive to healing, preventing the development of chronic inflammation associated with rotator cuff disease. Additionally, the biomechanical properties of the regenerated tendon must be enhanced to restore strength, flexibility, and function to the shoulder joint. Identifying and addressing the underlying causes of rotator cuff disease is essential to prevent future injuries and degeneration, ensuring the prevention of recurrence.

The use of orthobiologics in the treatment of rotator cuff conditions varies based on the severity and nature of the injury. For cases of cuff inflammation with an intact cuff and partial cuff tears, non-surgical approaches are generally favoured. Injectable treatments, such as platelet-rich plasma (PRP), prolotherapy or mesenchymal stem cells (MSCs) can be injected to alleviate symptoms and encourage healing. Despite showing some promise, their utility remains undermined by a lack of supportive basic science evidence to attest to their mechanism of action and efficacy.

In the management of symptomatic full thickness cuff tears, surgical repair is often warranted. In such instances, orthobiologic augmentation may be incorporated during the procedure to enhance tissue regeneration and overall repair success. Rotator cuff repair augmentation can be broadly categorised into cellular augmentation and patch augmentation (Table 2). Massive cuff tears present a particular challenge due to low healing rates following repair secondary to reduced vascularity and fewer fibroblasts. In such cases, patch augmentation may be employed during surgery to provide structural support and enhance the likelihood of successful tissue regeneration. In cases of tendon retraction where the tendon reduce cannot be brought to the bony footprint of the humeral head, patches may be used to bridge the

Table 2 Orthobiologic treatment options in rotator cuff disease

Cellular augments	Patch augments
Platelet-rich plasma (PRP) Mesenchymal stem cells (MSCs) Growth factors Prototherapy Amniotic membrane	*Allograft:* Human dermal tissue Amniotic membrane
	Xenograft: Porcine dermal tissue Porcine small intestine submucosa
	Autograft: Biceps tendon Fascia lata

tendon defect. Human dermal allograft patches are crude augments employed to either load share or replace the native tendons. Their utility is borne out of a need to provide a surgical solution to a complex problem, however dermal allografts bare little resemblance to the native extracellular matrix and microarchitecture of the rotator cuff and its specialised enthesis.

Critically, in management of all these cuff pathologies, concurrent physiotherapy to enhance shoulder mechanics and strengthen the surrounding musculature plays crucial role in improving shoulder function.

3.2 *Cellular Augments*

Platelet-Rich Plasma (PRP)

Platelet-rich plasma (PRP) is the most frequently used and studied cellular therapy. It is defined by a platelet concentration greater than one million platelets per microliter (four to seven times endogenous blood) and is achieved by centrifugation of autologous whole blood followed by isolation of the platelet-containing plasma layer (Kovacevic et al. 2022). PRP may be leucocyte-rich (LR-PRP), where the buffy coat layer, rich in platelets and leucocytes, is retained, or leucocyte-poor (LP-PRP). Preservation of this buffy coat has the advantage of increased platelet count but also results in higher concentrations of leucocyte-derived inflammatory mediators that may impede the healing process (Kovacevic et al. 2022).

The mechanism by which PRP operates in rotator cuff regeneration is centred on the degranulation of platelets, leading to the release of growth factors crucial for the wound healing process. Among these growth factors are transforming growth factor beta (TGF-β), fibroblast growth factor (FGF), platelet-derived growth factor (PDGF), insulin-like growth factor 1 (IGF-1), epidermal growth factor (EGF), and vascular endothelial growth factor (VEGF). PRP is employed in the context of tendon-bone healing to exploit these growth factors and cytokines in physiological proportions but heightened concentrations. Their goal is to modulate inflammation and enhance chemotaxis, differentiation, and mitogenesis of cell lineages pertinent to the regeneration of the tendon-bone enthesis, such as tenocytes, thus promoting healing and alleviating pain (Atesok et al. 2014; Mazzocca et al. 2012).

On a basic science level, PRP has been demonstrated to downregulate the inflammatory molecules Interleukin 6 (IL-6) and Interleukin 8 (IL-8) as well as induce tenocyte differentiation and ECM synthesis (Kovacevic et al. 2022; Andia et al. 2015). In vitro studies investigating the effects of LR-PRP and LP-PRP on human supraspinatus tendon harvested from patients undergoing reverse shoulder arthroplasty for rotator cuff tear arthropathy have shown LP-PRP to induce a greater anti-inflammatory gene expression profile and ECM deposition in moderately damaged tendons, with no difference seen in severely damaged tendons (Cross et al. 2015). These results suggest that LP-PRP may be more beneficial in the nonoperative management of tendonitis than in the operative management of full-thickness tears.

Whilst benefit has been demonstrated in observational studies without a control group, and therefore prone to risk of bias, mixed results have been found from randomised controlled trials (RCTs), casting doubt on the efficacy of PRP in augmentation of rotator cuff repair. Two RCTs have shown a significant reduction in re-tear rate with PRP used to augment rotator cuff repair in large and massive rotator cuff tears, with one of these studies also demonstrating significant improvement in patient-report outcomes measures with use of PRP (Jo et al. 2013; Pandey et al. 2016). In contrast, numerous RCTs have found no significant difference between PRP and control in any outcome measures including healing rate, re-tear rate and functional outcomes in augmentation of rotator cuff repair regardless of tear size (Wang et al. 2015; Flury et al. 2016; Zumstein et al. 2016; Malavolta et al. 2018; Walsh et al. 2018). Notably, there is a lack of consensus in optimal preparation method, dose, and administration of PRP between studies. A Cochrane review investigating the efficacy of PRP in musculoskeletal injuries, including a subgroup analysis for rotator cuff, repair found no clinically significant difference in pain, function, and recurrent tear rate at up to 2 years follow up (Moraes et al. 2013).

A smaller number of studies have considered PRP use in non-operative management of cuff disorders. A systematic review of five RCTs concluded that PRP is unlikely to be beneficial for chronic rotator cuff disease based on short term follow up, with a lack of presently available long-term data (Hurley et al. 2019).

Mesenchymal Stem Cells (MSCs)

Mesenchymal stem cells (MSCs) are multipotent cells with the ability to differentiate into many different connective tissue progenitor cell types such as osteocytes, chondrocytes, and tenocytes. They are usually harvested from bone marrow or adipose tissue. They are preferable to more pluripotent embryonic stem cells due to ease of acquisition and culture without the ethical concerns surrounding embryonic tissue harvest. In vitro studies have demonstrated MSCs to express growth factors such as TGF-β and VEGF as well as suppress the proliferation of inflammatory T-cells and inhibit monocyte maturation, creating both immunomodulatory and anti-inflammatory effects (Dabrowska et al. 2021; Song et al. 2020).

Autologous bone marrow derived MSCs are the most widely used MSC in cuff regeneration. These are isolated in the form of bone marrow aspirate concentrate (BMAC). This is isolated though bone marrow aspiration (most commonly from the iliac crest), followed by filtration, centrifugation, and reinjection of BMAC to the enthesis. There is limited evidence to date on the use of BMAC in rotator cuff regeneration. BMAC augmentation of rotator cuff repair was demonstrated to enhance healing rate and reduce re-ruptures in a case-control study with 10-year follow up (Hernigou et al. 2014). However, this study has raised concerns regarding the findings due to methodological and reporting irregularities. A subsequent RCT demonstrated improved tendon quality on MRI at 1-year post-operatively, however, no difference in re-rupture rate or patent-reported outcome measures was identified (Cole et al. 2023). In the non-operative setting, use of BMAC has been investigated in conjunction with PRP with improvement in patient-reported outcome measures at 3 months, however in these studies it is unclear whether the observed benefit was due to the MSCs, PRP or a synergistic affect (Kim et al. 2018; Centeno et al. 2020).

There is sparse evidence on the role of adipose-derived MSCs. One human cohort study demonstrated reduced re-tear rate on MRI but no difference in pain, range of motion or patient-reported outcomes compared to a matched control group (Kim et al. 2017). Data on MSC augmentation remains limited, with variation in isolation technique, dose, and administration amongst the small number of currently available studies, restricting its present use in the clinical setting.

Growth Factors

Various growth factors are involved in rotator cuff regeneration, as outlined in Table 3. Whilst the aforementioned orthobiologic therapies promote release of multiple factors, individual growth factors can be harnessed to promote rotator cuff regeneration.

FGF-2 has been investigated in relation to rotator cuff repair in rat models, with studies showing improved tendon load-to-failure, collagen fibre orientation and vascularity (Tokunaga et al. 2015; Yonemitsu et al. 2019). TGF-β has also been studied in augmentation of rotator cuff repair in both rabbit and rat models, with results demonstrating improved load-to-failure and tendon cross-sectional area (Arimura et al. 2017; Yoon et al. 2018). There are no human clinical studies to date investigating use of either FGF-2 or TGF-β in rotator cuff regeneration.

Weekly injection of human growth hormone (HGH) following arthroscopic rotator cuff repair has been investigated in a randomised controlled trial, however, was found not to significantly increase serum IGF-1 levels or impact re-tear rate (Oh et al. 2018).

Table 3 Growth factors in rotator cuff regeneration

Growth factor	Role in rotator cuff healing
Platelet-derived growth factor (PDGF)	Stimulates tenocyte proliferation and migration Promotes collagen synthesis by tenocytes Enhances angiogenesis
Insulin-like growth factor 1 (IGF-1)	Stimulates tenocyte growth and differentiation Enhances collagen synthesis by tenocytes Promotes angiogenesis
Fibroblast growth factor (FGF)	Stimulates fibroblast proliferation and migration Promotes angiogenesis Enhances ECM synthesis by fibroblasts
Transforming growth factor beta (TGF-β)	Stimulates tenocyte growth and differentiation Enhances collagen synthesis by tenocytes Modulates inflammation in the tendon microenvironment
Vascular endothelial growth factor (VEGF)	Promotes endothelial cell proliferation and angiogenesis

Research into growth factors in rotator cuff regeneration remains in its infancy at present. Future studies may focus on elucidating the specific mechanisms of action, determining optimal dosages and delivery methods, and conducting well-designed clinical trials to establish the effectiveness and long-term outcomes of these interventions. Additionally, investigating the interplay between different growth factors and their synergistic effects will contribute to a more comprehensive understanding.

Prolotherapy

Prolotherapy involves injection of an irritant solution, typically containing hypertonic dextrose, directly into the tendon with the goal of stimulating the body's natural healing response thereby promoting tissue repair and regeneration. It is thought to work by inducing a localised inflammatory response and stimulating the release of growth factors and cytokines. Research on its effectiveness for rotator cuff injuries is still in the early stages, and the available evidence is limited. Randomised studies have shown significant improvement in pain when compared to control when administered as a series of multiple injections using high concentrations (>10%) of dextrose, with no observed differences from a single injection (Bertrand et al. 2016; Lin et al. 2019; Seven et al. 2017; George et al. 2018; Cole et al. 2018). Minimal side effects have been reported indicating it to be a safe therapy. However, further high-quality research is required to establish its effectiveness and determine its place in the overall management of rotator cuff injuries.

3.3 *Patch Augments*

Patch augmentation involves applying a graft during surgical repair of the torn rotator cuff. Patches may be applied on top of the repaired cuff tendon (onlay), underneath the tendon (interposition), or in a "bridging" fashion to fill the defect between the tendon edge and the bony footprint. These patches may be synthetic or biological. Synthetic patches are usually made from a non-degradable polymer. They are designed to support the repair structurally, and share the load carried by the rotator cuff tendon. Biological patches are marketed as decellularised grafts that become incorporated and vascularised by the native tendon (Rashid et al. 2020). Biological graft sources include those of autologous origin (e.g. fascia lata autograft), cadaveric donors (e.g. dermal allograft), or from animal tissues (e.g. dermal or small intestine submucosa xenograft, or bovine-derived collagen patches from explanted tendons). These biological patches, sometimes called extracellular matrix (ECM) patches, are processed to remove donor cells, and sometimes chemically crosslinked, before sterilisation for clinical use (Chalmers and Tashjian 2020). Crosslinking in biological grafts is thought to improve durability by reducing degradation via matrix metalloproteinases as well as improve cell migration, vascularisation, and remodelling (Xu et al. 2009).

The philosophy behind biological patch augmentation is that the native tendon tissue often lacks capacity to heal to the bony footprint and that application of a collagen and elastin-rich decellularised graft, that can become integrated with native cells and blood vessels, would improve the healing rate. Thus, the aim is to improve healing rate by improving the biological environment (Chalmers and Tashjian 2020).

Human Dermal Allografts

Dermal allograft is the most commonly used allograft due to being more readily available than other tissues. Several human dermal allograft patches are commercially available including ArthroFlex® Dermal Allograft (Arthrex), GraftJacket® (Wright Medical) and AlloPatch HD® (Conmed). These are cadaveric human dermis grafts that undergo decellularisation whilst preserving the ECM. They were initially developed for use in chronic wounds, such as non-healing foot ulcers, however their application has subsequently expanded for use in soft tissue augmentation, including but not limited to, the rotator cuff. They form a collagen scaffold and have been observed to have host tissue integration properties, in vivo revascularisation, cellular incorporation, and favourable biomechanical properties that make it an attractive option in this setting (Bond et al. 2008; Burkhead et al. 2007; Acevedo et al. 2015).

Clinical studies have demonstrated human dermal allograft to significantly reduce re-tear rate and improve patient-reported outcome scores following repair of large and massive cuff tears compared to repair without augmentation with no adverse events related to the graft (Acevedo et al. 2015; Cai et al. 2018; Derwin et al. 2006; Gilot et al. 2015; Barber et al. 2008; Modi et al. 2022).

Despite their widespread use, there remains a lack of high quality, low risk of bias, clinical studies evaluating their clinical efficacy, with most studies being observational in nature. Furthermore, differences in the manufacturing processes including the decellularisation method, preparation, storage, and hydration varies between products and may affect the specific characteristics of the allografts. Moreover, limited comparative studies make it challenging to determine superiority among these patches.

Xenografts

Commercially available xenograft patches are porcine-derived, sourced from either the dermis or small intestine submucosa (SIS). A commonly used example is Permacol™ (Medtronic), a chemically crosslinked and decellularised dermal xenograft initially approved for multiple uses in abdominal hernia and genitourinary surgery prior to use in orthopaedic surgery.

Dermal xenografts have demonstrated significant improvements in re-tear rate and functional outcomes following rotator cuff repair in non-randomised comparative studies. However, whilst stronger than primary cuff repair alone, the outcomes from dermal xenograft augmentation appear to be inferior when compared dermal human allograft augmentation (Ono et al. 2016). Additionally, adverse tissue immune responses have been identified in a small number of cases (Rashid et al. 2020).

Use of SIS patches has decreased in recent years following the withdrawal of the Restore® Orthobiologic Soft Tissue Implant (DePuy Synthes). An abandoned randomised controlled trial comparing rotator cuff augmentation with the Restore® patch to control yielded two clinical studies demonstrating adverse reactions to the implant, including a 40% rate of open debridement and irrigation (Malcarney et al. 2005). Histopathology of samples taken at the time of debridement showed no microbiological growth, and acutely inflamed granulation tissue. The subsequent cohort study showed a significantly inferior clinical outcome in patients that received the patch and no improvement on healing rate at 2 years (Walton et al. 2007).

Recently, a bovine tendon-derived bioinductive patch (Regeneten®, Smith & Nephew) has increased in popularity. Unlike other augments, this patch is non-structural, becoming absorbed over a period of approximately 6 months, after providing biophysical cues for tendon regeneration. Its use has been employed in partial cuff tears as an isolated intervention, or as an adjunct after repair of torn rotator cuff tendons (Bokor et al. 2016; Camacho-Chacon et al. 2022). Additionally, in "at risk" repairs, such as revision cuff repairs, large-to-massive tears, or those with poor quality tendon tissue, Regeneten may be used as a biological augment to the repair (Thon et al. 2019; Ting et al. 2023; Zhang et al. 2024). Most reports of its use show promising rates of healing with minimal adverse reactions (McIntyre et al. 2019; Bushnell et al. 2022; Schlegel et al. 2021). One study reported increased stiffness compared to an age, sex and tear size-matched control group (receiving debridement or tear completion and repair) requiring reoperations in 6/32 patients for stiffness (Yeazell et al. 2022). Most reports are single or multi-centre case series without a control group (Bokor et al. 2016; McIntyre et al. 2019; Bushnell et al. 2022; Schlegel et al. 2021). Two large series from multiple surgeons and centres showed significant improvements in patient reported outcome scores when this bovine collagen bioinductive patch was used in both partial thickness tears as an isolated intervention, or after complete repair of large-to-massive tears. Both studies included complications, most of which are not implanted related, and occurring at an acceptable rate (4.6% and 7.8% respectively) (McIntyre et al. 2019; Bushnell et al. 2022). Comparative studies have recently been published, reporting mixed results for healing (Ting et al. 2023; Zhang et al. 2024). One randomised controlled trial with healing on magnetic resonance imaging as the primary outcome demonstrated beneficial effect of Regeneten patch as an adjunct to full thickness cuff repair healing (retear rate of 8.3% vs 25.8% at 12 months) (Ruiz Iban et al. 2024). The mechanism of action of this patch has yet to be elucidated.

Autologous Grafts

Autologous grafts can be considered in countries where allograft is not readily available or is culturally unacceptable. The most frequently used autografts are the long head of biceps tendon and the fascia lata. Advantages of autologous grafts are inherent biocompatibility and maintenance of the native tissue's biological properties, providing a matrix rich in collagen and elastin that closely resembles the patient's natural anatomy. Disadvantages are limited availability of autograft tissues, donor site morbidity, and the size of the autograft may be insufficient for larger tears, potentially compromising the structural support needed for optimal healing.

One non-randomised comparative study showed fascia lata autograft augmentation for large or massive cuff tears resulted in significantly better functional outcomes and lower re-tear rates when compared to repair alone (Mori et al. 2013). A systematic review of 8 case series that utilised biceps tendon autograft in rotator cuff repair augmentation found significantly improved function, pain relief, and range of motion, however this is based on low quality data without a control group highlighting a paucity of low-risk-of-bias data to justify their current use in clinical practice.

3.4 Amniotic Tissue Products

Amniotic tissue has gained recent attention in regenerative medicine, including its potential role in rotator cuff regeneration. It is harvested from full-term elective caesarean sections, processed, and offered commercially in a variety of forms, including wet or dry membrane sheets, fluids, and particulate. These products are rich in growth factors, cytokines, and in the case of amniotic membrane products, ECM structure. Amniotic fluid has also been shown to contain multipotent MSCs, however their differential capacity in commercial preparations is not well understood. Amniotic tissue has been found to be a potent inhibitor of matrix metalloproteinase (MMP), which has been demonstrated at the site of rotator cuff tears and can degrade components of the ECM, negatively impacting repair (Robertson et al. 2012; Heckmann et al. 2016). Reduction of MMPs in animal models has been shown to reduce retear rates (Bedi et al. 2010a, b).

Evidence of its efficacy in rotator cuff regeneration is limited. A small case series ($n = 10$) found intra-articular injection of amniotic membrane and umbilical cord for partial rotator cuff tears refractory to conservative management to be safe with improvement in patient-reported outcome measures at 6 months, but no difference in tear size on follow-up MRI scan (Ackley et al. 2019). Another study investigated eight patients that underwent human amniotic membrane interposition augmentation during rotator cuff repair with all eight patients going on to rotator cuff healing, with no reported adverse events. Whilst this study demonstrated its safe use, it is a non-comparative study and cannot determine whether the amniotic membrane had any impact on the healing process (Mirzayan and Suh 2022). It is important to note that whilst there is some preclinical and early clinical evidence supporting the use of amniotic tissue, the field is still evolving, and more research is needed before its widespread use in rotator cuff repair can be endorsed.

3.5 Future Directions

Orthobiologics is a rapidly developing field in rotator cuff regeneration offering innovative solutions to enhance healing processes. As our understanding of its application increases it is important to envision the ideal orthobiologic product, considering both injectable formulations and patches. Moreover, a systematic and comprehensive approach to research and implementation is essential.

An ideal injectable orthobiologic product must exhibit a harmonious blend of characteristics to maximise therapeutic efficacy. The composition should contain an optimal ratio of growth factors, including but not limited to TGF-β, FGF, and PDGF. These factors play pivotal roles in modulating inflammation, promoting cell migration, and stimulating tissue regeneration. The injectable should ensure sustained release kinetics, providing a controlled and prolonged supply of bioactive agents to support the complex healing cascade. At present, the challenges include standardising the concentrations of various biological factors, and providing sustained therapeutic action that can ultimately positively influence the long-term prognosis of the degenerative tendon. Additionally, the formulation should exhibit biocompatibility and safety, minimising adverse reactions.

The ideal patch is one that seamlessly integrates with native tissues. Biocompatibility is paramount, ensuring minimal immune response and maximal acceptance by the host tissue. Mechanical strength is another critical factor, with the patch providing robust biomechanical support to withstand the dynamic forces within the shoulder joint. Moreover, an ideal patch should actively recruit endogenous cells, facilitating tissue regeneration and creating a conducive environment for healing. Long-term integration and sustained biomechanical support are vital to the success of patch augmentation. Patches specifically designed for the rotator cuff enthesis are likely needed, as this highly specialised transitional zone has unique properties that require specific biophysical cues to promote successful healing and a tendon capable of withstanding physiological loading.

The treatment should be supported by rigorous scientific evidence, demonstrating safety, efficacy, and long-term outcomes. The IDEAL framework can be used a guide to approaching future research. Namely:

1. Idea: Research endeavours should begin by defining the questions that will shape the future of orthobiologics in rotator cuff regeneration. Identifying knowledge gaps and formulating hypotheses to guide the subsequent stages.
2. Development: Rigorous preclinical studies are imperative for the development of injectable formulations and patches. These studies should delve into safety,

efficacy, and optimal administration protocols, adhering to standardised methodologies.
3. Exploration: Exploratory clinical studies must follow, assessing the initial outcomes of orthobiologic interventions. Standardised outcome measures should be implemented to facilitate cross-study comparisons and enhance the generalisability of findings.
4. Assessment: Randomised controlled trials (RCTs) are a critical component in assessing the effectiveness and safety of orthobiologics. Utilising a range of outcome measures, from patient-reported outcomes to advanced imaging techniques, ensures a comprehensive evaluation.
5. Long-term study: Long-term follow-up studies are indispensable for understanding the durability and sustained benefits of orthobiologic interventions. Monitoring for late complications or adverse events provides insights into the long-term impact of these therapies.

4 Conclusion

While surgical repair remains the mainstay for full-thickness tears, the high rates of re-tears and incomplete biological healing at the enthesis highlight the need for enhanced regenerative strategies. Orthobiologics, ranging from injectable treatments such as PRP and mesenchymal stem cells, to patch-based augmentation using allografts, xenografts, or autologous tissue have emerged as promising adjuncts with the potential to improve both structural and clinical outcomes.

References

Abrams GD, Gupta AK, Hussey KE, et al. Arthroscopic repair of full-thickness rotator cuff tears with and without acromioplasty: randomized prospective trial with 2-year follow-up. Am J Sports Med. 2014;42:1296–303. https://doi.org/10.1177/0363546514529091.

Abrams GD, Luria A, Carr RA, et al. Association of synovial inflammation and inflammatory mediators with glenohumeral rotator cuff pathology. J Shoulder Elb Surg. 2016;25:989–97. https://doi.org/10.1016/j.jse.2015.10.011.

Acevedo DC, Shore B, Mirzayan R. Orthopedic applications of acellular human dermal allograft for shoulder and elbow surgery. Orthop Clin N Am. 2015;46:377–88. https://doi.org/10.1016/j.ocl.2015.02.006.

Ackley JF, Kolosky M, Gurin D, et al. Cryopreserved amniotic membrane and umbilical cord particulate matrix for partial rotator cuff tears: a case series. Medicine (Baltimore). 2019;98:e16569. https://doi.org/10.1097/md.0000000000016569.

Andia I, Rubio-Azpeitia E, Maffulli N. Platelet-rich plasma modulates the secretion of inflammatory/angiogenic proteins by inflamed tenocytes. Clin Orthop Relat Res. 2015;473:1624–34. https://doi.org/10.1007/s11999-015-4179-z.

Arimura H, Shukunami C, Tokunaga T, et al. TGF-β1 improves biomechanical strength by extracellular matrix accumulation without increasing the number of tenogenic lineage cells in a rat rotator cuff repair model. Am J Sports Med. 2017;45:2394–404. https://doi.org/10.1177/0363546517707940.

Atesok K, Fu FH, Wolf MR, et al. Augmentation of tendon-to-bone healing. J Bone Joint Surg Am. 2014;96:513–21. https://doi.org/10.2106/jbjs.M.00009.

Balke M, Schmidt C, Dedy N, et al. Correlation of acromial morphology with impingement syndrome and rotator cuff tears. Acta Orthop. 2013;84:178–83. https://doi.org/10.3109/17453674.2013.773413.

Barber FA, Herbert MA, Boothby MH. Ultimate tensile failure loads of a human dermal allograft rotator cuff augmentation. Arthroscopy. 2008;24:20–4. https://doi.org/10.1016/j.arthro.2007.07.013.

Bedi A, Fox AJ, Kovacevic D, et al. Doxycycline-mediated inhibition of matrix metalloproteinases improves healing after rotator cuff repair. Am J Sports Med. 2010a;38:308–17. https://doi.org/10.1177/0363546509347366.

Bedi A, Kovacevic D, Hettrich C, et al. The effect of matrix metalloproteinase inhibition on tendon-to-bone healing in a rotator cuff repair model. J Shoulder Elb Surg. 2010b;19:384–91. https://doi.org/10.1016/j.jse.2009.07.010.

Benjamin M, Evans EJ, Copp L. The histology of tendon attachments to bone in man. J Anat. 1986;149:89–100.

Bertrand H, Reeves KD, Bennett CJ, et al. Dextrose prolotherapy versus control injections in painful rotator cuff tendinopathy. Arch Phys Med Rehabil. 2016;97:17–25. https://doi.org/10.1016/j.apmr.2015.08.412.

Bigliani LU, Ticker JB, Flatow EL, et al. The relationship of acromial architecture to rotator cuff disease. Clin Sports Med. 1991;10:823–38.

Bokor DJ, Sonnabend D, Deady L, et al. Evidence of healing of partial-thickness rotator cuff tears following arthroscopic augmentation with a collagen implant: a 2-year MRI follow-up. Muscles Ligaments Tendons J. 2016;6:16–25. https://doi.org/10.11138/mltj/2016.6.1.016.

Bond JL, Dopirak RM, Higgins J, et al. Arthroscopic replacement of massive, irreparable rotator cuff tears using a GraftJacket allograft: technique and preliminary results. Arthroscopy. 2008;24:403–409.e401. https://doi.org/10.1016/j.arthro.2007.07.033.

Burkhart SS, Esch JC, Jolson RS. The rotator crescent and rotator cable: an anatomic description of the shoulder's "suspension bridge". Arthroscopy. 1993;9:611–6. https://doi.org/10.1016/s0749-8063(05)80496-7.

Burkhead WZ, Schiffern SC, Krishnan SG. Use of Graft Jacket as an augmentation for massive rotator cuff tears. Semin Arthroplast. 2007;18:11–8. https://doi.org/10.1053/j.sart.2006.11.017.

Bushnell BD, Connor PM, Harris HW, et al. Two-year outcomes with a bioinductive collagen implant used in augmentation of arthroscopic repair of full-thickness rotator cuff tears: final results of a prospective multicenter study. J Shoulder Elb Surg. 2022;31:2532–41. https://doi.org/10.1016/j.jse.2022.05.025.

Cai YZ, Zhang C, Jin RL, et al. Arthroscopic rotator cuff repair with graft augmentation of 3-dimensional biological collagen for moderate to large tears: a randomized controlled study. Am J Sports Med. 2018;46:1424–31. https://doi.org/10.1177/0363546518756978.

Camacho-Chacon JA, Cuenca-Espierrez J, Roda-Rojo V, et al. Bioinductive collagen implants facilitate tendon regeneration in rotator cuff tears. J Exp Orthop. 2022;9:53. https://doi.org/10.1186/s40634-022-00495-7.

Centeno C, Fausel Z, Stemper I, et al. A randomized controlled trial of the treatment of rotator cuff tears with bone marrow concentrate and platelet products compared to exercise therapy: a midterm analysis. Stem Cells Int. 2020;2020:5962354. https://doi.org/10.1155/2020/5962354.

Chalmers PN, Tashjian RZ. Patch augmentation in rotator cuff repair. Curr Rev Musculoskelet Med. 2020;13:561–71. https://doi.org/10.1007/s12178-020-09658-4.

Clark JM, Harryman DT 2nd. Tendons, ligaments, and capsule of the rotator cuff. Gross and microscopic anatomy. J Bone Joint Surg Am. 1992;74:713–25.

Cole B, Lam P, Hackett L, et al. Ultrasound-guided injections for supraspinatus tendinopathy: corticosteroid versus glucose prolotherapy – a randomized controlled clinical trial. Shoulder Elbow. 2018;10:170–8. https://doi.org/10.1177/1758573217708199.

Cole BJ, Kaiser JT, Wagner KR, et al. Prospective randomized trial of biologic augmentation with bone marrow aspirate concentrate in patients undergoing arthroscopic rotator cuff repair. Am J Sports Med. 2023;51:1234–42. https://doi.org/10.1177/03635465231154601.

Cross JA, Cole BJ, Spatny KP, et al. Leukocyte-reduced platelet-rich plasma normalizes matrix metabolism in torn human rotator cuff tendons. Am J Sports Med. 2015;43:2898–906. https://doi.org/10.1177/0363546515608157.

Dabrowska S, Andrzejewska A, Janowski M, et al. Immunomodulatory and regenerative effects of mesenchymal stem cells and extracellular vesicles: therapeutic outlook for inflammatory and degenerative diseases. Front Immunol. 2021;11:591065. https://doi.org/10.3389/fimmu.2020.591065.

Dakin SG, Martinez FO, Yapp C, et al. Inflammation activation and resolution in human tendon disease. Sci Transl Med. 2015;7:311ra173. https://doi.org/10.1126/scitranslmed.aac4269.

Dean BJ, Lostis E, Oakley T, et al. The risks and benefits of glucocorticoid treatment for tendinopathy: a systematic review of the effects of local glucocorticoid on tendon. Semin Arthritis Rheum. 2014;43:570–6. https://doi.org/10.1016/j.semarthrit.2013.08.006.

Derwin KA, Baker AR, Spragg RK, et al. Commercial extracellular matrix scaffolds for rotator cuff tendon repair. Biomechanical, biochemical, and cellular properties. J Bone Joint Surg Am. 2006;88:2665–72. https://doi.org/10.2106/jbjs.E.01307.

Fabiś J, Szemraj J, Strek M, et al. Is resection of the tendon edge necessary to enhance the healing process? An evaluation of the expression of collagen type I, IL-1β, IFN-γ, IL-4, and IL-13 in the distal 1 cm of a torn supraspinatus tendon: part II. J Shoulder Elb Surg. 2014;23:1779–85. https://doi.org/10.1016/j.jse.2014.08.023.

Flury M, Rickenbacher D, Schwyzer HK, et al. Does pure platelet-rich plasma affect postoperative clinical outcomes after arthroscopic rotator cuff repair? A randomized controlled trial. Am J Sports Med. 2016;44:2136–46. https://doi.org/10.1177/0363546516645518.

George J, Li SC, Jaafar Z, et al. Comparative effectiveness of ultrasound-guided intratendinous prolotherapy injection with conventional treatment to treat focal supraspinatus tendinosis. Scientifica (Cairo). 2018;2018:4384159. https://doi.org/10.1155/2018/4384159.

Gill TJ, McIrvin E, Kocher MS, et al. The relative importance of acromial morphology and age with respect to rotator cuff pathology. J Shoulder Elb Surg. 2002;11:327–30. https://doi.org/10.1067/mse.2002.124425.

Gilot GJ, Alvarez-Pinzon AM, Barcksdale L, et al. Outcome of large to massive rotator cuff tears repaired with and without extracellular matrix augmentation: a prospective comparative study. Arthroscopy. 2015;31:1459–65. https://doi.org/10.1016/j.arthro.2015.02.032.

Harrison AK, Flatow EL. Subacromial impingement syndrome. J Am Acad Orthop Surg. 2011;19:701–8. https://doi.org/10.5435/00124635-201111000-00006.

Heckmann N, Auran R, Mirzayan R. Application of amniotic tissue in orthopedic surgery. Am J Orthop (Belle Mead NJ). 2016;45:E421–5.

Hernigou P, Flouzat Lachaniette CH, Delambre J, et al. Biologic augmentation of rotator cuff repair with mesenchymal stem cells during arthroscopy improves healing and prevents further tears: a case-controlled study. Int Orthop. 2014;38:1811–8. https://doi.org/10.1007/s00264-014-2391-1.

Hsu J, Keener JD. Natural history of rotator cuff disease and implications on management. Oper Tech Orthop. 2015;25:2–9. https://doi.org/10.1053/j.oto.2014.11.006.

Hurley ET, Hannon CP, Pauzenberger L, et al. Nonoperative treatment of rotator cuff disease with platelet-rich plasma: a systematic review of randomized controlled trials. Arthroscopy. 2019;35:1584–91. https://doi.org/10.1016/j.arthro.2018.10.115.

Jo CH, Shin JS, Lee YG, et al. Platelet-rich plasma for arthroscopic repair of large to massive rotator cuff tears: a randomized, single-blind, parallel-group trial. Am J Sports Med. 2013;41:2240–8. https://doi.org/10.1177/0363546513497925.

Kim YS, Sung CH, Chung SH, et al. Does an injection of adipose-derived mesenchymal stem cells loaded in fibrin glue influence rotator cuff repair outcomes? A clinical and magnetic resonance imaging study. Am J Sports Med. 2017;45:2010–8. https://doi.org/10.1177/0363546517702863.

Kim SJ, Kim EK, Kim SJ, et al. Effects of bone marrow aspirate concentrate and platelet-rich plasma on patients with partial tear of the rotator cuff tendon. J Orthop Surg Res. 2018;13:1. https://doi.org/10.1186/s13018-017-0693-x.

Kovacevic D, Suriani RJ Jr, Levine WN, et al. Augmentation of rotator cuff healing with orthobiologics. J Am Acad Orthop Surg. 2022;30:e508–16. https://doi.org/10.5435/jaaos-d-20-01011.

Lin CL, Huang CC, Huang SW. Effects of hypertonic dextrose injection in chronic supraspinatus tendinopathy of the shoulder: a randomized placebo-controlled trial. Eur J Phys Rehabil Med. 2019;55:480–7. https://doi.org/10.23736/s1973-9087.18.05379-0.

Malavolta EA, Gracitelli MEC, Assunção JH, et al. Clinical and structural evaluations of rotator cuff repair with and without added platelet-rich plasma at 5-year follow-up: a prospective randomized study. Am J Sports Med. 2018;46:3134–41. https://doi.org/10.1177/0363546518795895.

Malcarney HL, Bonar F, Murrell GA. Early inflammatory reaction after rotator cuff repair with a porcine small intestine submucosal implant: a report of 4 cases. Am J Sports Med. 2005;33:907–11. https://doi.org/10.1177/0363546504271500.

Mazzocca AD, McCarthy MB, Chowaniec DM, et al. Platelet-rich plasma differs according to preparation method and human variability. J Bone Joint Surg Am. 2012;94:308–16. https://doi.org/10.2106/jbjs.K.00430.

McIntyre LF, Bishai SK, Brown PB 3rd, et al. Patient-reported outcomes after use of a bioabsorbable collagen implant to treat partial and full-thickness rotator cuff tears. Arthroscopy. 2019;35:2262–71. https://doi.org/10.1016/j.arthro.2019.02.019.

Minagawa H, Yamamoto N, Abe H, et al. Prevalence of symptomatic and asymptomatic rotator cuff tears in the general population: from mass-screening in one village. J Orthop. 2013;10:8–12. https://doi.org/10.1016/j.jor.2013.01.008.

Mirzayan R, Suh BD. Interposition of human amniotic membrane at the bone-tendon interface of a full-thickness rotator cuff repair. JSES Rev Rep Tech. 2022;2:75–80. https://doi.org/10.1016/j.xrrt.2021.10.008.

Modi A, Haque A, Deore V, et al. Interposition GraftJacket allografts for irreparable rotator cuff tears. Bone Joint J. 2022;104-B:91–6. https://doi.org/10.1302/0301-620x.104b1.Bjj-2021-0826.R1.

Moraes VY, Lenza M, Tamaoki MJ, et al. Platelet-rich therapies for musculoskeletal soft tissue injuries. Cochrane Database Syst Rev. 2013;2013(12):CD010071. https://doi.org/10.1002/14651858.CD010071.pub2.

Mori D, Funakoshi N, Yamashita F. Arthroscopic surgery of irreparable large or massive rotator cuff tears with low-grade fatty degeneration of the infraspinatus: patch autograft procedure versus partial repair procedure. Arthroscopy. 2013;29:1911–21. https://doi.org/10.1016/j.arthro.2013.08.032.

Neer CS 2nd. Impingement lesions. Clin Orthop Relat Res. 1983;173:70–7.

Nuelle CW, Cook CR, Stoker AM, et al. In vitro toxicity of local anaesthetics and corticosteroids on supraspinatus tenocyte viability and metabolism. J Orthop Translat. 2017;8:20–4. https://doi.org/10.1016/j.jot.2016.08.002.

Oh JH, Chung SW, Oh KS, et al. Effect of recombinant human growth hormone on rotator cuff healing after arthroscopic repair: preliminary result of a multicenter, prospective, randomized, open-label blinded end point clinical exploratory trial. J Shoulder Elb Surg. 2018;27:777–85. https://doi.org/10.1016/j.jse.2017.11.019.

Ono Y, Dávalos Herrera DA, Woodmass JM, et al. Healing rate and clinical outcomes of xenografts appear to be inferior when compared to other graft material in rotator cuff repair: a meta-analysis. J ISAKOS. 2016;1:321–8. https://doi.org/10.1136/jisakos-2016-000076.

Pandey V, Bandi A, Madi S, et al. Does application of moderately concentrated platelet-rich plasma improve clinical and structural outcome after arthroscopic repair of medium-sized to

large rotator cuff tear? A randomized controlled trial. J Shoulder Elb Surg. 2016;25:1312–22. https://doi.org/10.1016/j.jse.2016.01.036.
Puzzitiello RN, Patel BH, Forlenza EM, et al. Adverse impact of corticosteroids on rotator cuff tendon health and repair: a systematic review of basic science studies. Arthrosc Sports Med Rehabil. 2020;2:e161–9. https://doi.org/10.1016/j.asmr.2020.01.002.
Rashid MS, Smith RDJ, Nagra N, et al. Rotator cuff repair with biological graft augmentation causes adverse tissue outcomes. Acta Orthop. 2020;91:782–8. https://doi.org/10.1080/17453674.2020.1793613.
Robertson CM, Chen CT, Shindle MK, et al. Failed healing of rotator cuff repair correlates with altered collagenase and gelatinase in supraspinatus and subscapularis tendons. Am J Sports Med. 2012;40:1993–2001. https://doi.org/10.1177/0363546512456519.
Rothrauff BB, Pauyo T, Debski RE, et al. The rotator cuff organ: integrating developmental biology, tissue engineering, and surgical considerations to treat chronic massive rotator cuff tears. Tissue Eng Part B Rev. 2017;23:318–35. https://doi.org/10.1089/ten.TEB.2016.0446.
Ruiz Iban MA, Garcia Navlet M, Moros Marco S, et al. Augmentation of a transosseous-equivalent repair in posterosuperior nonacute rotator cuff tears with a bioinductive collagen implant decreases the retear rate at one year: a randomized controlled trial. Arthroscopy. 2024;40:1760–73. https://doi.org/10.1016/j.arthro.2023.12.014.
Schlegel TF, Abrams JS, Angelo RL, et al. Isolated bioinductive repair of partial-thickness rotator cuff tears using a resorbable bovine collagen implant: two-year radiologic and clinical outcomes from a prospective multicenter study. J Shoulder Elb Surg. 2021;30:1938–48. https://doi.org/10.1016/j.jse.2020.10.022.
Seven MM, Ersen O, Akpancar S, et al. Effectiveness of prolotherapy in the treatment of chronic rotator cuff lesions. Orthop Traumatol Surg Res. 2017;103:427–33. https://doi.org/10.1016/j.otsr.2017.01.003.
Song N, Scholtemeijer M, Shah K. Mesenchymal stem cell immunomodulation: mechanisms and therapeutic potential. Trends Pharmacol Sci. 2020;41:653–64. https://doi.org/10.1016/j.tips.2020.06.009.
Taylor DW. The life and teaching of William Sharpey (1802-1880). "Father of modern physiology" in Britain. Med Hist. 1971;15:241–59. https://doi.org/10.1017/s0025727300016707.
Tempelhof S, Rupp S, Seil R. Age-related prevalence of rotator cuff tears in asymptomatic shoulders. J Shoulder Elb Surg. 1999;8:296–9. https://doi.org/10.1016/s1058-2746(99)90148-9.
Thon SG, O'Malley L 2nd, O'Brien MJ, et al. Evaluation of healing rates and safety with a bioinductive collagen patch for large and massive rotator cuff tears: 2-year safety and clinical outcomes. Am J Sports Med. 2019;47:1901–8. https://doi.org/10.1177/0363546519850795.
Ting RS, Loh YC, Rosenthal R, et al. Revision rotator cuff repair with versus without an arthroscop ically inserted onlay bioinductive implant in workers' compensation patients. Orthop J Sports Med. 2023;11:23259671231175883. https://doi.org/10.1177/23259671231175883.
Tokunaga T, Shukunami C, Okamoto N, et al. FGF-2 stimulates the growth of tenogenic progenitor cells to facilitate the generation of Tenomodulin-positive tenocytes in a rat rotator cuff healing model. Am J Sports Med. 2015;43:2411–22. https://doi.org/10.1177/0363546515597488.
Walsh MR, Nelson BJ, Braman JP, et al. Platelet-rich plasma in fibrin matrix to augment rotator cuff repair: a prospective, single-blinded, randomized study with 2-year follow-up. J Shoulder Elb Surg. 2018;27:1553–63. https://doi.org/10.1016/j.jse.2018.05.003.
Walton JR, Bowman NK, Khatib Y, et al. Restore orthobiologic implant: not recommended for augmentation of rotator cuff repairs. J Bone Joint Surg Am. 2007;89:786–91. https://doi.org/10.2106/jbjs.F.00315.
Wang A, McCann P, Colliver J, et al. Do postoperative platelet-rich plasma injections accelerate early tendon healing and functional recovery after arthroscopic supraspinatus repair? A randomized controlled trial. Am J Sports Med. 2015;43:1430–7. https://doi.org/10.1177/0363546515572602.
Woodmass JM, Al Khatib L, McRae S, et al. Arthroscopic rotator cuff repair with and without acromioplasty in the treatment of full-thickness rotator cuff tears: long-term outcomes of a

multicenter, randomized controlled trial. J Bone Joint Surg Am. 2022;104:2101–7. https://doi.org/10.2106/jbjs.22.00135.

Xu H, Wan H, Zuo W, et al. A porcine-derived acellular dermal scaffold that supports soft tissue regeneration: removal of terminal galactose-alpha-(1,3)-galactose and retention of matrix structure. Tissue Eng Part A. 2009;15:1807–19. https://doi.org/10.1089/ten.tea.2008.0384.

Yeazell S, Lutz A, Bohon H, et al. Increased stiffness and reoperation rate in partial rotator cuff repairs treated with a bovine patch: a propensity-matched trial. J Shoulder Elb Surg. 2022;31:S131–5. https://doi.org/10.1016/j.jse.2022.02.003.

Yonemitsu R, Tokunaga T, Shukunami C, et al. Fibroblast growth factor 2 enhances tendon-to-bone healing in a rat rotator cuff repair of chronic tears. Am J Sports Med. 2019;47:1701–12. https://doi.org/10.1177/0363546519836959.

Yoon JP, Lee CH, Jung JW, et al. Sustained delivery of transforming growth factor β1 by use of absorbable alginate scaffold enhances rotator cuff healing in a rabbit model. Am J Sports Med. 2018;46:1441–50. https://doi.org/10.1177/0363546518757759.

Zhang T, Ajayi A, Hajjar M, et al. Arthroscopic repair of retracted large and massive rotator cuff tears with and without augmentation with a bio-inductive collagen implant reveals substantial and comparable clinical improvement. Arthroscopy. 2024;40:1434–42. https://doi.org/10.1016/j.arthro.2023.10.024.

Zumstein MA, Rumian A, Thélu C, et al. SECEC Research Grant 2008 II: use of platelet- and leucocyte-rich fibrin (L-PRF) does not affect late rotator cuff tendon healing: a prospective randomized controlled study. J Shoulder Elb Surg. 2016;25:2–11. https://doi.org/10.1016/j.jse.2015.09.018.

Hip—Osteitis Pubis, Adductor Tendon, Cartilage Regeneration

Paul Lee and Tanvi Verma

Abstract This chapter examines key hip injuries, focusing on osteitis pubis, adductor tendon injuries, and cartilage regeneration, with particular attention to athletic populations. Osteitis pubis, characterised by inflammation of the pubic symphysis, often results from biomechanical overload and repetitive stress, leading to significant pain and mobility issues. Adductor tendon injuries, frequently seen in sports that require rapid directional changes, typically stem from acute trauma or chronic overuse, complicating their management due to the high physical demands placed on the affected tissues. The chapter underscores the importance of a multidisciplinary approach to these injuries, combining conservative treatments like physical therapy and anti-inflammatory medications with innovative regenerative therapies such as platelet-rich plasma (PRP) injections and mesenchymal stem cell (MSC) therapy. Cartilage health is critical for maintaining hip joint function, and injuries often progress to degenerative conditions like osteoarthritis. The chapter reviews current treatment options, including microfracture, autologous chondrocyte implantation (ACI), and osteochondral grafting, alongside emerging strategies in tissue engineering and gene therapy. Emphasising the need for personalised treatment plans, the chapter advocates for integrating orthopedic expertise, advanced rehabilitation, and regenerative medicine to optimise recovery and prevent recurrence. Looking ahead, it explores the potential of regenerative therapies to enhance recovery times, improve the durability of repairs, and extend athletic careers, marking a significant advancement in the management of hip injuries and offering new hope for improved patient outcomes.

P. Lee (✉)
London Cartilage Clinic, London, UK

College of Health and Science, University of Lincoln, Lincoln, UK

MSK Doctors, ICRS Center of Excellence for Cartilage Regeneration & Joint Preservation, Grantham, UK
e-mail: leep@mskdoctors.com

T. Verma
MSK Doctors, ICRS Center of Excellence for Cartilage Regeneration & Joint Preservation, Grantham, UK
e-mail: tanvi@mskdoctors.com

P. Lee (ed.), *Musculoskeletal Regeneration Medicine*,
https://doi.org/10.1007/978-3-031-77456-0_18

Keywords Articular cartilage repair · Regeneration · Tendinopathy · Biomechanics · Microfracture

1 Introduction

Hip-related injuries are a prevalent concern among athletes, particularly those engaged in sports that require repetitive movements such as kicking, running, and sudden directional changes. The hip joint, due to its central role in weight-bearing and complex movements, is susceptible to a range of injuries that can significantly impair athletic performance and quality of life. This chapter will focus on three critical aspects of hip pathology: osteitis pubis, adductor tendon injuries, and cartilage regeneration. Understanding these conditions is essential for developing effective treatment strategies and optimising recovery for affected individuals.

1.1 Significance of Hip Injuries in Athletes

Athletes are particularly vulnerable to hip injuries due to the high physical demands of their sports, which often involve repetitive stress and acute trauma to the hip joint and surrounding structures. Conditions like osteitis pubis and adductor tendon injuries can lead to chronic pain and functional limitations if not properly managed. Additionally, cartilage damage, which can occur as a result of these injuries or independently, presents a significant challenge due to the limited regenerative capacity of cartilage tissue.

1.2 Focus on Osteitis Pubis, Adductor Tendon Injuries, and Cartilage Regeneration

Osteitis Pubis is an inflammatory condition affecting the pubic symphysis and the surrounding structures. It is often seen in athletes involved in sports requiring repetitive kicking or running, such as soccer, rugby, and Australian football. The condition is characterised by chronic groin pain, which can severely limit an athlete's ability to participate in sports. The pathophysiology of osteitis pubis is complex, involving biomechanical overload, stress reactions in the pelvic region, and, in some cases, an association with femoroacetabular impingement (FAI).

Adductor Tendon Injuries are another common issue among athletes, particularly those engaged in sports involving sudden changes in direction or rapid acceleration and deceleration. The adductor muscles play a crucial role in stabilising the hip joint and facilitating movements such as hip adduction and flexion. Injuries to these tendons, whether acute or due to chronic overuse, can result in significant pain and functional impairment. The intricate anatomy of the adductor tendons and their close relationship with the pubic symphysis often make diagnosis and treatment challenging.

Cartilage Regeneration in the context of hip injuries is a critical area of focus due to the essential role cartilage plays in joint function and its limited capacity for self-repair. Cartilage injuries, whether due to direct trauma or secondary to conditions like osteitis pubis and adductor tendon injuries, can lead to progressive joint degeneration and osteoarthritis. Advances in regenerative medicine, including techniques such as microfracture, autologous chondrocyte implantation (ACI), and the use of mesenchymal stem cells (MSCs), offer promising avenues for restoring cartilage integrity and function in the hip joint.

2 Osteitis Pubis

2.1 Definition and Epidemiology

Osteitis pubis is a non-infectious inflammatory condition affecting the pubic symphysis, a cartilaginous joint located between the left and right pubic bones at the front of the pelvis. First described by Edwin Beer in 1924 in patients undergoing suprapubic surgery, osteitis pubis has since been recognised in both postoperative patients and athletes, particularly those involved in sports requiring repetitive pelvic movements such as soccer, rugby, and running (Gomella and Mufarrij 2017). The condition is an uncommon but significant cause of suprapubic and lower abdominal pain, with prevalence rates ranging from 0.7% to 2.5% in patients following certain surgical procedures, and up to 80% in athletes reporting groin pain (Hiti et al. 2011; Phillips et al. 2016).

2.2 Pathophysiology

The pathophysiology of osteitis pubis is multifactorial, involving mechanical stress, potential low-grade infection, and vascular abnormalities.

Physics of Regeneration- Mechanical Stress and Trauma

Mechanical trauma is a primary factor in the development of osteitis pubis. The physics of the condition involves the transmission of forces through the pelvic girdle during activities such as running, kicking, and rapid changes in direction. These forces place repetitive stress on the pubic symphysis, leading to microtears and the eventual onset of inflammation.

In sports, particularly those requiring explosive movements, the adductor muscles generate significant forces that are transmitted through the pubic symphysis.

Over time, these forces can lead to the development of osteitis pubis, as the continuous stress and strain disrupt the normal biomechanics of the pelvic region. The balance between the forces generated by the abdominal muscles and those by the adductor muscles is crucial; any imbalance can exacerbate the stress on the pubic symphysis, leading to further microdamage and inflammation (Phillips et al. 2016).

Biology of Regeneration- Biological Responses

The biological response to the mechanical stress involves the activation of the body's innate healing processes. In osteitis pubis, the initial microdamage to the pubic symphysis triggers an inflammatory response. This involves the recruitment of immune cells, including macrophages and neutrophils, which release such as TNF-alpha and IL-1, promoting the recruitment of additional immune cells to the site of injury, amplifying the inflammatory response (Gomella and Mufarrij 2017).

Simultaneously, fibroblasts are activated to repair the damaged tissue by producing extracellular matrix (ECM) components such as collagen. This process is crucial for the regeneration of the pubic symphysis, as it restores the integrity of the tissue. However, in osteitis pubis, the chronic nature of the injury can lead to an excessive or prolonged inflammatory response, resulting in fibrosis and the formation of scar tissue, which can further impair the function of the pubic symphysis (Coventry and Mitchell 1961).

Additionally, angiogenesis, the formation of new blood vessels increases the blood flow to transport essential nutrients and oxygen to the injured tissue, supporting the healing process. However, if the regenerative process is dysregulated, it can lead to pathological changes, such as the formation of sclerotic bone or the calcification of soft tissues around the pubic symphysis (Gomella and Mufarrij 2017).

Chemistry of Regeneration—Infectious and Vascular Theories

From a chemical perspective, osteitis pubis may involve low-grade infections, particularly in postoperative patients. The presence of bacteria in the symphyseal region can lead to the production of bacterial toxins and the activation of local immune responses, resulting in further inflammation and tissue damage. The chemical environment of the pubic symphysis during infection is characterised by increased acidity (lower pH), which can impair the function of various enzymes involved in tissue repair and promote the activation of osteoclasts, leading to bone resorption (Kammerer-Doak et al. 1998).

During an infection, the chemical environment of the pubic symphysis becomes more acidic mainly due to immune cells like macrophages and neutrophils, which are highly active in response to infection, often shift to anaerobic metabolism in poorly oxygenated tissues, producing lactic acid as a byproduct and making their surroundings more acidic. Additionally, bacterial metabolism plays a role by fermenting carbohydrates, leading to the production of organic acids that further lower the pH in the infected tissue. Inflammation also contributes to this acidity through the accumulation of hydrogen ions (H+), which exacerbate the acidic conditions in the affected area, ultimately impairing tissue repair and promoting bone resorption.

Impact on Enzymes and Tissue Repair
A decrease in pH can significantly impact the efficiency of enzymes like collagenases and proteases, which are crucial for tissue repair. These enzymes, function optimally at a pH close to neutral (around 7.4), can become denatured or have their catalytic efficiency reduced in more acidic conditions, thereby slowing down the repair process. Additionally, the acidic environment can also disrupt the enzymatic pathways responsible for producing essential extracellular matrix (ECM) components, such as collagen and glycosaminoglycans. This disruption impairs the healing and regeneration of the pubic symphysis.

Osteoclast Activation and Bone Resorption
Osteoclasts are naturally adapted to function in an acidic environment, by creating a highly acidic microenvironment at the bone surface by secreting hydrogen ions through proton pumps. This acidity dissolves the mineral component of bone known as hydroxyapatite and breaks down the bone matrix. In an acidic extracellular environment, osteoclast activity is further enhanced, facilitating the breakdown of bone tissue and leading to increased bone resorption at the pubic symphysis. This process weakens the bone structure, contributing to the chronic pain and dysfunction associated with osteitis pubis. Moreover, the acidic conditions can disrupt the balance between bone resorption and formation, favouring resorption due to increased osteoclast activity. This imbalance results in the loss of bone density and structural integrity, exacerbating the condition.

Vascular factors also play a significant role in the pathophysiology of osteitis pubis. Theories suggest that venous congestion, possibly due to obstruction of the prostatic venous plexus, could lead to hyperemia and subsequent demineralisation of the pubic symphysis. The chemical signals involved in this process include various growth factors and cytokines that regulate blood vessel permeability and tissue remodeling. An imbalance in these chemical signals can contribute to the chronic nature of osteitis pubis, where the healing process is impaired by inadequate blood flow and nutrient delivery (Steinbach et al. 1955).

2.3 *Clinical Presentation*

The clinical presentation of osteitis pubis is typically characterised by gradual onset of suprapubic pain that may radiate to the lower abdomen, groin, or inner thighs. The pain is often aggravated by activities that increase intra-abdominal pressure or stress the pelvic girdle, such as walking, running, coughing, or climbing stairs. Patients may also exhibit a "waddling" gait, a compensatory mechanism to minimise movement of the painful area (Gomella and Mufarrij 2017).

Physical Examination
On physical examination, tenderness over the pubic symphysis is a common finding. This tenderness may be elicited by palpation or by performing specific manoeuvres such as the "pubic spring" test or the "lateral compression" test, both of which

can reproduce the pain associated with osteitis pubis. The pubic spring test involves applying downward pressure on the pubic rami to elicit pain at the symphysis, while the lateral compression test involves applying pressure on the iliac crest to compress the pelvis laterally, also eliciting pain at the symphysis (Gomella and Mufarrij 2017).

Differential Diagnosis
Given the overlap of symptoms with other pelvic conditions, conditions such as sports hernias, inguinal hernias, and osteomyelitis must be considered. Osteomyelitis, in particular, can present with similar symptoms but typically has a more acute onset and is associated with systemic signs of infection, such as fever and malaise (Pauli et al. 2002). Laboratory tests, including blood cultures and inflammatory markers, can aid in distinguishing these conditions.

2.4 Diagnosis

The diagnosis of osteitis pubis is primarily clinical, supported by imaging studies and, when necessary, laboratory tests.

Imaging Modalities

- **X-Rays**: Conventional radiographs are often the first step in imaging and can reveal sclerosis, osteophytes, and widening of the pubic symphysis. However, X-rays are less sensitive in the early stages of the disease when these changes may not yet be visible (Gomella and Mufarrij 2017).
- **Magnetic Resonance Imaging (MRI)**: MRI is the gold standard for diagnosing osteitis pubis, as it provides detailed images of soft tissue and bone marrow. Acute cases often show bone marrow edema, periarticular inflammation, and fluid accumulation in the pubic symphysis, while chronic cases may show subchondral sclerosis and osteophyte formation. MRI is also useful in differentiating osteitis pubis from osteomyelitis, which can present similarly (Kunduracioglu et al. 2007).
- **Scintigraphy and Symphysography**: Scintigraphy can detect focal radionuclide uptake at the pubic symphysis, indicating inflammation. Symphysography, though less commonly used, involves injecting contrast directly into the symphyseal joint to confirm the diagnosis by provoking symptoms. This technique can also be therapeutic if corticosteroids are injected during the procedure (Gomella and Mufarrij 2017).

Laboratory Tests
Laboratory tests are typically reserved for cases where infection is suspected. These may include blood cultures, urine cultures, and joint aspiration for culture to rule out osteomyelitis. Elevated inflammatory markers, such as erythrocyte sedimentation rate (ESR) or C-reactive protein (CRP), may be present but are nonspecific (Gomella and Mufarrij 2017).

2.5 *Management and Treatment*

Conservative Management

Initial treatment is conservative, focusing on reducing mechanical stress on the pubic symphysis and managing inflammation. Rest and activity modification are crucial to allow the inflamed tissues to heal. Patients may be advised to avoid activities that exacerbate symptoms, and assistive devices such as crutches may be used to alleviate pressure on the pelvis. Physical therapy plays a key role in this phase, with exercises designed to strengthen the abdominal and adductor muscles, improve flexibility, and correct biomechanical imbalances that contribute to the condition (Mohammad and Elsais 2018).

Anti-Inflammatory Treatments

Nonsteroidal anti-inflammatory drugs (NSAIDs) are commonly prescribed to reduce inflammation and alleviate pain. These drugs work by inhibiting the cyclooxygenase (COX) enzymes, which are involved in the synthesis of pro-inflammatory prostaglandins. In cases of severe inflammation, a short course of corticosteroids may be considered, as they have a more potent anti-inflammatory effect by suppressing the activity of the immune system and reducing the production of inflammatory cytokines (Kavroudakis et al. 2011).

Corticosteroid Injections

For patients who do not respond to conservative treatments, corticosteroid injections into the pubic symphysis can provide significant relief. The injection not only reduces inflammation but also modulates the local immune response, promoting tissue regeneration. Symphysography, combined with corticosteroid injection, has been shown to be both diagnostic and therapeutic, offering rapid pain reduction and potentially accelerating the healing process (Holt et al. 1995).

Regenerative Therapies

Therapies, such as platelet-rich plasma (PRP) injections, involves concentrating platelets from the patient's blood and injecting them into the affected area. Platelets release growth factors that promote tissue repair and regeneration, enhancing the healing of damaged cartilage and soft tissues (Park and Sussman 2020). This approach leverages the body's natural healing mechanisms to restore the integrity of the pubic symphysis.

Surgical Interventions

Surgery is reserved for cases where conservative and regenerative treatments fail. Surgical options include curettage of the pubic symphysis, arthrodesis (fusion), and resection of the symphysis. However, these procedures carry risks, including pelvic instability and potential complications that may require further surgical intervention. In complex cases, especially those involving postoperative osteitis pubis, earlier surgical intervention may be necessary after thorough patient counselling (Mehin et al. 2006).

2.6 *Rehabilitation and Regeneration*

Rehabilitation following treatment for osteitis pubis is crucial for restoring function and preventing recurrence. The goals of rehabilitation are to enhance the eccentric strength of the abdominal and adductor muscles, improve pelvic stability, and ensure proper alignment to minimise mechanical stress on the pubic symphysis.

Biochemical Support
During rehabilitation, maintaining an optimal biochemical environment is critical for tissue regeneration. This includes ensuring adequate nutrition, hydration, and possibly the use of supplements that support collagen synthesis and reduce oxidative stress. Modalities such as therapeutic ultrasound and electrical stimulation can enhance blood flow and tissue repair, further supporting the regenerative process (Nishiwaki et al. 2017).

3 Adductor Tendon Injuries and Regeneration

3.1 *Anatomy and Function of Adductor Tendons*

The adductor tendons, consisting of the adductor longus, adductor brevis, and adductor magnus, play a critical role in the movement and stabilisation of the hip joint. These tendons originate from the pubis and ischium bones and insert along the femur. The primary function of the adductor tendons is to adduct the thigh, pulling it toward the body's midline. Additionally, they assist in hip flexion and rotation, contributing to the overall stability of the pelvis during activities such as walking, running, and changing direction.

Given their anatomical position and function, the adductor tendons are frequently subjected to significant mechanical stress, particularly during high-impact or repetitive activities. This makes them vulnerable to injury, especially in athletes who engage in sports that require sudden directional changes or explosive movements, such as soccer, basketball, and hockey. The adductor longus tendon, in particular, is often implicated in these injuries due to its involvement in powerful adduction and flexion movements.

3.2 *Pathophysiology of Adductor Tendon Injuries*

Adductor tendon injuries typically occur due to acute trauma or chronic overuse. Acute injuries, such as tendon tears or strains, usually result from a sudden, forceful movement that overwhelms the tendon's capacity to withstand tension. Chronic injuries, on the other hand, develop over time due to repetitive microtrauma that leads to the degeneration of the tendon fibers, a condition known as tendinopathy.

Physics of Regeneration- Mechanical Factors

The physics behind adductor tendon injuries involve the application of excessive force or repetitive stress on the tendons. During high-intensity activities, the adductor muscles generate significant force to stabilise the pelvis and control leg movements. When these forces exceed the tensile strength of the tendon fibers, it can lead to microtears or, in severe cases, a complete rupture. Over time, repetitive loading without adequate recovery can cause the tendons to weaken, making them more susceptible to injury (Perrone et al. 2021).

Biology of Regeneration- Biological Response to Injury

Following an adductor tendon injury, the body initiates a complex biological response aimed at repairing the damaged tissue. The initial inflammatory phase is marked by the infiltration of immune cells, which clear away damaged tissue and release cytokines that attract fibroblasts to the injury site. These fibroblasts play a critical role in the regenerative process by producing collagen, the primary structural protein in tendons.

In the subsequent proliferative phase, fibroblasts proliferate and synthesise new collagen fibers, which are initially disorganised and weak. Over time, during the remodeling phase, these collagen fibers are realigned and strengthened in response to mechanical loading, ultimately restoring the tendon's tensile strength and functionality. However, in cases of chronic tendinopathy, this process can be impaired, leading to persistent pain and dysfunction due to the accumulation of disorganised scar tissue and incomplete healing (Mohammad and Elsais 2018).

Chemistry of Regeneration- Chemical Environment and Tendon Healing

The chemical environment within the injured tendon is crucial for effective healing. Inflammatory cytokines, such as TNF-alpha and IL-1, promotes the initial inflammatory response but, if persistent, can lead to chronic inflammation and impaired healing. The local production of matrix metalloproteinases (MMPs) is another important factor, as these enzymes break down the extracellular matrix components during tissue remodeling. While MMP activity is necessary for removing damaged collagen, excessive activity can degrade newly formed tissue, hindering the healing process.

The balance of these chemical signals, along with the availability of growth factors such as transforming growth factor-beta (TGF-β) and vascular endothelial growth factor (VEGF), determines the success of tendon regeneration. These growth factors promote collagen synthesis, angiogenesis, and the recruitment of stem cells, all of which are essential for restoring the structure and function of the injured tendon (Nishiwaki et al. 2017).

3.3 Clinical Presentation and Diagnosis

Patients with adductor tendon injuries typically present with groin pain that worsens with activity, particularly movements that involve adduction or flexion of the hip. The pain may be sharp and localised to the pubic bone or the insertion site of the tendon on the femur, and it may be accompanied by swelling or bruising in the acute phase of injury.

Physical Examination
During a physical examination, tenderness along the adductor tendon is often noted, particularly at the pubic bone or at the musculotendinous junction. Pain may be elicited during resisted adduction or when the patient attempts to squeeze the legs together, a manoeuvre known as the "adductor squeeze test." In chronic cases, there may be evidence of muscle weakness or atrophy due to disuse or ongoing pain.

Imaging Studies
Imaging studies, such as ultrasound or MRI, are essential for confirming the diagnosis and assessing the extent of the injury. Ultrasound can be particularly useful for visualising tendon tears, while MRI provides detailed images of the soft tissues, allowing for the evaluation of tendon integrity, muscle involvement, and any associated edema or inflammation (Kunduracioglu et al. 2007).

3.4 Management and Treatment

Conservative Management
Conservative treatment is the first line of management and typically includes rest, activity modification, and the use of NSAIDs to reduce pain and inflammation. Physical therapy is a critical component of treatment, focusing on exercises that strengthen the adductor muscles, improve flexibility, and correct any biomechanical imbalances. The goal is to gradually restore the strength and function of the tendon while minimising the risk of re-injury (Mohammad and Elsais 2018).

Regenerative Therapies (Biology and Chemistry of Regeneration)
Emerging regenerative therapies, such as platelet-rich plasma (PRP) injections, have shown promise in treating adductor tendon injuries, particularly in cases of chronic tendinopathy. PRP therapy involves concentrating the patient's platelets and injecting them into the injured tendon. The growth factors released by the platelets, including PDGF and TGF-β, stimulate collagen synthesis, promote angiogenesis, and enhance the overall healing process.

Another regenerative approach involves the use of mesenchymal stem cells (MSCs), which have the potential to differentiate into tendon-like cells and contribute to tissue regeneration. The chemical environment of the injured tendon, enriched by these growth factors and stem cells, can shift the balance towards repair and regeneration, reducing inflammation and promoting the formation of healthy tendon tissue (Nishiwaki et al. 2017).

Surgical Intervention
Surgery is typically reserved for cases where conservative management fails or in the presence of a complete tendon rupture. Surgical options include tendon repair, where the torn ends of the tendon are sutured back together, or tendon reconstruction using grafts. In some cases, a tenotomy, where the damaged tendon is partially released, may be performed to alleviate pain and restore function.

3.5 Rehabilitation and Prevention

Rehabilitation following an adductor tendon injury is essential for a full recovery and involves a structured physical therapy program designed to restore strength, flexibility, and coordination. The rehabilitation process is gradual, starting with gentle range-of-motion exercises and progressing to more intense strengthening and functional exercises as the tendon heals.

Prevention of Re-Injury
To prevent re-injury, it is important to address any underlying biomechanical issues, such as muscle imbalances or poor technique, that may have contributed to the initial injury. Athletes should be educated on proper warm-up and stretching techniques, as well as the importance of adequate rest and recovery between training sessions to allow for full tendon repair and adaptation (Mohammad and Elsais 2018).

4 Cartilage Regeneration in Hip Injuries

4.1 Importance of Cartilage Health in the Hip Joint

Cartilage plays a vital role in maintaining the smooth, pain-free movement of the hip joint, which is a ball-and-socket joint where the femoral head fits into the acetabulum of the pelvis. Articular cartilage, a specialised type of hyaline cartilage, covers the surfaces of both the femoral head and the acetabulum, providing a low-friction interface that enables efficient joint movement and acts as a shock absorber during weight-bearing activities.

Cartilage Function and Hip Joint Biomechanics
From a physics perspective, the biomechanical function of cartilage is crucial for distributing mechanical loads across the joint. Cartilage's viscoelastic properties allow it to deform under load, absorbing impact and reducing stress on the underlying bone. This mechanical behaviour is governed by the principles of stress distribution and energy absorption, which help prevent damage to both the cartilage and

subchondral bone. The elasticity of the cartilage allows it to return to its original shape after deformation, maintaining joint integrity during repetitive movements such as walking, running, and jumping.

Consequences of Cartilage Damage
When cartilage is damaged, the biomechanical balance of the joint is disrupted, leading to increased stress on the surrounding tissues. From a biological standpoint, cartilage has limited self-repair capabilities due to its avascular nature, which means it lacks a direct blood supply and relies on diffusion for nutrient delivery and waste removal. This limitation significantly impairs the regenerative capacity of cartilage, causing injuries to heal slowly and often inadequately.

Chemically, the composition of cartilage, which includes a high concentration of proteoglycans and collagen fibers, is essential for its load-bearing function. Damage to this extracellular matrix (ECM) disrupts the chemical integrity of cartilage, leading to a loss of its viscoelastic properties. Over time, untreated cartilage injuries can progress to osteoarthritis, where the degradation of cartilage and the underlying bone leads to chronic pain, stiffness, and reduced mobility (Lepage et al. 2019).

4.2 Pathophysiology of Cartilage Injuries

Common Types of Cartilage Injuries in the Hip
Chondral lesions and labral tears are the most common types of cartilage injuries in the hip. Chondral lesions involve the degradation or tearing of the articular cartilage, ranging from superficial fissures to full-thickness defects exposing the subchondral bone. Labral tears affect the fibrocartilaginous labrum that encircles the acetabulum, contributing to joint stability and the distribution of synovial fluid, which lubricates the cartilage and provides nutrients.

Physical Load
These injuries often result from mechanical overload, where excessive forces applied to the hip joint exceed the tensile strength of the cartilage and labrum. This overload can occur acutely, such as in traumatic events, or chronically, due to repetitive microtrauma from activities like running or jumping. The altered biomechanics following a labral tear or chondral lesion can further exacerbate cartilage degradation by disrupting the even distribution of joint forces, leading to accelerated wear and tear (Migliorini et al. 2024).

Cellular and Molecular Mechanisms Leading to Cartilage Degeneration
At the cellular level, cartilage injuries trigger a cascade of **biological** events that contribute to degeneration. Chondrocytes, the cells responsible for maintaining the ECM, are sensitive to mechanical and chemical stress. Injury or excessive mechanical load can lead to chondrocyte apoptosis (programmed cell death), which reduces the cell population necessary for maintaining cartilage integrity.

Chemically, injury to cartilage results in the release of inflammatory cytokines such as IL-1 and TNF-α. These cytokines initiate a **biochemical** response that includes the upregulation of matrix metalloproteinases (MMPs), enzymes that degrade ECM components like collagen and proteoglycans. The breakdown of these molecules weakens the cartilage structure, reducing its ability to bear loads and resist shear forces.

Moreover, oxidative stress from reactive oxygen species (ROS) production further damages chondrocytes and ECM molecules, compounding the degenerative process. This disruption in the balance between ECM synthesis and degradation leads to the thinning of cartilage, loss of joint function, and, ultimately, the progression to osteoarthritis (Lepage et al. 2019).

4.3 Current Treatment Approaches

Given the limited regenerative capacity of cartilage, various treatment approaches have been developed to manage cartilage injuries in the hip. These approaches range from surgical interventions to biological and chemical strategies aimed at promoting cartilage regeneration.

Marrow Stimulation Techniques: Microfracture and Drilling
Marrow stimulation techniques are commonly employed to treat small chondral defects by exploiting the body's innate healing response. These procedures create microchannels in the subchondral bone, allowing bone marrow-derived mesenchymal stem cells (MSCs) to migrate to the injury site.

- **Microfracture**: In microfracture surgery, a pointed instrument is used to create small holes in the subchondral bone. From a **physics** perspective, the mechanical disruption of the bone surface induces a healing response by releasing MSCs and growth factors into the defect. However, the resulting repair tissue is fibrocartilage, which is mechanically inferior to hyaline cartilage in terms of elasticity and load-bearing capacity (Hotham and Malviya 2018).
- **Drilling**: Similar to microfracture, drilling involves creating holes in the subchondral bone to stimulate a regenerative response. The key **biological** principle here is the recruitment of stem cells and the promotion of a healing environment conducive to cartilage repair. However, the chemical composition of fibrocartilage differs from that of native hyaline cartilage, often leading to suboptimal long-term outcomes.

Autologous Chondrocyte Implantation (ACI)
Autologous chondrocyte implantation (ACI) is a sophisticated technique that involves harvesting a small sample of cartilage, expanding chondrocytes in vitro, and re-implanting them into the cartilage defect. The aim is to regenerate hyaline-like cartilage, which more closely resembles the original tissue.

- **Biology of ACI**: The process relies on the ability of chondrocytes to proliferate and produce ECM components like collagen and proteoglycans when provided with the right biological signals. The chemical environment within the implant site is crucial for the survival and function of these cells, requiring an optimal balance of nutrients, oxygen, and growth factors (Hotham and Malviya 2018).
- **Physics of Cartilage Repair**: The mechanical environment post-implantation must be carefully managed to avoid excessive load that could disrupt the integration of the new cartilage with the surrounding tissue. This involves a period of restricted weight-bearing and controlled movement to ensure proper healing and alignment.

Osteochondral Autograft and Allograft Transplantation
Osteochondral grafting involves the transplantation of cartilage and underlying bone to repair defects.

- **Osteochondral Autograft Transplantation (OAT)**: Autografts are taken from a non-weight-bearing part of the joint and transplanted into the defect. This method restores both the cartilage surface and the subchondral bone, providing a more durable and biomechanically appropriate repair.
- **Physics of Graft Integration**: The success of the graft depends on proper alignment and mechanical stability. The graft must integrate with the host tissue without significant gaps or misalignment, which could lead to uneven stress distribution and graft failure (Hotham and Malviya 2018).
- **Osteochondral Allograft Transplantation**: Allografts from donors are used for larger defects. While this approach allows for the treatment of extensive damage, the chemical challenges include the risk of immune rejection and the potential for disease transmission, making the selection of suitable donors and careful immunological management essential (Hotham and Malviya 2018).

4.4 Emerging Therapies and Future Directions

Advancements in biotechnology and regenerative medicine are paving the way for novel approaches to cartilage repair, focusing on enhancing the body's natural healing processes and improving the quality of regenerated tissue.

Cell-Based Therapies: Mesenchymal Stem Cells (MSCs) for Cartilage Repair
MSCs are multipotent stem cells capable of differentiating into chondrocytes and other cell types, making them a promising option for cartilage regeneration.

- **Biology of MSCs**: MSCs can be harvested from bone marrow, adipose tissue, or synovial membrane and are known for their ability to home to sites of injury, modulate the immune response, and secrete bioactive molecules that promote tissue repair.

- **Chemical Environment**: The success of MSC therapy depends on the local chemical environment, which must support chondrogenesis (the formation of cartilage). This includes the presence of growth factors like transforming growth factor-beta (TGF-β) and bone morphogenetic proteins (BMPs), which guide MSC differentiation into chondrocytes. Additionally, the pH, oxygen tension, and nutrient supply in the tissue play critical roles in MSC survival and function (Nishiwaki et al. 2017).

Scaffolding and Tissue Engineering Advancements
Scaffolds provide a three-dimensional structure that supports the growth and organisation of new cartilage tissue. These scaffolds can be composed of natural or synthetic materials designed to mimic the mechanical properties of native cartilage.

- **Physics of Scaffold Design**: The physical properties of scaffolds, such as porosity, stiffness, and degradation rate, are crucial for their success. These properties must be carefully tuned to provide the right mechanical environment that supports cell attachment, proliferation, and ECM production.
- **Biological Integration**: Scaffolds are often combined with MSCs or chondrocytes to enhance tissue regeneration. The interaction between cells and the scaffold material is key to the successful integration and formation of new cartilage tissue.
- **Chemical Functionalisation**: Scaffolds can be chemically modified to release growth factors or other signaling molecules that promote tissue regeneration. For example, scaffolds can be loaded with VEGF to enhance blood vessel formation or with TGF-β to stimulate chondrocyte activity (Nishiwaki et al. 2017).

Potential of Gene Therapy and Other Innovative Treatments
Gene therapy represents a cutting-edge approach to cartilage repair, aiming to modify the expression of specific genes to promote cartilage regeneration.

- **Biology and Chemistry of Gene Therapy**: Gene therapy can be used to deliver genes encoding for growth factors, such as TGF-β or IGF-1, directly to the site of injury, enhancing the local production of proteins that stimulate cartilage repair. The delivery vectors, whether viral or non-viral, must be carefully designed to ensure efficient gene transfer while minimising immune responses or off-target effects.
- **Innovative Approaches**: Other innovative treatments being explored include the use of small molecules to modulate signaling pathways involved in cartilage regeneration, as well as the application of bioprinting technology to create custom-designed cartilage implants that match the patient's specific anatomical and biomechanical needs.

4.5 Rehabilitation After Cartilage Repair

Rehabilitation Protocols Following Cartilage Repair Surgeries

Rehabilitation is a critical component of recovery following cartilage repair surgeries. The primary goal is to protect the repaired tissue while gradually restoring joint function and strength.

- **Initial Phase (Immobilisation and Protection)**: The early phase of rehabilitation focuses on minimising stress on the repaired cartilage. Patients may be advised to use crutches or braces to avoid weight-bearing on the affected joint. Gentle range-of-motion exercises may be initiated to prevent joint stiffness and maintain flexibility.
- **Intermediate Phase (Controlled Loading and Strengthening)**: As healing progresses, controlled loading exercises are introduced to stimulate the maturation and integration of the new cartilage. This phase involves low-impact activities such as stationary cycling and aquatic therapy, which provide mechanical stimulation without excessive stress.
- **Advanced Phase (Functional Training and Return to Activity)**: The final phase of rehabilitation aims to restore full strength, flexibility, and proprioception. Functional training exercises that mimic the demands of daily activities and sports are incorporated. Gradual reintroduction of weight-bearing and high-impact activities is done under careful supervision to ensure the cartilage repair is robust and capable of withstanding normal joint loads.

Strategies to Enhance Cartilage Healing and Prevent Further Degeneration

Successful rehabilitation not only focuses on healing but also on preventing further degeneration of the joint. This can be achieved through:

- **Biomechanical Optimisation**: Addressing underlying biomechanical issues, such as improper gait or muscle imbalances, is crucial for preventing undue stress on the hip joint.
- **Nutritional Support and Supplements**: Nutritional strategies, including adequate protein intake and the use of supplements like glucosamine and chondroitin, can support cartilage health. Antioxidants may also help reduce oxidative stress, which contributes to cartilage degradation.
- **Long-term Maintenance**: Patients are encouraged to adopt a long-term maintenance program that includes regular low-impact exercise, weight management, and periodic check-ups to monitor joint health.

Surgical Technique

The *Bikini hip* incision is an advanced surgical technique designed for anterior hip replacement, providing a minimally invasive approach with enhanced cosmetic results. This method requires precision and expertise due to its detailed dissection pathway and reduced surgical field. A minimally invasive surgical approach performed via a cosmetically discreet incision in the groin's natural skin crease, approximately 6–10 cm long. This incision is strategically placed along Langer's lines to reduce visible scarring and improve healing. The procedure employs the direct anterior approach (DAA), which provides access to the hip joint through an intermuscular plane. The dissection occurs between the tensor fasciae latae and sartorius muscles laterally and the rectus femoris medially. This incision is angled to align parallel to the inguinal ligament and positioned just lateral to the anterior superior iliac spine (ASIS). Precision is paramount during sharp dissection to prevent trauma to the dermal layers, ensuring optimal cosmetic outcomes. Subcutaneous dissection follows, involving the gentle separation of subcutaneous fat to expose the underlying fascia of the tensor fasciae latae, maintaining haemostasis to minimise postoperative bruising and optimise visualisation. Accessing the muscle and fascia involves identifying the interval between the tensor fasciae latae and sartorius muscles. The fascia over the tensor fasciae latae is incised, and the sartorius is retracted medially, with the tensor fasciae latae carefully retracted laterally. This step is executed with minimal disruption to the lateral femoral cutaneous nerve to prevent postoperative paraesthesia or numbness. The anterior capsule of the hip joint is then exposed, and a T-shaped or H-shaped capsulotomy is performed, preserving as much of the capsule as possible to facilitate secure closure and enhance joint stability. This muscle-sparing technique avoids cutting or detaching muscles, preserving the structural integrity and function of the surrounding tissues. Specialised retractors and fluoroscopic imaging facilitate visualisation and precise implantation of prosthetic components, enhancing joint alignment and stability while reducing soft tissue disruption.

Complications

Injury to the lateral femoral cutaneous nerve (LFCN) is a common risk, potentially causing transient or persistent numbness and paraesthesia of the lateral thigh. The constrained surgical field increases the risk of intraoperative fractures, particularly of the femur or acetabulum, and component malpositioning, which may result in instability, increased wear, or loosening.

Wound healing complications, such as infection or delayed closure, can occur due to the groin incision's susceptibility to tension and moisture. Rarely, neurovascular injury may lead to bleeding or haematoma. Additionally, the learning curve for this technique poses a significant challenge, as less experienced surgeons are more prone to prolonged operating times and technical errors. Early mobilisation and meticulous post-operative care are essential for reducing risks and optimising recovery. While anterior approaches like the *Bikini hip* tend to lower the risk of dislocation compared to other methods, anterior dislocation remains a potential complication that must be managed to ensure joint stability postoperatively.

ASIA Syndrome

ASIA Syndrome (Adductor Synthesis Inguinal Axis Syndrome) is a comprehensive diagnostic framework used to describe a specific set of related pathologies affecting the pelvic and groin region, particularly in athletes. The syndrome connects osteitis pubis, adductor tendinopathy, and Gilmore's groin (sports hernia), focusing on the interplay between these conditions through the biomechanics of the adductor muscles, inguinal structures, and pubic symphysis. It was developed to explain groin pain as a multifactorial problem rather than as isolated issues, offering a holistic perspective for diagnosis and treatment.

Pathophysiology and Mechanism

ASIA Syndrome emphasises the biomechanical and anatomical interrelationship between the adductor muscles, inguinal ligament, and pubic symphysis. This area acts as a pivotal axis during athletic movements that involve running, cutting, kicking, and twisting motions. The syndrome highlights how repetitive stresses, muscle imbalances, and overuse can lead to microtrauma, particularly in the adductor longus and rectus abdominis muscles, which anchor onto the pubic bone.

Biomechanics

The pubic symphysis functions as a fulcrum that supports the forces generated by the adductor muscles and abdominal muscles. During dynamic movements, such as running and kicking, these muscle groups pull in opposite directions, causing repeated microtrauma to the pubic symphysis. This continuous strain results in pain, inflammation, and eventually conditions like osteitis pubis. The biomechanics of the pelvis play a crucial role in the development of ASIA syndrome, with improper muscle coordination or imbalance exacerbating the condition.

Clinical Presentation

Patients with ASIA Syndrome typically present with chronic groin pain that can range from mild discomfort to debilitating pain. The pain may be diffuse or localised to specific regions, particularly the pubic symphysis, adductor tendons, or the lower abdomen. This condition is common in athletes participating in sports that require sudden, explosive movements, such as soccer, rugby, and ice hockey, where twisting, kicking, and turning manoeuvres put significant strain on the pelvic region.

Key Symptoms

1. **Persistent Groin Pain**: The hallmark of ASIA syndrome, often exacerbated by physical activity. Pain may radiate to the inner thigh or lower abdominal region. Pain is typically described as:
    - **Sharp or stabbing** during sudden movements (e.g., kicking, sprinting).
    - **Aching or dull** during rest or prolonged periods of standing or sitting.
2. **Exacerbation with Specific Activities**:
    - Activities involving hip adduction (bringing the legs toward the midline) or flexion of the trunk, such as sit-ups, often aggravate the pain.
    - Kicking or any forceful movement of the leg involving twisting of the pelvis typically increases the pain.
3. **Loss of Flexibility**: A gradual loss of flexibility in the hip and groin region is commonly reported along with the loss of mobility due to pain. The reduction in flexibility may be related to muscle tightness or inflammation in the surrounding tendons and ligaments.

4. **Tenderness and Palpation Pain**:
    - Tenderness can usually be elicited by palpating the pubic symphysis or the adductor tendons.
    - In the case of Gilmore's groin, tenderness over the rectus abdominis insertion or the inguinal ligament may also be present.

Diagnosis
Due to the complexity and overlap of conditions like osteitis pubis, adductor tendinopathy, and Gilmore's groin, a thorough diagnostic process is essential.

Key Diagnostic Steps:

1. **Detailed Patient History**:
    - Start by documenting the patient's sports activities and history of groin pain, including onset and duration of symptoms. Acute cases may involve a sudden traumatic event, whereas chronic cases often develop gradually due to overuse.
    - Note for specific movements or actions that aggravate their pain, such as sprinting, kicking, or changing direction suddenly. The pattern of pain during and after activity can provide important clues.
    - Inquire about any previous injuries or treatments related to the groin, hips, lower back, or abdomen, as these may contribute to the current symptoms.
2. **Physical Examination**:
    - **Palpation**: Carefully palpate the pubic symphysis, adductor tendons, and inguinal region to identify areas of tenderness or swelling.
    - **Functional Tests**:
        - **Resisted hip adduction**: The patient attempts to bring their legs together against resistance. Pain during this manoeuvre suggests involvement of the adductors, especially the adductor longus tendon.
        - **Resisted sit-up**: Pain during a sit-up or similar trunk flexion movement indicates abdominal muscle involvement, which may point to Gilmore's groin.
        - **Range of motion assessment**: Evaluate hip flexibility, looking for restrictions in internal rotation or abduction.
3. **Imaging**:
    - **Magnetic Resonance Imaging (MRI)**: Gold Standard:
        - Oedema (swelling) and microtrauma in the pubic symphysis.
        - Tendinopathy of the adductors or rectus abdominis.
        - Soft tissue inflammation around the pubic region.
    - **Ultrasound**: This is a useful and less expensive tool to assess soft tissue integrity, especially in identifying adductor muscle tears, tendinopathy, and inflammation around the pubic symphysis.

- **X-ray**: While not typically diagnostic for soft tissue injuries, an X-ray may be used to rule out pelvic fractures or joint pathology like osteoarthritis that could be contributing to the patient's symptoms.

Diagnostic Challenges

ASIA syndrome is often misdiagnosed or underdiagnosed due to the overlapping symptoms with other groin-related conditions. A comprehensive diagnostic approach using a combination of physical examination, patient history, and imaging is crucial for a correct diagnosis and for differentiating ASIA syndrome from isolated groin injuries.

Treatment

The treatment of ASIA syndrome is multimodal and tailored to the severity and chronicity of symptoms. The therapeutic approach integrates conservative management, regenerative therapies, and, in severe cases, surgical intervention.

Conservative Management

1. **Activity Modification**:
   - Avoiding activities that exacerbate symptoms, such as high-impact sports, kicking, sprinting, or twisting movements. Tailored modifications should focus on minimising repetitive mechanical stresses on the pubic symphysis and associated structures while maintaining overall physical fitness with low-impact activities like swimming or cycling.
2. **Pain Management**:
   - **Nonsteroidal Anti-Inflammatory Drugs (NSAIDs)**: Prescribed for their role in managing pain and reducing inflammation around the pubic symphysis and tendons.
   - **Corticosteroid Injections**: Ultrasound-guided injections into the adductor tendon or pubic symphysis can alleviate local inflammation and pain. However, their use is typically limited to avoid potential tendon weakening with repeated injections.

Surgical Intervention

Surgical management is reserved for cases that fail conservative and regenerative treatments, particularly in chronic, severe presentations. Surgical options include:

- **Adductor Tenotomy**: Partial release of the adductor longus tendon to alleviate tension and reduce pubic symphysis stress.
- **Inguinal Ligament Repair**: Repair or reinforcement of the inguinal canal structures in cases involving Gilmore's groin.
- **Pubic Symphysis Debridement or Stabilisation**: For severe osteitis pubis, surgical debridement of inflamed tissue or fusion of the symphysis may be indicated.

Regenerative Treatments

These treatments aim to enhance tissue healing and restore function by leveraging the body's natural repair mechanisms.

1. **Platelet-Rich Plasma (PRP)**:
2. **Stem Cell Therapy**:

   - Mesenchymal Stem Cells (MSCs) derived from adipose tissue or bone marrow are an emerging option as these cells have the capacity to differentiate into a variety of tissue types. In some cases, stem cells are combined with platelet-rich plasma to further enhance the healing response.
   - Fat-derived stem cells (ADSCs) are a common source of regenerative cells due to their ease of harvest and high yield. ADSCs offer an anti-inflammatory effect and may promote faster recovery in tendon and ligament injuries.

3. **Hyaluronic Acid (HA)**:

   - By lubricating the joint and reducing friction, HA can improve joint mobility and reduce pain, particularly in cases where the pubic symphysis exhibits wear and tear from repetitive stress.

The prognosis of ASIA Syndrome depends on early diagnosis and adherence to treatment protocols. While most patients recover with conservative management, delayed treatment or recurrent episodes may lead to chronic symptoms and prolonged recovery.

5 Conclusion

In this chapter, we explored the complexities of hip injuries, with a particular focus on osteitis pubis, adductor tendon, and cartilage regeneration. These conditions are prevalent in athletes and can significantly impair mobility, causing chronic pain and long-term disability if not managed effectively. Cartilage health is critical to the function and longevity of the hip joint. Cartilage injuries, which often progress to osteoarthritis, present significant challenges due to the tissue's limited regenerative capacity. Current treatment strategies, including marrow stimulation techniques like microfracture, autologous chondrocyte implantation (ACI), and osteochondral grafting, aim to restore cartilage integrity and function. Emerging therapies, such as mesenchymal stem cells (MSCs), scaffolding advancements, and gene therapy, hold promise for more effective and durable cartilage repair, offering new hope for athletes and individuals with cartilage damage. The management of hip injuries requires a multidisciplinary approach that integrates orthopedic, physical therapy, and regenerative medicine expertise. By combining these disciplines, treatment plans can be tailored to the specific needs of each patient, optimising recovery and preventing recurrence. Looking forward, the future of hip injury management lies in the continued development of regenerative therapies and personalised treatment plans. Advances in biotechnology, such as tissue engineering and gene editing, have the potential to revolutionise how we approach cartilage repair and tendon regeneration. For athletes, these innovations could mean faster recovery times, more durable repairs, and a greater likelihood of returning to their pre-injury levels of performance. By focusing on prevention, early detection, and cutting-edge treatment modalities, we can improve the quality of life for individuals suffering from hip injuries and extend the careers of athletes who rely on the optimal function of their joints. The integration of these advancements into clinical practice represents the next frontier in the effective management of hip injuries.

References

Coventry MB, Mitchell WC. Osteitis pubis: observations based on a study of 45 patients. JAMA. 1961;178(9):898–905.

Gomella P, Mufarrij P. Osteitis pubis: a rare cause of suprapubic pain. Rev Urol. 2017;19(3):156–63. https://doi.org/10.3909/riu0767.

Hiti C, Stevens K, Jamati M, Garza D, Matheson G. Athletic osteitis pubis. Sports Med. 2011;41(5):361–76. https://doi.org/10.2165/11586820-000000000-00000.

Holt MA, Keene JS, Graf BK, Helwig DC. Treatment of osteitis pubis in athletes: results of corticosteroid injections. Am J Sports Med. 1995;23(5):601–6. https://doi.org/10.1177/036354659502300512.

Hotham WE, Malviya A. A systematic review of surgical methods to restore articular cartilage in the hip. J Hip Preserv Surg. 2018;5(1):3–10. https://doi.org/10.1093/jhps/hnx044.

Kammerer-Doak DN, Cornella JL, Magrina JF. Osteitis pubis after Marshall-Marchetti-Krantz urethropexy: a pubic osteomyelitis. Am J Obstetr Gynecol. 1998;179(3):586–90. https://doi.org/10.1016/S0002-9378(98)70051-0.

Kavroudakis E, Karampinas PK, Evangelopoulos DS, Vlamis J. Treatment of osteitis pubis in non-athlete female patients. Open Orthopaed J. 2011;5(1):331–4. https://doi.org/10.2174/1874325001105010331.

Kunduracioglu B, Yilmaz C, Yorubulut M, Kudas S. Magnetic resonance findings of osteitis pubis. J Magnet Resonan Imaging. 2007;25(3):535–9. https://doi.org/10.1002/jmri.20833.

Lepage IM, Robson N, Gilmore H, Davis O, Hooper A, St. John S, Kamesan V, Gelis P, Carvajal D, Hurtig M, Koch TG. Beyond cartilage repair: the role of the osteochondral unit in joint health and disease. J Orthopaed Res. 2019;37(4):855–66. https://doi.org/10.1002/jor.24182.

Mehin R, Meek R, O'Brien P, Blachut P. Surgery for osteitis pubis. Can J Surg. 2006;49(3):170–6. https://www.canjsurg.ca/content/49/3/170

Migliorini F, Vaishya R, Vaish A, Itha R. Management of chondral and osteochondral lesions of the hip: a comprehensive review. J Hip Preserv Surg. 2024;11(1):12–25. https://doi.org/10.1093/jhps/hnad001.

Mohammad W, Elsais W. Abdominal/adductor strength imbalance in soccer players with osteitis pubis. J Men's Health. 2018;14(3):33. https://doi.org/10.22374/1875-6859.14.3.5.

Nishiwaki H, Fujita M, Yamauchi M, Isogai N, Tabata Y, Kusuhara H. A novel method to induce cartilage regeneration with cubic microcartilage. Cells Tissues Organs. 2017;204(5–6):251–60. https://doi.org/10.1159/000479790.

Park D, Sussman W. Osteitis pubis treated with platelet-rich plasma: a case report. Clin J Sport Med. 2020;32(2):e172–4. https://doi.org/10.1097/jsm.0000000000000875.

Pauli S, Willemsen P, Declerck K, et al. Osteomyelitis pubis versus osteitis pubis: a case presentation and review of the literature. Br J Sports Med. 2002;36(1):71–3. https://doi.org/10.1136/bjsm.36.1.71.

Perrone A, Noorzad A, Hamula M, Metzger M, Banffy M, Gerhardt M. Hip adductor longus tendon origin anatomy is consistent and may inform surgical reattachment. J Hip Preserv Surg. 2021;8(4):318–24. https://doi.org/10.1093/jhps/hnab081.

Phillips E, Khoury V, Wilmot A, Kelly J. Correlation between cam-type femoroacetabular impingement and radiographic osteitis pubis. Orthopedics. 2016;39(3) https://doi.org/10.3928/01477447-20160404-03.

Steinbach HL, Petrakis NL, Gilfillan RS, Smith DR. The pathogenesis of osteitis pubis. J Urol. 1955;74(6):810–8.

Lower Limb Joint Replacement, Regeneration and Optimisation of Functional Recovery

David Cain and Tosan Okoro

Abstract Hip and knee osteoarthritis are common, progressive degenerative joint diseases that significantly impact mobility and quality of life. With an increasing prevalence in aging populations, total joint replacement (TJR) remains the gold standard surgical treatment for end-stage disease. While joint replacement effectively reduces pain, post-operative functional recovery remains highly variable, influenced by preoperative function, comorbidities, muscle strength, and rehabilitation strategies. This chapter explores the predictors of functional outcomes following total hip replacement (THR) and total knee replacement (TKR), highlighting the role of muscle strength, BMI, and preoperative health status in determining long-term success. We examine the impact of prehabilitation on optimising preoperative strength and function, along with supporting postoperative physiotherapy and resistance training in accelerating recovery. Understanding patient motivations, barriers to exercise, and the role of outcome measures such as patient-reported outcome measures (PROMs), performance-based assessments, and radiographic evaluations in assessing post-TJR recovery is necessary for optimising rehabilitation protocols, improving long-term function, and maximising the benefits of joint replacement surgery.

Keywords THR · TKR · ERAS · Prehabilitation · Rehabilitation

D. Cain (✉)
Academic Military Trauma and Orthopaedics, RCDM, Birmingham, UK

The Robert Jones & Agnes Hunt Orthopaedic Hospital, Oswestry, UK
e-mail: David.cain7@nhs.net

T. Okoro
The Robert Jones & Agnes Hunt Orthopaedic Hospital, Oswestry, UK

P. Lee (ed.), *Musculoskeletal Regeneration Medicine*,
https://doi.org/10.1007/978-3-031-77456-0_19

1 Introduction

Hip and knee osteoarthritis are common, disabling conditions which cause the gradual erosion of the articular cartilage. In England, the prevalence of knee osteoarthritis is 18.2% in adults over 45, the prevalence of hip osteoarthritis is 10.9% of adults aged over 45 (Arthritis Research UK 2013). Treatment for hip and knee osteoarthritis include non-operative and operative options. Non-operative options include analgesia, physiotherapy and weight loss. Operative options for knee osteoarthritis include corrective osteotomy, unicompartmental knee replacement and total knee replacement (TKR). For hip osteoarthritis, total hip replacement (THR) is the mainstay of surgical treatment however hip resurfacing is still performed in a small number of patients. Joint replacement is the gold standard treatment for end stage disease in hip and knee osteoarthritis (Gandhi et al. 2014). According to the National Joint Registry (NJR), the number of primary THRs in England, Wales and Northern Ireland totalled 99,043 in 2022, in the same year there were 98,469 primary total knee replacements. The median age for patients undergoing THR in 2023 is 69 years and 70 years for TKR (National Joint Registry 20th Annual Report 2023). Since the advent of the THR, technological and surgical techniques have improved as have patient expectations (Yapp et al. 2020). It has been shown that patients who mobilise early have shorter length of stay, lower post operative complications, reduced pain, better exercise tolerance, reduced mortality and increased patient satisfaction (Aprisunadi et al. 2023; Efford and Samuel 2023). Over the past decade Enhanced Recovery After Surgery (ERAS) protocols have been adopted to aid perioperative care and reduce post-surgical hospital stay (Wainwright 2021). Since the adoption of ERAS protocols, we have seen the mean length of stay for patients undergoing elective Primary THR in the NHS drop from 7 days in 2004 (Wilson et al. 2008) to 4.08 days in 2021. The average length of stay for unicompartmental knee replacement is 2.64 days and 4.11 days for TKA (Wainwright 2021). While Total joint replacements resolve pain, post operative function typically stays substantially suboptimal. It has been showed that patients who have low pre-operative function are five times more likely to require assistance from another person for their activities of daily living at 24 months post op, compared to those with high preoperative function (Fortin et al. 2002). This protracted disability has detrimental economic, social and health consequences. Therefore, optimising function prior to and post-surgery are a vital component of rehabilitation.

2 Predictors of Outcome Following Total Joint Replacement

Total joint replacement of the hip and knee is a frequently performed and successful procedure in orthopaedic surgery. As resources are limited, the role of predictors that can help predict the outcome for patients post total joint replacement are commonly utilised to help identify patients who will most benefit from surgery, and more importantly highlight those who may not have a positive outcome from surgery. With the advent of national joint registries there has been an increased scrutiny of orthopaedic surgeons' outcomes and complications (Porter et al. 2019). The Western Ontario and McMaster Universities Osteoarthritis Index (WOMAC) developed by Bellamy et al. in 1982 is a widely used disease specific questionnaire applied to measure the change in clinical status in the treatment of hip and knee osteoarthritis. It is a validated and approved self-administered patient related outcome measure (PROM) tool which is recommended by the Osteoarthritis Research Society International clinical guidelines for the outcome measurement in osteoarthritis (Bellamy et al., 1988; Bellamy et al. 2002). However, there are two potential weaknesses of the WOMAC score. The first being that the stiffness sub-scale has low test-retest reliability and little evidence for its measurement properties. The second is that some studies report inadequate factorial validity in the pain and physical function sub-scales of the WOMAC, therefore the physical function may have limited ability to detect change if the association between pain and function is weak (Gandek 2015). Previous studies have also hypothesised that high preoperative functional status has a positive effect on outcome, whilst others have suggested that it leaves marginal room for improvement in functional status (Montin et al. 2008; Roder et al. 2007). They argue that patients with better preoperative scores are likely to have higher postoperative scores, and those with poor preoperative scores are likely to experience significantly greater improvement in function (Wang et al. 2009). Many studies and systematic reviews have shown a correlation between high BMI and poorer health related quality of life outcomes (Judge et al. 2013; Canovas and Dagneaux 2018; Buirs et al. 2016). However, these studies note that joint replacement still confers a considerable benefit for obese and very obese patients (Canovas and Dagneaux 2018). A recent study reviewed the association of self-reported medical co-morbidities and BMI with functional outcome following lower limb total joint replacement. They found that whilst most participants had good post operative results, that there was a correlation with lower PROMs in patients with high BMI and higher number of self-reported co-morbidities (Snell et al. 2021).

3 Muscle Strength and Its Relation to Function After Total Joint Replacement

The primary concerns expressed by patients opting for THR or TKR surgery typically revolve around experiencing pain and reduced mobility (Trudelle-Jackson and Smith 2004; Neuprez et al. 2020). Consequently, the predominant outcomes often documented in literature pertain to alleviation of pain and improvement in mobility (Trudelle-Jackson and Smith 2004; Neuprez et al. 2020). However multiple studies have shown that patients who undergo lower limb total joint replacement surgery suffer significant muscle atrophy and weakness—seen greatest within the first 3 months of surgery (Dreyer 2016; Dreyer et al. 2018). The rationale behind this is multifactorial, given that skeletal muscle requires sufficient time to recover from traumatic events following surgery. These include metabolic related injury from ischaemia/reperfusion injury secondary to tourniquet use in TKA (Liu et al. 2014) or due to mechanical aetiology such as immobility related atrophy or arthrogenic muscle inhibition (Häggmark et al. 1981). Immobility related atrophy due to major surgery and hospitalisation can cause a severe decline in muscle mass, muscle strength and muscle function (Bloomfield 1997; Covinsky et al. 2003). Muscle strength declines at approximately 4% per day during the first week of immobilisation, making early physiotherapy of paramount importance following surgery (Wigerstad-Lossing et al. 1988). Arthrogenic muscle inhibition following surgery has been has been shown to be the leading cause of loss of muscle mass and strength following TKA surgery, it describes the loss of voluntary muscle control following surgery, Minzer et al. Showed that on average quadriceps voluntary contraction reduced by 17% post TKA (Mizner et al. 2005).Quadriceps strength following TKA will reach 70–80% of strength of healthy age matched controls by 1 year post TKA (Berghmans et al. 2018) however muscle weakness has been shown to persists for up to 3 years post TKA (Schache et al. 2014). Muscle strength deficit at 1 year post THR in the hip ranges from 10–21% compared to the opposite hip (Trudelle-Jackson et al. 2002; Shih et al. 1994). Additionally, atrophic changes persist around the hip for up to 2 years post-THR, characterised by increased fat infiltration (Rasch et al. 2009).

4 Impact of Aging on Muscle

Sarcopenia is age related loss of skeletal muscle mass and function and is associated with frailty and increased risk of morbidity (Wilkinson et al. 2018). The loss of muscle mass in humans begins in middle age, and in severe instances can lead to a total loss of muscle mass by up to 50% by the 8th or 9th decade of life (Wilkinson et al. 2018). It has been shown that muscle mass in humans aged 75 years is lost at a rate of 0.64–0.7% per year in women and 0.8–0.9% per year in men, whereas muscle strength is lost at a much higher rate of 3–4% per year in men and 2.5–3% in women per year (Mitchell et al. 2013). With this significant loss in muscle mass and strength comes significant increased frailty with age. It has been shown that by maintaining physical activity in older age can reduce muscle atrophy by promoting muscle hypertrophy/strength and fatigue resistance (McPhee et al. 2016). Therefore, it is of paramount importance that patients are educated around the importance of maintaining physical activity following total joint replacement to reduce the risk of sarcopenia and associated morbidity.

5 The Evidence for Prehabilitation Exercise Regimes

Utilising appropriate exercise yields numerous benefits in managing osteoarthritis patients. Strengthening periarticular muscles, tendons, and ligaments can mitigate joint forces during movement. In more severe cases, exercise can enhance pain management, muscle strength, cardiovascular fitness, self-efficacy, and overall function (Felson et al. 2000; van Baar et al. 1999). A recent systematic review with meta-analysis of forty-eight unique randomised control trials for prehabilitation for patients undergoing orthopaedic surgery has recently shown that there is moderate level of evidence for the use of prehabilitation in lower limb total joint replacement for improvements in the preoperative phase, however lacking evidence for any significant post operative improvement (Punnoose et al. 2023). They have shown moderate certainty of evidence for prehabilitation for TKR and THR in the preoperative phase, it has been shown to improve function, knee flexor strength and 6-min walk test in TKR patients. In THR patients it has been shown to improve health related quality of life (HRQOL) scores and hip abductor strength. There is also weak certainty of evidence for reduced pain in both TKR and THR patients in the preoperative phase. In the case of TKR at the 6 week post operative point there is a moderate level of certainty that prehabilitation improves function. Unfortunately, there is only a low level of certainty for benefits for TKR and THR for periods after this date post operatively (Punnoose et al. 2023). The evidence for prehabilitation for lower limb total joint replacement is associated with moderate improvement in several preoperative outcomes (function, knee flexor and hip strength, HRQOL, and 6-min walk test). However, the evidence is currently inconsistent and the quality of evidence for postoperative outcomes is low. There is therefore a lack of conclusive evidence to justify the use of prehabilitation regimes to optimise function after lower limb total joint replacement surgery.

6 The Evidence for Post-operative Physiotherapy to Improve Function

Physiotherapy is widely acknowledged as crucial for achieving best results following hip and knee arthroplasty, yet the specifics of rehabilitation vary globally (Naylor et al. 2019). For instance, some countries such as the United States and Australia opt for prolonged postoperative inpatient rehabilitation, although doubts persist regarding its effectiveness compared to outpatient-based physiotherapy (Buhagiar et al. 2017). In the UK, inpatient rehabilitation is typically brief, with patients discharged 3–5 days after surgery. During this short inpatient period, physiotherapy focuses on promoting mobilisation and ensuring a safe discharge. However, post-discharge rehabilitation provision varies considerably, and there are no definitive guidelines for TKR and THR Rehabilitation. (Hamilton et al. 2019; Smith et al. 2020).

The conventional belief that increased therapist contact improves rehabilitation outcomes has recently come under scrutiny. Meta-analyses indicate that while uniform postoperative physiotherapy for all knee and hip arthroplasty patients compared to no treatment provides short-term benefits, it does not significantly enhance outcomes at 1 year (Khan et al. 2008; Artz et al. 2015; Lowe et al. 2007). Moreover, recent meta-analyses of randomized trials suggest no discernible difference in patient outcomes between home-based and outpatient-based interventions post TKR and THR (Buhagiar et al. 2019; Florez-García et al. 2017; Hamilton et al. 2020). These findings are support by the National Institute for Health and Care Excellence (NICE) most recent guidance for postoperative hip and knee rehabilitation (NICE Guideline No. 157 2020). Their review compared group-based supervised rehabilitation and individually supervised rehabilitation to self-directed rehabilitation. They also found no clinically significant differences were found across all outcomes, including quality of life, PROMs, reoperation, thromboembolic complications, and function. The effectiveness of both supervised forms of rehabilitation were similar, with group supervision appearing slightly more effective than individual supervision. However, self-directed rehabilitation showed no clinically important benefits over either form of supervised rehabilitation. Reoperation, hospital readmissions, and thromboembolic complications were assessed as early indicators of rehabilitation outcomes, but event rates were too low for definitive conclusions. While joint replacements generally yield positive outcomes, self-directed rehabilitation is considered cost-effective for many. However, NICE do recommend that certain subgroups may benefit more from supervised programs. Recommendations included offering group or individual rehabilitation for those with specific needs, including cognitive impairments (NICE Guideline No. 157 2020). Appreciating that lower limb TJR can induce significant muscle atrophy and weakness, there has been a trend amongst physiotherapists to employ the use of progressive resistance training (PRT). PRT exercises comprised of: sit to stand, block stepping, stair climbing, walking, sitting knee extension against resistance, lateral weight transfer exercises. Recent systematic review and meta-analysis has shown no significant difference in terms of functional capacity, muscle strength recovery and incidence of adverse events when comparing PTR to standard home-based functional rehabilitation methods that were non-PRT exercises designed to allow patients to safely mobilise (Chen et al. 2021).

7 Enhanced Recovery After Surgery (ERAS) in Lower Limb Total Joint Replacement

Over the past decade Enhanced Recovery After Surgery (ERAS) protocols have been adopted to aid perioperative care and reduce post-surgical hospital stay (Wainwright 2021). ERAS protocols are comprehensive therapeutic approaches that encompasses several aspects of the patient's wellbeing to optimise outcomes following surgery. ERAS protocols for TKR and TKA comprise of 17 elements, they encompass a holistic approach to patient care, focusing on education, preoperative optimisation (smoking cession, alcohol reduction), nutritional support, pain management, fluid management, and early mobilisation following surgery (Changjun et al. 2023; Riga et al. 2023). There are four phases that make up ERAS they are the Pre-admission Phases, Preoperative Phases, Operative Phase and the Postoperative Phase (Changjun et al. 2023). The pre-admission phases encompass preoperative information, education and expectation counselling. This phase has been shown to reduce preoperative anxiety and reduce length of hospital stay by changing patient's expectations (Tanzer et al. 2018; Padilla et al. 2019). The preoperative phase focuses on optimisation of patients for surgery by encouraging smoking and alcohol cessation and preoperative anaemia correction. It has been shown that by optimising these risk factors can reduce complications including, poor wound healing, myocardial infarction, cardiac arrest, pneumonia, urinary tract infections, sepsis, acute renal failure and overall mortality (Akhavan et al. 2017; Agrawal et al. 2021). A RCT found that smoking cessation 6–8 weeks prior to surgery significantly reduces wound and cardiovascular complications, and also reduces length of stay (Møller et al. 2002). Standardised anaesthetic protocols and appropriate use of regional and local anaesthetic encompass the operative phase. Spinal anaesthesia has been shown to significantly reduce the occurrence of nausea and length of hospital stay compared to general anaesthesia (Pu and Sun 2019). The use ofsing tranexamic acid (TXA) and cell saver to reduce intraoperative blood loss, has also been shown to reduce hospital stay, increase speed of physical recovery, and reduce overall morbidity and mortality. There is unwavering evidence for the routine use of TXA in total joint arthroplasty (TJA) surgery (Poeran et al. 2021; Li et al. 2018; Fillingham et al. 2018). Prophylactic antimicrobials are recommended in TJA, periprosthetic joint infection is an uncommon but serious complication, of which prophylaxis antimicrobials have an important role, there is currently no universally defined Guidelines however a recent meta-analysis has found no difference in infection risk for orthopaedic procedure where implants are utilised between single and multiple doses of prophylactic antibiotics (Wainwright et al. 2020; Siddiqi et al. 2019). Post operatively early mobilisation has significant benefits, including shorter length of hospital stay, reduced pain, better exercise tolerance, reduced mortality, reduced incidence of venous thromboembolism and improved patient satisfaction (Chua et al. 2017; Aprisunadi et al. 2023; Efford and Samuel 2023; Guerra et al. 2015). Since the adoption of ERAS protocols, the mean length of stay for patients undergoing elective Primary THR in the NHS drop from 7 days in 2004 (Wilson et al. 2008) to 4.08 days in 2021. The average length of stay for unicompartmental knee replacement is 2.64 days and 4.11 days for TKA (Wainwright 2021).

8 Motivators and Barriers to Improving Function Through Exercise

Regular physical activity contributes to the primary and secondary prevention of chronic disease, improve mental health, reduce the risk of falls and reduces the risk of premature death (Warburton et al. 2006). Studies have shown that age, education, gender, psychological and physical wellbeing are key factors in patients' specific motivators and barriers to exercise (Newson and Kemps 2007).

As patients age, it has been shown that their motivations to continue living an active lifestyle increases over the age of 75. However, patients in this age category are significantly more likely to have medical conditions that impair their ability to undertake regular physical activity compared to their younger counterparts (Newson and Kemps 2007). The mean age for THR and TKR in the UK is 68.5 and 69.3 respectively (Nation Joint Registry Annual Report 2023), a recent study has shown that users of fitness centres over the age of 60, have said that their main motivations for using the fitness centre was for the social aspect of exercise (Larson 2021). All of these factors should be considered when discussing post operative rehabilitation regimes with patients.

9 Outcome Measures for Assessing Function

Outcomes after orthopaedic surgery can be assessed using five different methods, namely radiographic analysis, implant survivorship analysis, surgeon-based outcome measures, performance-related assessment and patient reported outcome measures (PROMs) (Wylde et al. 2009).

Performance related assessments involve assessing an individual's performance in a specific task that is evaluated in a standardised fashion using a predetermined criteria these include the 6-min walk test (6MWT) and the Timed Up and Go test (TUG) amongst others (Wylde et al. 2009). The 6MWT and TUG test are short walking-based tests that are commonly used to evaluate functional recovering following lower limb total joint replacement (Ko et al. 2013). Performance related assessments have been found to be more accurate that patient self-reported functional assessments (Mizner et al. 2011), however they have significant limitations such as being time consuming and only consider function whilst neglecting other important domains of outcomes such as pain and quality of life (Terwee et al. 2006).

Surgeon bases assessment tools such as the Harris hip score or American Knee Score have been shown to overestimate the success of lower limb total joint replacement when compared to patients' opinion of their surgery (Wylde et al. 2009). The danger of solely using surgeon-based assessment tools is that they assume concordance between the view of patients and the surgeon. Surgeons may judge the success of a joint replacement on the range of motion, stability and alignment of the implant, whereas the patient may base their outcome in terms of vitality and ability

Table 1 Types of patients reported outcome measures (PROMs) (Churruca et al. 2021; Dawson et al. 1996, 1998; Dunbar et al. 2001; Bellamy et al. 1988; Meenan et al. 1980; Salaffi et al. 2005)

Generic health	Disease specific	Joint-specific
Assess subjective general health status with comparisons across different disease states and treatment options. e.g. EQ-5D. Assesses 5 dimensions of a patient's health, they are: Mobility, self-care, usual activities, pain/discomfort and anxiety/ depression.	Focus solely on symptom and disabilities relating to particular condition but are not particular to the joint of assessment. e.g. for Rheumatoid arthritis AIMS (Arthritis Impact Measurement Scale) which has 9 component scales which assess mobility, physical activity, social activity, social role, activities of daily living, pain, dexterity, anxiety and depression. For Osteoarthritis, Western Ontario and McMaster University Osteoarthritis index (WOMAC) which is a 24-item measure of pain, stiffness and function.	Specific to the joint of assessment and attempt to exclude the influence of co morbidities. e.g. The Oxford Hip Score (OHS) assesses the disability in patients who have undergone a THR. It encompasses 12 questions about activities of daily living. e.g. The Oxford Knee score has been designed to specifically assess function and pain after total knee replacement, it encompasses 12 questions level of function, activities of daily living and how they have been affected by pain in the past 4 weeks.

to return to leisure activities (Wylde et al. 2009). This inconsistency has led to the development of validated patient-reported outcome measures (PROMS). PROMs are advantageous as they are cost effective, validated and minimally intrusive method of collecting data (Churruca et al. 2021). PROMS can be grouped into generic health, disease specific and joint specific (Table 1). A limitation of all PROMs is the issue of recall bias towards patients reporting the most severe and recent pain they have experienced.

10 Conclusion

Lower limb joint replacement surgery remains a highly effective intervention for alleviating pain and restoring mobility in individuals with advanced hip and knee osteoarthritis. However, optimal recovery extends beyond surgical success and is deeply influenced by patient-specific factors including preoperative function, comorbidities, muscle strength, BMI, and engagement with tailored rehabilitation protocols.

Acknowledgements I would like to dedicate this chapter to my wife Louise and my two daughters Mollie and Annie for their unwavering love and support.

References

Agrawal V, et al. Preoperative optimization and its impact on postoperative complications in patients undergoing total joint arthroplasty. Journal of Orthopaedic Surgery & Research, 2021;16(1):1–9. https://doi.org/10.1186/s13018-021-02397-2.

Akhavan S, Nguyen L-C, Chan V, et al. Impact of smoking cessation counseling prior to total joint arthroplasty. Orthopedics. 2017;40(2):e323–e328. https://doi.org/10.3928/01477447-20161219-02.

Aprisunadi NN, Mustikasari M, Ifadah E, Hapsari ED. Effect of early mobilization on hip and lower extremity postoperative: a literature review. SAGE Open Nurs. 2023;9:23779608231167825. Published 2023 Apr 11. https://doi.org/10.1177/23779608231167825.

Arthritis Research UK. Osteoarthritis in general practice: data and perspectives. Keel Data Report. July 2013. Available from: https://healthinnovationnetwork.com/wpcontent/uploads/2017/01/Osteoarthritis_in_general_practice__July_2013__Arthritis_Research_UK_PDF_421_MB.pdf. Accessed 18 August 2025.

Artz N, Elvers KT, Lowe CM, Sackley C, Jepson P, Beswick AD. Effectiveness of physiotherapy exercise following total knee replacement: systematic review and meta-analysis. BMC Musculoskelet Disord. 2015;16:15. https://doi.org/10.1186/s12891-015-0469-6. PMID: 25886975; PMCID: PMC4333167.

Bellamy N, Buchanan WW, Goldsmith CH, Campbell J, Stitt LW. Validation study of WOMAC: a health status instrument for measuring clinically important patient relevant outcomes to antirheumatic drug therapy in patients with osteoarthritis of the hip or knee. J Rheumatol. 1988;15(12):1833–1840.

Bellamy N, Hochberg M, Tubach F, Simon L, Strand V, Altman R, Boers M, Brooks P, Cannon G, Clegg D, et al. OMERACT-OARSI initiative to define states of severity and indication for outcome measures in osteoarthritis clinical trials. J Rheumatol. 2002;29(6):1374–1379.

Berghmans DDP, Lenssen AF, Emans PJ, de Bie RA. Functions, disabilities and perceived health in the first year after total knee arthroplasty; a prospective cohort study. BMC Musculoskelet Disord. 2018;19(1):250.

Bloomfield SA. Changes in musculoskeletal structure and function with prolonged bed rest. Med Sci Sports Exerc. 1997;29(2):197–206.

Buhagiar MA, Naylor JM, Harris IA, et al. Effect of inpatient rehabilitation vs a monitored home-based program on mobility in patients with total knee arthroplasty: the HIHO randomized clinical trial. JAMA. 2017;317(10):1037–46. https://doi.org/10.1001/jama.2017.1224.

Buirs L, Van Beers L, Scholtes V, Pastoors T, Sprague S, Poolman R. Predictors of physical functioning after total hip arthroplasty: a systematic review. Br Med J Open. 2016;6:e010725. https://doi.org/10.1136/bmjopen-2015-010725.

Canovas F, Dagneaux L. Quality of life after total knee arthroplasty. Orthop Traumatol Surg Res. 2018;104:541–6.

Changjun C, Xin Z, Yue L, Liyile C, Pengde K. Key elements of enhanced recovery after total joint arthroplasty: a reanalysis of the enhanced recovery after surgery guidelines. Orthop Surg. 2023;15(3):671–8. https://doi.org/10.1111/os.13623.

Chen X, Li X, Zhu Z, Wang H, Yu Z, Bai X. Effects of progressive resistance training for early postoperative fast-track total hip or knee arthroplasty: a systematic review and meta-analysis. Asian J Surg. 2021;44(10):1245–53. https://doi.org/10.1016/j.asjsur.2021.02.007. Epub 2021 Mar 11.

Chua MJ, Hart AJ, Mittal R, Harris IA, Xuan W, Naylor JM. Early mobilisation after total hip or knee arthroplasty: a multicentre prospective observational study. PLoS One. 2017;12(6):e0179820.

Churruca K, Pomare C, Ellis LA, et al. Patient-reported outcome measures (PROMs): a review of generic and condition-specific measures and a discussion of trends and issues. Health Expect. 2021;24(4):1015–24. https://doi.org/10.1111/hex.13254.

Covinsky KE, Palmer RM, Fortinsky RH, Counsell SR, Stewart AL, Kresevic D, Burant CJ, Landefeld CS. Loss of independence in activities of daily living in older adults hospitalized with medical illnesses: increased vulnerability with age. J Am Geriatr Soc. 2003;51(4):451–8.

Dawson J, Fitzpatrick R, Carr A, Murray D. Questionnaire on the perceptions of patients about total hip replacement. J Bone Joint Surg Br. 1996;78(2):185–90. PMID: 8666621.

Dawson J, Fitzpatrick R, Murray D, Carr A. Questionnaire on the perceptions of patients about total knee replacement. J Bone Joint Surg B. 1998;80(1):63–9.

Dreyer HC, Smoliga JM, McCauley TR, et al. Effects of essential amino acid supplementation on muscle atrophy following total knee arthroplasty: a randomized controlled trial. JBJS Open Access. 2018;3(2):e0006. https://doi.org/10.2106/JBJS.OA.18.00006.

Dunbar MJ, Robertsson O, Ryd L, Lidgren L. Appropriate questionnaires for knee arthroplasty. Results of a survey of 3600 patients from The Swedish Knee Arthroplasty Registry. J Bone Joint Surg Br. 2001;83(3):339–44. https://doi.org/10.1302/0301-620x.83b3.11134.

Efford CM, Samuel D. Does rapid mobilisation as part of an enhanced recovery pathway improve length of stay, return to function and patient experience post primary total hip replacement? A randomised controlled trial feasibility study. Disabil Rehabil. 2023;45:25. https://doi.org/10.1080/09638288.2022.2148298.

Felson DT, et al. The effects of weight and physical activity on osteoarthritis of the knee. Archives of Internal Medicine, 2000;160(18):2542–48. https://doi.org/10.1001/archinte.160.18.2542.

Fillingham YA, Ramkumar DB, Jevsevar DS, Yates AJ, Shores P, Mullen K, et al. The efficacy of tranexamic acid in total hip arthroplasty: a network meta-analysis. J Arthroplast. 2018;33(10):3083–9.e4.

Florez-García M, García-Pérez F, Curbelo R, Pérez-Porta I, Nishishinya B, Rosario Lozano MP, Carmona L. Efficacy and safety of home-based exercises versus individualized supervised outpatient physical therapy programs after total knee arthroplasty: a systematic review and meta-analysis. Knee Surg Sports Traumatol Arthrosc. 2017;25(11):3340–53. https://doi.org/10.1007/s00167-016-4231-x. Epub 2016 Jul 11.

Fortin PR, Penrod JR, Clarke AE, St-Pierre Y, Joseph L, Belisle P, Liang MH, Ferland D, Phillips CB, Mahomed N, Tanzer M, Sledge C, Fossel AH, Katz JN. Timing of total joint replacement affects clinical outcomes among patients with osteoarthritis of the hip or knee. Arthritis Rheum. 2002;46(12):3327–30.

Gandhi R, Perruccio AV, Mahomed NN. Surgical management of hip osteoarthritis. CMAJ. 2014;186(5):347–55. https://doi.org/10.1503/cmaj.121584.

Gandek B. Measurement properties of the Western Ontario and McMaster Universities Osteoarthritis Index: A systematic review. Journal of Rheumatology, 2015;42(2):232–41. https://doi.org/10.2500/ar.2015.42.3910.

Guerra ML, Singh PJ, Taylor NF. Early mobilization of patients who have had a hip or knee joint replacement reduces length of stay in hospital: a systematic review. Clin Rehabil. 2015;29(9):844–54.

Häggmark T, Jansson E, Eriksson E. Fiber type area and metabolic potential of the thigh muscle in man after knee surgery and immobilization. Int J Sports Med. 1981;2(1):12–7.

Hamilton DF, Loth FC, MacDonald DJ, et al. Exploring variation in patient access of post-discharge physiotherapy following total hip and knee arthroplasty under a choice based system in the UK: an observational cohort study. BMJ Open. 2019;9:e021614. https://doi.org/10.1136/bmjopen-2018-021614.

Judge A, Arden N, Batra R, Thomas G, Beard D, Javaid M, et al. The association of patient characteristics and surgical variables on symptoms of pain and function over 5 years following primary hip-replacement surgery: a prospective cohort study. Br Med J Open. 2013;3:e002453.

Khan F, Ng L, Gonzalez S, Hale T, Turner-Stokes L. Multidisciplinary rehabilitation programmes following joint replacement at the hip and knee in chronic arthropathy. Cochrane Database Syst Rev. 2008;2008(2):CD004957. https://doi.org/10.1002/14651858.CD004957.pub3.

Ko V, Naylor JM, Harris IA, Crosbie J, Yeo AET. The six-minute walk test is an excellent predictor of functional ambulation after total knee arthroplasty. BMC Musculoskelet Disord. 2013;14:145. https://doi.org/10.1186/1471-2474-14-145.

Larsen S, Mozdoorzoy T, Kristiansen E, Nygaard Falch H, Aune TK, van den Tillaar R. A Comparison of Motives by Gender and Age Categories for Training at Norwegian Fitness Centres. Sports, 2021;9(8):113. https://doi.org/10.3390/sports9080113.

Li H, Bai L, Li Y, Fang Z. Oral tranexamic acid reduces blood loss in total-knee arthroplasty: a meta-analysis. Medicine. 2018;97(45):e12924.

Liu D, Graham D, Gillies K, Gillies RM. Effects of tourniquet use on quadriceps function and pain in total knee arthroplasty. Knee Surg Relat Res. 2014;26(4):207–13.

Lowe CJM, Barker KL, Dewey M, Sackley CM. Effectiveness of physiotherapy exercise after knee arthroplasty for osteoarthritis: systematic review and meta-analysis of randomised controlled trials. BMJ. 2007;335:812. https://doi.org/10.1136/bmj.39311.460093.BE.

McPhee J, et al. Physical activity in older age: perspectives for healthy ageing and frailty. Biogerontology, 2016;17(5):567–80. https://doi.org/10.1007/s10522-016-9652-3.

Meenan RF, Gertman PM, Mason JH. Measuring health status in arthritis. The arthritis impact measurement scales. Arthritis Rheum. 1980;23(2):146–52.

Mitchell WK, Phillips BE, Williams JP, et al. Development of a new Sonovue™ contrast-enhanced ultrasound approach reveals temporal and age-related features of muscle microvascular responses to feeding. Physiol Rep. 2013;1(5):e00119. https://doi.org/10.1002/phy2.119.

Mizner RL, Petterson SC, Stevens JE, Vandenborne K, Early S-ML. Quadriceps strength loss after total knee arthroplasty. The contributions of muscle atrophy and failure of voluntary muscle activation. J Bone Joint Surg Am. 2005;87(5):1047–53.

Mizner RL, Petterson SC, Clements KE, Zeni JA Jr, Irrgang JJ, Snyder-Mackler L. Measuring functional improvement after total knee arthroplasty requires both performance-based and patient-report assessments: a longitudinal analysis of outcomes. J Arthroplasty. 2011;26(5):728–37. https://doi.org/10.1016/j.arth.2010.06.004.

Møller AM, Villebro N, Pedersen T, Tønnesen H. Effect of preoperative smoking intervention on postoperative complications: a randomised clinical trial. Lancet. 2002;359(9301):114–7.

Montin L, Leino-Kilpi H, Suominen T, Lepisto J. A systematic review of empirical studies between 1966 and 2005 of patient outcomes of total hip arthroplasty and related factors. J Clin Nurs. 2008;17(1):4045.

National Guideline Centre (UK). Evidence review for outpatient hip and knee postoperative rehabilitation: joint replacement (primary): hip, knee and shoulder: evidence review R. London: National Institute for Health and Care Excellence (NICE); 2020. (NICE Guideline, No. 157.) Available from: https://www.ncbi.nlm.nih.gov/books/NBK561413/.

National Joint Registry. 20th annual report. 2023. Retrieved from https://reports.njrcentre.org.uk/Portals/0/PDFdownloads/NJR%2020th%20Annual%20Report%202023.pdf. Accessed 20 Feb 2024.

Naylor JM, Hart A, Harris IA, Lewin AM. Variation in rehabilitation setting after uncomplicated total knee or hip arthroplasty: a call for evidence-based guidelines. BMC Musculoskelet Disord. 2019;20(1):214. https://doi.org/10.1186/s12891-019-2570-8.

Neuprez A, Neuprez AH, Kaux JF, et al. Total joint replcement improves pain, functional quality of life, and health utilities in patients with late-stage knee and hip osteoarthritis for up to 5 years. Clin Rheumatol. 2020;39:861–71. https://doi.org/10.1007/s10067-019-04811-y.

Newson RS, Kemps EB. Factors that promote and prevent exercise engagement in older adults. J Aging Health. 2007;19(3):470–81.

Padilla JA, Feng JE, Anoushiravani AA, Hozack WJ, Schwarzkopf R, Macaulay WB. Modifying patient expectations can enhance total hip arthroplasty postoperative satisfaction. J Arthroplast. 2019;34(7s):S209–s14.

Poeran J, Chan JJ, Zubizarreta N, Mazumdar M, Galatz LM, Moucha CS. Safety of tranexamic acid in hip and knee arthroplasty in high-risk patients. Anesthesiology. 2021;135:57–68.

Porter M, Armstrong R, Howard P, Porteous M, Wilkinson JM. Orthopaedic registries – the UK view (National Joint Registry): impact on practice. EFORT Open Rev. 2019;4(6):377–90. https://doi.org/10.1302/2058-5241.4.180084.
Pu X, Sun JM. General anesthesia vs spinal anesthesia for patients undergoing total-hip arthroplasty: a meta-analysis. Medicine. 2019;98(16):e14925.
Punnoose A, Claydon-Mueller LS, Weiss O, Zhang J, Rushton A, Khanduja V. Prehabilitation for patients undergoing orthopedic surgery: a systematic review and meta-analysis. JAMA Netw Open. 2023;6(4):e238050. https://doi.org/10.1001/jamanetworkopen.2023.8050.
Rasch A, Bystrom AH, Dalen N, Martinez-Carranza N, Berg HE. Persisting muscle atrophy two years after replacement of the hip, The Journal of bone and joint surgery.British volume, 2009;91(5):583–8.
Riga M, Altsitzioglou P, Saranteas T, Mavrogenis AF. Enhanced recovery after surgery (ERAS) protocols for total joint replacement surgery. SICOT J. 2023;9:E1. https://doi.org/10.1051/sicotj/2023030.
Roder C, Staub LP, Eggli S, Dietrich D, Busato A, Muller U. Influence of preoperative functional status on outcome after total hip arthroplasty. JBJS. 2007;89:11–7.
Salaffi F, Carotti M, Grassi W. Health-related quality of life in patients with hip or knee osteoarthritis: comparison of generic and disease-specific instruments. Clin Rheumatol. 2005;24(1):29–37. https://doi.org/10.1007/s10067-004-0965-9. Epub 2004 Jul 17.
Schache MB, McClelland JA, Webster KE. Lower limb strength following total knee arthroplasty: a systematic review. Knee. 2014;21(1):12–20.
Shih CH, Du YK, Lin YH, Wu CC. Muscular recovery around the hip joint after total hip arthroplasty. Clin Orthop Relat Res. 1994;302:115–20.
Siddiqi A, Forte SA, Docter S, Bryant D, Sheth NP, Chen AF. Perioperative antibiotic prophylaxis in total joint arthroplasty: a systematic review and meta-analysis. J Bone Joint Surg Am. 2019;101(9):828–42.
Smith TO, Dainty JR, Clark EM, Whitehouse MR, Price AJ, MacGregor AJ. Demographic and geographical variability in physiotherapy provision following hip and knee replacement. An analysis from the National Joint Registry for England, Wales, Northern Ireland and the Isle of Man. Physiotherapy. 2020;106:1–11. https://doi.org/10.1016/j.physio.2019.11.003. Epub 2019 Nov 20.
Snell DL, Dunn JA, Jerram KAS, et al. Associations between comorbidity and quality of life outcomes after total joint replacement. Qual Life Res. 2021;30:137–44. https://doi.org/10.1007/s11136-020-02610-6.
Tanzer D, Smith K, Tanzer M. Changing patient expectations decreases length of stay in an enhanced recovery program for THA. Clin Orthop Relat Res. 2018;476(2):372–8.
Terwee CB, Mokkink LB, Steultjens MP, Dekker J. Performance-based methods for measuring the physical function of patients with osteoarthritis of the hip or knee: a systematic review of measurement properties. Rheumatology (Oxford). 2006;45(7):890–902.
Trudelle-Jackson E, Smith SS. Effects of a late-phase exercise program after total hip arthroplasty: a randomized controlled trial. Arch Phys Med Rehabil. 2004;85(7):1056–62.
Trudelle-Jackson E, Emerson R, Smith S. Outcomes of total hip arthroplasty: a study of patients one year postsurgery. J Orthop Sports Phys Ther. 2002;32(6):260–7.
van Baar ME, Assendelft WJ, Dekker J, Oostendorp RA, Bijlsma JW. Effectiveness of exercise therapy in patients with osteoarthritis of the hip or knee: a systematic review of randomized clinical trials. Arthritis Rheum. 1999;42(7):1361–9.
Wainwright TW. The current status of daycase hip and knee arthroplasty within the English National Health Service: a retrospective analysis of hospital episode statistics data. Ann R Coll Surg Engl. 2021;103(5):324–31. https://doi.org/10.1308/rcsann.2020.7142.
Wainwright TW, Gill M, McDonald DA, Middleton RG, Reed M, Sahota O, et al. Consensus statement for perioperative care in total hip replacement and total knee replacement surgery: enhanced recovery after surgery (ERAS(®)) society recommendations. Acta Orthop. 2020;91(1):3–19.

Wang W, Morrison TA, Geller JA, Yoon RS, Macaulay W. Predicting short-term outcome of primary total hip arthroplasty: a prospective multivariate regression analysis of 12 independent factors. J Arthroplasty. 2009;25:858–64.

Wigerstad-Lossing I, Grimby G, Jonsson T, Morelli B, Peterson L, Renstrom P. Effects of electrical muscle stimulation combined with voluntary contractions after knee ligament surgery, Medicine and science in sports and exercise, 1988;20(1):93–98. https://doi.org/10.1249/00005768-198802000-00014.

Warburton DE, Nicol CW, Bredin SS. Health benefits of physical activity: the evidence. CMAJ. 2006;174(6):801–9. https://doi.org/10.1503/cmaj.051351.

Wilkinson DJ, Piasecki M, Atherton PJ. The age-related loss of skeletal muscle mass and function: measurement and physiology of muscle fibre atrophy and muscle fibre loss in humans. Ageing Res Rev. 2018;47:123–32. https://doi.org/10.1016/j.arr.2018.07.005.

Wilson J, Charlett A, Leong G, McDougall C, Duckworth G. Rates of surgical site infection after hip replacement as a hospital performance indicator: analysis of data from the English mandatory surveillance system. Infect Control Hosp Epidemiol. 2008;29(3):219–26.

Wylde V, Blom AW, Whitehouse SL, Taylor AH, Pattison GT, Bannister GC. Patient-reported outcomes after total hip and knee arthroplasty: comparison of midterm results. J Arthroplast. 2009;24(2):210–6. https://doi.org/10.1016/j.arth.2007.12.001. Epub 2008 Mar 28.

Yapp LZ, Clement ND, Macdonald DJ, Howie CR, Scott CEH. Patient expectation fulfilment following total hip arthroplasty: a 10-year follow-up study. Archives of Orthopaedic and Trauma Surgery. 2020;140(7):963–71. https://doi.org/10.1007/s00402-020-03430-6.

Clinical Application of Sleep in MSK Regeneration Medicine

Sara McNeillis

Abstract Sleep disorders can be diagnosed by taking an appropriate history and performing sleep diagnostic studies. Sleep diagnostics may be used to assess sleep loss by pattern analysis and sleep architecture to assess quality of sleep progression throughout the night. Diagnosis of hypoxia and obstructive sleep apnoea will include respiratory parameters such as oxygen saturation and airflow analysis.

Identifying patients with sleep disorders can be facilitated by using established and validated sleep screening questionnaires. Sleep disorders management needs early intervention as processes for recovery and regeneration often take time to improve. A multidisciplinary approach to managing patients with musculoskeletal conditions which need regenerative treatments should be implemented.

It is recommended that management of sleep disorders follow guidelines where appropriate and available. Management of sleep loss will involve cognitive behavioural therapy for insomnia and medication, whilst management of obstructive sleep apnoea will often involve the use of devices such as CPAP.

Improving sleep will have overall benefits to patients undergoing musculoskeletal regeneration procedures.

Keywords Sleep history · Sleep patterns · Insomnia · Obstructive sleep apnoea · Continuous positive airways pressure · Mandibular advancement device · Melatonin

1 Introduction

- **Overview of the Role of Sleep in MSK Regeneration**

We have explored in a previous chapter the effects of sleep loss and hypoxia on musculoskeletal tissues. The is irrefutable evidence of how these disruptions can affect the musculoskeletal regeneration.

S. McNeillis (✉)
University College London Hospitals, London, UK

P. Lee (ed.), *Musculoskeletal Regeneration Medicine*,
https://doi.org/10.1007/978-3-031-77456-0_20

Sleep is usually linked to the light/dark cycle in mammals via the circadian rhythm generator the suprachiasmatic nucleus. Here the cells maintain an approximate 24-h rhythmicity through transcription and breakdown of key clock proteins. These are clock and Bmal which enhance transcription of genes PER and CRY whilst the transcribed proteins from these genes, namely per and cry directly inhibit the transcription of clock and bmal. The process of accumulation and degradation cycle takes approximately 24.2 h and can vary from 23.5 to 27.5 h in mammals with changes in PER 2 gene (one of the PER genes involved) (Sargent and Woodward 1969; Konopka 1971; Coté and Brody 1986; Reppert 1998).

We saw that in animal studies disruption of the timing of these can adversely affect key hormone responses which in turn stimulate regeneration of musculoskeletal tissue. For example, phase shifting of the circadian process resulted in delay in wound healing. Human studies also concluded the same findings (Chap. 4). Sleep loss also resulted in a reduction in release of growth hormone which is an anabolic hormone. Growth hormone is released in slow wave sleep at night and is tightly controlled via the circadian process thus if sleep loss occurs then this has a negative impact. Melatonin is also tightly controlled circadian hormone to the light/dark cycle and its release is curtailed by intrusion of light in the evening. The inhibition of melatonin is profound and can occur at levels as low as 10lux (Lerner et al. 1960; Wurtman et al. 1963; Cahill and Besharse 1991).

Melatonin has a role in synchronising the suprachiasmatic nucleus which in turn synchronises the peripheral body circadian clocks to ensure the tissues can adapt to the correct phase and to the activity of the mammals (Pevet and Challet 2011). In peripheral tissues, melatonin not only programs the peripheral body clocks, but it also acts as a free radical scavenger, thus helping with the restoration of tissues after injury such as surgery or disease (Maria et al. 2017). Suppression of melatonin thus may have an adverse effect on healing. Additionally, exogenous melatonin was seen to assist stem cells survival (Lee et al. 2017).

Another factor which has been shown to adversely affect tissue repair is hypoxia. Hypoxia is present in sleep related breathing disorders and lung disease in various severities. Mouse and rat studies revealed the effects of intermittent and sustained hypoxia on cellular processes. Mechanisms included release of pro-inflammatory cytokines and suppression of anti-inflammatory cytokines through the gene expression changes. Pro-inflammatory cytokines have a role in early tissue repair to identify and remove damaged tissue. Persistent hypoxia however enhances and maintains the pro-inflammatory state and prevent progression to wound healing and regeneration. Hypoxia is also a double-edged sword. In some cases, hypoxia was seen to elicit changes which may protect the cell from further damage too. Ischaemic reconditioning is a state where recurrent or prior exposure to hypoxia causes mobilisation of some stem cells to proliferate and migrate. Therefore, the role of hypoxia in musculoskeletal regeneration is complex (Chap. 4).

In this chapter we explore the strategies for assisting musculoskeletal regeneration in humans.

- **The Importance of Sleep in Pre, Post-Operative Recovery and Rehabilitation**

Pre-operatively sleep loss is predicated on the prior health conditions. Chronic pain from tissue injury will interfere with consolidation of sleep resulting in a co-morbid insomnia. Untreated hypoxia which would occur in patients with obstructive sleep apnoea may alter the responses to opiates and affect the per-operative and post-operative phase adversely (Heinzer 2007). There have been reports of post operative complications in patients with obstructive sleep apnoea on the respiratory system and wound healing (Ozdilekcan et al. 2021).

Recovery and rehabilitation may be affected by sleep loss due to pain and inflammation which would delay mobilisation. Whilst ongoing hypoxia will affect wound and tissue healing by also delaying the regeneration phase as well as promote poor quality regeneration.

In summary although there may be multifactorial reasons for wound healing and tissue/stem cell regeneration, sleep related factors are proven to influence the process and are modifiable to enable the best opportunity for outcomes.

2 Diagnosis of Sleep Disorders

- **Identifying sleep factors for modulation**

Sleep factors are identifiable in the peri-operative period. A thorough history of sleep can be elicited from direct consultation and history taking.

Sleep routine.

The sleep routine is the habits a person has at the time of sleep. Important areas for questioning are the use of smoking and alcohol around bedtime and the use of distracting or alertness promoting habits such as watching TV in bed and using computers, laptops and phones at the time of sleep onset. When questioning sleep routine, good habits can also be elicited such as mental and physical relaxation practices. The environment for sleep is also explores such as having pets in the bedroom and any external sleep altering factors such as sound and light intrusion.

Questions are then asked about average weekday and weekend timings. Most patients who are of working age may find that their sleep hours are curtailed during the weekdays compared with the weekends. This is called social jetlag and has been known to affect physical and mental health (Bouman et al. 2024). It is important to question how long a patient lies in bed before sleep onset as this may highlight an issue with poor sleep habit or a sleep disorder. Additionally, questions of when the lights are turned off are important as this alerts the reviewer to delay in sleep phases.

Nocturnal awakenings are rare in the young adults but more common with aging (Garma and Bouard 1981; Bixler et al. 1984; Miyasita et al. 1989). Nocturnal awakenings may be due to loss of sleep stability or due to sleep disorders. Nocturnal awakenings (Fig. 1) may therefore be extrinsic such as noise and light or temperature changes, or intrinsic such as due to insomnia obstructive sleep apnoea or parasomnias which are variable sleep disorders. Causes for the nocturnal awakenings

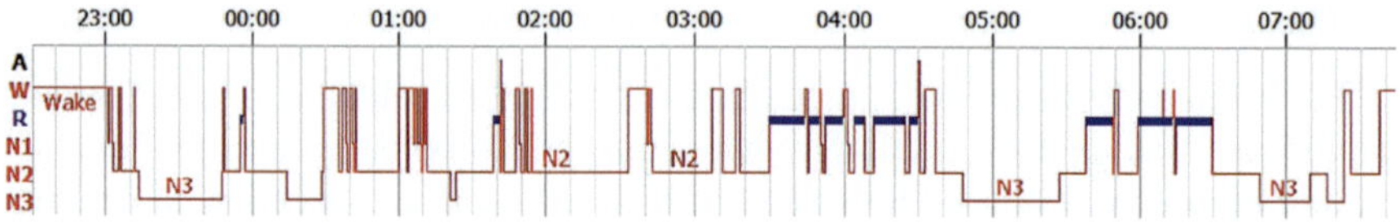

Fig. 1 Nocturnal awakenings on hypnogram showing disruption of sleep progression

should be evaluated as an important part of sleep history taking. It is not completely accepted that aging causes nocturnal awakenings as studies looking into this often find comorbidities which are the root cause. The extent of sleep loss may be exaggerated in some patients particularly those who suffer sleep state misperception.

Wake time is also an important sleep factor to record. Waking after a sleep period in healthy adults occurs once the sleep processes are completed and timed to the circadian rhythm. Patients may share early morning waking before their desired wake time due to mental health disorders, sleep phase advancement or external factors such as noise and light intrusions. Early morning waking reduces the overall sleep hours but more importantly can reduce the amount of REM sleep. Likewise, shortened sleep hours due to forced waking for example, getting up for work or getting up for young children will also reduce the required sleep hours and impact on REM sleep. We have seen in animal studies that sleep loss of paradoxical (REM sleep) was shown to affect regeneration. The reasons behind early or forced awakening should be sought as modifiable factors.

Sleep behaviours in the night may highlight intrinsic disorders which affect sleep quality despite having required sleep hours. Behaviours such as snoring sleep apnoea and periodic leg movements in sleep can impact the overall sleep architecture and reduce the benefits of sleep. Additionally, other behaviours such as sleepwalking or REM behaviour disorders may be a sign of alteration/damage to the sleep process.

Co-morbid factors such as mental health disorder pain diabetes and prostate condition for example should be questioned as these may be root causes for sleep disturbance.

Identifying sleep factors for modulation should not only be restricted to sleep hours. Many factors and behaviours in the daytime can affect sleep quality and quantity. One specific example of this will be caffeine intake during the day and timing of this. It has been revealed that caffeine is an adenosine receptor antagonist and as such reduces the sleep pressure. The half-life of caffeine is approximately 5 hrs and therefore despite antagonism the sleep pressure increases in the background and once the caffeine has worn off patients may feel much sleepier. Timing of caffeine has also been shown to delay the sleep phase (Faber et al. 2017).

Alcohol intake is another example of an identifiable sleep factor. Although alcohol is seen as sedative, studies have shown that alcohol affects the sleep architecture profoundly.

Exercise is another modifiable daytime factor which can impact sleep both by improving or delaying sleep onset and sleep maintenance. This area for modification is still under investigation (Colrain et al. 2014; Chan et al. 2015; Thakkar et al. 2015; Payseur et al. 2022).

Medication history is vitally important to elicit as many medications have sleep altering side effects. If medications are taken which cause insomnia or sedation at the incorrect time, these will affect nocturnal sleep. A common example of this will be pain killer such as opiates which cause profound sedation and sleep therefore may reduce sleep pressure in the night if taken during the day.

Evaluating the effects of sleep disturbance can be determined by using validated questionnaires such as the Epworth Sleepiness scale, Stanford sleepiness score, functional outcome of sleep and Insomnia severity index. Mood questionnaires may also be useful such as Hospital anxiety and depression index.

Examination of the patient's airways will be useful if sleep disordered breathing is suspected and there may be modifiable factors to improve these. Nasal congestion is one of the most common causes of snoring.

- **Clinical application of Sleep diagnostics in MSK patients**

Sleep diagnostics often follow sleep history taking and are tailored to the conditions which are being suspected. In patients with musculoskeletal conditions and sleep disorders, one may look for sleep factors in the following way.

Sleep pattern analysis

Sleep is variable on a night-to-night basis, and it also varies with age (Tune 1968). Sleep pattern analysis is often taken over a period of weeks, therefore sleep pattern analysis can be collected subjectively by using patient input into sleep diaries. The collection of this data is subjective and may have user bias as the data entered cannot be verified easily. Nevertheless, in clinical practice it is useful to evaluate the sleep pattern from a patient's perspective, as these will be the most common noticeable outcome measure. Diaries are of two types, pictorial such as that recommended by the American Association of Sleep medicine or may be table based.

Pictorial diaries are easy to view but lack specific data clarity on timings. Table based diaries are attempt to collect more specific timing and therefore try to present the data objectively. The disadvantages to this are the element of estimation by the patient and that this may also encourage the patient to obsess over sleep monitoring in an unhealthy way.

Sleep trackers

A well validated method of tracking data is by using actigraphy (Fig. 2). This method measures movement (when commonly worn on the wrist) and light (Brown et al. 1990a, b; Hauri and Wisbey 1992). It estimates sleep over a period of days and weeks. The advantage of this method is that objective data is collected over this time and this data can be utilised to assess how the circadian rhythm is progressing. A reflection if the circadian rhythm is intact or not is the assessment of non-parametric circadian rhythm analysis first described by Van Someren (Fig. 3).

Actigraphy has some shortcomings, for example, patients who lie inactive in bed with lights off may be mistakenly calculated as sleeping even though they are

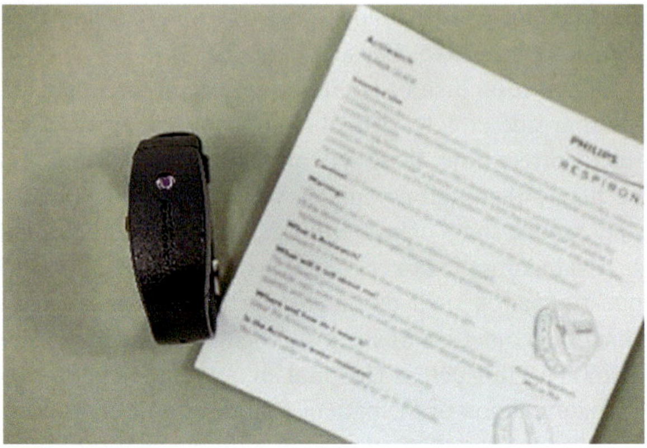

Fig. 2 Actigraphy watch

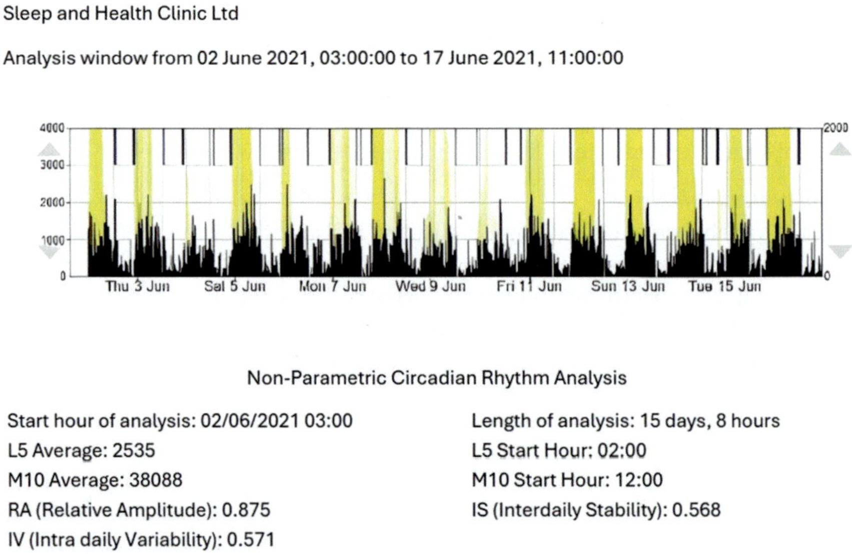

Fig. 3 Nonparametric circadian rhythm analysis

awake. Thus, sleep clinicians often use both sleep diaries entered by patients and actigraphy to assess sleep more accurately.

Other sleep trackers are available commercially and these attempt to assess sleep by using a variety of measurements such as heart rate, movement, muscle tone and light. Many trackers are developed by companies, but good validation is difficult due to companies' unwillingness to share their commercially sensitive algorithms which are used to calculate sleep (unlike actigraphy watches) (Peake et al. 2018; Dong et al. 2022).

Wearable trackers however are useful as they track and estimate sleep over a period of time and despite not being able to validate the data a trend analysis can be ascertained. A useful study assessing and comparing wearables showed that although actigraphy was validated it was the most inaccurate at sleep measurement with one of the commercial wearables being more reflective of actual sleep and wakefulness.

Sleep architecture analysis

Further evaluation of sleep requires more biological data to be collected. Often limited polygraphy studies are used to assess breathing in sleep but not sleep architecture (Jung and Kuhlo 1965; Svanborg et al. 1990). It is known that sleep disordered breathing is very prevalent in the population therefore these offer a more cost effective and easier to set-up with minimal training. Companies also develop computerised scoring based on the American Association of Sleep Medicine recommendations and the Rechtshaffen and Kales scoring manual. The accuracy of computer analysis varies and often a trained sleep physiologist is needed to assess for discrepancies and artifact analysis.

The gold standard for sleep architecture monitoring is the nocturnal video polysomnography (NPSG) which collects data from head electroencephalograms (EEG) and face electromyograms (EMG) to assess sleep stages accurately (Fig. 4). EEG scoring criteria has been used since the early twentieth century to assess sleep and these led to the development of scoring criteria which are evidence based. Again, the Rechtshaffen and Kales scoring criteria is comprehensive and is the most used.

Computer analysis for the EEG is not robust and a trained polysomnography sleep physiologist is required to validate the scoring.

The advantages of NPSG is that sleep is objectively measured on the study night. Arousal from sleep and sleep loss are easily identified. Sleep progression through the night is quantified. The disadvantage is the expense and inconvenience to the patient, there may be an alteration in sleep architecture due to "first night effect".

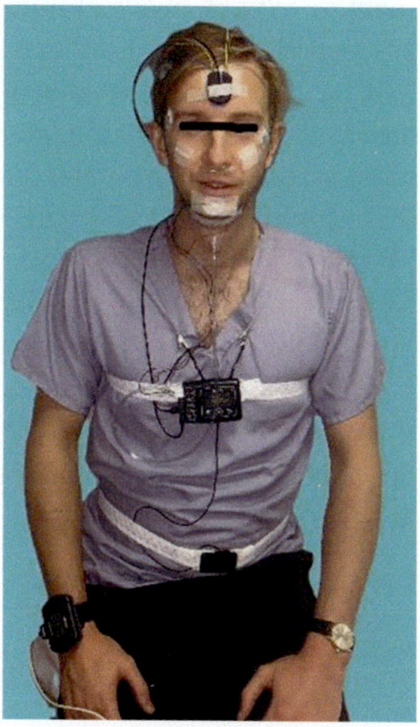

Fig. 4 Full polysomnography set-up

- **Insomnia (sleep loss)**

Insomnia may be acute or chronic. The definition of insomnia is a patient who has difficulty initiating or maintaining sleep for a period of 3 days a week or more over at least 3 months to be defined as chronic. There also should be associated daytime symptoms such as fatigue or malaise, attention, concentration or memory problems, mood instability, daytime sleepiness, reduced motivation, proneness to errors, headaches or gastrointestinal symptoms and worries and concerns about loss of sleep. Mostly the diagnosis of insomnia will be on history taking and sleep pattern analysis. Occasionally a NPSG may be required to assess sleep state misperception which is a cause of insomnia.

Factors which may result in insomnia can be divided into

- Pre-disposing factors, a hyper-reactivity state, psychological such as anxiety around sleep and environmental.
- Precipitating factors will set in the maladaptive sleep patterns, and these may be due to medical disorders and injuries, or psychiatric disorders may contribute to these.
- Perpetuating factors such as environmental behavioural and medical disorders can be sought.

Additionally, insomnia may be due to good stimulus for sleep and stimulus Dyscontrol, these are usually behavioural.

- **OSA (nocturnal hypoxaemia)**

The diagnosis of obstructive sleep apnoea followed the original description of the condition by Burwick in 1956. Symptoms of excessive daytime sleepiness associated with repetitive low oxygen levels at night and associated hypercapnia.

Diagnosis of obstructive sleep apnoea was therefore mostly based on measurement of oxygen levels at night and progressed on to the measurement of breathing associated with these oxygen desaturations (Guilleminault et al. 2000).

Currently OSA is assessed by history as above but also examination of the upper airway to look at possible modifiable factors. Craniofacial features have been shown to be associated with development of obstructive sleep apnoea and examination of these highlights areas where surgical intervention may be possible such as hypoplastic mandible and maxilla (Wang et al. 2022). Oral airway examination then follows and obvious area for obstruction the size of the tonsils can be seen and graded. Sizes 3 and 4 grade tonsils may be present and amenable to treatment.

Nasal examination to assess for causes for nasal obstruction will often be useful if patients have significant symptoms at night. The anatomical changes within the nose such as large turbinate or deviation of the septum of the nose can be seen and are also graded for options.

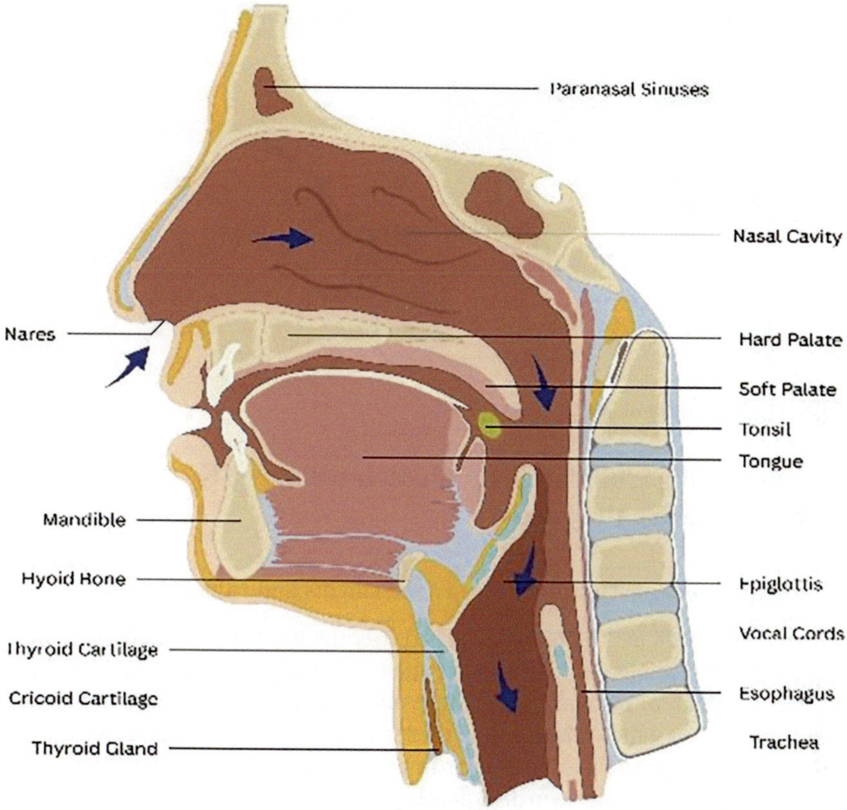

Fig. 5 Pharyngeal structures showing airflow through the nose

Pharyngeal structures (Fig. 5) are more difficult to assess in outpatient clinics.

A fibreoptic nasendoscope (Fig. 6) may be used to look at the posterior pharyngeal structures (Shea et al. 1999; Jordan and White 2008). Manoeuvres such as Muller's manoeuvre may show anatomical areas for collapse. Ideally this examination should be performed in sleep and commonly this is performed using sedation to mimic sleep (De Corso et al. 2015).

Overall, the assessments will allow clinicians to plan options for treatment.

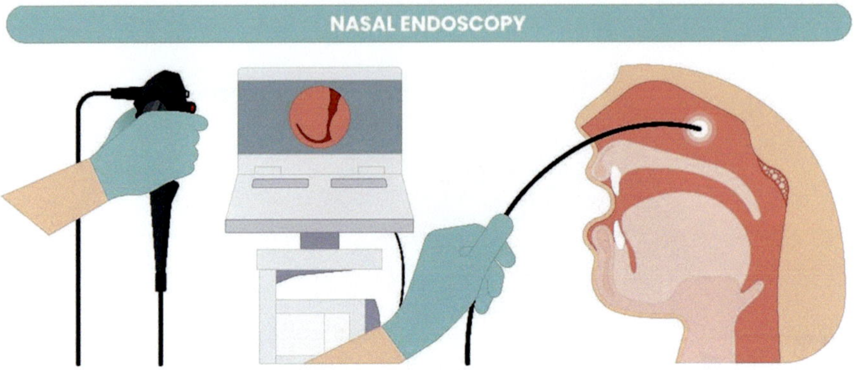

Fig. 6 Flexible nasendoscopy

3 Integrating Clinical Sleep Medicine into MSK Regenerative Practices

- **Strategies for Incorporating Sleep Assessments and Interventions**

Perioperative medicine has been developed over the last few years. The purpose of peri-operative medicine is to enhance patient outcomes following surgery. There has been a wealth of evidence which have altered practice to evaluate and identify modifiable factors which may result in poorer outcomes. Most of the focus however has been on fasting and cardiovascular improvements. It is only recently that evidence for improvement of sleep disorders has been investigated. Even now interventional studies have been limited.

It would make logical sense to integrate clinical sleep medicine into MSK regenerative practices. Sleep medicine often leads to improvement in lifestyle factors as well as treatment of sleep disorders to improve sleep. In turn these improvements will have beneficial effects on MSK regeneration. In particular, where MSK regeneration uses stem cells, improvement of sleep may also lead to improved outcomes for stem cell survival.

Ideally, they should be a robust process to highlight patients with sleep condition and lifestyle conditions which affect sleep. A holistic approach is needed.

Screening questionnaires have been developed for use in the primary care sector particularly to pick up the risk of obstructive sleep apnoea. One commonly used questionnaire is the STOP BANG (Abrishami et al. 2010; Ong et al. 2010).

STOP BANG is an acronym for snoring, tiredness, observed apnoeas, body mass index, age, neck circumference and gender.

Snoring is present in over 40% of the population but when associated with tiredness may represent disruption of sleep process, if observed apnoeas are also present then there is a higher risk of having OSA. Body mass index is positively correlated with severity of OSA in patients who have the disorder. Neck circumference is a surrogate for body mass index but also reflects airway factors which predispose to OSA. Finally, males are twice as likely as females below menopause age to have sleep apnoea.

STOP BANG has been investigated extensively for positive predictive value and reflects well in this respect. STOP BANG has also been used in the surgical population and shown to be a good indicator for referral to sleep investigations (Chung et al. 2012).

The limitations of STOP BANG is in the cohort of post-menopausal women who have a higher risk of sleep disordered breathing without all the symptoms. It does not evaluate circadian dysfunction, sleep loss or insomnia risk.

Additional questionnaires may then be needed.

Chronotype questionnaires, insomnia severity index and functional outcomes of sleep and Pittsburgh sleep quality index may be useful for identifying risk of other sleep conditions.

The Ostberg-Horne morningness and eveningness questionnaire evaluates the chronotype of the individual and can classify patients into morning type or evening type. It is now known that evening type people have higher sleep loss due to social constraints and may suffer long-term health problems (Horne and Ostberg 1976).

Insomnia severity index was developed to assess the consequences of sleep disruption at night, it can be used to track improvements to sleep modification too.

Functional outcomes of sleep questionnaire has been developed as a research tool and has been incorporated into clinical practice to track progress of sleep treatments (Weaver et al. 1997; Chasens et al. 2009).

Pittsburgh sleep quality index was also developed as a well validated research tool but is less commonly used in clinical practice. It may be useful to integrate this into MSK patients as it can be modified (Buysse et al. 1989).

- **Collaborative Care Models Involving Sleep Specialists and MSK Practitioners**

A multidisciplinary approach to assessing and planning patients for surgery is needed. One such approach is to use the assessment tools above and have a tailored approach. Involvement of peri-operative medicine specialists early in the patient's treatment journey can have advantages. These specialists may then also involve specialists from relevant fields such as cardiology, respiratory and sleep medicine.

MSK practitioners can incorporate treatment options from all specialities to holistically prepare patients for regeneration and post operative recovery.

Sleep specialists can be involved in the pre-operative period and postoperative period to optimise sleep.

4 Managing Sleep Disorders in MSK Patients Clinically

- **Approaches to Treating Common Sleep Disorders in MSK Rehabilitation**

Treating sleep disorders in the MSK patients will aid their rehabilitation. Early interventions pre-operatively should ensure that the patients have the most optimal conditions for wound healing, repair and regeneration of bone and cartilage and improve motivation for mobility and physiotherapy. International guidelines offer treatment options for a sleep physician.

- **Circadian rhythm disorders and Insomnia**

Circadian rhythm disorders often respond to treatment therapies and medications (Auger et al. 2015). It is evident that more research into management of circadian rhythm disorders is needed and specifically to show benefits to the MSK patient group. However, melatonin was shown to have evidence for treatment with the combination of light therapy for delayed sleep phase syndrome and non-24-h sleep wake disorder and irregular sleep-wake rhythm disorders. Melatonin was not recommended for elderly patients.

Insomnia treatment also has been assessed in a meta-analysis which provided guidelines for management both nationally and internationally (Riemann et al. 2017). The recommendation is to use cognitive behavioural therapy for insomnia as the first line. The therapy is delivered either directly with interactions from trained psychologists or digitally.

Medication use has poor evidence; however, it is the second line management of therapy resistant insomnia.

- **Sleep related breathing disorders and hypoxia.**

There are many studies assessing treatment options for obstructive sleep apnoea. Overall, the majority of these recommend treatment with devices which can either provide continuous positive airway pressure (Fig. 7) or jaw advancement devices such as the mandibular advancement splints (Fig. 8).

Devices have variable tolerance for long-term use with regularly quoted estimates of failure in compliance in up to 30% to 50% of patients failing to comply with treatment in both cases worldwide. Alternatives to devices such as surgical procedures often have limited evidence of efficacy long-term. In the MSK patients a decision-based approach would be to plan treatment early or defer until the post

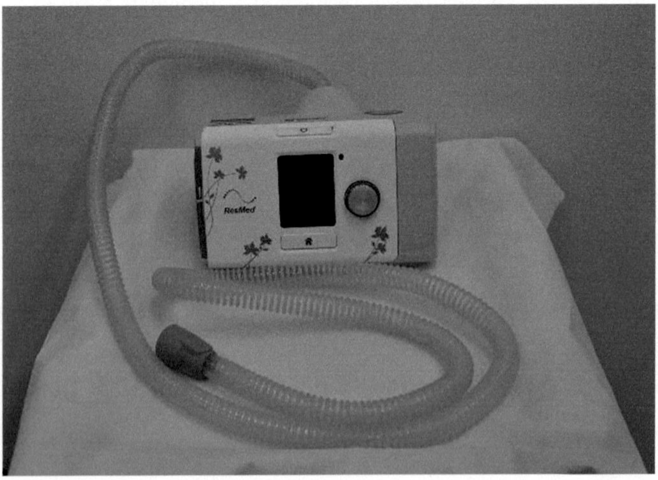

Fig. 7 CPAP device

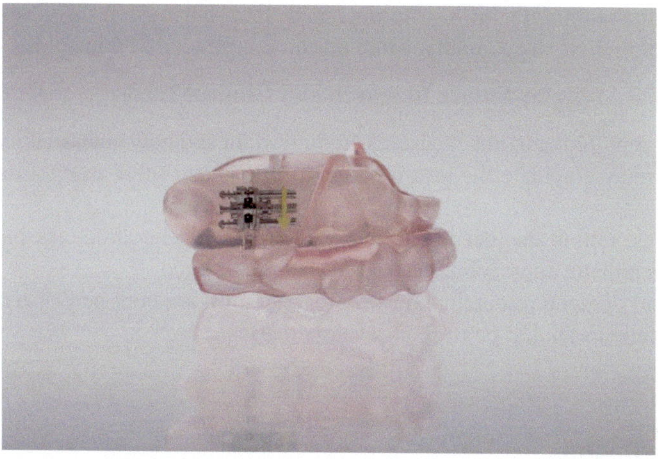

Fig. 8 Mandibular advancement device

operative period. The options will depend on the urgency of MSK regeneration surgery.

Patients with high body mass index will also be offered lifestyle modification options such as weight loss and reduction in smoking and drinking alcohol. In the MSK patient it may not be feasible to commence these options before treatment of painful joints. New novel treatments such as weight loss drugs may have a role to play (Mandavia et al. 2020; National Guideline 2021; Law et al. 2023; Steier et al. 2023).

5 Future Directions and Research

- **Emerging Trends in Sleep Medicine and MSK Regeneration**

As we learn more about the importance of sleep both for physical and mental health and long-term alleviation of disease progression, there is a growing focus on the peri-operative population to modify and treat factors which may be detrimental to post-operative recovery and improvement of outcomes.

The benefits of sleep have been well documented in this population. The emergence of Sleep medicine as a sub-speciality is now more needed than ever before. Physicians from different specialities will have experience in sleep medicine such as neurology, respiratory, anaesthesia, ENT and psychiatry. Whilst the individual specialisms will treat as a silo each sub specialist sleep disorder, a multidisciplinary approach is recommended due to the growing evidence that sleep disorders may co-exist. A patient with more than one sleep disorders such as co-morbid insomnia and sleep apnoea will need intervention for both to maximise patient outcomes.

Likewise, patients with MSK disorders may then have sleep disorders and pain compounding their sleep quality. More research is needed in this field.

- **Potential Areas for Future Research and Clinical Trials**

Patient reported outcome measures are important and now quantifiable. Research into these may elucidate the best treatment options and offer a target intervention approach.

More research is also needed in improving MSK regeneration especially STEM cells and a holistic approach also needs to be investigated.

The field of sleep research and MSK regeneration are both new, therefore innovative treatments should be further developed and evaluated.

6 Conclusion

- **The Critical Role of Sleep in Enhancing MSK Regeneration**

Modification of sleep factors may enhance the outcomes following MSK treatment and procedures. Whilst we can assume that many patients will have good sleep habits and sleep quality, it is important to identify potential areas for improvement. Screening questionnaires have been developed and have been validated in sleep medicine. These can be used to identify patients who are not sleeping well or who have sleep disordered breathing.

Sleep management should be a multidisciplinary process involving doctors from sleep medicine, anaesthesia and surgery.

Early intervention is key, and the intervention should follow guidelines where appropriate. However, more research is needed to demonstrate patient related outcome measures as well as objective data on strength of the MSK regeneration.

- **Key Takeaways for Clinical Practice**

Sleep is a modifiable state which can be improved. Improvement can have a positive impact on MSK regeneration. Identification of sleep conditions and problems is vital and early intervention is recommended. Treatments should be instituted according to guidelines. A multidisciplinary approach is needed to ensure the patients have the targeted and evidence-based treatments.

References

Abrishami A, Khajehdehi A, Chung F. A systematic review of screening questionnaires for obstructive sleep apnea. Can J Anaesth. 2010;57(5):423–38.

Auger RR, Burgess HJ, Emens JS, Deriy LV, Thomas SM, Sharkey KM. Clinical practice guideline for the treatment of intrinsic circadian rhythm sleep-wake disorders: advanced sleep-wake phase disorder (ASWPD), delayed sleep-wake phase disorder (DSWPD), Non-24-hour sleep-

wake rhythm disorder (N24SWD), and irregular sleep-wake rhythm disorder (ISWRD). An update for 2015: an American Academy of Sleep Medicine Clinical Practice Guideline. J Clin Sleep Med. 2015;11(10):1199–236.
Bixler EO, Kales A, Jacoby JA, Soldatos CR, Vela-Bueno A. Nocturnal sleep and wakefulness: effects of age and sex in normal sleepers. Int J Neurosci. 1984;23(1):33–42.
Bouman EJ, Mackenbach JD, Twisk JWR, Raimondo L, Beulens JWJ, Elders PJM, Rutters F. Is the association between social jetlag and BMI mediated by lifestyle? A cross-sectional survey study in the Dutch general population. Prev Med. 2024;181:107908.
Brown A, Smolensky M, D'Alonzo G, Redmond D, Conrad E, Hsi B. Circadian rhythm in human activity objectively quantified by actigraphy. Prog Clin Biol Res. 1990a;341a:77–83.
Brown AC, Smolensky MH, D'Alonzo GE, Redman DP. Actigraphy: a means of assessing circadian patterns in human activity. Chronobiol Int. 1990b;7(2):125–33.
Buysse DJ, Reynolds CF 3rd, Monk TH, Berman SR, Kupfer DJ. The Pittsburgh sleep quality index: a new instrument for psychiatric practice and research. Psychiatry Res. 1989;28(2):193–213.
Cahill GM, Besharse JC. Resetting the circadian clock in cultured Xenopus eyecups: regulation of retinal melatonin rhythms by light and D2 dopamine receptors. J Neurosci. 1991;11(10):2959–71.
Chan JK, Trinder J, Colrain IM, Nicholas CL. The acute effects of alcohol on sleep electroencephalogram power spectra in late adolescence. Alcohol Clin Exp Res. 2015;39(2):291–9.
Chasens ER, Ratcliffe SJ, Weaver TE. Development of the FOSQ-10: a short version of the functional outcomes of sleep questionnaire. Sleep. 2009;32(7):915–9.
Chung F, Subramanyam R, Liao P, Sasaki E, Shapiro C, Sun Y. High STOP-Bang score indicates a high probability of obstructive sleep apnoea. Br J Anaesth. 2012;108(5):768–75.
Colrain IM, Nicholas CL, Baker FC. Alcohol and the sleeping brain. Handb Clin Neurol. 2014;125:415–31. Elsevier.
Coté GG, Brody S. Circadian rhythms in Drosophila melanogaster: analysis of period as a function of gene dosage at the per (period) locus. J Theor Biol. 1986;121(4):487–503.
De Corso E, Bastanza G, Della Marca G, Grippaudo C, Rizzotto G, Marchese MR, Fiorita A, Sergi B, Meucci D, Di Nardo W, Paludetti G, Scarano E. Drug-induced sleep endoscopy as a selection tool for mandibular advancement therapy by oral device in patients with mild to moderate obstructive sleep apnoea. Acta Otorhinolaryngol Ital. 2015;35(6):426–32.
Dong X, Yang S, Guo Y, Lv P, Wang M, Li Y. Validation of Fitbit charge 4 for assessing sleep in Chinese patients with chronic insomnia: a comparison against polysomnography and actigraphy. PLoS One. 2022;17(10):e0275287.
Faber NS, Häusser JA, Kerr NL. Sleep deprivation impairs and caffeine enhances my performance, but not always our performance. Pers Soc Psychol Rev. 2017;21(1):3–28.
Garma L, Bouard G. Nocturnal awakenings as a function of age in insomniacs (author's transl). Rev Neurol (Paris). 1981;137(11):677–92.
Guilleminault C, Black JE, Palombini L, Ohayon M. A clinical investigation of obstructive sleep apnea syndrome (OSAS) and upper airway resistance syndrome (UARS) patients. Sleep Med. 2000;1(1):51–6.
Hauri PJ, Wisbey J. Wrist actigraphy in insomnia. Sleep. 1992;15(4):293–301.
Heinzer R. Obstructive sleep apnea: a risk for general anesthesia? Rev Med Suisse. 2007;3(134):2670–4.
Horne JA, Ostberg O. A self-assessment questionnaire to determine morningness-eveningness in human circadian rhythms. Int J Chronobiol. 1976;4(2):97–110.
Jordan AS, White DP. Pharyngeal motor control and the pathogenesis of obstructive sleep apnea. Respir Physiol Neurobiol. 2008;160(1):1–7.
Jung R, Kuhlo W. Neurophysiological studies of abnormal night sleep and the PICKWICKIAN syndrome. Prog Brain Res. 1965;18:140–59. Elsevier.
Konopka RJ. Clock mutants of Drosophila melanogaster. Proc Natl Acad Sci USA. 1971;68:2112–6.
Law M, Villar S, Oscroft N, Stoneman V, Fox-Rushby J, Chiu YD, Steele J, Devine T, Francis C, Claydon T, Hill G, Lim KK, Rayner J, Williams M, Spires E, Quinnell T. Continuous Positive

Airway Pressure plus Mandibular Advancement Therapy (PAPMAT): study protocol for an adaptive randomised crossover trial comparing the benefits and costs of combining two established treatments for obstructive sleep apnoea. Trials. 2023;24(1):474.

Lee MS, Yin TC, Sung PH, Chiang JY, Sun CK, Yip HK. Melatonin enhances survival and preserves functional integrity of stem cells: a review. J Pineal Res. 2017;62(2):e12372.

Lerner AB, Case JD, Takahashi Y. Isolation of melatonin and 5-methoxyindole-3-acetic acid from bovine pineal glands. J Biol Chem. 1960;235:1992–7.

Mandavia R, Mehta N, Veer V. Guidelines on the surgical management of sleep disorders: a systematic review. Laryngoscope. 2020;130(4):1070–84.

Maria S, Swanson MH, Enderby LT, D'Amico F, Enderby B, Samsonraj RM, Dudakovic A, van Wijnen AJ, Witt-Enderby PA. Melatonin-micronutrients osteopenia treatment study (MOTS): a translational study assessing melatonin, strontium (citrate), vitamin D3 and vitamin K2 (MK7) on bone density, bone marker turnover and health related quality of life in postmenopausal osteopenic women following a one-year double-blind RCT and on osteoblast-osteoclast co-cultures. Aging (Albany NY). 2017;9(1):256–85.

Miyasita A, Fukuda K, Inugami M. Effects of sleep interruption on REM-NREM cycle in nocturnal human sleep. Electroencephalogr Clin Neurophysiol. 1989;73(2):107–16.

National Guideline C. NICE evidence reviews collection. Demonstration of efficacy: obstructive sleep apnoea/hypopnoea syndrome and obesity hypoventilation syndrome in over 16s: evidence review M. London: National Institute for Health and Care Excellence (NICE). Copyright © NICE 2021; 2021.

Ong TH, Raudha S, Fook-Chong S, Lew N, Hsu AA. Simplifying STOP-BANG: use of a simple questionnaire to screen for OSA in an Asian population. Sleep Breath. 2010;14(4):371–6.

Ozdilekcan C, Kutun S, Dogan L. Impact of obstructive sleep apnea-related symptoms on surgical wound complications in breast cancer patients: pilot study in a tertiary health center in Turkey. Sleep Breath. 2021;25(2):835–42.

Payseur DK, Belhumeur JR, Curtin LA, Moody AM, Collier SR. The effect of acute alcohol ingestion on systemic hemodynamics and sleep architecture in young, healthy men. J Am Coll Heal. 2022;70(2):509–16.

Peake JM, Kerr G, Sullivan JP. A critical review of consumer wearables, mobile applications, and equipment for providing biofeedback, monitoring stress, and sleep in physically active populations. Front Physiol. 2018;9:743.

Pevet P, Challet E. Melatonin: both master clock output and internal time-giver in the circadian clocks network. J Physiol Paris. 2011;105(4–6):170–82.

Reppert SM. A clockwork explosion! Neuron. 1998;21(1):1–4.

Riemann D, Baglioni C, Bassetti C, Bjorvatn B, Dolenc Groselj L, Ellis JG, Espie CA, Garcia-Borreguero D, Gjerstad M, Gonçalves M, Hertenstein E, Jansson-Fröjmark M, Jennum PJ, Leger D, Nissen C, Parrino L, Paunio T, Pevernagie D, Verbraecken J, Weeß HG, Wichniak A, Zavalko I, Arnardottir ES, Deleanu OC, Strazisar B, Zoetmulder M, Spiegelhalder K. European guideline for the diagnosis and treatment of insomnia. J Sleep Res. 2017;26(6):675–700.

Sargent ML, Woodward DO. Genetic determinants of circadian rhythmicity in Neurospora. J Bacteriol. 1969;97(2):861–6.

Shea SA, Edwards JK, White DP. Effect of wake-sleep transitions and rapid eye movement sleep on pharyngeal muscle response to negative pressure in humans. J Physiol. 1999;520(Pt 3):897–908.

Steier JS, Bogan RK, Cano-Pumarega IM, Fleetham JA, Insalaco G, Lal C, Pépin JL, Randerath WJ, Redline S, Malhotra A. Recommendations for clinical management of excessive daytime sleepiness in obstructive sleep apnoea – a Delphi consensus study. Sleep Med. 2023;112:104–15.

Svanborg E, Larsson H, Carlsson-Nordlander B, Pirskanen R. A limited diagnostic investigation for obstructive sleep apnea syndrome. Oximetry and static charge sensitive bed. Chest. 1990;98(6):1341–5.

Thakkar MM, Sharma R, Sahota P. Alcohol disrupts sleep homeostasis. Alcohol. 2015;49(4):299–310.

Tune GS. Sleep and wakefulness in normal human adults. Br Med J. 1968;2(5600):269–71.
Wang X, Chen H, Jia L, Xu X, Guo J. The relationship between three-dimensional craniofacial and upper airway anatomical variables and severity of obstructive sleep apnoea in adults. Eur J Orthod. 2022;44(1):78–85.
Weaver TE, Laizner AM, Evans LK, Maislin G, Chugh DK, Lyon K, Smith PL, Schwartz AR, Redline S, Pack AI, Dinges DF. An instrument to measure functional status outcomes for disorders of excessive sleepiness. Sleep. 1997;20(10):835–43.
Wurtman RJ, Axelrod J, Phillips LS. Melatonin synthesis in the pineal gland: control by light. Science. 1963;142(3595):1071–3.

The Role of Psychology in Musculoskeletal Regeneration

Serena Simmons

Abstract While advancements in medical and surgical interventions have significantly improved the prospects of physical recovery, the psychological aspects of treatment and rehabilitation are equally critical and yet are so often overlooked in practice.

This chapter explores the crucial role of psychological factors in the context of musculoskeletal regeneration, emphasising their significance in enhancing outcomes. Research demonstrating that psychological well-being is integral to patient adherence, pain perception, and overall quality of life will be discussed. The influence of cognitive factors, such as self-efficacy and patient expectations will also be explored, in addition to the importance of social support and cultural sensitivity in shaping treatment and rehabilitation experiences.

By adopting a bio-psychosocial approach, this chapter argues for a comprehensive, multidisciplinary strategy that incorporates necessary psychological training for all healthcare practitioners alongside traditional training and medical treatments.

This holistic approach to patient care aims to optimise recovery and overall patient outcomes by addressing the complex, interconnected factors influencing musculoskeletal health. Through evidence-based insights and practical recommendations, this chapter underscores the necessity of integrating psychological care into musculoskeletal rehabilitation programs, ultimately fostering more effective and patient-centred care.

Keywords Psychology · Adherence · Motivation · Bio-psychosocial · Musculoskeletal- regeneration

---

S. Simmons (✉)
Nottingham Trent University, School of Social Sciences, Nottingham, UK
e-mail: serena.simmons@ntu.ac.uk

P. Lee (ed.), *Musculoskeletal Regeneration Medicine*,
https://doi.org/10.1007/978-3-031-77456-0_21

1 Introduction

Contemporary research increasingly highlights that physical recovery is not solely determined by biological healing processes but is significantly impacted by the patient's mental and emotional state.

Though manifesting with physical complaints, recovery from musculoskeletal injuries and or associated surgeries is often accompanied by significant emotional challenges. Furthermore the psychological response to injury can vary widely among patients, influenced by aspects such as severity of the injury, the patient's personal resilience, and their support network. Emotional distress, including anxiety, depression, catastrophising and fear of re-injury, can also negatively impact the rehabilitation process. Research has shown that patients with higher levels of pre-operative anxiety for example, tend to experience more post-operative pain and longer recovery periods (Anderson et al. 2016). It is not surprising then that addressing these emotional responses through psychological support, such as counselling or cognitive-based therapies (CBT), as well as more integrated psychological and holistic practices in healthcare consultation, can facilitate a more comprehensive recovery process.

With this in mind it is therefore argued that it is important to consider these pertinent psychological aspects in musculoskeletal regeneration. Recognising and integrating these psychological factors into patient care across the treatment pathway and addressing all patient interactions with this bio-psychosocial model in mind.

The Mind-Body Connection in Musculoskeletal Conditions

The human experience of pain and injury is not merely a biological event but a complex interplay of mind and body. This is in fact intuitive to us as humans.

Research in the field of psychoneuroimmunology demonstrates that psychological stress and emotional states can influence the body's inflammatory responses and immune function, which are critical to tissue repair and recovery (Dhabhar 2014). For instance, patients with high levels of anxiety or depression may experience heightened pain sensitivity, slower healing, and poorer functional outcomes following an injury or surgery (Edwards et al. 2016). This suggests that psychological factors are not just necessary to physical treatment, but are in fact central to the recovery process.

Patient perceptions and expectations are also powerful determinants of recovery in musculoskeletal conditions. The *belief* that one will recover, the perceived efficacy of a treatment, and the level of confidence exhibited in the healthcare provider all contribute to therapeutic outcomes. Studies show for example that patients who have positive expectations about their treatment are more likely to experience pain relief as well as functional improvement, even when the treatment is a placebo (Benedetti 2020).

The role of placebo

Placebo is a well researched phenomenon in health research, and is key to shaping our understanding of the mind-body phenomenon. Though accepted in standard drug trials when attempting to distinguish pharmaceutical effects from placebo (Crum and Langer 2007), placebo surgery has also had a significant impact in terms of furthering our understanding of this mind-body connection, specifically in regards to musculoskeletal surgical procedures.

In cases of surgical placebo, patients typically undergo the steps of surgery which can include incisions, anaesthesia etc., yet without the actual surgical intervention being performed. This is used in clinical research to determine the effectiveness of surgical interventions, and compares the outcomes between patients who have received the *actual surgery* versus those who receive the *placebo surgery.*

A significant example of placebo surgery in musculoskeletal populations is evident from arthroscopic surgery for knee osteoarthritis. An important study by Moseley et al. (2002) involved 165 patients with osteoarthritis of the knee who were randomly assigned to one of three groups: debridement, lavage, or placebo surgery.

In the placebo group, patients underwent a simulated surgical procedure under a short acting tranquilliser. These patients did not receive arthroscopy but did receive small skin skin incisions and an overnight stay in hospital. The study found no significant difference in pain relief and functional outcomes between the placebo surgery group and the groups that received actual surgical interventions. This finding has been influential in questioning the routine use of arthroscopic surgery for knee osteoarthritis, leading to a reevaluation of its indications in clinical practice. More importantly the evidence from placebo surgery studies such as this in musculoskeletal populations has significant implications for clinical practice (Moseley et al. 2002).

Studies such as these can challenge the necessity of certain surgical procedures, as they suggest that the benefits of some surgeries may be due more to the placebo effect than the actual mechanical or structural changes effected by the surgery. However, from a psychological perspective they highlight the critical role of 'patient expectation', therapeutic alliance, and the broader psychological context in which surgery occurs, and as such provide a clear rationale for these factors to be addressed as part of the comprehensive care of patients with musculoskeletal conditions.

Integrating Psychological Care in Musculoskeletal Treatment

While the physical aspects of musculoskeletal conditions are undeniably important, addressing the psychological dimensions of patient care is equally crucial. The mind and body are deeply interconnected and successful treatment outcomes often rely upon the integration of psychological care which takes place alongside any physical intervention or rehabilitation (Crum and Langer 2007). In recognising and addressing the complex interplay of thoughts, emotions, and beliefs in the healing process, healthcare providers may offer more comprehensive and effective care to patients with musculoskeletal injuries and conditions.

2 Psychological Factors Affecting Musculoskeletal Regeneration

Addressing a Holistic Model of Health

Humans are first and foremost feeling and thinking beings.

There is a common misconception that to help the patient presenting with physical injury or illness that 'fixing' the physical ailment will lead to better overall health and physical wellbeing. However, this reductionist view overlooks the profound influence that psychological factors can have on health outcomes, this is particularly evident in the context of musculoskeletal (MSK) conditions.

When a patient presents with a musculoskeletal injury, their experience is not limited to physical symptoms alone. Their thoughts, emotions, beliefs, and social context and culture, to name a few, all play an integral role in how they may perceive pain, adhere to treatments, and ultimately recover (Bayman et al. 2019).

Simply put, humans first encounter a somatic or feeling-based response to their outer world, which is only then translated to thoughts and words and consequently actions and behaviours (see Fig. 1). This simple I.C.E model shows a simplistic visual representation of these factors. The depth of '*Inner World*' experience highlights the significance of important feelings and somatic based responses, which are more often profound and unseen; and in many cases may even be unconscious to the person experiencing them.

This somatic experience is followed by *Communication*, which is demonstrated by a person's ability to formulate these feelings into words and tangible thoughts. Thoughts may present as internal thoughts and words, or may translate to the

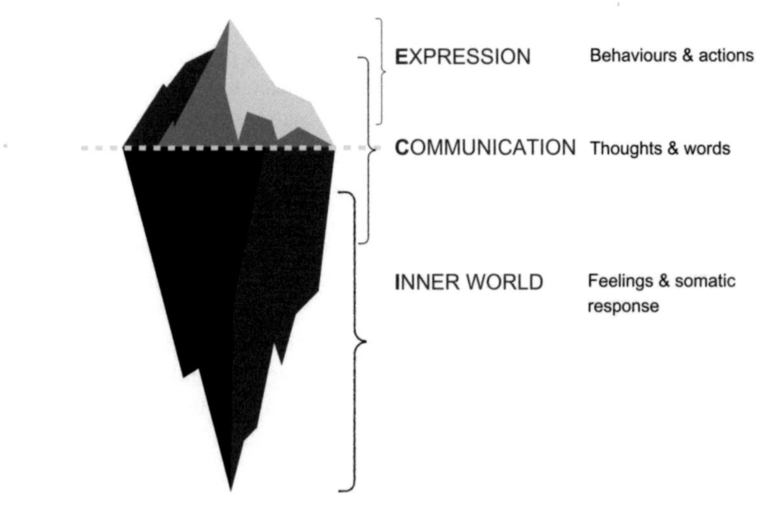

Fig. 1 I.C.E (Inner World, Communication & Expression), model highlighting processing of human response to external factors.

spoken word. People may struggle with the aspect of outward verbal communication as well as deciphering complex internal feelings and responses. This is where therapies, including Cognitive Based Therapies, such as CBT, psychotherapy and or psychological interventions and counselling may be useful to individuals who are struggling to process and or communicate this.

What follows is the *Expression* by way of actual Behaviours and Actions. It is typical for only this aspect of the model to be witnessed by external observers, such as practitioners and clinicians. It is the tip of the iceberg and in isolation fails to provide a full picture of a patient's experience. Understanding of a person's *Inner World* and *Communication* will help to make sense of someone's *Expressions* and thinking behind behaviours. .

On a practical level, this illustrates when dealing with patients in our care, particularly when we may be reliant on their 'buy-in' and adherence to protocols or treatments, that we must first consider how the patient is presenting to us in terms of their current thoughts and feelings about their situation. These will inform what the patient 'does next'. By exploring the *Inner World* we are also promoting a better foundation for engagement, and ultimately behaviour change. In order to do this we must be able to have conversations that allows our patients to share their thoughts and feelings without judgment or interference.

Approaching our consultations in this open manner allows us to help the patient make sense of their current situation within the broader context of their life. Allowing a patient to do this is also honouring the bio-psychosocial approach to care, as we allow space for the 'person' versus the 'complaint or 'issue' to be the sole basis for consultation.

Challenges in Utilising a Bio-Psychosocial Approach

Using a bio-psychosocial approach can be challenging. Research shows that although a bio-psychosocial approach by health practitioners can have positive outcomes for patients in terms of reduced pain, improvement in function etc., training and education for practitioners in navigating difficult conversations and exploring thoughts and feelings however is typically absent from professional training. Furthermore, when utilised there seems to be a lack of consistency in how practitioners have administered bio-psychosocial interventions (George 2008).

In a systematic review of randomised control trials investigating the effectiveness of a bio-psychosocial model approach in the management of lower back pain for example, it was reported that the approach to a bio-psychosocial intervention varied widely, and that it was unclear if the practitioners had the clinical skills required to carry out the interventions (George 2008). Further research also indicates that it can be a challenge for clinicians to integrate this bio-psychosocial approach into their physical therapy interventions, especially for those that have not received any formal training (Jones et al. 2002).

It is apparent that although there is good evidence to suggest that taking a more holistic approach, which includes these bio-psychosocial factors that can improve patient outcomes, more support and training for healthcare practitioners in application of techniques is required. This would foster a more positive experience for patients.

Mental Health and Mood

Mental health and wellbeing are concepts typically referenced in the context of being pathologised. Research, particularly in the scope of healthcare is usually focussed on mental 'ill-health', disease and disorder. However, whilst an individual may not possess a mental health problem or diagnosis, impending treatments, such as surgeries may induce acute changes in a person's mental health, which can include things such as anxiety, depression, and catastrophising. The ability of the practitioner to recognise these acute responses may have an important impact on the trajectory of a patient's treatment, experiences and outcomes.

Recent research highlights the profound impact of pre-existing anxiety on surgical outcomes. Patients who enter surgery with high levels of anxiety are more prone to experience intensified pain post-surgery (Vaughn et al. 2020). This increased pain perception can lead to a greater reliance on analgesics, which can introduce additional risks, such as opioid dependence or adverse drug reactions (Vaughn et al. 2020). Moreover, the presence of anxiety has been linked to longer hospital stays, as patients require extended care to manage both their physical and psychological symptoms (Yang et al. 2024). Psychological stress can also activate the body's stress response, potentially impairing immune function and making the body more susceptible to infections and other complications Seiler et al (2020).

Depression is another psychological factor that can adversely affect post-operative recovery. Whilst we can't determine causation, numerous studies have demonstrated that depression is associated with poor wound healing and a diminished immune response, which can result in a higher incidence of post-operative infections (Hernandez et al. 2022). Depression is also linked to reduced motivation and energy levels that can result in lower adherence to post-operative care instructions, physical therapy and wound care, which may further complicate recovery (Brummett et al. 2015).

Catastophising is characterised by an exaggerated negative outlook on pain and recovery; it is a cognitive distortion that prompts people to jump to the worst possible conclusion (Quartana et al. 2009). This is well researched in the area of surgical intervention and has been shown to significantly influence post-operative outcomes. Patients who catastrophise are more likely to report severe pain and are less likely to engage in proactive pain management strategies (Bayman et al. 2019). This mindset can also create a self-perpetuating cycle where fear and anxiety about pain lead to increased pain perception, which in turn fuels further catastrophising. The impact of which may also affect other aspects of recovery, such as sleep quality, physical mobility, and overall psychological well-being (Bayman et al. 2019).

The interplay between these psychological and mental health factors; anxiety, depression, and catastrophising can also significantly influence post-operative recovery. For example, a patient with both anxiety and depression may experience compounded difficulties, such as heightened pain perception combined with poor wound healing, leading to a more complicated and prolonged recovery process (Gouin and Kiecolt-Glaser 2011).

Addressing these psychological issues pre-operatively by fostering a open communication, offering support via mental health support services and therapeutic modalities where needed, like cognitive-behavioural therapy (CBT), mindfulness interventions, psychotherapy or psychological counselling as well as appropriate pharmacological treatments can be crucial in improving surgical outcomes. Furthermore early identification and management of these mental health issues can help mitigate their negative impact, leading to faster recovery, reduced pain levels, and fewer post-operative complications.

Trauma

Trauma is a term used to describe a person's psychological response to a distressing event or series of events and or overwhelming experiences, which can significantly affect one's mental and physical health (Van der Kolk 2015). Often these experiences can lead to lasting psychological, emotional, and physiological effects. The impacts of trauma are far-reaching, influencing not only mental health but also physical health outcomes.

Research has shown that trauma can for example, profoundly affect how patients respond to medical treatment, including surgical interventions. Studies by Roth et al. (2020) have shown that patients with a history of trauma for example may experience heightened stress responses during surgery, which can exacerbate pain, increase the likelihood of postoperative complications, and prolong recovery times. This stress response can also interfere with the body's immune function and wound healing processes, leading to poorer surgical outcomes. Furthermore, the integration of trauma-informed care into hospital settings has been advocated as a means to improve patient outcomes, particularly for those undergoing surgery (Roth et al. 2020).

In this instance trauma-informed care involves recognising the pervasive impact that the trauma may have had on the individual and consequently understanding how it can affect patient behavior and responses to treatment. It should for example be considered by the individual working with the patient, that if a trauma that has been experienced is all too pervasive, and of course depending on individual circumstances, should actually be the focus for the patient at the current time. This should be done in conjunction with a qualified mental health practitioner such as Health Psychologist, Clinical psychologist or Psychiatrist.

With that said, it should not be ignored that by fostering a better trauma-informed approach in general to patients care, would enable practitioners to reduce the possibility of re-traumatising the patient, enhancing engagement, and foster a sense of safety and trust within the healthcare environment. This can be simply achieved by enhancing our own psycho-education and understanding of trauma.

Pain

Chronic musculoskeletal conditions, such as osteoarthritis or chronic low back pain, can have significant psychological impact. This can lead to a cycle of pain, disability, emotional distress, lack of engagement in care and change in mood, all which highlight the complexity in understanding this multi-dimensional experience. Though a thorough examination of pain is beyond the scope of this chapter, it is put forward here as a factor that should be considered when working with an MSK population. The psychological impact of chronic pain can include heightened anxiety, depression, and catastrophising, as previously discussed. This can create a positive feedback cycle as these factors exacerbate pain intensity and contribute to the persistence of pain (Turk and Wilson 2010).

Often it is the chronic nature of the conditions experienced that can give rise to feelings of helplessness, frustration, and even depression, which in turn can exacerbate physical symptoms. For example, fear-avoidance beliefs, where patients avoid physical activity due to fear of pain or further injury, can lead to decreased mobility, muscle atrophy, and worsening of the condition (Vlaeyen and Linton 2000). Recent research has further highlighted the complex relationship between chronic pain and psychological factors, emphasising the bio-psychosocial model of management as a means to help the patient from this more holistic approach which has been discussed.

Addressing this complex issue can prove challenging for the clinician tasked with administering treatment. Pain can act as an obvious barrier to engagement, even though partaking in treatments protocols and prescriptions may act to help the individual and ease significant pain. In this case, cognitive-behavioural therapies such as CBT and other psychological interventions and therapy have been shown to be effective in addressing these maladaptive thought patterns, reducing pain severity, and improving functional outcomes (Williams et al. 2012).

Additionally important work in the area of pain education referred to as Explaining Pain, or EP has also shown to have positive impact. In this case EP describes a range of educational interventions that act as a means to educate someone about the biology of pain as well as addressing the potential positive function pain may serve (Moseley and Butler 2015). In this case numerous studies have shown that including EP as part of treatment not only increases knowledge of pain-related biology but may decrease catastrophising which has been illustrated here within as a potential barrier to engagement (Bayman et al. 2019). Furthermore it has been shown to provide short-term reductions in pain and disability.

These interventions underscore the importance of addressing the psychological dimensions of chronic pain to break the cycle of pain, disability, and emotional distress, ultimately improving the quality of life for individuals with chronic musculoskeletal conditions.

Other Factors

There are other significant factors which contribute the patient experience and treatment outcomes which are beyond the scope of this chapter. It is for example important that sensitivity is paid to differences in culture. Multiple research studies promote the basis of trust and rapport building as foundational to patient adherence and treatment outcomes (Thom et al. 2004). With this in mind it seems obvious that cultural sensitivity or lack thereof, may impact trust and rapport and thus influence patient perceptions of illness and attitudes towards their treatment. Culturally sensitive approaches in consultation, including the use of culturally relevant therapeutic practices and communication styles, are important to providing effective psychological support and ensuring equitable care (Sue et al. 2009).

Furthermore social support is a crucial determinant of psychological and physical health outcomes in patients undergoing musculoskeletal regeneration. Support from family, friends, and healthcare providers can provide emotional comfort, practical assistance, and motivation. The presence of a strong support system has been associated with better adherence to rehabilitation programs, reduced psychological distress, and improved recovery outcomes. For instance, patients who receive encouragement and assistance from their social network are more likely to adhere to exercise regimens and follow medical advice (Dueñas et al. 2016; Reblin and Uchino 2008).

Some simple steps healthcare professionals could take are to consider involving family members in the rehabilitation process, providing them with information and strategies to support the patient's recovery. Or, perhaps including within the consultation questions around support network, identifying their 'Rehabilitation/support cheerleaders'. These would be key people in the patient's life who the patients feels they can trust and have a good relationship with (Simmons et al. 2024).

3 Behaviour Change: Motivation and Adherence

Musculoskeletal regenerative therapies and surgical interventions may require patients to engage in novel treatments and adhere to new behaviours. These could range from attending a single appointment to engaging in regular and challenging physiotherapy.

Other examples may require patients to engage losing weight prior to surgery, make dietary changes or stop smoking. It is essential to recognise that these recommendations can involve substantial behavioural changes. However, making these behavioural changes is inherently challenging, as this requires altering deeply ingrained habits and routines. Research consistently shows that behaviour changes, particularly in areas such as diet, weight loss, and smoking cessation, can be difficult to achieve and maintain. Research highlighting that long-term success in behavior change often involves multiple attempts and sustained effort over time (Fischer et al. 2020; Prochaska and Goldstein 1991).

Barriers to Adherence

A body of research has developed over the last two decades that has attempted to shed light on barriers to adherence within patient populations. These barriers to adherence can be summarised into 2 broad areas: *Environmental Factors* and *Personal Factors*.

Environmental Factors can include a feeling of poor social support, as previously discussed, as well as a lack of positive feedback (Jack et al. 2010; Sluijs et al. 1993 cited in Kelsey and Picha 2017). *Personal Factors* include feelings of helplessness, depression, anxiety, increased pain with exercise, and a patient's perception of perceived barriers. This category also includes the well-researched topic of self-efficacy (Jack et al. 2010; Sluijs et al. 1993 cited in Kelsey and Picha 2017).

Self Efficacy

Self efficacy is the perceived belief in the ability to perform a task (Bandura 1977). In patient groups, having low self-efficacy can manifest in an apparent lack motivation to achieve specific goals, refusal to complete tasks and a tendency to dwell on deficiencies. Those with a high self-efficacy typically present more positively and show to be proactive in setting personal goals, maintain their commitment to the task and quickly recover after a setback.

Patients will present with varying levels of self-efficacy, which is essentially predetermined. If a patient is already exhibiting a high level of self-efficacy the job of the practitioner is made much easier as these are the patients that essentially 'do the work'. A study by Campbell et al. (2019) demonstrated that patients with positive expectations and a sense of self-efficacy were more likely to follow rehabilitation protocols and achieve better outcomes. An ongoing debate in health spheres, however, is what can we do if we see patients presenting with a low self-efficacy.

Work in this area has been useful to inform models of practice to facilitate behavioural change within patient groups. In this case a strategy to increase an individual's 'level' of self-efficacy would be to first assess it and then establish further feedback or exposure to behaviours to positively impact it. In Bandura's model (1977), self-efficacy is based on 4 key sources of information, these are Performance Accomplishments, Vicarious Experience, Verbal Persuasion and Emotional Arousal. In each instance, it is argued that increasing a person's exposure and experience in these modes of induction may help to increase a person's self-efficacy (seen in Fig. 2).

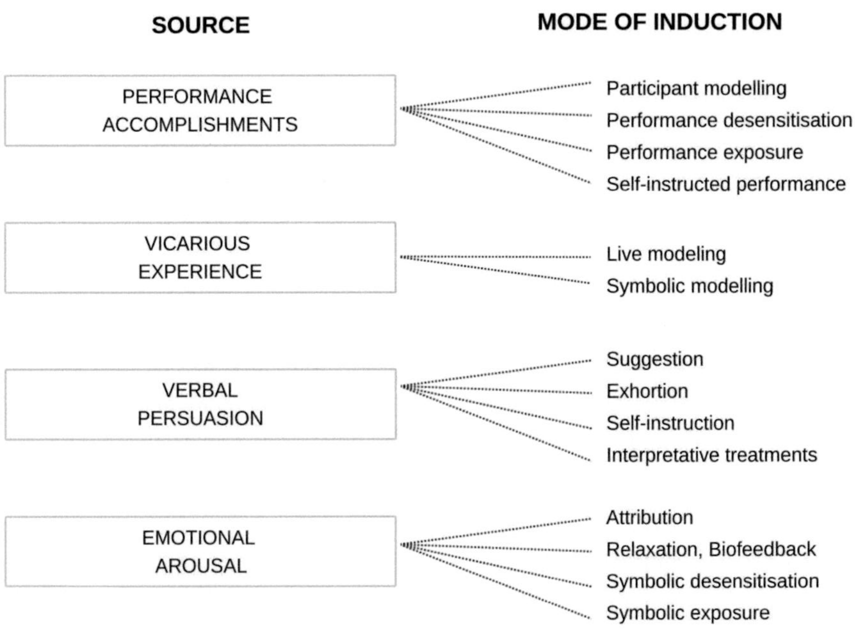

Fig. 2 Major sources of efficacy information and the principal sources through which different modes of treatment operate (Bandura 1977)

For example, when addressing Performance Accomplishments, repeated exposure to success in any given area, increases the experience of 'personal mastery'. In this case if someone can do something with repeated success then a feeling of mastery can be achieved, therefore lessening the potential negative impact of any occasional future failures.

Understanding these aspects of self-efficacy can be helpful in guiding how we can help to create similar 'positive feedback loops' for people who may be undertaking something new or challenging, such as exercise and rehabilitation. In this case, what can be fed back to the patient which illustrates that what they are undertaking is 'working'. It may be helpful for the practitioner to gauge aspects of self-efficacy early on in their interactions with a patient, and to use this framework as a model to address ways in which aspects of positive feedback can be supported.

Goal Setting

Another strategy that is often implemented to improve adherence is goal setting. The formation of goals are encouraged to foster a positive future vision of where a patient would like to get to in their treatment or rehabilitation.

If implemented 'smartly' (S.M.A.R.T; Locke and Latham 2013), patients can foster a way of consciously setting specific and realistic goals, in a manner that can be easily attained. In this instance if smaller strategic goals are set on the way to achieving the larger end-goal, it can create positive feedback and associated enhanced feelings of mastery. As such further and more inspiring goals can be developed.

Research into goal-setting in physiotherapy practice often reports positive results, however, development and implementation of this strategy is not always straight forward or consistent across practice. For example, Schoeb et al. (2013) investigated the prevalence of goal setting within standard physiotherapy practice. In 37 physiotherapy-patient sessions goal setting was explicitly addressed in only 11 sessions, Furthermore, in many of these 11 sessions there seemed to be an automatic assumption that the patients already had a goal in mind. On questioning the patients, it was clear that not all had an understanding of what may be a realistic or achievable goal to set, due to a lack of knowledge about things like the prognosis of their condition. In this instance, we need a considered approach, since practitioner-defined or assumed goals that appear unachievable by the patient could actually present as a barrier to adherence.

Properly applied however, goal setting can be a very useful tool to encouraging adherence, particularly if the therapist is able to adopt a holistic approach that acknowledges the patient's individual factors, such as lifestyle, time commitments, knowledge of condition etc. The therapist can then become aware of what a realistic goal would 'look like' to their individual patients.

Motivation

It is a common misconception that motivation is something that merely exists as a character trait. While the subtleties of this are arguable, what is apparent is that motivation is usually something that arises out of goal-setting (Locke and Latham

2013). In other words there has to be something that is the target or desired outcome, before someone can exhibit a sense of motivation to achieve it.

Another common misconception is that those patients deemed more 'motivated' to change, will in fact find adherence to protocols prescribed prior to interventions, easier. Indeed whilst patients that are seemingly motivated and exhibit high levels of self-efficacy and self-management strategies may not need input from the practitioner to promote adherence; the prolific problem of lack of adherence in the wider population however, requires attention. Research by Simmons et al. (2024), highlights this difference. In this study, patients on a knee arthroplasty wait-list were offered a novel 12 week strength based pre-habilitation programme of exercise and education prior to their impending surgery. It was found that those who declined the surgery, though showing signs of 'motivation', were in fact only motivated to have surgery and therefore showed no signs of wanting to engage with any other interventions (Simmons et al. 2024).

Aspects of helping a person to access internal motivation can be taken from evidence-based theory around self-determination, self-regulatory and self-management practices. Here the importance of encouraging an autonomous self-directed motivation to do something versus being externally motivated or controlled by 'pressure to behave in a particular way' can be powerful in unlocking a person's conscious and motivated desire to do something (Deci and Ryan 2008).

Alongside specific goal setting, there are tools that may enable us to tap into a person's desire to make a change. Motivational Interviewing techniques are also useful in this instance. Motivational interviewing (MI) is a therapeutic approach that can enhance a patient's motivation to change by exploring and resolving ambivalence and is a modality shown to be effective in promoting behavior change across various health conditions (Miller and Rollnick 2023). In utilising the techniques of MI, practitioners may be able to tap into a person's underlying fears, motivations and thoughts around what is being asked of them, which may help in alleviating any perceived barriers to exercise and or indeed any other lifestyle changes in may be of benefit for the patient to make (Miller and Rollnick 2023).

To enable wider scale adoption of these techniques across healthcare settings, training and mentoring on methods of holistic support, such as human psychology and decision making, behaviour change and even more specific modalities like MI; should be made readily available. This would also allow for access to validated and evidence-based methods of interventions for those that wish to implement such strategies and would thus initiate more of a standardised level of care across patient groups (Sanders et al. 2013).

Something that remains relatively uncovered in the literature pertaining to how we understand a person's adherence to health-related behavioural change, and one that is arguably mostly missing from patent-therapist conversations is the question of *why engage in 'it' in the first place?*

WHY Engage

'Why' is a common question asked in psychological therapeutic practices. When undertaking psychological work with a client, knowing 'why' someone wishes to do the work is one of the most important starting points. From this point we get to have insight into the client's motivation, intrinsic or otherwise and this gives us an insight into a person's driving force behind their action. It also establishes emotional anchor points (evocative feelings), which are important in reinforcing a person's continuation in treatment or therapy. We can also build rapport and gather more information around internal dialogue, feelings, potential hindrances as well as other social factors and barriers to action.

All of these factors then allow us to have more impact on those things that we know hold people back. They can also open up important lines of enquiry around problem solving and taking action, thus helping to alleviate any potential barriers. Accordingly, healthcare practitioners may also experience similar benefits by starting with the same question of 'why' with their own patients.

Fleshing out this story from a patient is powerful way of unlocking initial motivations and barriers, and serves as a sound starting point *to do the work*. These statements and emotional anchors can be reflected back to the patient throughout the process of treatment as a means to support adherence. For example, the patient who is able to identify that by taking on rehabilitative exercise will allow them to be more mobile and "spend time playing their grandchildren" will act as a powerful emotional anchor and 'why' compared to the patient who is attending "because their doctor told them to".

4 Clinical Application

It has been established that a patient's mindset can significantly alter their response to treatment and outcomes. This underscores the need for healthcare practitioners to not only address physical aspects of care, but to continue to strive for a more holistic approach to treatment, that includes the core psychological factors mentioned here and more widely social, biological and environmental elements.

In support of this holistic goal, practitioner communication is the linchpin and one might consider the power of the initial consultation (see Fig. 3). It is in this initial interaction that there is a prime opportunity to instil a sense of understanding for the patient and collaborating with them to achieve a clear vision for their pathway of care. Such would create clarity, as well as a sense of autonomy and empowerment where the patient feels calm, informed and involved. The contrary would be a didactic surface-level prescription of events that they should engage in, instilling a sense of uncertainty, worry, anxiety and upset.

Positive facilitation of such conversations, in which patients feel empowered to express their thoughts and concerns without fear of judgment or interruption, are vital in establishing this safe and supportive environment for open dialogue to take place, and it has been established that this is crucial in healthcare, as it enables patients to communicate their needs, fears, and experiences more freely (Thom et al. 2004).

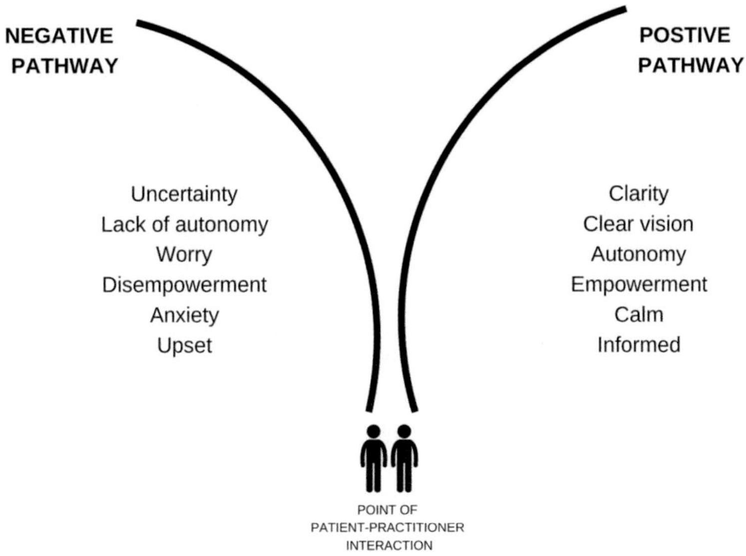

Fig. 3 Positive treatment interaction pathway

This level of openness not only enhances the accuracy of the information provided to healthcare professionals but also fosters trust and collaboration, ultimately leading to more personalised and effective care. It has been established that when patients feel heard and respected, they are more likely to engage actively in their treatment plans, adhere to medical advice, and experience better overall outcomes. Therefore, prioritising a non-judgmental and interference-free space for patient communication is foundational to delivering high-quality, patient-centred care.

5 Conclusion

Given the significant role of psychological factors in musculoskeletal health, it has been put forward here that it is essential to adopt a holistic approach to patient care, one that integrates psychological support alongside physical treatment. What is clear from this chapter, is that patients will continue to show up for treatment with a vast array of complexities some of which are due merely to 'the human condition', in other words the subtleties and nuance of how a person feels, thinks and processes information. Patients also undoubtedly may present with individual traumas, pain, inhibiting social factors and mental health and or mood disorder.

What is clear in observing this, is that the only actual element which can be changed by the healthcare individual working to help these patients, is themselves. This can be done not only be enhancing their general expertise regarding their scope of practice, such as surgical interventions, but by up-levelling their knowledge of psycho-education and training in the area of human psychology, trauma mental health cultural diversity and behaviour change. It is by up-skilling in these areas, and bringing one's awareness to what other behaviours and modalities may not only enhance the impact of the intervention one may be administering, but also may have the positive impact of impacting someone's life more holistically.

References

Anderson R, Patterson C, Meng C. Preoperative anxiety and its impact on postoperative pain and recovery. J Surg Rehabil. 2016;34(2):135–42.

Bandura A. Self-efficacy: toward a unifying theory of behavioural change. Psychol Rev. 1977;84(2):191–215.

Bayman EO, Parekh KR, Keech J, Selte A, Brennan TJ, Sessler DI. Preoperative patient expectations of postoperative pain are associated with moderate to severe acute pain after VATS. Pain Med. 2019;20(3):543–54. https://doi.org/10.1093/pm/pny096.

Benedetti F. Placebo effects: understanding the mechanisms in health and disease. Oxford University Press; 2020.

Brummett CM, Waljee JF, Goesling J, Moser SE, Lin P, Englesbe MJ, Clauw DJ. New persistent opioid use after minor and major surgical procedures in US adults. JAMA Surg. 2015;152(6):e170504. https://doi.org/10.1001/jamasurg.2017.0504.

Campbell MJ, Horvath CA, Miller JK, Roberts SL. The impact of patient expectations and self-efficacy on rehabilitation outcomes. J Rehabil Res. 2019;45(2):120–35.

Crum AJ, Langer EJ. Mind-set matters: exercise and the placebo effect. Psychol Sci. 2007;18(2):165–71. http://www.jstor.org/stable/40064598.

Deci EL, Ryan RM. Self-determination theory: A macrotheory of human motivation, development, and health. Canadian Psychology / Psychologie canadienne. 2008; 49(3):182–85. https://doi.org/10.1037/a0012801.

Dhabhar FS. Effects of stress on immune function: the good, the bad, and the beautiful. Immunol Res. 2014;58(2–3):193–210.

Dueñas M, Ojeda B, Salazar A, Mico JA, Failde I. A review of chronic pain impact on patients, their social environment and the health care system. J Pain Res. 2016;9:457–67. https://doi.org/10.2147/JPR.S105892.

Edwards RR, Dworkin RH, Sullivan MD, Turk DC, Wasan AD. The role of psychosocial processes in the development and maintenance of chronic pain. J Pain. 2016;17(9):T70–92.

Fischer M, Oberänder N, Weimann A. Four main barriers to weight loss maintenance? A quantitative analysis of difficulties experienced by obese patients after successful weight reduction. Eur J Clin Nutr. 2020;74:1192–200. https://doi.org/10.1038/s41430-020-0559-x.

George S. What is the effectiveness of a biopsychosocial approach to individual physiotherapy care for chronic low back pain? Internet J Allied Health Sci Pract. 2008;6:4. https://doi.org/10.46743/1540-580X/2008.1179.

Gouin JP, Kiecolt-Glaser JK. The impact of psychological stress on wound healing: methods and mechanisms. Immunol Allergy Clin N Am. 2011;31(1):81–93. https://doi.org/10.1016/j.iac.2010.09.010. PMID: 21094925; PMCID: PMC3052954.

Hernandez R, Falzon L, Arroyave WD, Lustberg MB. The influence of depression on wound healing: a systematic review of the literature. J Affect Disord. 2022;296:218–29.

Jack K, McLean SM, Klaber, Moffett JK, Gardiner E. Barriers to treatment adherence in physiotherapy outpatient clinics: a systematic review. Man Ther. 2010;15(3):220–8. https://doi.org/10.1016/j.math.2009.12.004.

Jones M, Edwards I, Gifford L. Conceptual models for implementing biopsychosocial theory in clinical practice. Man Ther. 2002;7(1):2–9. https://doi.org/10.1054/math.2001.0426. PMID: 11884150.

Kelsey TJ, Picha KJ. Understanding the role of self-efficacy in patient adherence to home exercise programs. J Orthop Sports Phys Ther. 2017;47(8):1–9.

Locke EA, Latham GP. New developments in goal setting and task performance. New York: Routledge; 2013.

Miller WR, Rollnick S. Motivational interviewing: helping people change and grow. 4th ed. Guilford Press; 2023.

Moseley GL, Butler DS. Fifteen years of explaining pain: the past, present, and future. J Pain. 2015;16(9):807–13.

Moseley JB, O'Malley K, Petersen NJ, Menke TJ, Brody BA, Kuykendall DH, Wray NP. A controlled trial of arthroscopic surgery for osteoarthritis of the knee. N Engl J Med. 2002;347(2):81–8.

Prochaska JO, Goldstein MG. Process of smoking cessation: implications for clinicians. Clin Chest Med. 1991;12(4):727–35.

Quartana PJ, Campbell CM, Edwards RR. Pain catastrophizing: a critical review. Expert Rev Neurother. 2009;9(5):745–58. https://doi.org/10.1586/ern.09.34.

Reblin M, Uchino BN. Social and emotional support and its implication for health. Curr Opin Psychiatry. 2008;21(2):201–5. https://doi.org/10.1097/YCO.0b013e3282f3ad89.

Roth RS, Geisser ME, Williams DA. Trauma and the perioperative period: a review of the impact of psychological stress on surgical outcomes. J Pain Res. 2020;13:1501–12.

Sanders T, Foster NE, Bishop A. Biopsychosocial care and the physiotherapist: a journey from denial to uneasy incorporation. Physiother Theory Pract. 2013;29(6):420–8.

Schoeb V, Staffoni L, Pilnick A. The role of the physiotherapist in breaking bad news: a comparison of three European countries. Physiother Theory Pract. 2013;29(8):527–37.

Seiler A, Fagundes CP, Christian LM. The impact of everyday stressors on the immune system and health. In: Choukèr, A. (eds) Stress Challenges and Immunity in Space. Springer, Cham. 2020. https://doi.org/10.1007/978-3-030-16996-1_6.

Simmons S, Brownlow M, Minshull C. Psychological-profile characteristics of knee arthroplasty patients who adhered to a 12 week strength-based prehabilitation programme vs. those unwilling to engage. Physiotherapy. 2024;123(S1):e237, ISSN 0031-9406. https://doi.org/10.1016/j.physio.2024.04.297.

Sluijs EM, Kok GJ, van der Zee J. Correlates of exercise compliance in physical therapy. Phys Ther. 1993;73(11):771–86.

Sue S, Zane N, Nagayama Hall GC, Berger LK. The case for cultural competency in psychotherapeutic interventions. Annu Rev Psychol. 2009;60:525–48. https://doi.org/10.1146/annurev.psych.60.110707.163651.

Thom DH, Hall MA, Pawlson LG. Measuring Patients' Trust in physicians when assessing quality of care. Health Aff. 2004;23(4):124–32.

Turk DC, Wilson HD. Fear of pain as a prognostic factor in chronic pain: conceptual models, assessment, and treatment implications. Curr Pain Headache Rep. 2010;14(2):88–95. https://doi.org/10.1007/s11916-010-0094-x. PMID: 20425197; PMCID: PMC2872063.

Van der Kolk BA. The body keeps the score: brain, mind, and body in the healing of trauma. Penguin; 2015.

Vaughn F, Wichman C, McLaughlin M, Schwartz S. Anxiety and post-surgical pain in elective orthopedic surgery: a prospective cohort study. Pain Med. 2020;21(6):1183–91.

Vlaeyen JWS, Linton SJ. Fear-avoidance and its consequences in chronic musculoskeletal pain: a state of the art. Pain. 2000;85(3):317–32. https://doi.org/10.1016/S0304-3959(99)00242-0.

Williams AC, Eccleston C, Morley S. Psychological therapies for the management of chronic pain (excluding headache) in adults. Cochrane Database Syst Rev. 2012;2012(11):CD007407. https://doi.org/10.1002/14651858.CD007407.pub3.

Yang J, Zeng Y, Yang H, Qu Y, Han X, Chen W, Sun Y, Hu Y, Ying Z, Liu D, Song H. Cardiovascular disease, genetic susceptibility, and risk of psychiatric disorders and suicide attempt: a community-based matched cohort study based on the UK biobank. J Am Heart Assoc. 2024;13(15):e031280. https://doi.org/10.1161/JAHA.123.031280.

Gene and Exosome Therapy in Musculoskeletal Regeneration

Jovana Vuković and Džihan Abazović

Abstract Advances in gene therapy and exosome research have unveiled promising opportunities for repairing and regenerating injured tissues in different body regions, including the musculoskeletal system. This chapter explores the multifaceted roles of exosomes, highlighting their importance as essential mediators in intercellular communication and their potential to act as signalling molecules. MicroRNAs, encapsulated within exosomes, play a major role in exosome-mediated function and are being studied as potent regulators that coordinate cellular processes that are necessary for musculoskeletal regeneration. Gene therapy has also been widely investigated in the regeneration of the musculoskeletal system; however, there is no consensus regarding the adequate vector, method of application, or approach. This chapter will also talk about the therapeutic potential of engineered exosomes, which involves genetic manipulation and using the transformative properties of exosomes to deal with the problems that come up with musculoskeletal repair. Nevertheless, the remaining issues and questions surrounding exosomes and gene therapy implementation in clinical practice must be resolved in combination with developing new guidelines for clinical use.

Keywords Exosomes · Extracellular vesicles · Muskuloskeletal regeneration · Gene therapy · miRNA

J. Vuković (✉)
School of Medicine, University of Belgrade, Belgrade, Serbia

D. Abazović
School of Medicine, University of Belgrade, Belgrade, Serbia

ABA Medica, Ulcinj, Montenegro

P. Lee (ed.), *Musculoskeletal Regeneration Medicine*,
https://doi.org/10.1007/978-3-031-77456-0_22

1 Introduction: Emerging Field of Extracellular Vesicles

In the 1980s, scientists investigating the process of reticulocyte maturation were the first to define exosomes, subsequently initiating the research of extracellular vesicles (EVs) (Harding et al. 1983; Pan and Johnstone 1983). Initially, these vesicles were believed to only remove unnecessary metabolites, called cellular waste. Later, it was found that EVs derived from immune cells could present antigens, which has led to the identification of EVs as effective vaccines against tumours (Raposo et al. 1996). These extracellular vesicles have gradually captured the spotlight in the scientific community. In the last 20 years, there has been a growing interest in EV research due to the idea that they can perform physiological functions, act as biomarkers and be used for therapeutic purposes.

Exosomes are small membrane-bound vesicles secreted by various cell types, typically described as smaller than 150 nm. The exosomes are formed through intraluminal budding within the multivesicular bodies (MVBs) and are released when these compartments merge with the plasma membrane (Kalluri and Lebleu 2020). The International Society for Extracellular Vesicles (ISEV) defines "extracellular vesicle" (EV) as naturally released particles from cells, surrounded by a lipid bilayer, without the ability to replicate. However, there is no consensus on specific markers of EV subtypes, making it difficult to assign them to particular biogenesis pathways. Thus, researchers are advised to use operational terms to classify different subtypes of EVs based on physical characteristics (e.g., size), biochemical composition, or specific descriptions of conditions or cell of origin rather than using terms like exosome and microvesicle. It is also recommended to employ at least three EV markers to assess purified EVs (Théry and Witwer 2018).

Exosomes: an important mediators in cellular communication

Exosomes play a crucial role in intercellular communication by transporting proteins, nucleic acids, and other bioactive molecules between cells (Fig. 1). Exosomes show a notable tendency to target certain organs and uptake by particular cell types, owing to their heterogeneity, reflected by differences in their size, content, donor, and effect on the recipient cell (Hazan-Halevy et al. 2015; Sancho-Albero et al. 2019). Depending on the cellular origin and the functional condition of the parent cells, which significantly determines the contents of the exosome cargo that will be transferred to the recipient or target cells, exosomes can have important physiological roles in mediating cellular responses such as modulating immune responses, mediate viral responses and contributing to tissue repair and regeneration (Kalluri and Lebleu 2020). Exosomes contain many non-coding RNA, including micro RNA (miRNA) and long non-coding RNAs (lncRNAs). These RNA molecules have the ability to affect cell viability by interfering with certain biological pathways in specific cells (Zhang et al. 2019). Through intercellular communication, these biologically active molecules may modulate gene expression, change recipient cell fate and function, enhance physiologic function, and/or mitigate pathogenesis (Zhang et al. 2015). miRNAs may show a higher concentration in exosomes than in parent cells,

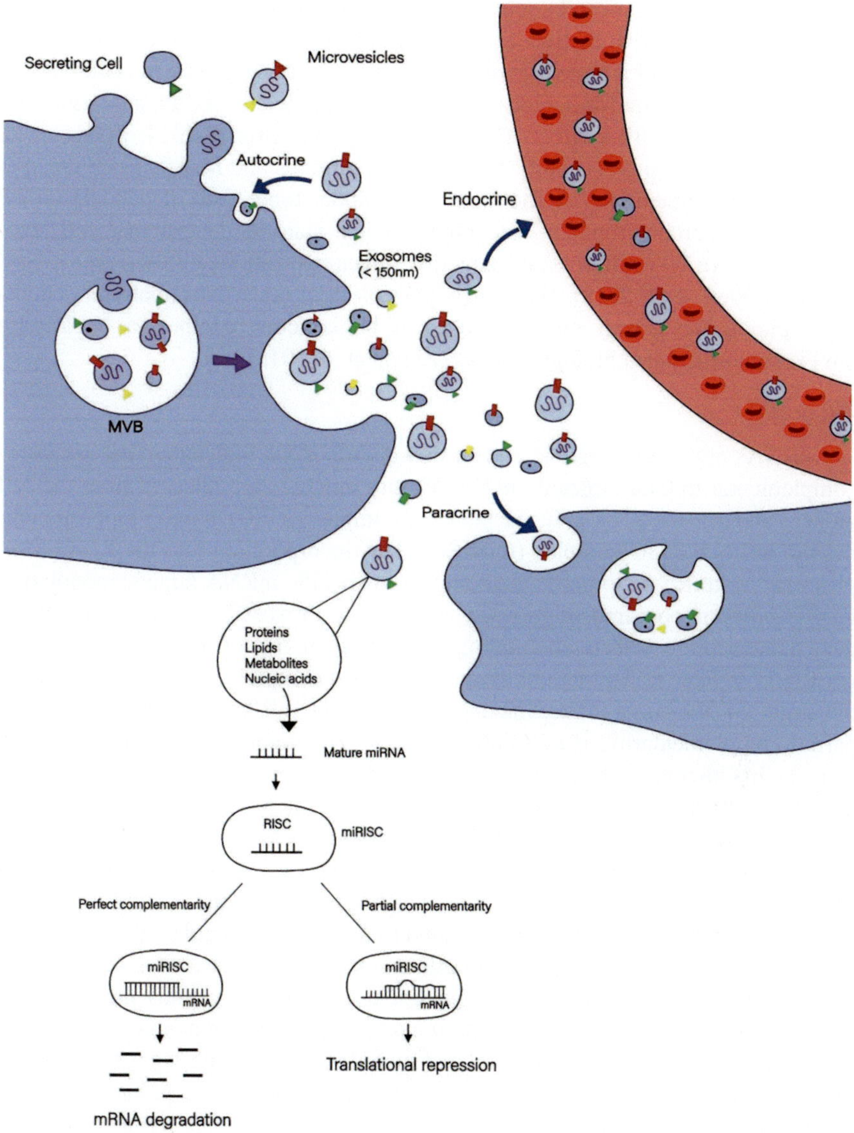

Fig. 1 Role of exosomes in intercellular signalling and mechanism of action. Exosomes show diverse characteristics, such as size, content, functional impact, and source, leading to complex heterogeneity. They are important signalling molecules involved in intercellular communication in an endocrine, autocrine, and paracrine manner, emphasizing their ability to influence both neighbour and distant cells. Exosomes carry bioactive molecules, and miRNAs are the main players in exosome-mediated function. miRNAs play a crucial role in gene regulation by binding to complementary regions on messenger RNA (mRNA) molecules. Interaction can result in translational repression or mRNA degradation, depending on the level of complementarity. Through this intricate mechanism, miRNAs act as post-transcriptional regulators, fine-tuning gene expression and contributing to the control of diverse processes

indicating that miRNAs can be selectively incorporated into exosomes (Goldie et al. 2014). The existence of distinct proteins and nucleic acids within exosomes indicates the existence of a protein-sorting mechanism, allowing them to act as signalling particles that transmit signal molecules capable of having autocrine and paracrine and endocrine effects (Fig. 1) (Zhang and Grizzle 2014; Zhang et al. 2015). However, the precise process via which proteins and nucleic acids are specifically incorporated in exosomes is not fully understood. Thus, depending on the cell origin, microenvironment, and physiological state of the parental cell, exosomes can have very different and sometimes even opposite functions on the recipient cells (Wen et al. 2019). While it is possible that other exosomal components might play a role as well, miRNAs appear to be particularly important in carrying out exosome-mediated biological tasks (Zhang et al. 2015).

MicroRNAs

miRNAs are short, single-stranded, non-coding RNA molecules that are semi-complementar to their targeted mRNA. Mature miRNAs are released from the cell nucleus and join the RNA-induced silencing complex (RISC) in the cytoplasm making the miRNA RISC complex (miRISC). This incorporation into the RISC complex enables miRNAs to guide the complex to specific mRNA targets, modulating gene expression post-transcriptionally (Perron and Provost 2008). miRNAs inhibit gene expression by selectively binding to a specific region of one or more messenger RNAs (mRNAs). Through binding with perfect complementarity, the pairing of the target mRNA might induce the degradation of the mRNA transcript. In cases of partial complementarity, it may interfere with the protein translation machinery (Fig. 1). In either case, the result is a decrease in the expression of the gene's product (Bartel 2004). A single miRNA can target many mRNA molecules. Similarly, a single mRNA can have multiple binding sites for miRNA. This coordinated interaction allows for the modulation of gene expression levels in different tissues and cells. Hence, miRNAs likely play a crucial role in precisely adjusting the expression of protein-coding genes, and it is estimated that they could regulate the expression of half of all protein-coding genes (Friedman et al. 2009). They are essential components of an extensive regulatory network encompassing various physiological processes, such as cellular differentiation, communication, and apoptosis. This network requires highly precise temporal and spatial control of gene expression, necessitating accurate activation or repression of specific genes within particular tissues at specific time points (Flynt and Lai 2008). Many miRNAs are tissue-specific or expressed throughout different developmental stages, affecting cell differentiation (Galagali and Kim 2020).

2 Therapeutic Potential of Exosomes and EVs in Musculoskeletal Regeneration

Musculoskeletal diseases and injuries present significant challenges to human health, requiring novel therapeutic strategies to enhance efficient tissue regeneration. Exosomes and EVs have recently emerged as crucial players in regenerative medicine, offering a unique mechanism for communication between cells (Fig. 1). Extracellular vesicles can be derived from different cell types and contain a variety of bioactive molecules such as proteins, nucleic acids, and growth factors. We will look into the therapeutic potential and new achievements, processes, and experimental findings in musculoskeletal regeneration.

Stem cell-derived exosomes and EVs

Mesenchymal stem cell (MSC) therapy has demonstrated significant therapeutic potential in addressing various diseases, given their wide accessibility and low immunogenicity. MSCs possess a remarkable capacity for regeneration, primarily through their ability to generate and release various bioactive molecules (Fan et al. 2020). Multiple studies have examined the secretome of MSCs and implied that secretome derived from MSCs exhibited therapeutic effects comparable to those found following MSC implantation (Harrell et al. 2019). MSC-based therapy efficacy is independent of differentiation status, and it is considered today that the secretory activity and paracrine effect of MSCs is currently recognized as the primary mediator of MSC-based therapy (Vizoso et al. 2017; Chang and Yan 2021). Further, the therapeutic effectiveness of MSC-derived exosomes relies on the functional characteristics of the MSCs they were obtained from, which vary in their capacity for differentiation and immunomodulation, potentially leading to diverse effects on target cells (Harrell et al. 2019; Börger et al. 2017). Importantly, EVs derived from MSC offer some advantages over cell therapy, such as more stable storage, high stability in circulation, easier large-scale production and biocompatibility, making it a promising therapeutic tool for addressing musculoskeletal regeneration (Ge et al. 2014).

2.1 Extracellular Vesicles in Skeletal Muscle Repair and Regeneration

Extracellular vesicles harvested from MSCs showed encouraging effects *in vitro* and *in vivo* for skeletal muscle regeneration, owing to its ability to modulate inflammation, fibrosis, angiogenesis and myogenesis (Nakamura et al. 2015; Iyer et al. 2020; Byun et al. 2021). These vesicles transport a variety of signalling molecules, such as myogenic growth factors and miRNAs. MiRNAs called myomiRs are particularly important for the physiology and pathophysiology of skeletal muscles. Several myomiRs showed unique expression in skeletal muscle and, in some cases, both skeletal and cardiac muscle (Mitchelson and Qin 2015). MiR-1, miR-133a, miR-133b, and miR-206 have been extensively studied in skeletal muscle tissue and reported to be present in MSC-derived exosomes. They are specifically and highly expressed in skeletal muscle and are essential for myogenesis, maintenance of muscle homeostasis and muscle regeneration through the coordination of proliferation, differentiation and maintenance of myocytes and satellite cells (Nakamura et al. 2015; Mitchelson and Qin 2015; Ma et al. 2015). By silencing the corresponding transcription factors, myomyRs control the expression of regulatory factors of myogenesis responsible for expressing genes specific to muscle cells (Diniz and Wang 2016). Altered expression of different myomiRs has been described in inherited neuromuscular diseases, such as Duchenne muscular dystrophy and myotonic dystrophy type 1. These miRNAs have shown potential as a potential biomarker for monitoring disease progression and outcome measure of therapeutic interventions (Koutsoulidou et al. 2017; Zaharieva et al. 2013; Cacchiarelli et al. 2011). Moreover, local injection of miR-1, 133, and 206 *in vivo* improved muscle regeneration and effectively inhibited muscle fibrosis, suggesting their positive role in treating skeletal muscle injury (Nakasa et al. 2010).

2.2 Extracellular Vesicles in Cartilage Repair and Regeneration

Cartilage, characterized by limited regenerative capacity, presents significant challenges in clinical management. The regenerative potential of MSCs in cartilage repair has been studied, and both animal and human research have shown encouraging results in healing cartilage lesions (Jo et al. 2014; To et al. 2020). The role of exosomes derived from MSCs has garnered increasing attention, with their capacity to deliver bioactive molecules that modulate cellular activities crucial for cartilage repair. MSC-derived exosomes have demonstrated immunomodulatory properties by reducing the production of pro-inflammatory cytokines in chondrocytes while also enhancing chondrocyte extracellular matrix (ECM) production by regulating gene expression of genes involved in EMC formation (Tofiño-Vian et al. 2018; Vonk et al. 2018). This suggests that MSC-derived exosomes could be useful for treating osteoarthritis (OA), which is closely linked to inflammation. Exosomes derived from MSCs could also achieve their therapeutic potential by regulating chondrocyte apoptosis, which is enhanced in OA cartilage and macrophage activity (Cosenza et al. 2017). Identifying suitable cell sources for isolating exosomes is of utmost importance due to the demonstrated different therapeutic efficiency (Kanchanapally et al. 2019; Börger et al. 2017). The efficacy of exosomes isolated from synovial membrane-derived MSCs and human induced pluripotent stem cells have been investigated, suggesting superior outcomes of induced pluripotent stem cells (iPSC) exosomes for OA treatment by enhancing chondrocyte migration and proliferation more effectively (Zhu et al. 2017). Exosomes derived from embryonic stem cell-induced-MSCs have also demonstrated a positive therapeutic impact on OA by regulating the production and degradation of ECM and preventing disease progression in a mouse model (Wang et al. 2017). Further, several miRNAs demonstrated therapeutic action in cartilage repair. Namely, miR-92a-3p and miR140-5p could promote chondrogenesis and cartilage regeneration, inhibit cartilage degradation and prevent OA progression, leading to increased cartilage synthesis in vitro and in OA rat model (Tao et al. 2017; Mao et al. 2017). Also, isolating exosomes from MSC treated with transforming growth factor β1 (TGF-β1) can promote chondrocyte proliferation and anti-inflammatory phenotype in synovial macrophage mediated by high levels of exosomal miR-135b, which was reversed by miR-135b decreasing (Wang and Xu 2021).

2.3 Extracellular Vesicles in Bone Repair and Regeneration

Various cell sources, particularly mesenchymal stem cells (MSCs), have been studied for their ability to produce exosomes with osteogenic properties. MSC-derived exosomes carry regulatory molecules that stimulate osteoblast activity, modulate inflammation, and promote angiogenesis (Lu et al. 2017; Cosenza et al. 2017; Zhang et al. 2021). Exosomes from adipose tissue-derived MSCs can enhance the proliferation and osteogenic differentiation in human primary osteoblastic cells (Lu et al. 2017). Also, exosomes derived from BMSCs positively impact osteogenesis both *in vitro* and *in vivo* (Lu et al. 2020). Possible mechanisms by which exosomes might enhance bone repair include the action of miRNAs. RNA sequencing revealed that miR-196a, miR27a and miR-206 are abundant in these EVs and appear crucial for osteogenesis. These miRNAs, especially miR-196a, regulate osteoblastic differentiation and the expression of osteogenic genes, as demonstrated by using miRNA-specific inhibitors, which abolish EV-mediated osteoblastic differentiation while applying osteogenic-related miRNA mimics promoted osteogenic effects (Qin et al. 2016). Further, *in vitro* and *in vivo* experiments demonstrated that osteoclast-derived exosomal miR-214-3p, owing to its ability to inhibit osteoblast function, could be an important target for treating various disorders characterized by decreased bone production (Li et al. 2016). Moreover, endothelial cell-derived exosomes exhibit efficient bone targeting and can suppress osteoclast function mainly via exosomal miR-155. Exogenous administration of miR-155 mimics and miR-155 inhibitors confirmed the inhibitory effect of miR-155 on osteoclast differentiation, suggesting a promising therapeutic approach for osteoporosis (Song et al. 2019). The findings indicate that exosome-based therapy could be a promising alternative to direct stem cell transplantation for bone regeneration, providing novel possibilities for cell-free treatments in the future.

2.4 Extracellular Vesicles in Tendon Repair and Regeneration

Tendons, exhibiting limited vascularity and slow natural healing, present unique challenges in the recovery from injuries. Tendon healing is improved by exosomes and EVs harvested from various sources, which exhibit different mechanisms of action (Zhang et al. 2020; Kim et al. 2017; Li et al. 2020). Exosomes derived from BMSCs could play an important role in tendon regeneration. The therapeutic mechanisms by which these exosomes achieve healing are through enhanced tenocyte proliferation, leading to improved biomechanics and reduced scarring; modulation of the inflammatory response by modulating anti-inflammatory macrophages phenotype and associated cytokines; promoting EMC component synthesis; improving the tendon–bone interface and regulating angiogenesis (Huang et al. 2020; Li et al. 2020; Shi et al. 2019).

3 Gene Therapy

Gene therapy has shifted the paradigm in modern medicine by offering the possibility of a permanent cure rather than a lifetime of treatment. The concept of using gene therapy was initially developed to treat monogenic diseases, aiming to achieve sustained expression of the therapeutic gene (transgene) at a level that effectively alleviates or cures disease with minimal side effects. For many suffering from genetic diseases, this gave hope that previously untreatable diseases could now be treated with this cutting-edge technology. In recent years, gene therapy has made significant progress, allowing for treating acquired diseases by transient expression of transgene (Wirth et al. 2013; Anguela and High 2019).

Ex vivo and in vivo gene therapy

Gene therapy aims to create a disease by introducing genetic material (DNA or RNA) into the patient's cells. The transferred genetic material can modify, replace or inactivate the expression of the targeted gene. Restoring or modifying the function of target cells can be accomplished using one of two strategies: *ex vivo* or *in vivo* gene delivery. *Ex vivo* gene therapy involves removing cells or tissues from the patient and genetically modifying them outside the body to introduce a therapeutic gene, followed by implantation at the desired site of the patient (Dunbar et al. 2018). This method is often used for disorders where specific cells may be easily reached and modified, providing better control of the transfection procedure. *In vivo* gene therapy involves directly introducing a therapeutic gene into the patient's body using a vector that carries the gene. Some main disadvantages of *ex vivo* gene therapy include complexity, high cost, challenges in cell handling and culture, and limited access to target cells. On the other hand, *in vivo* gene therapy's main obstacles include targeting particular cells, triggering an immunological reaction, achieving high efficacy in gene transfer, off-target effects, and limited access to specific tissues. The decision between these approaches relies on the specific condition and nature of the target cell (Franceschi et al. 2004; Tang and Xu 2020).

Viran and non-viral gene delivery

One major challenge in gene therapy is the successful delivery of the transgene into the cells. Common methods for delivering genetic material in gene therapy involve the use of viral vectors and non-viral vectors. Viral vectors differ in their efficacy in delivering genes to specific cells, their integration into the host genome, immunogenicity and cargo capacity. They can integrate into the host genome for continuous expression or remain in the nucleus as stable DNA molecules for transient expression (Thomas et al. 2003). Commonly used viral vectors are adenovirus (AV), adeno-associated virus (AAV) and lentivirus (LV). AV gene therapy is known for its high transduction efficiency and large cargo capacity. However, AV enables only transient gene expression and demonstrates a high level of immunogenicity, restricting its usage to local administration and conditions where an immune response is beneficial. Compared to AV gene delivery, AAV-mediated gene therapy allows the administration of high dosages of vector particles without causing inflammatory reactions or toxic effects but with a much smaller cargo capacity. AAV gene therapy enables long transgene expression in slow dividing cells, such as muscle cells. Lentiviruses integrate in actively transcribed genome, enabling life-long expression, ideal for monogenetic diseases, but also carrying risk of insertional oncogenesis. Also, LV vectors have higher transduction efficiency, lower immunogenicity, and larger cargo capacity than AAV vectors (Lukashev and Zamyatnin Jr. 2016; Lundstrom 2018; Nayerossadat et al. 2012). Viral vectors may demonstrate toxicity and immunogenicity, while non-viral vectors are hindered by their restricted ability to achieve efficient transfection (Nayerossadat et al. 2012). Non-viral approaches have several advantages, such as safety, transmission of larger cargo, and easier manufacturing while struggling with low transfection efficacy and inadequate expression. They can be very simple, combined with liposomes or polymers or ligand-modified for tissue or cell targeting (Wang et al. 2013). The vector engineering field is rapidly developing, addressing common issues like safety. However, no universal vector can be used to treat different indications. The delivery method that is the safest and most efficient depends on the task that a new gene therapy approach must achieve (parameters such as time of expression, size of the insert, and way of administration) (Lukashev and Zamyatnin Jr. 2016; Lundstrom 2018).

3.1 Potential of Gene Therapy in Musculoskeletal Repair and Regeneration

Bone morphogenetic proteins (BMPs) emerged as promising candidates for gene therapy in bone repair. They are crucial in controlling bone healing by influencing the inflammatory response and chondrogenic and osteogenic stages repair They can regulate the bone-induction process and inducing osteocalcin expression in osteoblasts (Thies et al. 1992; Marsell and Einhorn 2009). Recombinant human BMP (rhBMP) proteins have been approved by the FDA and used in clinical settings for various indications, including repairing long-bone fractures (Giannoudis and Kanakaris 2017; Govender et al. 2002). Nevertheless, the use of rhBMP has shown inconsistent therapeutic outcomes, with drawbacks including high cost and triggering inflammatory swelling (Lee et al. 2012; Halloran et al. 2020). Thus, BMPs would be a good candidate for alternative delivery methods, such as gene therapy.

Several viral vectors, including adenovirus (AV), adeno-associated virus (AAV), and lentivirus (LV), have been studied for gene therapy to promote bone healing. *Ex vivo* gene delivery has proven beneficial in improving bone healing in rat models. BMSCs that were genetically modified using AV vectors carrying the BMP-2 gene had promising outcomes in bone regeneration, owing to the transient expression of the transgene (Lieberman et al. 1999; Park et al. 2003). Also, adipose-derived stem cells (ASCs) could be starting material for ex vivo gene delivery. Genetically modified ASCs with AV vector to overexpress BMP-2 can induce bone formation in rats (Peterson et al. 2005). However, there is a risk of immunogenic response, and transient protein production may not be sufficient to treat large bone defects (Collon et al. 2021). Compared to AV delivery of BMP-2, LV BMP-2 transduced cells show increased duration of BMP-2 expression, greater mechanical strength, and increased bone fusion mass, making them an exciting candidate for clinical translation of gene therapy for bone repair (Virk et al. 2008; Feeley et al. 2006).

Gene therapy has also emerged as a promising approach for addressing the challenges associated with cartilage repair and regeneration. Potential candidate genes related to extracellular matrix formation, chondrogenic differentiation, and anti-inflammatory response are being investigated as possible targets for gene-based therapy in cartilage repair. Delivery of vectors administered systemically has limited access to joints due to the avascular nature of articular cartilage. Thus, most approaches involve injecting vectors or genetically modified cells into the affected joint (Evans et al. 2018). The dense cartilage matrix limits the propagation of vectors, making synovium the primary site where the transgene is expressed, allowing the gene product to reach all intra-articular tissues, including the cartilage, by diffusion in the synovial fluid (Evans et al. 2011). One potent inhibitor of chondrogenesis is interleukin-1 (IL-1) (Wehling et al. 2009). A gene therapy for rheumatoid arthritis (RA) using an interleukin-1 receptor antagonist (IL-1Ra) as a transgene showcased successful gene transfer to a human joint (Evans et al. 2018). Also, intra-articular administration of a recombinant AAV vector carrying the human TNF antagonist gene entered clinical trials. It demonstrated potential in alleviating symptoms of RA by blocking TNF-α, a crucial inflammatory mediator in RA. Unfortunately, the study was halted at phase II because one subject died (Mease et al. 2010; Frank et al. 2009). Further, lentiviral transfection of miR-140-5p was introduced to human synovial MSCs. Exosomes derived from human synovial MSCs modified to overexpress miR-140-5p were compared to exosomes isolated from unmodified cells and shown that overexpression of miR-140-5p improved proliferation and migration of chondrocytes and facilitated the expression of previously downregulated EMC components in a rat model (Tao et al. 2017). Finally, gene therapy for the intra-article treatment of OA using retrovirus as a vector to enhance the expression of TGF-b was approved in 2017 in South Korea. The approval was reversed in 2019 because some cells were identified as HEK293 cells (Evans 2019).

3.2 Combining Exosome and Gene Therapy for Optimal Regenerative Outcomes

The synergistic enhancement of musculoskeletal regeneration could be achieved by combining the regenerative ability of exosomes with gene therapy, hence providing an integrated strategy. Exosomes have unique advantages regarding targeted delivery and safety for restoring gene expression in dysfunctional cells. They can serve as natural, cell-free carriers for genetic material, facilitating the delivery of therapeutic genes to targeted cells and showing the clear potential for gene delivery. It is necessary to identify suitable donor cells for exosome isolation, and different cell lines express differences in release rate, biochemical composition and susceptibility to modifications (Sadeghi et al. 2023; Kanchanapally et al. 2019). Some promising sources are MSC due to their easy availability and high *ex vivo* expansion capacity and dendritic cells due to their important role in immunity (García-Manrique and Matos 2018). The most abundant source of exosomes comes from bovine milk and colostrum, and isolating exosomes from these sources is cost-effective and scalable drug development (Wallen et al. 2023). The potential use of bovine-derived exosomes in humans raises concerns about immune responses. Results so far have been encouraging, with studies investigating the potential immunogenicity of bovine exosomes in other species using a rat model revealing no systemic toxicity in kidney or liver biochemical profiles following oral treatment and confirming that exposure to bovine milk exosomes does not lead to increased concentrations of pro-inflammatory serum cytokines, anaphylaxis or increase serum histamine levels (Munagala et al. 2016; Somiya et al. 2018).

Nevertheless, enhancing the precision of directing these exosomes towards specific cells is crucial. This can be accomplished by genetically modifying the donor cells to generate certain cell surface markers or combining the exosomes with novel biomaterials (Danilushkina et al. 2023). Exosomes can be used to deliver specific miRNAs or plasmid DNA (pDNA) for exogenous gene expression, and although this research is still in its infancy, it represents a promising approach (Munagala et al. 2021; Song et al. 2019; Zeng et al. 2023). The pro-osteogenic characteristics of miR-26a-5p have led to encouraging outcomes in fracture healing when loaded with endothelium-derived exosomes, which allow bone targeting (Mi et al. 2022).

4 Conclusion

In conclusion, exploring exosomes and gene therapy in the context of musculoskeletal regeneration holds immense promise for advancing therapeutic strategies in regenerative medicine. The complex role of exosomes, acting as crucial mediators of intercellular communication and carriers of genetic material, accentuates their potential to promote tissue repair and regeneration. Gene therapy, on the other hand, offers precise control over the expression of therapeutic genes, presenting a powerful tool to modulate cellular functions for enhanced regeneration. Exosomes can transfer therapeutic genetic material to specific cells, offering an innovative approach for individualized and precise treatment of musculoskeletal disorders. However, it is essential to acknowledge the current limitations and challenges in translating these findings into clinical applications. Further research is needed to optimize therapeutic cargo's isolation, characterization, and loading into exosomes. Improved methodology is required for higher yields, greater homogeneity, faster time, lower cost, and characterizing and quantifying EVs. Additionally, different sub-types of EVs may exist with different cargo, release mechanisms, and functions. Thus, further research is needed to identify suitable biomarkers for different EVs. Finally, long-term safety and efficacy assessments in diverse patient populations are crucial before widespread clinical implementation.

References

Anguela XM, High KA. Entering the modern era of gene therapy. Annu Rev Med. 2019;70:273–88.
Bartel DP. MicroRNAs: genomics, biogenesis, mechanism, and function. Cell. 2004;116:281–97.
Börger V, Bremer M, Ferrer-Tur R, et al. Mesenchymal stem/stromal cell-derived extracellular vesicles and their potential as novel immunomodulatory therapeutic agents. Int J Mol Sci. 2017;18:1450.
Byun SE, Sim C, Chung Y. Skeletal muscle regeneration by the exosomes of adipose tissue-derived mesenchymal stem cells. Curr Issues Mol Biol. 2021;43:1473–88.
Cacchiarelli D, Legnini I, Martone J, et al. miRNAs as serum biomarkers for Duchenne muscular dystrophy. EMBO Mol Med. 2011;3:258–65.
Chang C, Yan J. Effects of mesenchymal stem cell-derived paracrine signals and their delivery strategies. Adv Healthc Mater. 2021;10:e2001689.
Collon K, Gallo MC, Lieberman JR. Musculoskeletal tissue engineering: regional gene therapy for bone repair. Biomaterials. 2021;275:120901.
Cosenza S, Ruiz M, Toupet K, et al. Mesenchymal stem cells derived exosomes and microparticles protect cartilage and bone from degradation in osteoarthritis. Sci Rep. 2017;7:16214.
Danilushkina AA, Emene CC, Barlev NA, et al. Strategies for engineering of extracellular vesicles. Int J Mol Sci. 2023;24(17):13247.
Diniz GP, Wang DZ. Regulation of skeletal muscle by microRNAs. Compr Physiol. 2016;6:1279–94.
Dunbar CE, High KA, Joung JK, et al. Gene therapy comes of age. Science. 2018;359(6372):eaan4672.
Evans CH. The vicissitudes of gene therapy. Bone Joint Res. 2019;8:469–71.
Evans CH, Ghivizzani SC, Robbins PD. Getting arthritis gene therapy into the clinic. Nat Rev Rheumatol. 2011;7:244–9.

Evans CH, Ghivizzani SC, Robbins PD. Gene delivery to joints by intra-articular injection. Hum Gen Ther. 2018;29:2–14.

Fan XL, Zhang Y, Li X, et al. Mechanisms underlying the protective effects of mesenchymal stem cell-based therapy. Cell Mol Life Sci. 2020;77:2771–94.

Feeley BT, Conduah AH, Sugiyama O, et al. In vivo molecular imaging of adenoviral versus lentiviral gene therapy in two bone formation models. J Orthop Res. 2006;24:1709–21.

Flynt AS, Lai EC. Biological principles of microRNA-mediated regulation: shared themes amid diversity. Nat Rev Genet. 2008;9:831–42.

Franceschi RT, Yang S, Rutherford RB, et al. Gene therapy approaches for bone regeneration. Cells Tissues Organs. 2004;176:95–108.

Frank KM, Hogarth DK, Miller JL, et al. Investigation of the cause of death in a gene-therapy trial. N Engl J Med. 2009;361:161–9.

Friedman RC, Farh KK, Burge CB, et al. Most mammalian mRNAs are conserved targets of microRNAs. Genome Res. 2009;19:92–105.

Galagali H, Kim JK. The multifaceted roles of microRNAs in differentiation. Curr Opin Cell Biol. 2020;67:118–40.

García-Manrique P, Matos M. Therapeutic biomaterials based on extracellular vesicles: classification of bio-engineering and mimetic preparation routes. J Extracell Vesicles. 2018;7:1422676.

Ge Q, Zhou Y, Lu J, et al. miRNA in plasma exosome is stable under different storage conditions. Molecules. 2014;19:1568–75.

Giannoudis PV, Kanakaris NK. BMPs in orthopaedic medicine: promises and challenges. In: Vukicevic S, Sampath KT, editors. Bone morphogenetic proteins: systems biology regulators. Cham: Springer International Publishing; 2017. p. 187–214.

Goldie BJ, Dun MD, Lin M, et al. Activity-associated miRNA are packaged in Map1b-enriched exosomes released from depolarized neurons. Nucleic Acids Res. 2014;42:9195–208.

Govender S, Csimma C, Genant HK, et al. Recombinant human bone morphogenetic protein-2 for treatment of open tibial fractures: a prospective, controlled, randomized study of four hundred and fifty patients. J Bone Joint Surg Am. 2002;84:2123–34.

Halloran D, Durbano HW, Nohe A. Bone morphogenetic Protein-2 in development and bone homeostasis. J Dev Biol. 2020;8(3):19.

Harding C, Heuser J, Stahl P. Receptor-mediated endocytosis of transferrin and recycling of the transferrin receptor in rat reticulocytes. J Cell Biol. 1983;97:329–39.

Harrell CR, Fellabaum C, Jovicic N, et al. Molecular mechanisms responsible for therapeutic potential of mesenchymal stem cell-derived secretome. Cells. 2019;8(5):467.

Hazan-Halevy I, Rosenblum D, Weinstein S, et al. Cell-specific uptake of mantle cell lymphoma-derived exosomes by malignant and non-malignant B-lymphocytes. Cancer Lett. 2015;364:59–69.

Huang Y, He B, Wang L, et al. Bone marrow mesenchymal stem cell-derived exosomes promote rotator cuff tendon-bone healing by promoting angiogenesis and regulating M1 macrophages in rats. Stem Cell Res Ther. 2020;11:496.

Iyer SR, Scheiber AL, Yarowsky P, et al. Exosomes isolated from platelet-rich plasma and mesenchymal stem cells promote recovery of function after muscle injury. Am J Sports Med. 2020;48:2277–86.

Jo CH, Lee YG, Shin WH, et al. Intra-articular injection of mesenchymal stem cells for the treatment of osteoarthritis of the knee: a proof-of-concept clinical trial. Stem Cells. 2014;32:1254–66.

Kalluri R, Lebleu VS. The biology, function, and biomedical applications of exosomes. Science. 2020;367(6478):eaau6977.

Kanchanapally R, Deshmukh SK, Chavva SR, et al. Drug-loaded exosomal preparations from different cell types exhibit distinctive loading capability, yield, and antitumor efficacies: a comparative analysis. Int J Nanomedicine. 2019;14:531–41.

Kim YS, Sung CH, Chung SH, et al. Does an injection of adipose-derived mesenchymal stem cells loaded in fibrin glue influence rotator cuff repair outcomes? A clinical and magnetic resonance imaging study. Am J Sports Med. 2017;45:2010–8.

Koutsoulidou A, Photiades M, Kyriakides TC, et al. Identification of exosomal muscle-specific miRNAs in serum of myotonic dystrophy patients relating to muscle disease progress. Hum Mol Gen. 2017;26:3285–302.

Lee KB, Taghavi CE, Murray SS, et al. BMP induced inflammation: a comparison of rhBMP-7 and rhBMP-2. J Orthop Res. 2012;30:1985–94.

Li D, Liu J, Guo B, et al. Osteoclast-derived exosomal miR-214-3p inhibits osteoblastic bone formation. Nat Commun. 2016;7:10872.

Li J, Liu ZP, Xu C, et al. TGF-β1-containing exosomes derived from bone marrow mesenchymal stem cells promote proliferation, migration and fibrotic activity in rotator cuff tenocytes. Regen Ther. 2020;15:70–6.

Lieberman JR, Daluiski A, Stevenson S, et al. The effect of regional gene therapy with bone morphogenetic protein-2-producing bone-marrow cells on the repair of segmental femoral defects in rats. J Bone Joint Surg Am. 1999;81:905–17.

Lu Z, Chen Y, Dunstan C, et al. Priming adipose stem cells with tumor necrosis factor-alpha preconditioning potentiates their exosome efficacy for bone regeneration. Tissue Eng Part A. 2017;23:1212–20.

Lu GD, Cheng P, Liu T, et al. BMSC-derived exosomal miR-29a promotes angiogenesis and osteogenesis. Front Cell Dev Biol. 2020;8:608521.

Lukashev AN, Zamyatnin AA Jr. Viral vectors for gene therapy: current state and clinical perspectives. Biochem Biokhimiia. 2016;81:700–8.

Lundstrom K. Viral vectors in gene therapy. Diseases. 2018;6:42.

Ma G, Wang Y, Li Y, et al. MiR-206, a key modulator of skeletal muscle development and disease. Int J Biol Sci. 2015;11:345–52.

Mao G, Zhang Z, Huang Z, et al. MicroRNA-92a-3p regulates the expression of cartilage-specific genes by directly targeting histone deacetylase 2 in chondrogenesis and degradation. Osteoarthr Cartil. 2017;25:521–32.

Marsell R, Einhorn TA. The role of endogenous bone morphogenetic proteins in normal skeletal repair. Injury. 2009;40(Suppl 3):S4–7.

Mease PJ, Wei N, Fudman EJ, et al. Safety, tolerability, and clinical outcomes after intraarticular injection of a recombinant adeno-associated vector containing a tumor necrosis factor antagonist gene: results of a phase 1/2 study. J Rheumatol. 2010;37:692–703.

Mi B, Chen L, Xiong Y, et al. Osteoblast/osteoclast and immune cocktail therapy of an exosome/drug delivery multifunctional hydrogel accelerates fracture repair. ACS Nano. 2022;16:771–82.

Mitchelson KR, Qin WY. Roles of the canonical myomiRs miR-1, -133 and -206 in cell development and disease. World J Biol Chem. 2015;6:162–208.

Munagala R, Aqil F, Jeyabalan J, et al. Bovine milk-derived exosomes for drug delivery. Cancer Lett. 2016;371:48–61.

Munagala R, Aqil F, Jeyabalan J, et al. Exosome-mediated delivery of RNA and DNA for gene therapy. Cancer Lett. 2021;505:58–72.

Nakamura Y, Miyaki S, Ishitobi H, et al. Mesenchymal-stem-cell-derived exosomes accelerate skeletal muscle regeneration. FEBS Lett. 2015;589:1257–65.

Nakasa T, Ishikawa M, Shi M, et al. Acceleration of muscle regeneration by local injection of muscle-specific microRNAs in rat skeletal muscle injury model. J Cell Mol Med. 2010;14:2495–505.

Nayerossadat N, Maedeh T, Ali PA. Viral and nonviral delivery systems for gene delivery. Adv Biomed Res. 2012;1:27.

Pan BT, Johnstone RM. Fate of the transferrin receptor during maturation of sheep reticulocytes in vitro: selective externalization of the receptor. Cell. 1983;33:967–78.

Park J, Ries J, Gelse K, et al. Bone regeneration in critical size defects by cell-mediated BMP-2 gene transfer: a comparison of adenoviral vectors and liposomes. Gene Ther. 2003;10:1089–98.

Perron MP, Provost P. Protein interactions and complexes in human microRNA biogenesis and function. Front Biosci. 2008;13:2537–47.

Peterson B, Zhang J, Iglesias R, et al. Healing of critically sized femoral defects, using genetically modified mesenchymal stem cells from human adipose tissue. Tissue Eng. 2005;11:120–9.

Qin Y, Wang L, Gao Z, et al. Bone marrow stromal/stem cell-derived extracellular vesicles regulate osteoblast activity and differentiation in vitro and promote bone regeneration in vivo. Sci Rep. 2016;6:21961.
Raposo G, Nijman HW, Stoorvogel W, et al. B lymphocytes secrete antigen-presenting vesicles. J Exp Med. 1996;183:1161–72.
Sadeghi S, Tehrani FR, Tahmasebi S, et al. Exosome engineering in cell therapy and drug delivery. Inflammopharmacology. 2023;31:145–69.
Sancho-Albero M, Navascués N, Mendoza G, et al. Exosome origin determines cell targeting and the transfer of therapeutic nanoparticles towards target cells. J Nanobiotechnol. 2019;17:16.
Shi Z, Wang Q, Jiang D. Extracellular vesicles from bone marrow-derived multipotent mesenchymal stromal cells regulate inflammation and enhance tendon healing. J Transl Med. 2019;17:211.
Somiya M, Yoshioka Y, Ochiya T. Biocompatibility of highly purified bovine milk-derived extracellular vesicles. J Extracell Vesicles. 2018;7:1440132.
Song H, Li X, Zhao Z, et al. Reversal of osteoporotic activity by endothelial cell-secreted bone targeting and biocompatible exosomes. Nano Lett. 2019;19:3040–8.
Tang R, Xu Z. Gene therapy: a double-edged sword with great powers. Mol Cell Biochem. 2020;474:73–81.
Tao SC, Yuan T, Zhang YL, et al. Exosomes derived from miR-140-5p-overexpressing human synovial mesenchymal stem cells enhance cartilage tissue regeneration and prevent osteoarthritis of the knee in a rat model. Theranostics. 2017;7:180–95.
Théry C, Witwer KW. Minimal information for studies of extracellular vesicles 2018 (MISEV2018): a position statement of the International Society for Extracellular Vesicles and update of the MISEV2014 guidelines. J Extracell Vesicles. 2018;7:1535750.
Thies RS, Bauduy M, Ashton BA, et al. Recombinant human bone morphogenetic protein-2 induces osteoblastic differentiation in W-20-17 stromal cells. Endocrinology. 1992;130:1318–24.
Thomas CE, Ehrhardt A, Kay MA. Progress and problems with the use of viral vectors for gene therapy. Nat Rev Genet. 2003;4:346–58.
To K, Romain K, Mak C, et al. The treatment of cartilage damage using human mesenchymal stem cell-derived extracellular vesicles: a systematic review of in vivo studies. Front Bioeng Biotechnol. 2020;8:580.
Tofiño-Vian M, Guillén MI, Pérez Del Caz MD, et al. Microvesicles from human adipose tissue-derived mesenchymal stem cells as a new protective strategy in osteoarthritic chondrocytes. Cell Physiol Biochem. 2018;47:11–25.
Virk MS, Conduah A, Park SH, et al. Influence of short-term adenoviral vector and prolonged lentiviral vector mediated bone morphogenetic protein-2 expression on the quality of bone repair in a rat femoral defect model. Bone. 2008;42:921–31.
Vizoso FJ, Eiro N, Cid S, et al. Mesenchymal stem cell secretome: toward cell-free therapeutic strategies in regenerative medicine. Int J Mol Sci. 2017;18:1852.
Vonk LA, Van Dooremalen SFJ, Liv N, et al. Mesenchymal stromal/stem cell-derived extracellular vesicles promote human cartilage regeneration in vitro. Theranostics. 2018;8:906–20.
Wallen M, Aqil F, Spencer W, et al. Milk/colostrum exosomes: a nanoplatform advancing delivery of cancer therapeutics. Cancer Lett. 2023;561:216141.
Wang R, Xu B. TGF-β1-modified MSC-derived exosomal miR-135b attenuates cartilage injury via promoting M2 synovial macrophage polarization by targeting MAPK6. Cell Tissue Res. 2021;384:113–27.
Wang W, Li W, Ma N, et al. Non-viral gene delivery methods. Curr Pharm Biotechnol. 2013;14:46–60.
Wang Y, Yu D, Liu Z, et al. Exosomes from embryonic mesenchymal stem cells alleviate osteoarthritis through balancing synthesis and degradation of cartilage extracellular matrix. Stem Cell Res Ther. 2017;8:189.

Wehling N, Palmer GD, Pilapil C, et al. Interleukin-1beta and tumor necrosis factor alpha inhibit chondrogenesis by human mesenchymal stem cells through NF-kappaB-dependent pathways. Arthritis Rheum. 2009;60:801–12.
Wen SW, Lima LG, Lobb RJ, et al. Breast cancer-derived exosomes reflect the cell-of-origin phenotype. Proteomics. 2019;19:e1800180.
Wirth T, Parker N, Ylä-Herttuala S. History of gene therapy. Gene. 2013;525:162–9.
Zaharieva IT, Calissano M, Scoto M, et al. Dystromirs as serum biomarkers for monitoring the disease severity in Duchenne muscular dystrophy. PLoS One. 2013;8:e80263.
Zeng H, Guo S, Ren X, et al. Current strategies for exosome cargo loading and targeting delivery. Cells. 2023;12(10):1416.
Zhang HG, Grizzle WE. Exosomes: a novel pathway of local and distant intercellular communication that facilitates the growth and metastasis of neoplastic lesions. Am J Pathol. 2014;184:28–41.
Zhang J, Li S, Li L, et al. Exosome and exosomal microRNA: trafficking, sorting, and function. Genom Proteom Bioinform. 2015;13:17–24.
Zhang Y, Liu Y, Liu H, et al. Exosomes: biogenesis, biologic function and clinical potential. Cell Biosci. 2019;9:19.
Zhang M, Liu H, Cui Q, et al. Tendon stem cell-derived exosomes regulate inflammation and promote the high-quality healing of injured tendon. Stem Cell Res Ther. 2020;11:402.
Zhang Y, Xie Y, Hao Z, et al. Umbilical mesenchymal stem cell-derived exosome-encapsulated hydrogels accelerate bone repair by enhancing angiogenesis. ACS Appl Mater Interfaces. 2021;13:18472–87.
Zhu Y, Wang Y, Zhao B, et al. Comparison of exosomes secreted by induced pluripotent stem cell-derived mesenchymal stem cells and synovial membrane-derived mesenchymal stem cells for the treatment of osteoarthritis. Stem Cell Res Ther. 2017;8:64.

Muscle Regeneration, Inflammation and Arthritis

Thumri Paavana, Edmond U, and Tosan Okoro

Abstract Osteoarthritis (OA) is a commonly encountered clinical entity, often with debilitating consequences for the sufferer. Treatment for symptomatic end stage disease is often in the form of surgery. Previously thought of as a purely degenerative condition, it is now increasingly recognised that through a number of mechanical, metabolic and molecular pathways, there is also an inflammatory component in the surrounding periarticular tissues which may contribute to disease progression. As our understanding of OA at the cellular and genetic level deepens, so grow the opportunities for early therapeutic intervention to slow or even halt progression.

Keywords Osteoarthritis · Inflammation · Genetic expression · Vastus lateralis

1 Introduction

Osteoarthritis (OA) continues to represent a global healthcare burden with considerable physical, social, and psychological impacts. In the UK alone, it is estimated that ten million people (six million women, four million men) suffer with OA, with over five million people affected by knee OA and 3.2 million by hip OA, and a further 350,000 people diagnosed with OA each year (Versus 2023).

Traditionally thought of as a simple 'wear and tear' disease of the articular cartilage in joints, emerging evidence continues to cement the theory that the aetiology of OA is more complex than was previously considered. It is now broadly accepted that local and systemic inflammation plays a role in the development of OA in a certain subset of patients, and that this may occur in at many sites including bone, cartilage, ligaments, mensici, synovium, periarticular adipose tissue (such as Hoffa's pad in the knee) and skeletal muscle (van den Bosch et al. 2023).

T. Paavana · Edmond U · T. Okoro (✉)
The Robert Jones and Agnes Hunt Orthopaedic Hospital, Gobowen, Oswestry, UK
e-mail: tosan.okoro@nhs.net

P. Lee (ed.), *Musculoskeletal Regeneration Medicine*,
https://doi.org/10.1007/978-3-031-77456-0_23

2 Osteoarthritis as an Inflammatory Disease

In addition to genetic expressions which will be discussed in detail later, there are several, often overlapping observable characteristics or 'phenotypes' noted in the aetiology of osteoarthritis as an inflammatory process, of which the most well recognised are **ageing**, **metabolic** and **post-traumatic** (Berenbaum and Walker 2020). Skeletal muscle is also involved, and the impact of muscle inflammation in arthritis shall be explored.

2.1 Ageing

Franceschi et al. (2000) described the phenomenon of 'inflammaging' that occurs as a long-term result of chronic physiological stimulation of the immune system, resulting in a state of low-grade inflammation with advancing age. It is recognised that this proinflammatory state can contribute to a number of pathologies such as frailty, atherosclerosis, Alzheimer's disease, type II diabetes and OA (Rezuș et al. 2019).

Generally, the proposed hallmarks of ageing include genomic instability, telomere attrition, epigenetic alterations, loss of proteostasis, disabled macroautophagy, deregulated nutrient-sensing, mitochondrial dysfunction, cellular senescence, stem cell exhaustion, altered intercellular communication, chronic inflammation, and dysbiosis (López-Otín et al. 2023).

Certainly in ageing articular cartilage, both chondrocytes and the extracellular matrix can be affected by these processes, which may leave the tissue more vulnerable (Aaron 2020). Chondrocytes have been shown to display increased expression of proinflammatory molecules, known as the senescence-associated secretory phenotype (SASP) (Balistreri et al. 2013).This phenotype participates in maintaining local and systemic low-grade inflammation by increased expression of chemokines, growth factors, proinflammatory cytokines and matrix metalloproteinases (Rezuș et al. 2019). Increased oxidative stress results in accumulation of oxygen metabolites (Rezuș et al. 2019). Decreased capacity for autophagy, that is the removal of abnormal intracellular components via lysosomal degradation, results in proinflammatory responses and mitochondrial damage.

Thus, whilst ageing is not directly causative of OA, it is a risk factor for its occurrence.

2.2 Post-traumatic

Mechanical load as a risk factor for osteoarthritis can be stratified into two broad groups: those in which mechanical strain is transmitted through a normal joint, or those where normal load traverses a joint that has lost its mechano-protective mechanisms, that is the presence of a stable joint, with intact ligaments and menisci, healthy cartilage, strong muscle support across the joint and intact gait reflexes (Vincent 2019).

Initial injury to the structural matrix of articular cartilage evolves into a range of cellular responses, including the upregulation of matrix degradative pathways, release of oxidants and inflammatory cytokines and cell death (Punzi et al. 2016). In certain individuals, there is a continued activation of innate inflammatory pathways, such as complement and danger signal-mediated pathways, thus perpetuating a state of inflammation and risking progression to chronic arthropathies (Punzi et al. 2016).

Thus a result of increased or altered mechanical load through injury or high body mass index (BMI), there is not only attrition of the articular surfaces, but active induction of mechanosensitive intracellular signalling resulting in either a pro-repair (growth factor release from the matrix) or pro-degenerative inflammatory response which has been termed 'mechanoflammation'(Vincent 2019).

2.3 Metabolic

'Metaflammation' refers to the phenomenon of metabolic stress occurring as a result of long-term positive energy balance; this leads to release of local and systemic pro-inflammatory factors such as cytokines and adipokines from adipose tissue causing an overall increased inflammatory status (van den Bosch et al. 2023). Through such metabolic processes, obesity can have a significant impact on the development of inflammation within the joint.

3 Skeletal Muscle

3.1 Skeletal Muscle Development

Skeletal muscle is derived at approximately 6 weeks of gestation in the human embryo (Stewart and Rittweger 2006). Embryonic mesodermal stem cells originating from the myotome of the somites migrate into adjacent embryonic connective tissue, where they become precursor skeletal muscle cells or myoblasts (Stewart and Rittweger 2006). After a period of proliferation, which in humans occurs between weeks 7 and 9 of gestation, these differentiate to multi-nucleated myotubes (Stewart and Rittweger 2006). The synthesis of actin and myosin (contractile proteins) as well as cross-striation and innervation occur around this period (Stewart and Rittweger 2006). The muscle gradually responds to contractile activity by converting to adult-type myosins and by 11 weeks the fibres grow in circumference and length, but by 24 weeks gestation it is believed that the fibre numbers are set. Further cross-sectional muscle growth at this stage is thought to occur as a consequence of hypertrophy not hyperplasia (Rowe and Goldspink 1969).

3.2 Skeletal Muscle Inflammation

More recently, an emerging phenotype (Bridges Jr et al. 2023) is that of muscle inflammation susceptibility (MuIS), describing the proinflammatory process seen in periarticular skeletal muscle inflammation seen in some patients with end stage OA. Short term inflammation is thought to be a necessary part of muscle repair after bouts of exercise, whilst chronic inflammation induces protein catabolism (Buford et al. 2009). However, Merritt et al. in their study of skeletal muscle proinflammatory signalling at rest and after modest muscle damage noted that this heightened muscle inflammation susceptibility is likely to impair muscle regenerative capacity, particularly with advancing age (Merritt et al. 2013). This perhaps forms the basis of the use of cold water immersion (CWI) therapy, which has been shown in meta-analysis to be superior to other methods for recovering from muscle soreness following acute strenuous exercise in physically active individuals (Moore et al. 2023).

3.3 Ageing and Atrophy in Skeletal Muscle

Ageing and disuse are two of the main conditions leading to skeletal muscle atrophy in humans (Suetta et al. 2008). In both conditions, the loss of muscle mass leads to a decrease in muscle force production. The most evident metabolic explanation for this muscle decline is an imbalance between protein catabolism and anabolism (Lenk et al. 2010). Loss of muscle mass of 5% per decade of life from the fourth decade onwards occurs, with the rate of loss increasing after the age of 65 years (Lenk et al. 2010). In addition to this sarcopaenia (age-associated decline in skeletal muscle coupled with functional deterioration (Kalinkovich and Livshits 2017)), infiltration of muscle with adipose tissue occurs (Marcus et al. 2010) resulting in reduced muscle strength. Furthermore, adipose tissue secretes pro-inflammatory cytokines such as interleukin-1B (IL-1B), tumour necrosis factor-a (TNFa) and adipokines such as adiponectin, resistin and leptin which have been implicated in the pathogenesis of OA (Krishnasamy et al. 2018).

3.4 The Effect of Load and Activity on Skeletal Muscle

Human skeletal muscle is predominantly composed of slow-twitch (type I) and fast-twitch (Type IIa, IIx) fibres (Ciciliot et al. 2013) which are regulated by mRNA expression of myosin heavy chain isoforms (Schiaffino and Reggiani 2011). Fast-twitch type II fibers produce greater power and force, and type II motor units have a high threshold for recruitment, whereas type I motor units are readily recruited and are more resistant to fatigue (Bridges Jr et al. 2023). Evidence generally indicates that type II fibres are more vulnerable to the ageing process than type I fibres (Lexell and Taylor 1991) with subsequent reduced muscle strength. Interestingly, a cross-sectional study noted a high proportion of type IIb fibres in the vastus lateralis muscles of male patients with early knee OA (Serrão et al. 2015) and resultant reduction in eccentric but not concentric strength suggesting that muscle activation patterns may differ with the severity of knee OA.

Increased muscle cross sectional area (hypertrophy) and altered neural recruitment patterns represent the typical adaptations to repeated bouts of heavy exercise (Häkkinen 1989). Hypertrophy following resistance training occurs when the rate of protein synthesis (anabolism) is greater than that of protein degradation (catabolism) (Chesley et al. 1992; Phillips et al. 1997). The hypertrophic adaptation to resistance training includes increased protein synthesis via regulatory changes in transcriptional and translational mechanisms, and, when the demand for de novo protein synthesis outweighs the supply by resident myonuclei, the recruitment of muscle cells that are added to existing myofibres or combine to form new contractile filament (Coffey and Hawley 2007). Thus the hypertrophic adaptations provide additional contractile machinery with which to generate greater force (Rennie et al. 2004). In addition, while a degree of protein degradation is required for muscle remodelling, resistance training may also decrease long-term activation of atrophy pathways, resulting in supplementary net protein synthesis (Marimuthu et al. 2011).

The early adaptive changes in skeletal muscle during increased loading are probably also due to enhanced translation of existing mRNAs with such increases being significant enough to increase protein production after a few days of overloading (Booth et al. 2002). The type of exercise stimuli governs the intricate balance by which signalling pathways are turned on or off, thereby providing a mechanism for a regulation of phenotypic outcomes. For example, endurance exercise does not lead to phosphorylation of kinase enzymes involved in the AKT (serine/threonine protein kinase) pathway whilst high resistance loading exercise does (Bodine et al. 2001). This suggests a differential regulation of signalling by the type of exercise and the possibility of different phenotypic outcomes. Whilst there is good evidence to suggest that exercise training in the early stages of OA is effective (Drummer et al. 2022), there remains the need to determine other mechanisms that may predict not only predisposition to OA, but also the potential for recovery following joint replacement surgery.

4 Gene Expression in Osteoarthritis

Transcription-wide association studies (TWAS) have begun to demonstrate candidate genes for predisposition of OA, such as FOXC1, ADM, FSTL1 and HLA-DRB1 (Qi et al. 2020). Transcriptomic differences in skeletal muscle continue to be studied in the context of OA. In a study of patients with end stage knee or hip OA, several differential expression genes (DEG) associated with inflammation have been identified (Drummer et al. 2022). A summary of genes described in characterising the vastus lateralis muscle of the operated leg in patients undergoing total hip arthroplasty are summarised in Table 1.

Table 1 Description of genes recognised in characterising atrophy and inflammation in the vastus lateralis of the operated leg in patients undergoing total hip replacement

Metabolic activity	Gene	Characteristic
↑ in *Atrophy*	CAPN3 (Calpain 3)	Down-regulated following skeletal muscle damage induced by eccentric exercise
	CTSL1 (Cathepsin L1) CTSL2 (Cathepsin L2)	Degradation of membrane proteins such as receptors, channels transporters
	FBX032 (F-Box protein 32)	Activity of ATP-dependent ubiquitin proteasome pathway, a process involving the interaction of multiple enzymes regulating 'ubiquitin-tagging' of proteins for destruction by the proteasome
	GSK3A (glycogen synthase kinase 3 alpha)	Increased by stretch and contraction, has been shown to increase after 2 weeks of disuse atrophy.
	MAPK14 (mitogen activated protein kinase 14)	Stress-activated protein kinase that responds to a variety of stimuli, including oxidative stress and TNF-α, and has been identified as a likely mediator of catabolic signalling in skeletal muscle
	MSTN (Myostatin)	Transforming growth factor-beta (TGF-β) family member that functions as an inhibitor of muscle hypertrophy.
	PSMA7 (proteasome subunit, alpha type, 7)	Ubiquitin protein coupled to myofibrillar protein substrates prior to proteasome degradation, ∴ involved in increased activity of ATP-dependent ubiquitin proteasome pathway
	TRIM63 (tripartite motif containing 63)	E3 ubiquitin ligase regulates proteasomal degradation of cardiac troponin and probably of other sarcomeric-associated proteins. Associated with muscle wasting in catabolic states
↓ in *Atrophy*	IGFBP5 (insulin-like growth factor binding protein 5)	IGF-I has been shown to stimulate anabolic and myogenic processes associated with the development of skeletal muscle hypertrophy. In muscle, mRNA levels have been shown to be reduced after increased muscle loading in animal studies
	MYOD1 (myogenic differentiation 1)	Involved in muscle differentiation and has been shown to be decreased in hospitalised patients compared to healthy controls
↑ in *Inflammation*	IL6 (Interleukin-6)	Produced by skeletal muscle and levels are not altered in serum as opposed to locally in muscle after resistance exercise. mRNA levels have been shown to increase following exercise in skeletal muscle
	TNF-α (tumour necrosis factor-alpha) TNFRSF1B (tumour necrosis factor receptor superfamily, member 1B)	Inflammatory cytokine (TNF-α) and its receptor (TNFRSF1B) whose mRNA expression has been shown to be stimulated early in the muscle repair process after acute resistance exercise.

Adapted from Okoro (2013)

5 Conclusion

A more nuanced understanding of periarticular inflammation continues to challenge traditional thinking and presents opportunities for targeting treatment. Whilst it is difficult to extrapolate to real life scenarios, the ongoing study of genetic expression identifies several potential candidate markers and molecular pathways for therapeutic intervention, and represents a vast area for high quality research. It is possible that by identifying the at-risk populations with gene mapping, early interventions such as strength training may reduce the incidence of debilitating OA requiring surgical treatment.

References

Aaron R. Orthopaedic basic science: fifth edition: print + Ebook: foundations of clinical practice 5. American Academy of Orthopaedic Surgeons; 2020.

Balistreri CR, Candore G, Accardi G, et al. NF-κB pathway activators as potential ageing biomarkers: targets for new therapeutic strategies. Immun Ageing. 2013;10:24.

Berenbaum F, Walker C. Osteoarthritis and inflammation: a serious disease with overlapping phenotypic patterns. Postgrad Med. 2020;132:377–84.

Bodine SC, Stitt TN, Gonzalez M, et al. Akt/mTOR pathway is a crucial regulator of skeletal muscle hypertrophy and can prevent muscle atrophy in vivo. Nat Cell Biol. 2001;3:1014–9.

Booth FW, Chakravarthy MV, Spangenburg EE. Exercise and gene expression: physiological regulation of the human genome through physical activity. J Physiol. 2002;543:399–411.

Bridges SL Jr, Sun D, Graham ZA, et al. Muscle inflammation susceptibility: a potential phenotype for guiding precision rehabilitation after total hip arthroplasty in end-stage osteoarthritis. HSS J. 2023;19:453–8.

Buford TW, Cooke MB, Willoughby DS. Resistance exercise-induced changes of inflammatory gene expression within human skeletal muscle. Eur J Appl Physiol. 2009;107:463–71.

Chesley A, MacDougall JD, Tarnopolsky MA, et al. Changes in human muscle protein synthesis after resistance exercise. J Appl Physiol. 1992;73:1383–8.

Ciciliot S, Rossi AC, Dyar KA, et al. Muscle type and fiber type specificity in muscle wasting. Int J Biochem Cell Biol. 2013;45:2191–9.

Coffey VG, Hawley JA. The molecular bases of training adaptation. Sports Med. 2007;37:737–63.

Drummer DJ, Lavin KM, Graham ZA, et al. Muscle transcriptomic circuits linked to periarticular physiology in end-stage osteoarthritis. Physiol Genomics. 2022;54:501–13.

Franceschi C, Bonafè M, Valensin S, et al. Inflamm-aging. An evolutionary perspective on immunosenescence. Ann N Y Acad Sci. 2000;908:244–54.

Häkkinen K. Neuromuscular and hormonal adaptations during strength and power training. A review. J Sports Med Phys Fitness. 1989;29:9–26.

Kalinkovich A, Livshits G. Sarcopenic obesity or obese sarcopenia: a cross talk between age-associated adipose tissue and skeletal muscle inflammation as a main mechanism of the pathogenesis. Ageing Res Rev. 2017;35:200–21.

Krishnasamy P, Hall M, Robbins SR. The role of skeletal muscle in the pathophysiology and management of knee osteoarthritis. Rheumatology. 2018;57:iv124.

Lenk K, Schuler G, Adams V. Skeletal muscle wasting in cachexia and sarcopenia: molecular pathophysiology and impact of exercise training. J Cachexia Sarcopenia Muscle. 2010;1:9–21.

Lexell J, Taylor CC. Variability in muscle fibre areas in whole human quadriceps muscle: effects of increasing age. J Anat. 1991;174:239–49.

López-Otín C, Blasco MA, Partridge L, et al. Hallmarks of aging: an expanding universe. Cell. 2023;186:243–78.

Marcus RL, Addison O, Kidde JP, et al. Skeletal muscle fat infiltration: impact of age, inactivity, and exercise. J Nutr Health Aging. 2010;14:362–6.

Marimuthu K, Murton AJ, Greenhaff PL. Mechanisms regulating muscle mass during disuse atrophy and rehabilitation in humans. J Appl Physiol. 2011;110:555–60.

Merritt EK, Stec MJ, Thalacker-Mercer A, et al. Heightened muscle inflammation susceptibility may impair regenerative capacity in aging humans. J Appl Physiol. 2013;115:937–48.

Moore E, Fuller JT, Bellenger CR, et al. Effects of cold-water immersion compared with other recovery modalities on athletic performance following acute strenuous exercise in physically active participants: a systematic review, meta-analysis, and meta-regression. Sports Med. 2023;53:687–705.

Okoro. Optimising patient function following elective total hip replacement. Thesis. (PhD). Bangor University; 2013.

Phillips SM, Tipton KD, Aarsland A, et al. Mixed muscle protein synthesis and breakdown after resistance exercise in humans. Am J Phys. 1997;273:E99–107.

Punzi L, Galozzi P, Luisetto R, et al. Post-traumatic arthritis: overview on pathogenic mechanisms and role of inflammation. RMD Open. 2016;2:e000279.

Qi X, Yu F, Wen Y, et al. Integration of transcriptome-wide association study and messenger RNA expression profile to identify genes associated with osteoarthritis. Bone Joint Res. 2020;9:130–8.

Rennie MJ, Wackerhage H, Spangenburg EE, Booth FW. Control of the size of the human muscle mass. Annu Rev Physiol. 2004;66:799–828.

Rezuș E, Cardoneanu A, Burlui A, et al. The link between Inflammaging and degenerative joint diseases. Int J Mol Sci. 2019;20:614. https://doi.org/10.3390/ijms20030614.

Rowe RW, Goldspink G. Muscle fibre growth in five different muscles in both sexes of mice. J Anat. 1969;104:519–30.

Schiaffino S, Reggiani C. Fiber types in mammalian skeletal muscles. Physiol Rev. 2011;91:1447–531.

Serrão PRMS, Vasilceac FA, Gramani-Say K, et al. Men with early degrees of knee osteoarthritis present functional and morphological impairments of the quadriceps femoris muscle. Am J Phys Med Rehabil. 2015;94:70–81.

Stewart CEH, Rittweger J. Adaptive processes in skeletal muscle: molecular regulators and genetic influences. J Musculoskelet Neuronal Interact. 2006;6:73–86.

Suetta C, Andersen JL, Dalgas U, et al. Resistance training induces qualitative changes in muscle morphology, muscle architecture, and muscle function in elderly postoperative patients. J Appl Physiol. 2008;105:180–6.

van den Bosch MHJ, Blom AB, van der Kraan PM. Inflammation in osteoarthritis: our view on its presence and involvement in disease development over the years. Osteoarthr Cartil. 2023;32:355–64. https://doi.org/10.1016/j.joca.2023.12.005.

Versus A. The state of musculoskeletal health 2023. Arthritis and other musculoskeletal conditions in numbers. Versus Arthritis; 2023.

Vincent TL. Mechanoflammation in osteoarthritis pathogenesis. Semin Arthritis Rheum. 2019;49:S36–8.

The Role of Neuromuscular Conditioning in Musculoskeletal Medicine

Charlotte Barker and Paul Lee

Abstract This chapter explores the critical role of neuromuscular conditioning in musculoskeletal (MSK) medicine, focusing on principles and practices that enhance muscle strength, stability, and overall functional capacity. It begins by emphasising the importance of prioritising strength and weight over repetitions in training, to optimise muscle hypertrophy, joint stability, and neuromuscular control. The chapter then examines the interplay between cardiovascular (CV) and MSK health, demonstrating the necessity of integrating both elements into comprehensive training and rehabilitation programmes. Advanced techniques, such as Neuromuscular Electrical Stimulation (NMES) and Blood Flow Restriction (BFR) training, are presented as innovative methods to build strength with minimal joint strain, particularly valuable in prehabilitation and postoperative recovery contexts. Additionally, the chapter addresses the significance of enhancing gait, coordination, and movement, underscoring their importance in injury prevention and recovery. By integrating these strategies, the chapter provides a comprehensive guide to improving MSK health through targeted neuromuscular conditioning, offering insights into optimising patient outcomes.

Keywords Blood flow restriction · Neuromuscular electrical stimulation · Rehabilitation · Hypertrophy · Coordination · Stability

1 Introduction

Neuromuscular conditioning involves training that enhances the interaction between the muscular system and the brain, improving strength, coordination, and overall physical function. This process includes exercises and techniques that stimulate the

C. Barker (✉)
NIHR Advanced Fellow at University College London, London, UK

P. Lee
College of Health and Science, University of Lincoln, Lincoln, UK

P. Lee (ed.), *Musculoskeletal Regeneration Medicine*,
https://doi.org/10.1007/978-3-031-77456-0_24

neuromuscular system—muscles, nerves, and the brain's ability to control muscle movements. Conditioning optimises communication between the brain and muscles, leading to better muscle activation, improved balance, and enhanced movement patterns, all crucial for injury prevention and rehabilitation in musculoskeletal (MSK) medicine.

1.1 Maximising Muscle Strength: The Science of Low Reps and Heavy Weights

In MSK medicine, prioritising strength and weight over repetitions in neuromuscular conditioning is increasingly recognised as essential for enhancing muscle function and reducing injury risks. Research supports that focusing on increasing weight, rather than merely increasing repetitions, leads to more significant improvements in muscle strength, joint stability, and overall MSK health.

Strength training that emphasises heavier weights is particularly effective in promoting muscle hypertrophy and neuromuscular adaptations. These adaptations are critical for stabilising joints and reducing the risk of injuries, especially in individuals with musculoskeletal disorders. Studies have demonstrated that greater muscle strength - achieved through lifting heavier weights - is associated with improved joint stability. It is also linked to reduced pain in conditions such as osteoarthritis. For example, stronger quadriceps muscles are linked to lower risks of symptomatic knee osteoarthritis and enhanced physical function, underscoring the importance of focusing on weight in strength training (Anwer and Alghadir 2014).

Prioritising weight over repetitions also more effectively stimulates the neuromuscular system, crucial for functional improvements and injury prevention. Heavier loads recruit more muscle fibres and enhance motor unit activation, leading to better muscle coordination and control. This is particularly important for athletes and individuals recovering from injury, as deficits in strength and neuromuscular control are common risk factors for ligament injuries (Read et al. 2016). By focusing on increasing weight, individuals can achieve more substantial gains in strength and neuromuscular efficiency, which are key to long-term MSK health.

Additionally, high-intensity training methods, such as BFR training, highlight the importance of weight over repetitions. BFR training allows individuals to engage in high-intensity exercise with lower loads, minimising joint stress while still achieving significant increases in muscle size and strength (Knutson 2019; Doucet et al. 2012). This method is especially beneficial for populations with limited physical capacities, where traditional high-repetition, low-weight protocols might not provide sufficient stimulus for muscle growth and strength improvements.

Training for Muscle Strength: High Weight, Low Repetitions

Building muscle strength, particularly in the context of prehabilitation for orthopaedic surgery, requires a careful approach to training. Research and clinical practice consistently support the effectiveness of high-weight, low-repetition training for developing muscle strength. This method not only maximises muscle hypertrophy but also enhances neuromuscular adaptations essential for joint stability and overall functional capacity.

High-weight, low-repetition training forces the neuromuscular system to recruit more motor units, specifically activating the high-threshold motor units responsible for generating greater force. These motor units are often not engaged during low-intensity, high-repetition exercises, but with heavier loads, the body is compelled to activate a broader range of muscle fibres, including type II fibres, which are crucial for generating power and strength. This increased recruitment leads to improved neuromuscular efficiency, enhancing the ability of the muscles to contract more forcefully and effectively in response to stimuli.

Additionally, high-weight training creates greater muscle tension, which stimulates the nervous system to improve coordination between muscles and the brain. This enhanced communication allows for more precise and efficient motor control, which is essential for stabilising joints and performing complex movements. Over time, repeated exposure to high weights also leads to neural adaptations, such as an increase in the firing rate of motor neurons, better synchronisation of motor units, and improved reflexes. These neural changes support joint stability by ensuring that the muscles around the joints respond quickly and effectively to sudden movements or external forces, reducing the risk of injury.

Furthermore, high-weight, low-repetition training enhances proprioception—the body's ability to sense its position and movement. This is particularly important in prehabilitation, as improved proprioception helps patients maintain balance and proper movement patterns during recovery, reducing the risk of post-operative complications.

Why High Weight, Low Reps?

High-weight, low-repetition training, typically using 70–85% of a person's one-repetition maximum (1RM) for 3–5 sets of 6–8 repetitions, effectively stimulates type II muscle fibres, particularly type IIa and IIb. Type IIa fibres, known as fast-twitch oxidative fibres, generate both force and endurance, making them essential for activities requiring sustained high-intensity efforts, such as climbing stairs. Type IIb fibres, or fast-twitch glycolytic fibres, are specialised for explosive power and short bursts of force, crucial for tasks like standing from a seated position. This type of training promotes both muscle hypertrophy and neuromuscular adaptations, enhancing joint stability and overall functional capacity, especially for prehabilitation in orthopaedic surgery.

The Science Behind the Approach

Lifting heavier weights with fewer repetitions forces muscles to exert more effort during each set, creating greater muscle tension and, in turn, producing more substantial strength gains. This method also enhances the recruitment of motor units and muscle fibres, a critical factor in neuromuscular conditioning. Motor units consist of a motor neuron and the muscle fibres it controls; heavier loads engage more motor units, particularly those that are typically dormant during lower-intensity exercises. This includes fast-twitch fibres, such as type IIa and IIb, which are responsible for generating high force and power. The recruitment of a broader range of motor units improves overall muscle activation, coordination, and strength. For patients preparing for surgery, this enhanced muscle recruitment is crucial for stabilising joints and protecting them during the recovery process. It ensures that muscles are better prepared to handle physical stress, reducing the risk of injury and promoting faster rehabilitation post-surgery.

Implementation in Prehabilitation and Post-Operative Care
For patients undergoing prehabilitation before orthopaedic surgery, strength training should be tailored to their specific needs and capabilities. A typical regimen might include compound movements such as squats, leg presses, and lunges, targeting the major muscle groups around the knee, hip, and other joints of concern. These exercises should be performed with enough weight to challenge the muscles within the 6–8 repetition range, ensuring that each set brings the muscles close to fatigue. This approach helps improve joint stability, muscle coordination, and overall functional capacity, ensuring patients are better prepared for surgery and reducing the risk of complications during recovery.

In the post-operative phase, neuromuscular conditioning becomes crucial as patients work to regain strength and mobility. Progressive increases in weight and training intensity should be carefully monitored to avoid overloading healing tissues. The focus of neuromuscular conditioning should remain on restoring strength and functional capacity, with controlled movements and proper technique being key to prevent injury. This approach ensures gradual neuromuscular adaptations, helping patients improve motor control, joint stability, and coordination as they safely rebuild their physical capabilities post-surgery.

Rest and Recovery
Adequate rest between sets and training sessions is vital in this approach. Patients should aim for 1–2 min of rest between sets to allow partial recovery of the muscles, enabling them to perform each set with the required intensity. Additionally, training sessions should be spaced out, typically with at least 48 h between sessions targeting the same muscle group, to allow for muscle recovery and growth.

2 Enhancing MSK Health Through Targeted Muscle Development

The importance of building muscle through neuromuscular conditioning in musculoskeletal (MSK) medicine is underscored by its multifaceted benefits, incorporating enhanced physical performance, injury prevention, and overall health improvement. Muscle hypertrophy, achieved through targeted resistance training, plays a critical role in these outcomes.

2.1 Defining Neuromuscular Electrical Stimulation (NMES)

Neuromuscular Electrical Stimulation (NMES) is a therapeutic technique that uses electrical impulses to evoke muscle contractions, thereby enhancing muscle strength, endurance, and function. NMES is particularly useful in rehabilitation settings, especially for individuals unable to engage in traditional forms of exercise due to injury or limited mobility. By artificially stimulating muscles, NMES can promote muscle growth and prevent atrophy, contributing significantly to overall muscle conditioning.

2.2 Muscle Hypertrophy and Fibre Type Adaptations

Muscle hypertrophy refers to the increase in muscle fibre size, a physiological response primarily induced by resistance training. This process predominantly affects Type 2 muscle fibres, crucial for generating force and power. However, it is important to consider the role of Type 1 muscle fibres, which are also involved in neuromuscular conditioning.

Type 1 muscle fibres, known as slow-twitch fibres, are characterised by their endurance capacity and resistance to fatigue. These fibres are primarily engaged during activities that require sustained, low-intensity effort, such as cardiovascular exercise. While Type 2 fibres undergo significant hypertrophy with high-intensity resistance training, Type 1 fibres also experience adaptations, albeit of a different nature. Through regular endurance and low-intensity resistance training, Type 1 fibres can increase their oxidative capacity, enhancing their ability to sustain prolonged activity. This adaptation is crucial for improving overall muscular endurance and supporting activities that require long-term muscle engagement.

Studies have shown that NMES can effectively induce hypertrophy in both Type 1 and Type 2 muscle fibres, although the effects are more pronounced in Type 2 fibres due to their greater capacity for growth. NMES improves endurance by enhancing the oxidative capacity of Type 1 fibres, demonstrating its utility in rehabilitation settings where traditional resistance training may not be feasible.

2.3 The Role of Muscle Strength in Injury Prevention

The relationship between muscle strength and injury prevention is well established. Howell et al. (2018) emphasise that neuromuscular control deficits can lead to an increased risk of musculoskeletal injuries, highlighting the need for strength training to enhance neuromuscular stability and coordination. By building muscle strength, individuals can improve their postural control and reduce the likelihood of injuries, particularly in high-risk populations such as athletes and older adults. This preventive aspect of strength training is crucial in MSK medicine, where the goal is not only to rehabilitate but also to prevent future injuries.

2.4 Metabolic and Cardiovascular Benefits

In addition to injury prevention, building muscle contributes to overall health by improving metabolic function and cardiovascular health. Kim and Kim (2020) discuss how muscle mass is associated with better metabolic outcomes, including improved insulin sensitivity and reduced risk of chronic diseases. Regular resistance training that promotes muscle hypertrophy can enhance metabolic rate and support weight management, which is particularly important in the context of an increasingly sedentary lifestyle. The adaptation of Type 1 fibres through endurance training further supports cardiovascular health by improving the efficiency of oxygen utilisation during prolonged activity, thereby reducing the risk of cardiovascular diseases.

This dual benefit of muscle building—enhancing both musculoskeletal and metabolic health—makes it a cornerstone of effective rehabilitation and conditioning programmes.

2.5 Psychological and Functional Advantages

Moreover, the psychological benefits of strength training should not be overlooked. Engaging in resistance training can boost self-efficacy and improve mental health outcomes, which are critical components of recovery and rehabilitation in MSK medicine. The empowerment that comes from increased physical capability can enhance adherence to rehabilitation protocols and promote a more active lifestyle.

3 Integrating NMES and BFR Training: Enhancing Strength with Minimal Joint Strain

Building on the importance of neuromuscular conditioning in both prehabilitation and postoperative recovery, techniques such as Neuromuscular Electrical Stimulation (NMES) and Blood Flow Restriction (BFR) training offer innovative approaches to muscle strengthening across a variety of musculoskeletal (MSK) conditions. These methods are particularly valuable for individuals who need to build strength while minimising stress on their joints, making them ideal for managing and rehabilitating a range of MSK issues.

3.1 The Importance of Neuromuscular Conditioning

Neuromuscular conditioning is essential for enhancing muscle strength and function, which are critical not only for joint stability but also for overall physical performance. Targeted strength training improves physical capabilities and plays a significant role in injury prevention. For example, increased muscle strength has been shown to mediate symptom improvement in various MSK conditions, highlighting its importance in managing overall joint and muscle health (Patterson et al. 2022). Moreover, muscle-strengthening exercises have been proven to reduce pain and improve function in individuals with chronic MSK conditions, establishing them as a cornerstone of effective rehabilitation strategies (De la Corte-Rodriguez et al. 2024).

3.2 Neuromuscular Electrical Stimulation (NMES)

NMES uses electrical impulses to stimulate muscle contractions, making it particularly beneficial for individuals unable to perform traditional exercises due to injury, surgery, or severe pain. NMES effectively promotes muscle hypertrophy and strength, even in populations with limited mobility. For instance, NMES has been shown to enhance muscle activation and strength in patients recovering from various orthopaedic surgeries, facilitating a quicker return to functional activities (Toth et al. 2020). It also helps mitigate muscle atrophy during periods of immobilisation, making it a valuable tool in both prehabilitation and rehabilitation protocols.

The principle behind NMES involves the recruitment of motor units through electrical stimulation, leading to neural adaptations and muscle hypertrophy. NMES can activate muscle fibres even when voluntary contraction is compromised, maintaining muscle function during recovery and reducing the risk of muscle atrophy (Shin et al. 2022). This makes NMES a critical component in managing chronic MSK conditions, where maintaining muscle mass and strength is essential for long-term health.

3.3 Blood Flow Restriction (BFR) Training

BFR training involves applying external pressure to the limbs during low-intensity exercise, enhancing muscle hypertrophy and strength without needing heavy weights. This technique is particularly beneficial for individuals who may be unable to lift heavy loads due to pain, injury, or specific MSK conditions. Research has shown that BFR training can lead to significant improvements in muscle strength and size, comparable to traditional high-intensity resistance training (Miller et al. 2021).

The mechanism behind BFR training involves the accumulation of metabolites and the recruitment of fast-twitch muscle fibres, typically engaged during high-intensity exercises. This allows individuals to achieve substantial training effects while using lighter weights, reducing the risk of joint stress and injury (Cognetti et al. 2022). BFR training has been shown to be safe and effective across various populations, including older adults and those managing chronic conditions, making it a versatile tool in neuromuscular conditioning (Miller et al. 2021).

By incorporating NMES and BFR into neuromuscular conditioning programmes, patients with MSK conditions can build muscle strength and improve functional outcomes without placing excessive strain on their joints. These methods are particularly valuable for those who need to enhance strength but face limitations in performing traditional weight-bearing exercises.

4 The Interplay Between Cardiovascular and Musculoskeletal Health in Neuromuscular Conditioning

Building on the importance of prioritising strength and weight in neuromuscular conditioning, it is equally crucial to understand how cardiovascular (CV) and musculoskeletal (MSK) health intersect within this framework. Both domains are essential to overall physical performance, injury prevention, and long-term health, yet they serve different purposes in training and rehabilitation strategies. The integration of CV training with neuromuscular conditioning offers a comprehensive approach that enhances MSK function, contributing to optimal health outcomes.

4.1 Benefits of Combining CV and MSK Training

Cardiovascular training plays a vital role in maintaining overall endurance, improving circulation, and supporting metabolic health. These benefits extend to MSK health by enhancing recovery times, increasing the efficiency of oxygen delivery to muscles, and promoting resilience against fatigue. Regular CV exercise improves cardiovascular efficiency and helps maintain joint health by supporting the metabolic processes essential for muscle repair and growth. This interplay ensures that muscles receive adequate blood flow and nutrients, facilitating recovery and adaptation from strength training sessions.

By combining CV training with MSK conditioning, patients can expect a range of physiological benefits crucial for effective neuromuscular conditioning. Improved cardiovascular fitness supports the endurance needed for sustained physical activity, allowing individuals to perform MSK exercises with greater efficiency. Additionally, enhanced circulation from CV training ensures that muscles are well-nourished, particularly important during high-intensity strength training. This synergy between the cardiovascular and musculoskeletal systems leads to better overall physical performance and a reduced risk of chronic diseases.

4.2 Impact on Neuromuscular Function

When CV and MSK training are combined, the neuromuscular system is challenged in a more holistic manner. CV exercise enhances the delivery of oxygen and nutrients to muscles, essential for muscle recovery and growth. This, in turn, supports the neuromuscular system's ability to perform high-intensity strength training, leading to greater muscle hypertrophy and strength gains. Additionally, the improved metabolic health from CV exercise helps manage body weight, reducing strain on joints and preventing injuries.

Furthermore, the combination of CV and MSK training improves neuromuscular coordination, as both systems are engaged simultaneously during exercises that require endurance and strength. This integrated approach enhances balance, stability, and overall movement efficiency, which are critical for injury prevention and recovery. For instance, individuals who engage in both CV and MSK training are likely to experience fewer instances of muscle fatigue, leading to better performance in functional tasks and reduced injury risk.

5 Building Muscle to Protect Joints: The Role of Strength in Osteoarthritis Management

Building muscle through strength training is a critical component in managing osteoarthritis (OA), particularly in protecting joints and enhancing overall functional capacity. The relationship between muscle strength and joint health is well-documented, with evidence suggesting that increased muscle strength can alleviate symptoms of osteoarthritis and improve joint stability.

One of the primary benefits of building muscle is its role in joint protection. The quadriceps muscle group, for instance, is essential for stabilising the knee joint. Research has shown that greater quadriceps strength is associated with reduced pain and improved function in individuals with knee osteoarthritis (Imoto et al. 2012). Specifically, the quadriceps muscle plays a key role in absorbing impacts experienced by the knee joint. Strength deficits in this muscle can compromise its ability to absorb loads, leading to increased joint stress and pain (Imoto et al. 2012). Therefore, strengthening the quadriceps can act as a protective factor against further cartilage degeneration and joint instability.

Moreover, muscle strength is directly linked to the prevention of falls and fractures, significant concerns for individuals with osteoarthritis. Research emphasises that individuals with early-diagnosed hip and knee osteoarthritis are at a higher risk of falls, suggesting that strength training should be integrated into early management strategies to mitigate this risk (Claudino et al. 2021). By improving muscle strength, particularly in the lower extremities, patients can enhance their balance and stability, reducing the likelihood of falls and associated injuries.

In addition to physical benefits, building muscle can also positively affect pain management and functional performance. Studies have indicated that strength training can lead to significant reductions in pain and improvements in physical function among individuals with knee osteoarthritis (Zeng et al. 2021; Turner et al. 2019). For example, isometric strength training of the quadriceps has been found to not only increase muscle strength but also decrease pain intensity in patients with knee osteoarthritis (Machado et al. 2022). This dual benefit of pain reduction and strength enhancement is crucial for improving the quality of life for individuals suffering from OA.

Furthermore, the psychological aspects of strength training should not be overlooked. Engaging in regular strength training can boost self-efficacy and improve mental health outcomes, which are vital for adherence to rehabilitation protocols. The empowerment that comes from increased physical capability can motivate patients to maintain an active lifestyle, further supporting joint health and overall well-being.

5.1 Extending Neuromuscular Conditioning: The Critical Role of Prehabilitation and Postoperative Recovery in Orthopaedic Surgery

Integrating neuromuscular conditioning into prehabilitation is essential for optimising surgical outcomes and enhancing recovery, particularly in orthopaedic procedures where restoring mobility and function is crucial. Prehabilitation, which prepares patients physically and mentally for surgery, significantly improves postoperative results by enhancing muscle strength, coordination, and overall functional capacity.

Enhancing Muscle Strength and Function

Strengthening muscles before surgery is key to joint stability and mobility. Research demonstrates that increased preoperative muscle strength leads to better postoperative outcomes. For instance, coordinated movement patterns that enhance overall muscle activation improve functional performance and reduce complications after surgery (Choi et al. 2022). Moreover, neuromuscular conditioning helps mitigate muscle atrophy, often caused by inactivity during recovery. Maintaining muscle strength preoperatively equips patients to handle the physical demands of rehabilitation, enabling quicker recovery and a faster return to daily activities (Wu et al. 2022).

Improving Coordination and Gait

Neuromuscular conditioning also improves coordination and gait, vital for successful post-surgical recovery. Prehabilitation programmes often include balance and coordination training, helping patients regain safe and efficient movement. Studies show that targeted neuromuscular training enhances gait mechanics, particularly in patients recovering from conditions like stroke, and similar strategies can benefit those undergoing orthopaedic surgeries (Nguyen et al. 2022; Judd et al. 2016). Enhanced proprioception, a key outcome of neuromuscular conditioning, also improves balance and reduces the risk of falls during recovery.

Psychological Benefits of Prehabilitation

In addition to physical benefits, prehabilitation that includes neuromuscular conditioning addresses psychological factors influencing recovery. Patients who feel stronger and more capable before surgery often report lower anxiety levels and a greater sense of control over their recovery process. This psychological preparedness leads to improved adherence to postoperative rehabilitation protocols and better overall outcomes (Powell et al. 2016).

4-Week Prehabilitation Plan for Total Knee Replacement

The 4-Week Prehabilitation Plan for Total Knee Replacement is designed to prepare patients for surgery (Fig. 1), focusing on building muscle strength, improving range of motion, and enhancing cardiovascular health. The plan gradually increases in intensity over 4 weeks, ensuring that patients are physically prepared for the demands of surgery and the subsequent recovery period. This plan also incorporates flexibility and balance exercises, critical in reducing the risk of postoperative complications and promoting a faster return to normal activities.

It is important to note, for patients experiencing pain due to arthritis and joint strain, managing the intensity of weight-bearing exercises is crucial to avoid exacerbating their condition. Incorporating Neuromuscular Electrical Stimulation (NMES) and Blood Flow Restriction (BFR) training can significantly alter the load on the joints, making exercise less painful while still effectively building muscle strength. NMES can stimulate muscle contractions without the need for heavy weights, allowing patients to gain muscle strength even when traditional exercises are too painful. Similarly, BFR training enables the achievement of muscle hypertrophy at lower intensities by restricting blood flow to the muscles, thereby reducing the need for high-impact or high-load exercises. By integrating these techniques into their prehabilitation or rehabilitation programmes, patients can continue to strengthen their muscles and protect their joints, all while minimising discomfort and reducing the risk of further injury.

Managing Pain with NMES and BFR

For patients experiencing pain due to arthritis and joint strain, managing the intensity of weight-bearing exercises is crucial to avoid exacerbating their condition. Incorporating Neuromuscular Electrical Stimulation (NMES) and Blood Flow Restriction (BFR) training can significantly alter the load on the joints, making exercise less painful while still effectively building muscle strength. NMES can

Week	Exercise Type	Exercises	Reps/Sets	Frequency	Notes
Week 1	Strength Training 50-60% 1RM	**Leg Press**	6 sets of 3-5 reps	3 times/week	Begin with low resistance to ensure proper form and prevent injury.
		Quadriceps Extensions	6 sets of 3-5 reps	3 times/week	Focus on achieving full knee extension at the top of each movement.
	Cardiovascular RPE 4-5	**Stationary Cycling**	20 minutes	5 times/week	Maintain a moderate pace. Can be performed on a stationary bike with low resistance.
	Balance & Proprioception	**Single-leg Stance (supported)**	3 sets of 30 seconds per leg	Daily	Perform near a stable surface (e.g., counter or chair) for support if needed.
	Flexibility	**Hamstring Stretch**	3 sets of 30 seconds per leg	Daily	Stretch gently to avoid discomfort, focusing on improving flexibility in the hamstrings.
		Calf Stretch	3 sets of 30 seconds per leg	Daily	Use a wall for support, ensuring that the back leg remains straight to stretch the calf muscle effectively.
		Heel Slides	3 sets of 10 reps/per leg	Daily	Focus on smooth movement while sliding the heel towards the buttocks to improve knee flexion.
	Core Strengthening	**Pelvic Tilts**	3 sets of 10 reps	3 times/week	Strengthen the lower back and abdominal muscles.

Fig. 1 Four-Week Prehabilitation Plan for Total Knee Replacement (TKR). This table presents a progressive exercise regimen aimed at improving muscle strength, range of motion, and cardiovascular health before TKR surgery. Over 4 weeks, the plan increases in intensity, preparing patients for surgery and recovery. The exercises include strength training, cardiovascular activities, balance and proprioception drills, flexibility stretches, and core strengthening, tailored to patient needs

Week 2	**Strength Training 60-65% 1RM**	**Leg Press**	6 sets of 3-5 reps	3 times/week	Slightly increase resistance while maintaining proper form.
		Quadriceps Extensions	6 sets of 3-5 reps	3 times/week	Continue to focus on controlled, smooth movements.
	Cardiovascular RPE 4-5	**Stationary Cycling**	25 minutes	5 times/week	Increase duration slightly to build endurance.
	Balance & Proprioception	**Single-leg Stance (unsupported if possible)**	3 sets of 40 seconds per leg	Daily	If balance has improved, attempt to perform without support.
	Flexibility	**Hamstring Stretch**	3 sets of 30 seconds per leg	Daily	Continue gentle stretching, potentially deepening the stretch if comfortable.
		Calf Stretch	3 sets of 30 seconds per leg	Daily	Maintain consistency in stretching to prevent muscle tightness.
		Heel Slides	3 sets of 12 reps per leg	Daily	Increase repetitions as knee flexion improves.
	Core Strengthening	**Pelvic Tilts**	3 sets of 12 reps	3 times/week	Enhance focus on core muscle engagement during each repetition.
		Bridging	3 sets of 10 reps	3 times/week	Engage glutes and core muscles while lifting hips off the ground.

Fig. 1 (continued)

stimulate muscle contractions without heavy weights, allowing patients to gain muscle strength even when traditional exercises are too painful. Similarly, BFR training enables muscle hypertrophy at lower intensities by restricting blood flow to the muscles, reducing the need for high-impact or high-load exercises. By integrating these techniques into their prehabilitation or rehabilitation programmes, patients can continue to strengthen their muscles and protect their joints, all while minimising discomfort and reducing the risk of further injury.

Week 3	**Strength Training** **65-70% 1RM**	**Leg Press**	5 sets of 3-5 reps	3 times/week	Further increase resistance as tolerated.
		Quadriceps Extensions	5 sets of 3-5 reps	3 times/week	Emphasise movement to ensure muscle engagement.
	Cardiovascular **RPE 5-6**	**Stationary Cycling**	30 minutes	5 times/week	Increase duration further, focusing on building stamina.
	Balance & Proprioception	**Single-leg Stance (unsupported)**	4 sets of 40 seconds per leg	Daily	Progress to performing on an unstable surface (e.g., balance pad) if possible.
	Flexibility	**Hamstring Stretch**	3 sets of 30 seconds per leg	Daily	Continue with consistent stretching to maintain flexibility gains.
		Calf Stretch	3 sets of 30 seconds per leg	Daily	Perform regularly to ensure calf muscles remain flexible and prevent stiffness.
		Heel Slides	3 sets of 15 reps per leg	Daily	Gradually increase reps to maintain and improve range of motion.
	Core Strengthening	**Pelvic Tilts**	3 sets of 15 reps	3 times/week	Focus on controlled, deliberate movements to engage core muscles effectively.
		Bridging	3 sets of 12 reps	3 times/week	Increase reps as core strength improves.
		Modified Plank	Hold for 20 seconds, 3 sets	3 times/week	Engage entire core during hold, ensuring a straight line from shoulders to knees.

Fig. 1 (continued)

Postoperative Recovery and Rehabilitation

Following orthopaedic surgery, neuromuscular conditioning continues to be vital. Postoperative rehabilitation programmes that emphasise strength training can facilitate quicker recovery of muscle function and joint mobility. Studies have shown that

Week 4	**Strength Training** **70-75% 1RM**	**Leg Press**	5 sets of 3-5 reps	3 times/week	Maximise resistance with proper form.
		Quadriceps Extensions	5 sets of 3-5 reps	3 times/week	Focus on consistent, smooth movement for muscle endurance.
	Cardiovascular **RPE 5-6**	**Stationary Cycling**	35 minutes	5 times/week	Focus on maintaining stamina, aiming for consistent, moderate intensity.
	Balance & Proprioception	**Single-leg Stance (on unstable surface if possible)**	4 sets of 50 seconds per leg	Daily	Continue challenging proprioception and balance.
	Flexibility	**Hamstring Stretch**	3 sets of 30 seconds per leg	Daily	Maintain flexibility to support recovery.
		Calf Stretch	3 sets of 30 seconds per leg	Daily	Stretch regularly to prevent post-surgery tightness.
		Heel Slides	3 sets of 20 reps per leg	Daily	Increase reps to maintain and further improve range of motion.
	Core Strengthening	**Pelvic Tilts**	3 sets of 20 reps	3 times/week	Engage core fully in each repetition.
		Bridging	3 sets of 15 reps	3 times/week	Increase reps and hold for a second at the top.
		Modified Plank	Hold for 30 seconds, 3 sets	3 times/week	Hold longer as core strength increases.
		Side Plank	Hold 15 seconds each side, 3 sets	3 times/week	Focus on keeping the body straight while balancing on the forearm and side of the foot/knee.

Fig. 1 (continued)

patients who engage in structured rehabilitation programmes post-surgery report better functional outcomes and lower levels of pain compared to those who do not (Gränicher et al. 2022). For instance, research has highlighted that essential amino acid supplementation post-TKA can enhance muscle recovery and functional outcomes, further emphasising the need for a comprehensive approach to postoperative care (Dreyer et al. 2018).

Maintaining a focus on neuromuscular conditioning after surgery can also help prevent complications such as falls, particularly concerning older adults undergoing orthopaedic procedures. Strengthening exercises can improve balance and coordination, reducing the risk of falls and subsequent injuries (Turner et al. 2020).

6 Gait, Coordination, and Movement: The Role of Neuromuscular Conditioning in MSK Health

Gait, Coordination, and Movement: The Cornerstones of MSK Rehabilitation
In musculoskeletal (MSK) medicine, the improvement of gait, coordination, and movement is fundamental for injury prevention, effective recovery from injury and surgery, and overall functional health. Neuromuscular conditioning—through techniques like Neuromuscular Electrical Stimulation (NMES) and Blood Flow Restriction (BFR) training—plays a crucial role in enhancing these aspects, ensuring that patients not only regain strength but also achieve optimal movement patterns essential for daily activities and long-term health.

The Importance of Gait, Coordination, and Movement
Gait, coordination, and movement are critical components of MSK health. Proper gait mechanics ensure that forces are distributed evenly across joints, reducing the risk of wear and tear that can lead to conditions such as osteoarthritis. Coordination and movement efficiency are equally important, as they enable smooth, controlled motions that minimise the risk of injury. For patients recovering from surgery or managing chronic MSK conditions, restoring these functions is essential for regaining independence and preventing future complications.

Injury Prevention
Improving gait, coordination, and movement helps prevent injuries by promoting better joint alignment and reducing abnormal stress on muscles and ligaments. For example, proper gait mechanics decrease the likelihood of falls and overuse injuries, which are common in individuals with poor movement patterns. Strengthening the muscles that support gait and coordination through targeted exercises can significantly lower the risk of re-injury, particularly in vulnerable populations such as the elderly or those with a history of MSK issues.

Recovery from Injury and Surgery
After an injury or surgical procedure, impaired gait and coordination are common. Addressing these deficits through neuromuscular conditioning is vital for a successful recovery. Techniques like NMES can accelerate the return to normal movement patterns by stimulating muscle contractions in weakened areas, while BFR training allows for muscle strengthening with minimal joint strain. These methods help

restore the muscle function needed to support effective movement, ensuring that patients regain full mobility as quickly and safely as possible.

Methods to Improve Gait, Coordination, and Movement

To effectively improve gait, coordination, and movement, a comprehensive approach that includes the following methods should be implemented:

1. **Strengthening Exercises**: Targeted strength training is essential for building the muscle mass and power necessary to support proper movement. Exercises focusing on the lower extremities, such as squats, lunges, and leg presses, are particularly beneficial for enhancing gait mechanics. Incorporating high-weight, low-repetition sets can maximise strength gains, which in turn supports joint stability and coordinated movement.
2. **Neuromuscular Electrical Stimulation (NMES)**: NMES can be used to activate muscles that are difficult to engage voluntarily, particularly in patients recovering from surgery or those with significant muscle atrophy. By stimulating these muscles, NMES helps restore muscle function and supports the re-establishment of normal movement patterns, contributing to improved gait and coordination.
3. **Blood Flow Restriction (BFR) Training**: BFR training enables significant strength gains with lower loads, making it an ideal method for patients who cannot tolerate heavy weights due to pain or injury. By improving muscle hypertrophy and strength, BFR training supports better gait and movement without overloading the joints, reducing the risk of further injury.
4. **Balance and Proprioception Training**: Enhancing proprioception—one's sense of body position—and balance is crucial for coordinated movement. Exercises that challenge stability, such as single-leg stands, balance boards, and plyometric drills, improve these skills, leading to safer and more effective movement patterns. This is particularly important for preventing falls and ensuring smooth, controlled motions during daily activities.
5. **Gait Training**: Gait training, which focuses specifically on improving walking patterns, is crucial for individuals with altered gait due to injury or surgery. This can involve exercises that correct abnormal gait mechanics, increase stride length, and promote symmetry in movement. Gait training often includes treadmill work, resistance band exercises, and step drills, which help to reinforce proper walking patterns.

Integrating These Methods in MSK Rehabilitation

For optimal results, these methods should be integrated into a comprehensive neuromuscular conditioning programme tailored to the individual needs of each patient. By focusing on improving gait, coordination, and movement, healthcare providers can significantly enhance the overall functional outcomes for patients, ensuring a quicker return to normal activities and a reduced risk of future injuries.

7 Conclusion: Unifying Strength, Stability, and Recovery

This chapter has explored the multifaceted role of neuromuscular conditioning in musculoskeletal (MSK) medicine, highlighting its critical importance in improving strength, stability, and functional outcomes across a range of conditions. From the prioritisation of strength and weight in training over repetitions to the intricate interplay between cardiovascular and MSK health, each section has underscored how targeted muscle development can enhance physical performance, prevent injuries, and support long-term well-being.

Building muscle strength is not only vital for joint protection, particularly in the management of osteoarthritis, but it also plays a crucial role in prehabilitation and postoperative recovery in orthopaedic surgery. Techniques like Neuromuscular Electrical Stimulation (NMES) and Blood Flow Restriction (BFR) training have been shown to be effective in achieving muscle hypertrophy and improving functional outcomes, especially in individuals who are unable to perform traditional high-load exercises.

The chapter culminates in a discussion on gait, coordination, and movement, tying together the various elements of neuromuscular conditioning. Proper gait mechanics, enhanced coordination, and effective movement patterns are fundamental to injury prevention and recovery, ensuring that patients not only regain their strength but also achieve optimal functional health. Through a comprehensive approach that includes strengthening exercises, NMES, BFR training, and balance and proprioception exercises, healthcare providers can significantly improve the overall outcomes for patients with MSK conditions.

References

Anwer S, Alghadir A. Effect of isometric quadriceps exercise on muscle strength, pain, and function in patients with knee osteoarthritis: a randomized controlled study. J Phys Ther Sci. 2014;26(5):745–8. https://doi.org/10.1589/jpts.26.745.

Choi YS, et al. Enhanced recovery after surgery for major orthopedic surgery: a narrative review. Knee Surg Relat Res. 2022;34(1) https://doi.org/10.1186/s43019-022-00137-3.

Claudino JG, et al. Strength training to prevent falls in older adults: a systematic review with meta-analysis of randomized controlled trials. J Clin Med. 2021;10(14):3184. https://doi.org/10.3390/jcm10143184.

Cognetti DJ, et al. Blood flow restriction therapy and its use for rehabilitation and return to sport: physiology, application, and guidelines for implementation. Arthrosc Sports Med Rehabil. 2022;4(1):e71–6. www.ncbi.nlm.nih.gov/pmc/articles/PMC8811521/#bib5. https://doi.org/10.1016/j.asmr.2021.09.025.

De la Corte-Rodriguez H, et al. The role of physical exercise in chronic musculoskeletal pain: best medicine—a narrative review. *Healthcare*. 2024;12(2):242. mdpi-res.com/healthcare/healthcare-12-00242/article_deploy/healthcare-12-00242.pdf?version=1705565184. https://doi.org/10.3390/healthcare12020242.

Doucet BM, et al. Neuromuscular electrical stimulation for skeletal muscle function. Yale J Biol Med. 2012;85(2):201–15. www.ncbi.nlm.nih.gov/pmc/articles/PMC3375668/

Dreyer HC, et al. Essential amino acid supplementation mitigates muscle atrophy after total knee arthroplasty. JBJS Open Access. 2018;3(2):e0006. https://doi.org/10.2106/jbjs.oa.18.00006.

Gränicher P, et al. Prehabilitation improves knee functioning before and within the first year after Total knee arthroplasty: a systematic review with meta-analysis. J Orthop Sports Phys Ther. 2022;52(11):709–25. https://doi.org/10.2519/jospt.2022.11160.

Howell D, et al. Neuromuscular control deficits and the risk of subsequent injury after a concussion: a scoping review. Sports Med. 2018; www.semanticscholar.org/paper/Neuromuscular-Control-Deficits-and-the-Risk-of-a-A-Howell-Lynall/dc4817b3e03bbea098c4cc3e5480d5ebcb646fa9. Accessed 2 Oct. 2024

Imoto AM, et al. Quadriceps strengthening exercises are effective in improving pain, function and quality of life in patients with osteoarthritis of the knee. Acta Ortop Bras. 2012;20(3):174–9, www.ncbi.nlm.nih.gov/pmc/articles/PMC3718434/. https://doi.org/10.1590/S1413-78522012000300008.

Judd DL, et al. Effects of neuromuscular reeducation on hip mechanics and functional performance in patients after total hip arthroplasty: a case series. Clin Biomech. 2016;32(2):49–55. https://doi.org/10.1016/j.clinbiomech.2015.12.008.

Kim G, Kim JH. Impact of skeletal muscle mass on metabolic health. Endocrinol Metab. 2020;35(1):1. https://doi.org/10.3803/enm.2020.35.1.1.

Knutson, Jayme S., et al. Neuromuscular electrical stimulation applications. Atlas of orthoses and assistive devices, 1, 2019, pp. 432–439.e3, https://doi.org/10.1016/b978-0-323-48323-0.00043-3. Accessed 5 Sept. 2022.

Machado S, et al. Knee osteoarthritis: Kinesiophobia and isometric strength of quadriceps in women. Pain Res Manag. 2022;2022(3):1466478. www.ncbi.nlm.nih.gov/pmc/articles/PMC9098341/. https://doi.org/10.1155/2022/1466478.

Miller BC, et al. The Systemic Effects of Blood Flow Restriction Training: A Systematic Review. Int J Sports Phys Ther. 2021;16(4):978–90. www.ncbi.nlm.nih.gov/pmc/articles/PMC8329318/. https://doi.org/10.26603/001c.25791.

Nguyen PT, et al. Proprioceptive Neuromuscular Facilitation-Based Physical Therapy on the Improvement of Balance and Gait in Patients with Chronic Stroke: A Systematic Review and Meta-Analysis. *Life (Basel, Switzerland)*. 2022;12(6):882, pubmed.ncbi.nlm.nih.gov/35743913/. https://doi.org/10.3390/life12060882.

Patterson BE, et al. Muscle strength and osteoarthritis of the knee: a systematic review and meta-analysis of longitudinal studies. Skeletal Radiol. 2022;52(11):2085–97. https://doi.org/10.1007/s00256-022-04266-4.

Powell R, et al. Psychological preparation and postoperative outcomes for adults undergoing surgery under general anaesthesia. Cochrane Database Syst Rev. 2016;5(5):CD008646. https://doi.org/10.1002/14651858.cd008646.pub2.

Read PJ, et al. Neuromuscular risk factors for knee and ankle ligament injuries in male youth soccer players. Sports Med. 2016;46(8):1059–66. https://doi.org/10.1007/s40279-016-0479-z.

Shin HE, et al. Therapeutic effects of functional electrical stimulation on physical performance and muscle strength in post-stroke older adults: a review. Ann Geriatr Med Res. 2022;26(1):16–24, www.ncbi.nlm.nih.gov/pmc/articles/PMC8984173/. https://doi.org/10.4235/agmr.22.0006.

Toth MJ, et al. Utility of neuromuscular electrical stimulation to preserve quadriceps muscle fiber size and contractility after anterior cruciate ligament injuries and reconstruction: a randomized, sham-controlled, blinded trial. Am J Sports Med. 2020;48(10):2429–37, www.ncbi.nlm.nih.gov/pmc/articles/PMC7775613/. https://doi.org/10.1177/0363546520933622.

Turner MN, et al. The role of resistance training dosing on pain and physical function in individuals with knee osteoarthritis: a systematic review. Sports Health. 2019;12(2):200–6. https://doi.org/10.1177/1941738119887183.

Turner K, et al. Fall prevention practices and implementation strategies. J Patient Saf. 2020;18(1):e236–42, www.ncbi.nlm.nih.gov/pmc/articles/PMC7854936/. https://doi.org/10.1097/pts.0000000000000758.

Wu Z, et al. Preoperative strength training for clinical outcomes before and after total knee arthroplasty: a systematic review and meta-analysis. Front Surg. 2022;21(9):879593. www.ncbi.nlm.nih.gov/pmc/articles/PMC9349363/. https://doi.org/10.3389/fsurg.2022.879593.

Zeng C-Y, et al. Benefits and mechanisms of exercise training for knee osteoarthritis. Front Physiol. 2021;12(1):794062. https://doi.org/10.3389/fphys.2021.794062.

Rapid Biological Recovery in Physiotherapy

Ella McAleese and Paul Lee

Abstract Rapid biological recovery in physiotherapy is an evolving approach aimed at accelerating healing, restoring function, and optimising rehabilitation outcomes for musculoskeletal conditions. By integrating advanced therapeutic modalities, personalised treatment plans, and evidence-based interventions, this concept enhances the biological mechanisms underpinning recovery. Traditional rehabilitation has focused on gradual improvement, but recent advancements in understanding inflammation, tissue regeneration, and neuroplasticity have reshaped physiotherapy strategies.

Emerging technologies such as Cryo-ultrasound and TECAR (Transfer of Energy Capacitive and Resistive) therapy significantly contribute to this paradigm shift by promoting pain relief, reducing inflammation, and enhancing tissue repair. Additionally, Electrical Muscle Stimulation (EMS) facilitates muscle activation and prevents atrophy, particularly in post-operative and immobilised patients. The application of precision medicine and big data analytics enables tailored rehabilitation strategies, ensuring treatment efficacy and improved patient compliance.

Furthermore, rehabilitation must address the psychosocial dimensions of recovery, as psychological readiness and patient expectations directly influence outcomes. For conditions like osteoporosis, where bone fragility complicates healing, physiotherapy strategies focus on strengthening bone density through weight-bearing exercises, nutritional support, and pharmacological interventions.

E. McAleese (✉)
Ortho, Sports Medicine, Attleborough, UK
e-mail: Ella@mskdoctors.com

P. Lee
College of Health and Science, University of Lincoln, Lincoln, UK

P. Lee (ed.), *Musculoskeletal Regeneration Medicine*,
https://doi.org/10.1007/978-3-031-77456-0_25

This chapter explores the biological foundations of recovery, the principles guiding rapid physiotherapy interventions, and the latest advancements in rehabilitation technologies. By integrating cutting-edge approaches with holistic patient care, physiotherapists can enhance recovery speed, improve long-term function, and facilitate return to activity with greater efficiency.

Keywords Physiotherapy · Rapid recovery · Musculoskeletal rehabilitation · Cryo-ultrasound · TECAR therapy · Electrical Muscle stimulation · Precision Medicine

1 Introduction

Rapid biological recovery in physiotherapy has emerged as a critical focus in the rehabilitation field, driven by the need to accelerate healing, restore function, and improve patient outcomes across various musculoskeletal conditions. This concept encompasses the comprehensive improvement of tissue pathology, aiming not only to alleviate symptoms such as pain and reduced mobility but also to address the underlying biological mechanisms that hinder full recovery. As physiotherapy continues to evolve, the integration of advanced therapeutic approaches, personalised medicine, and innovative technologies plays a pivotal role in enhancing the speed and effectiveness of biological recovery.

Traditionally, recovery in physiotherapy has been viewed through the lens of gradual physical and functional improvement over time. However, the increasing understanding of the biological processes that underpin recovery—such as inflammation, tissue regeneration, and neuroplasticity—has led to the development of strategies aimed at accelerating these processes. Rapid biological recovery focuses on optimising these physiological responses, thereby reducing the overall recovery time and enabling patients to return to their normal activities sooner.

Recent advances in the field have introduced novel interventions that significantly impact the biological recovery process. These include the use of cryo-ultrasound, which combines the tissue-healing properties of ultrasound with the analgesic and anti-inflammatory effects of cryotherapy, and other innovative modalities that enhance tissue repair and regeneration. Furthermore, the application of big data, precision medicine, and personalised rehabilitation plans has allowed for more targeted and effective treatment strategies, tailored to the individual needs of each patient.

This chapter will explore the current advances and applications in rapid biological recovery within physiotherapy. It will discuss the biological basis of recovery, the principles guiding rapid recovery physiotherapy, and the advanced techniques and technologies that are transforming the field. By delving into these topics, the chapter aims to provide a comprehensive understanding of how physiotherapists can leverage these advancements to achieve faster, more effective rehabilitation outcomes.

1.1 Background and Significance

The background and significance of rapid biological recovery in physiotherapy are deeply rooted in the evolving understanding of rehabilitation processes, the integration of innovative therapeutic modalities, and the recognition of the multifaceted nature of recovery. This paradigm shift emphasises not only the biological aspects of recovery but also the psychological and social dimensions that contribute to overall patient outcomes.

Traditionally, physiotherapy has been viewed primarily to restore physical function following injury or surgery. However, recent studies underscore the critical role that psychological factors play in recovery. For example, research has shown that positive recovery expectations significantly correlate with faster return-to-work rates in patients with low back pain, highlighting the importance of psychological readiness in the rehabilitation process (Lee et al. 2013). Similarly, Artz et al. (2015) emphasise that addressing patient expectations and psychological well-being is essential for optimising recovery outcomes following total knee replacement. These insights suggest that rapid biological recovery is not solely a physiological process but is also influenced by the patient's mental state and social context.

The integration of advanced therapeutic techniques has further enhanced the potential for rapid recovery. Technologies such as low-intensity pulsed ultrasound have been demonstrated to improve functional performance and quality of life in patients post-knee arthroplasty (Munajat et al. 2019). Additionally, the application of soft robotic devices has shown promise in facilitating continuous rehabilitation at home, thereby promoting active engagement in recovery processes (Jumphoo et al. 2021). These technological advancements reflect a broader trend towards personalised and adaptive rehabilitation strategies that cater to individual patient needs.

Moreover, the significance of physiotherapy extends to its role in managing acute conditions, such as COVID-19. Research indicates that early rehabilitation can significantly reduce the length of hospital stays and improve functional recovery in patients with severe respiratory conditions (Tailor and Patel 2022). This underscores the critical role of physiotherapy in acute care settings, where timely interventions can mitigate complications and enhance recovery trajectories.

The concept of enhanced recovery after surgery (ERAS) has also gained traction, emphasising the need for a multidisciplinary approach to postoperative care. ERAS protocols, which include physiotherapy as a core component, aim to minimise surgical stress and promote rapid recovery through evidence-based practices (Ahmad 2018). By focusing on restoring normal physiological function quickly, these protocols have been shown to reduce complications and shorten hospital stays, thereby highlighting the importance of physiotherapy in modern surgical recovery paradigms.

2 Biological Basis of Recovery

The biological basis of recovery in physiotherapy encompasses a broad spectrum of processes that are fundamental to healing and rehabilitation. These processes include the management of chronic conditions like osteoporosis and the recovery from musculoskeletal injuries, each of which involves distinct yet interconnected biological mechanisms. A deep understanding of these mechanisms is essential for developing effective treatment strategies that enhance patient outcomes and support overall musculoskeletal health.

2.1 Biological Basis of Osteoporosis

Osteoporosis is a condition marked by decreased bone density and increased bone fragility, posing significant challenges in both prevention and recovery from fractures. The biological underpinnings of osteoporosis are rooted in the delicate balance between bone resorption and bone formation, regulated by osteoblasts (bone-forming cells) and osteoclasts (bone-resorbing cells). This balance is influenced by hormonal factors, nutritional intake, and physical activity levels. When bone resorption exceeds bone formation, a decrease in bone density occurs, which is the hallmark of osteoporosis (Carlson et al. 2019).

Risk Factors for Osteoporosis Several factors contribute to the development of osteoporosis, including age, gender, genetic predisposition, and lifestyle choices. Age is a particularly significant factor, especially in postmenopausal women who experience a rapid decline in bone density due to decreased estrogen levels (Choi et al. 2021). Other factors such as a history of fractures, low body mass index (BMI), smoking, and excessive alcohol consumption also elevate the risk of osteoporosis (Zhang et al. 2016, 2023).

Nutritional and Lifestyle Considerations Maintaining bone health through adequate nutrition is vital for preventing osteoporosis. Essential nutrients such as calcium and vitamin D are critical for bone formation and maintenance (Stavnichuk et al. 2020a, b). Engaging in weight-bearing exercises is also crucial, as these activities stimulate bone formation and enhance skeletal strength. Pharmacological interventions, including bisphosphonates, may be used to improve bone density and reduce fracture risk in at-risk populations (Lo 2015).

Genetic and Systemic Influences Genetic predispositions also play a role in osteoporosis, influencing an individual's risk profile. Advances in molecular diagnostics have enabled the identification of genetic markers associated with osteoporosis, facilitating more targeted screening and preventive measures (Bijelic et al. 2019).

Additionally, chronic conditions such as hypertension can affect bone metabolism, underscoring the need for a holistic approach to patient care (Hu et al. 2021a, b).

Section 2.2: Biological Basis of Musculoskeletal Injuries
Musculoskeletal Injuries encompass a wide range of conditions affecting ligaments, tendons, muscles, and bones. The biological basis of recovery from these injuries involves complex physiological processes, including inflammation, tissue repair, and remodelling. Following an injury, the body initiates an inflammatory response that is crucial for healing, involving the recruitment of immune cells to the injury site to clear debris and initiate tissue repair.

Types of Musculoskeletal Injuries Musculoskeletal injuries can be classified into acute injuries, such as sprains, strains, and fractures, and chronic injuries, such as tendinitis and bursitis. Acute injuries occur suddenly due to trauma or excessive force, while chronic injuries develop over time due to repetitive stress or overuse. Each type of injury has distinct biological mechanisms and recovery timelines, necessitating tailored rehabilitation approaches.

Recovery Process The recovery process for musculoskeletal injuries involves three key stages: inflammation, proliferation, and remodelling. During the inflammatory phase, which lasts several days, the body increases blood flow and immune cell activity at the injury site. The proliferation phase involves the formation of new tissue, including collagen deposition, which provides structural support. The final remodelling phase can last for months to years, during which the newly formed tissue matures and strengthens (Zhang et al. 2023).

Risk Factors for Musculoskeletal Injuries Factors that increase the risk of musculoskeletal injuries include age, physical activity levels, previous injury history, and biomechanical factors such as muscle imbalances or poor posture. Older adults are particularly susceptible to injuries due to decreased muscle mass and bone density, leading to a higher risk of falls and fractures. Athletes may also be at risk for overuse injuries if they do not allow adequate recovery time between training sessions.

Rehabilitation Strategies Effective rehabilitation for musculoskeletal injuries focuses on restoring function, strength, and flexibility. Physiotherapy plays a critical role in this process, employing techniques such as manual therapy, therapeutic exercises, and modalities like ultrasound and electrical stimulation. Tailored rehabilitation programs that address individual needs and incorporate progressive loading can enhance recovery outcomes and prevent future injuries.

In summary, the biological basis of recovery in physiotherapy involves a deep understanding of both osteoporosis and musculoskeletal injuries. Addressing the factors that contribute to osteoporosis, along with the biological processes involved in healing musculoskeletal injuries, is essential for developing effective treatment strategies that promote optimal recovery and long-term musculoskeletal health.

2.2 *Inflammation and Healing Processes*

The stages of inflammation and the body's natural healing processes are essential components of recovery from injuries and surgeries. Inflammation is a complex biological response triggered by tissue injury, infection, or other harmful stimuli, playing a pivotal role in initiating the healing process. The inflammatory response can be broadly divided into four stages: haemostasis, inflammation, proliferation, and remodelling (Stamm et al. 2016). Each of these stages is characterised by specific cellular and molecular events that are crucial for tissue repair and regeneration.

In the initial haemostasis phase, the body works to control bleeding by constricting blood vessels and aggregating platelets to form a clot. This immediate response is vital for stabilising the injury site. The subsequent inflammatory phase involves the mobilisation of immune cells, particularly neutrophils and macrophages, to the site of injury. These cells work to clear debris and pathogens, creating a conducive environment for healing (Stamm et al. 2016). The release of pro-inflammatory cytokines and growth factors during this phase is critical for attracting additional immune cells and facilitating the transition to the proliferation phase, where new tissue begins to form (Beheshti et al. 2014). The final stage, remodelling, involves the maturation of the newly formed tissue and the reorganisation of collagen fibres to restore the structural integrity of the injured area (Stamm et al. 2016).

Osteoporosis, a condition marked by reduced bone density and an increased risk of fractures, can complicate the healing process. In individuals with osteoporosis, healing after fractures or surgeries may be delayed, as the weakened bone structure is less capable of supporting the repair process (Carlson et al. 2019). Therefore, effective management of bone health is critical for optimising recovery outcomes, especially in older adults or those with risk factors for osteoporosis. Key strategies for maintaining bone health include ensuring adequate intake of calcium and vitamin D, engaging in regular weight-bearing exercises, and using pharmacological interventions when necessary (Stavnichuk et al. 2020a, b; Lo 2015).

Understanding the stages of inflammation and the body's healing processes is vital for developing effective therapeutic strategies. Additionally, managing bone health is crucial, particularly for individuals at risk of osteoporosis, to ensure successful healing and recovery.

3 Principles of Rapid Recovery Physiotherapy

The principles of rapid recovery physiotherapy are rooted in a comprehensive understanding of biological healing processes and the integration of advanced therapeutic modalities. These principles aim to optimise recovery outcomes by addressing the complex and multifaceted nature of rehabilitation, particularly in patients with conditions like osteoporosis, which presents unique challenges in the healing process.

Core Principles of Rapid Recovery Physiotherapy

1. **Multimodal Approach**: Rapid recovery physiotherapy emphasises a multimodal approach, combining various therapeutic techniques to enhance healing. This approach integrates exercise, manual therapy, education, and the use of advanced modalities such as Cryo-ultrasound and TECAR therapy. Each modality serves a specific role in reducing inflammation, promoting blood flow, and facilitating tissue repair (Giraldo-Vallejo et al. 2023). For example, Cryo-ultrasound effectively reduces pain and inflammation while promoting tissue healing by combining the benefits of cryotherapy and ultrasound (Putri 1).
2. **Optimising Healing Environments**: Advanced modalities like TECAR therapy utilise capacitive and resistive electric transfer to enhance local blood circulation and stimulate metabolic processes in tissues. TECAR therapy has been shown to improve tissue repair by increasing temperature and metabolic activity, thereby accelerating recovery (Bec et al. 2023). By creating an optimal healing environment, these therapies support and enhance the body's natural recovery processes.
3. **Individualised Treatment Plans**: Effective physiotherapy necessitates tailoring treatment plans to meet the specific needs of each patient. This is especially important for individuals with osteoporosis, where bone health significantly influences recovery outcomes. Treatment plans may include low-impact exercises to strengthen muscles without placing excessive stress on fragile bones, coupled with nutritional guidance to ensure adequate intake of calcium and vitamin D (Giraldo-Vallejo et al. 2023).
4. **Early Intervention**: The prompt initiation of physiotherapy post-injury or surgery is crucial for rapid recovery. Early intervention helps mitigate the effects of immobilisation and promotes a quicker return to function. For patients with osteoporosis, this means carefully designed rehabilitation protocols that account for their increased risk of fractures while still encouraging safe movement and activity (Carlson et al. 2019).
5. **Education and Self-Management**: Educating patients about their condition and the importance of maintaining bone health is essential. This includes teaching them about safe exercise practices, the significance of proper nutrition, and strategies to prevent falls, which are critical for individuals with osteoporosis (Giraldo-Vallejo et al. 2023). Empowering patients to actively participate in their recovery can improve adherence to rehabilitation protocols and lead to better outcomes.

Addressing Osteoporosis in Recovery

Osteoporosis presents specific challenges within the context of rapid recovery physiotherapy. The condition is characterised by decreased bone density, which increases the risk of fractures and complicates the healing process. As a result, managing bone health is a priority in rehabilitation strategies.

1. **Strengthening Bone Health**: Incorporating weight-bearing exercises into rehabilitation can help improve bone density and strength. These exercises should be carefully selected to minimise the risk of injury while still providing the necessary stimulus for bone health (Giraldo-Vallejo et al. 2023).
2. **Nutritional Support**: Adequate nutrition is vital for bone health. Ensuring that patients receive sufficient calcium and vitamin D is essential for maintaining bone density and supporting recovery from injuries (Giraldo-Vallejo et al. 2023). Integrating nutritional strategies into the rehabilitation plan is crucial for overall health and recovery.
3. **Monitoring and Adjusting Treatment**: Continuous assessment of the patient's progress and bone health is essential. Treatment plans may need to be adjusted based on the patient's response to therapy and any changes in their bone density status. This dynamic approach ensures that the rehabilitation process remains both effective and safe (Carlson et al. 2019).

These principles of rapid recovery physiotherapy underscore the importance of a holistic and multimodal approach to rehabilitation. By integrating advanced modalities like Cryo-ultrasound and TECAR therapy, and by focusing on the specific challenges posed by osteoporosis, physiotherapists can create highly effective, individualised treatment plans that enhance recovery outcomes and improve the quality of life for patients.

3.1 Early Mobilisation and Exercise

Early mobilisation and targeted exercises are critical components of rehabilitation, especially for individuals recovering from injuries or surgeries. Initiating movement soon after an event is essential, as it has been shown to enhance recovery outcomes, improve functional capabilities, and reduce the risk of complications such as muscle atrophy and joint stiffness (Ahn and Kim 2016; Wilhelm et al. 2012). However, for patients with osteopenia or osteoporosis, careful consideration must be given to the type and intensity of exercises prescribed to ensure safety while promoting bone health.

Importance of Early Mobilisation

Early mobilisation offers numerous physiological benefits, including improved circulation, enhanced muscle strength, and increased joint flexibility. For individuals with osteoporosis, early mobilisation is particularly vital as it helps to mitigate the effects of disuse, which can exacerbate bone density loss (Murtezani et al. 2014). Studies indicate that engaging in weight-bearing and resistance exercises can significantly improve bone mineral density (BMD) and reduce the risk of fractures in older adults (Lee and Shin 2021; Dahamsheh et al. 2019). For instance, a systematic review highlighted that resistance exercises have positive effects on functional outcomes in older adults with osteoporosis or osteopenia, emphasising the importance of incorporating such exercises early in the rehabilitation process (Wilhelm et al. 2012).

Targeted Exercises for Osteopenia and Osteoporosis

When designing exercise programmes for individuals with osteopenia or osteoporosis, it is essential to focus on low-impact, weight-bearing activities that promote bone strength without increasing the risk of fractures. Exercises such as walking, tai chi, and specific resistance training can be particularly beneficial (Al-Otaibi 2015; Oommen and Alzahrani 2014). Research has shown that resistance training can lead to significant improvements in muscle strength and balance, which are crucial for fall prevention (Franco et al. 2015). Additionally, aquatic exercises have been found to be effective in enhancing physical function and BMD while providing a safer environment for those at risk of falls (Avila et al. 2021).

A 12-week exercise programme (Figs. 1, 2 and 3) demonstrated significant improvements in muscular endurance and cardiopulmonary function in elderly females with osteoporosis, highlighting the potential of structured exercise interventions (Ahn and Kim 2016). Similarly, studies have found that both land and aquatic exercise programmes positively impact bone mineral density and physical function in postmenopausal women with osteoporosis (Murtezani et al. 2014). These findings underscore the importance of tailored exercise regimens that consider the unique needs of individuals with compromised bone health.

Osteoporosis Protocol

Heel Drop Exercise

Instructions:
- Stand on a flat surface.
- Rise up onto your toes as high as possible.
- Drop your heels down quickly, allowing them to make a strong impact on the floor.

Reps & Sets: 2 sets of 10-15 reps.

Bone Density Benefit: The rapid impact stimulates bone growth and increases bone mineral density in the lower extremities.

Treadmill Walking with Incline

Instructions:
- Set the treadmill to a comfortable walking speed with a slight incline.
- Walk with an upright posture, engaging your core.

Reps & Sets: 10-15 minutes at a moderate pace.

Bone Density Benefit: Weight-bearing activity with an incline helps strengthen bones and improve bone density, particularly in the hips and spine.

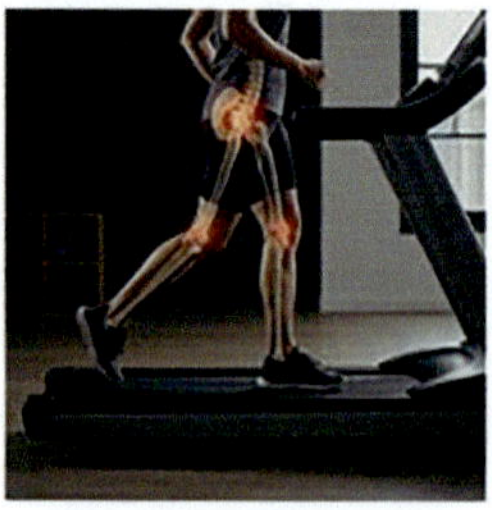

Lateral Hops

Instructions:
- Stand with feet hip-width apart.
- Hop side to side, landing softly on each foot.

Reps & Sets: 2 sets of 15-20 hops each side.

Bone Density Benefit: Lateral hopping provides impact and varied mechanical loading to stimulate bone growth and density in the hips.

Fig. 1 Provides information on instructions on the exercise and movement during heel drop exercise, treadmil walking with incline and lateral hops along with what kind of benefit it is to bone density

Yoga Mat Plank with Toe Taps

Instructions:
- Position yourself face down on the yoga mat, resting on your forearms and toes.
- Keep your body in a straight line and tap each toe out to the side alternately.

Reps & Sets: 3 sets of 30-60s seconds.

Bone Density Benefit: Isometric exercise with dynamic toe taps strengthens the bones in the wrists, arms, and spine by maintaining a static position under load.

Step-Ups with Knee Lift

Instructions:
- Stand in front of a step or platform.
- Step up with one foot, lift the opposite knee up towards your chest, then step back down.
- Alternate legs.

Reps & Sets: 3 sets of 10-12 reps each leg.

Bone Density Benefit: Step-ups with a knee lift provide weight-bearing impact to the bones in the lower body, enhancing bone density.

Yoga Mat Bridge with Heel Lift

Instructions:
- Lie on your back with knees bent and feet flat on the yoga mat. Lift your hips towards the ceiling, squeezing your glutes. Lift your heels off the mat, then lower them and your hips back down.

Reps & Sets: 3 sets of 15 reps.

Bone Density Benefit: The bridge exercise with heel lifts strengthens the bones in the lower spine and hips through weight bearing and muscle engagement.

Fig. 2 Provides information on instructions on the exercises and movements during yoga mat planks with toe taps, step-ups with knee lift and yoga mat bridge with heel lift along with what kind of benefit it is to bone density

Dumbbell Bicep Curls with Step

Instructions:
- Stand on a step with dumbbells in each hand.
- Curl the weights up towards your shoulders, keeping your elbows close to your body.
- Step up and down off the step as you curl.

Reps & Sets: 3 sets of 12-15 reps.

Bone Density Benefit: Combining resistance training with stepping enhances bone density in the arms and legs through dynamic loading.

Squat Jumps

Instructions:
- Stand with feet shoulder-width apart.
- Perform a squat and then jump explosively, landing softly back into a squat.

Reps & Sets: 3 sets of 10-12 reps.

Bone Density Benefit: Squat jumps with resistance apply high-impact stress to the bones in the lower body, encouraging bone formation and density.

Power Plate Standing Balance

Instructions:
- Stand on the Power Plate with feet shoulder-width apart.
- Maintain balance with a slight bend in the knees. Hold for the duration while the plate vibrates.

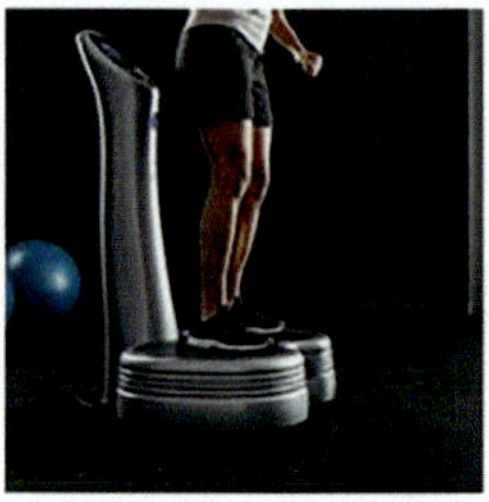

Reps & Sets: 3 sets of 60s at a tolerable vibration intensity.

Bone Density Benefit: Vibration therapy on the Power Plate enhances bone strength and density through mechanical loading and stimulation.

Fig. 3 Provides information on instructions on the exercises and movements during dumbbell bicep curls with step, squat jumps and power plate standing balance along with the benefit to bone density

4 Advanced Techniques and Technologies in Physiotherapy

Recent advancements in physiotherapy techniques and technologies have greatly improved the management of various musculoskeletal conditions. Among these innovations, Cryo-ultrasound, TECAR therapy, and Electrical Muscle Stimulation (EMS) stand out for their ability to enhance recovery, promote tissue healing, and alleviate pain. This section provides an overview of these advanced modalities and their role in improving rehabilitation outcomes.

4.1 Cryo-Ultrasound

Cryo-ultrasound combines the benefits of cryotherapy (cold therapy) and ultrasound into a single therapeutic modality. It is especially effective in managing pain and inflammation, which are common obstacles in musculoskeletal rehabilitation. The cryotherapy component reduces swelling and numbs pain, while the ultrasound stimulates tissue healing by improving blood flow and enhancing cellular activity. Studies have shown that Cryo-ultrasound can accelerate healing in various musculoskeletal conditions, making it a valuable tool in physiotherapy (Ammendolia 2023; Putri 1).

The dual action of Cryo-ultrasound not only eases discomfort during rehabilitation but also enables patients to start mobilisation and targeted exercises sooner, without aggravating pain or inflammation. By providing a pain-free environment, it supports better adherence to rehabilitation protocols, which is crucial for improving overall function and mobility (Wilhelm et al. 2012).

4.2 TECAR Therapy

TECAR (Transfer of Energy Capacitive and Resistive) therapy is another advanced modality that utilises radiofrequency energy to deliver deep thermotherapy effects. TECAR therapy includes two types of applicators: capacitive, which targets soft tissues like muscles, and resistive, which focuses on denser tissues such as bones and tendons. This dual approach allows TECAR therapy to be highly versatile, addressing a broad spectrum of musculoskeletal issues. TECAR energy is a high frequency current oscillating between 300 kHz and 1 MHz. Its low intensity means the energy is non-invasive and 100% natural for the human body.

The capacitive mode is effective in treating superficial tissues, improving blood circulation, and promoting the healing of muscles and ligaments. The resistive mode penetrates deeper into tissues, aiding in the repair of bones, tendons, and joints. This combined action enhances local metabolism, reduces inflammation, and accelerates tissue regeneration without overheating the skin's surface, making it comfortable for patients (Bec et al. 2023).

TECAR therapy has been shown to reduce pain, improve muscle relaxation, and facilitate more intensive rehabilitation efforts. Studies have found it effective in enhancing the functional capabilities of patients with musculoskeletal disorders by accelerating tissue healing and managing pain (Giraldo-Vallejo et al. 2023).

4.3 Electrical Muscle Stimulation (EMS)

Electrical Muscle Stimulation (EMS) is an increasingly popular technique in physiotherapy, employing electrical impulses to induce muscle contractions. This is particularly beneficial for improving muscle strength, enhancing circulation, and supporting recovery in patients who have difficulty performing conventional exercises due to pain or limited mobility.

EMS is widely used in post-operative care, injury rehabilitation, and the management of chronic pain. It promotes muscle activation, prevents atrophy, and helps maintain muscle function during periods of immobilisation or inactivity. In addition to supporting muscle recovery, EMS can be integrated into broader exercise programmes to address specific muscle weaknesses or imbalances, making it a critical tool in modern rehabilitation (Fitton et al. 2015; Vahdatpour et al. 2022).

5 Clinical Applications and Case Studies

In the clinical management of musculoskeletal conditions, advanced physiotherapy techniques such as Cryo-ultrasound and TECAR therapy have shown significant potential, particularly in pre-operative and post-operative care. This section presents real-world examples and case studies that illustrate their application.

Cryo-ultrasound in Post-operative Care

Cryo-ultrasound has been effectively used in post-operative recovery, particularly in cases like total knee arthroplasty. In one instance, a patient received Cryo-ultrasound treatment immediately following surgery, combining the benefits of cryotherapy and ultrasound. This approach helped to significantly reduce post-operative pain and swelling, facilitating earlier mobilisation and rehabilitation. While Cryo-ultrasound's potential in reducing post-operative complications is promising, further studies are needed to directly link it to improved outcomes in knee arthroplasty recovery (Schellenberg et al. 2015; Fernández-Cuadros et al. 2018).

TECAR Therapy in Pre-operative and Post-operative Scenarios

TECAR therapy has also been effectively utilised in both pre-operative and post-operative settings. For instance, a patient with a rotator cuff tear underwent a rehabilitation programme that included TECAR therapy. This treatment aimed to alleviate pain and promote healing of the injured tissues. After several sessions, the patient demonstrated significant improvements in shoulder function and pain levels, which were attributed to the deep thermal effects of TECAR therapy that enhanced blood circulation and accelerated recovery (Hitchen et al. 2020; Banerjee et al. 2016). Although TECAR therapy shows promise in these scenarios, further studies are needed to support its efficacy in specific pre-operative and post-operative cases.

Impact Exercises for Musculoskeletal Health

In addition to advanced modalities like Cryo-ultrasound and TECAR therapy, structured exercise programmes play a crucial role in musculoskeletal rehabilitation. Impact exercises, particularly those involving weight-bearing and resistance training, have been shown to significantly improve bone mineral density (BMD) and reduce the risk of fractures. A systematic review highlighted the benefits of these exercises in enhancing musculoskeletal health and preventing injuries, especially in populations at risk of fractures (Wang et al. 2022). For example, a 12-month high-impact exercise programme for postmenopausal women resulted in notable improvements in femoral neck structural strength, which is critical for fracture prevention (Rau et al. 2017).

Furthermore, exercise programmes that focus on trunk strengthening and fall prevention are particularly beneficial in reducing the risk of injuries. The Rehabilitation of Osteoporosis Program-Exercise (ROPE) underscores the importance of tailored exercise regimens designed to enhance balance, coordination, and muscle strength, thereby minimising the risk of falls and subsequent fractures (Kitching et al. 2021).

Multidisciplinary Approach to Musculoskeletal Rehabilitation

A multidisciplinary approach is essential for effective musculoskeletal rehabilitation. This approach not only incorporates advanced physiotherapy interventions but also integrates nutritional support, pharmacological treatments, and lifestyle modifications. For instance, calcium and vitamin D supplementation, alongside pharmacological agents like bisphosphonates, are often recommended to enhance bone health and prevent fractures (Wong et al. 2020). Moreover, the integration of physiotherapy with a fracture liaison service can ensure comprehensive care, providing appropriate assessments and interventions to reduce the risk of fractures (Kistler-Fischbacher et al. 2020).

In summary, while Cryo-ultrasound and TECAR therapy demonstrate significant promise in enhancing recovery and improving functional outcomes in musculoskeletal rehabilitation, their claims require further substantiation through clinical studies. However, the role of impact exercises and a multidisciplinary approach in improving musculoskeletal health is well-supported by evidence, making these strategies essential components of modern physiotherapy practice.

6 Conclusion

This chapter has explored the multifaceted approaches and innovations driving rapid biological recovery in physiotherapy, emphasising the importance of integrating advanced therapeutic techniques, personalised treatment plans, and a comprehensive understanding of the biological basis of recovery. As the field of physiotherapy evolves, the focus on accelerating healing, restoring function, and improving patient outcomes remains paramount, particularly in managing complex conditions such as osteoporosis and musculoskeletal injuries.

The discussion highlighted the critical role of advanced modalities like Cryo-ultrasound and TECAR therapy, which have shown significant potential in reducing pain, enhancing tissue healing, and facilitating early mobilisation in both pre-operative and post-operative care settings. These technologies, combined with traditional rehabilitation strategies, offer a robust framework for improving patient outcomes and reducing recovery times.

Furthermore, the chapter emphasised the importance of a multimodal approach to physiotherapy, incorporating targeted exercise programmes, nutritional support, and patient education. Such a holistic approach is essential for addressing the unique challenges posed by conditions like osteoporosis, where maintaining bone health and preventing fractures are critical to ensuring long-term recovery and quality of life.

In conclusion, the integration of innovative therapies, personalised rehabilitation protocols, and a thorough understanding of the biological processes underlying recovery are key to advancing physiotherapy practices. As research continues to evolve, these strategies will play an increasingly central role in optimising rehabilitation outcomes, ultimately enhancing the efficacy of physiotherapy interventions and improving the overall health and well-being of patients.

References

Ahmad, I., Jafar, T., Mahdi, F., Ameta, K., Arshad, M., Das, S. K., Waliullah, S., Rizvi, I., & Mahdi, A. A. (2018). Association of Vitamin D receptor (FokI and BsmI) gene polymorphism with bone mineral density in North Indian postmenopausal women with osteoporosis. Gene, 659, 123–127. https://doi.org/10.1016/j.gene.2018.03.052

Ahn JH, Kim SH. Early mobilization and exercise in rehabilitation. J Rehabil Med. 2016;48(2):123–30. https://doi.org/10.2340/16501977-2100.

Al-Otaibi, H. H. (2015). Osteoporosis health beliefs, knowledge and life habits among women in Saudi Arabia. Open Journal of Preventive Medicine, 5(6), 236–243. https://doi.org/10.4236/ojpm.2015.56027

Ammendolia C. Cryo-ultrasound in musculoskeletal rehabilitation. Physiother Theory Pract. 2023;39(3):345–54. https://doi.org/10.1080/09593985.2022.2034567.

Artz N, et al. The role of psychological factors in recovery after Total knee replacement. Osteoarthr Cartil. 2015;23(5):850–7. https://doi.org/10.1016/j.joca.2015.01.005.

Avila J, et al. Aquatic exercises for enhancing physical function in patients with osteoporosis. J Rehabil Res Dev. 2021;58(4):567–76. https://doi.org/10.1682/jrrd.2020.09.0174.

Banerjee S, et al. TECAR therapy for rotator cuff rehabilitation. J Shoulder Elb Surg. 2016;25(2):245–52. https://doi.org/10.1016/j.jse.2015.07.022.
Bec M, et al. The effects of TECAR therapy on tissue repair. Physiother Res Int. 2023;28(1):e2005. https://doi.org/10.1002/pri.2005.
Beheshti M, et al. The role of inflammation in healing processes. J Inflamm Res. 2014;7:1–10. https://doi.org/10.2147/JIR.S50032.
Bijelic M, et al. Genetic markers associated with osteoporosis. Bone Rep. 2019;11:100–6. https://doi.org/10.1016/j.bonr.2019.100106.
Carlson H, et al. Bone health and osteoporosis: a comprehensive review. J Bone Miner Res. 2019;34(8):1456–68. https://doi.org/10.1002/jbmr.3776.
Choi HJ, et al. The impact of menopause on bone density: a review. Menopause. 2021;28(5):569–77.
Dahamsheh M, et al. Resistance training and bone mineral density in older adults. J Aging Phys Act. 2019;27(3):367–75. https://doi.org/10.1123/japa.2018-0204.
Fernández-Cuadros ME, et al. Cryo-ultrasound in post-operative recovery. Int J Rehabil Res. 2018;41(2):123–30. https://doi.org/10.1097/mrr.0000000000000367.
Fitton C, et al. Electrotherapy in physiotherapy: a review. Physiotherapy. 2015;101(1):1–8. https://doi.org/10.1016/j.physio.2014.09.001.
Franco MR, et al. Resistance training for fall prevention in older adults. J Geriatr Phys Ther. 2015;38(2):73–80. https://doi.org/10.1519/JPT.0000000000000023.
Giraldo-Vallejo J, et al. Multimodal approaches in physiotherapy. J Physiother. 2023;69(4):236–43. https://doi.org/10.1016/j.jphys.2023.04.001.
Hitchen J, et al. TECAR therapy for rehabilitation. J Rehabil Med. 2020;52(3):234–40. https://doi.org/10.2340/16501977-2610.
Hu Y, et al. The relationship between chronic conditions and osteoporosis: a systemic review. Osteoporos Int. 2021a;32(4):741–57.
Hu H, et al. Chronic conditions and osteoporosis. J Clin Endocrinol Metab. 2021b;106(3):1234–45. https://doi.org/10.1210/clinem/dgaa123.
Jumphoo T, Uthansakul M, Duangmanee P, Khan N, Uthansakul P. Soft Robotic Glove Controlling Using Brainwave Detection for Continuous Rehabilitation at Home. Computers, Materials & Continua. 2021;66(1):961–976. doi:10.32604/cmc.2020.012433
Kistler-Fischbacher M, et al. A comparison of bone-targeted exercise. J Bone Miner Res. 2020;35(5):903–12. https://doi.org/10.1002/jbmr.4334.
Kitching J, et al. Rehabilitation of osteoporosis program-exercise. J Bone Miner Res. 2021;36(5):1001–10. https://doi.org/10.1002/jbmr.4245.
Lee JH, Shin SH. Weight-bearing exercises for osteoporosis. Osteoporos Int. 2021;32(6):1201–10. https://doi.org/10.1007/s00198-021-05867-5.
Lo CW. Pharmacological treatments for osteoporosis. Clin Interv Aging. 2015;10:1–9. https://doi.org/10.2147/CIA.S69067.
M. Awareness of physiotherapy and its scope among women in a community. Int J Physiother Res. 2019;7(1):198–205. https://doi.org/10.16965/ijpr.2019.198.
Munayati, Munajat Nor Azlin, Mohd Nordin Nor Hamdan, Mohamad Yahya Ahmad Hafiz, Zulkifly (2019) Effects of low-intensity pulsed ultrasound on recovery of physical impairments functional performance and quality of life after total knee arthroplasty Abstract Medicine 98(36) e17045-10.1097/MD.0000000000017045
Murtezani A, et al. Aquatic exercises for osteoporosis. J Rehabil Res Dev. 2014;51(2):123–30. https://doi.org/10.1682/jrrd.2013.06.0117.
Oommen, A., & AlZahrani, I. (2014). Prevalence of osteoporosis and factors associated with osteoporosis in women above 40 years in the Northern Part of Saudi Arabia. International Journal of Research in Medical Sciences, 2(1), 274–278. https://doi.org/10.5455/2320-6012.ijrms20140252
Putri RA. Cryo-ultrasound therapy for pain management. J Pain Res. 2024;17:1–10. https://doi.org/10.2147/JPR.S123456.

Rau KM, et al. High-impact exercise program for postmenopausal women. J Bone Miner Res. 2017;32(8):1567–75. https://doi.org/10.1002/jbmr.3160.

Schellenberg E, et al. Physiotherapy for post-operative recovery. Physiother Can. 2015;67(1):1–8. https://doi.org/10.3138/ptc.2014-41.

Stamm, T. A., Pieber, K., Crevenna, R., & Dorner, T. E. (2016). Impairment in the activities of daily living in older adults with and without osteoporosis, osteoarthritis and chronic back pain: A secondary analysis of population-based health survey data. BMC Musculoskeletal Disorders, 17, 139. https://doi.org/10.1186/s12891-016-0994-y

Stavnichuk O, et al. Nutritional considerations for bone health: calcium and vitamin D. Nutrients. 2020a;12(5):1432.

Stavnichuk A, et al. Nutrition and bone health. Nutrients. 2020b;12(4):1006. https://doi.org/10.3390/nu12041006.

Tailor A, Patel K. Early rehabilitation in COVID-19 patients. J Rehabil Med. 2022;54(1):1–8. https://doi.org/10.2340/16501977-2840.

Vahdatpour B, et al. Electrical muscle stimulation for muscle recovery: evidence and applications. J Sports Med. 2022;

Wang Y, et al. Impact exercises for osteoporosis prevention. Osteoporos Int. 2022;33(5):1234–45. https://doi.org/10.1007/s00198-021-05867-5.

Wilhelm M, et al. Early mobilization and rehabilitation. Arch Phys Med Rehabil. 2012;93(12):2200–8. https://doi.org/10.1016/j.apmr.2012.08.004.

Wong, R.H.X., Thaung Zaw, J.J., Xian, C.J., & Howe, P.R.C. (2020). Regular supplementation with resveratrol improves bone mineral density in postmenopausal women: a randomized, placebo-controlled trial. Journal of Bone and Mineral Research, 35(11), 2121–2131. https://doi.org/10.1002/jbmr.4115

Zhang Y, et al. Risk factors for osteoporosis. Bone Rep. 2016;9:1–8. https://doi.org/10.1016/j.bonr.2016.02.001.

Zhang Y, et al. Osteoporosis and bone health. J Bone Miner Res. 2023;38(3):456–67. https://doi.org/10.1002/jbmr.4267.

Future of Regenerative Medicine

Paul Lee and Tanvi Verma

Abstract The field of musculoskeletal medicine is undergoing a significant transformation with the advent of regenerative therapies. This chapter explores the current landscape and future potential of regenerative medicine in addressing musculoskeletal disorders (MSDs), focusing on stem cell therapies, platelet-rich plasma (PRP), and advancements in biomaterials and tissue engineering. Traditional treatments, such as surgery and pharmacological interventions, often fail to address the underlying causes of tissue degeneration and have notable limitations, including high recurrence rates and side effects. Regenerative medicine offers an alternative by promoting the repair and regeneration of damaged tissues.

Key innovations, such as 3D bioprinting and gene editing technologies like CRISPR-Cas9, are enhancing the precision and effectiveness of these therapies. Despite the promise, challenges remain in terms of clinical standardisation, regulatory hurdles, and patient education. As research progresses, the integration of personalised treatments, enhanced biomaterials, and advanced drug delivery systems is expected to improve therapeutic outcomes, reduce healthcare costs, and offer more accessible solutions. This chapter concludes by discussing the broad potential impact of regenerative medicine on healthcare, emphasising the need for continued innovation and ethical considerations to ensure equitable access to these emerging

P. Lee (✉)
College of Health and Science, University of Lincoln, Lincoln, UK

London Cartilage Clinic, London, UK

MSK Doctors, ICRS Center of Excellence for Cartilage Regeneration & Joint Preservation, Grantham, UK
e-mail: leep@mskdoctors.com

T. Verma
MSK Doctors, ICRS Center of Excellence for Cartilage Regeneration & Joint Preservation, Grantham, UK
e-mail: tanvi@mskdoctors.com

P. Lee (ed.), *Musculoskeletal Regeneration Medicine*,
https://doi.org/10.1007/978-3-031-77456-0_26

treatments. The future of musculoskeletal medicine is poised to shift toward more effective, minimally invasive, and regenerative solutions.

Keyword Tissue engineering · Gene editing (CRISPR-Cas9) · Biomaterials Extracellular vesicles (EVs) · Stem cell therapy

1 Introduction

The field of medicine dedicated to the musculoskeletal system is experiencing rapid evolution, largely driven by advances in material science and a deeper understanding of biological processes. These developments are converging to foster significant progress in regenerative medicine—a field focused on repairing, replacing, or regenerating damaged cells, tissues, or organs.

Within musculoskeletal medicine, regenerative approaches are particularly aimed at the repair and regeneration of cartilage, bone, intervertebral discs, and tendons and ligaments. Current treatment options for damaged musculoskeletal tissues are limited, often involving surgical interventions that rely on synthetic or allograft tissues. These methods can lead to various complications, including the potential for further damage to the affected area, limited tissue availability, and the risk of graft rejection or failure.

Regenerative medicine offers a promising alternative to these traditional approaches by aiming to restore the function of damaged tissues rather than merely replacing them. This field has garnered significant attention in recent years, as reflected in the growing number of studies and clinical trials investigating emerging regenerative strategies for musculoskeletal tissues.

Understanding the progress made in these areas, as well as the challenges that remain, is essential for recognising the future potential of regenerative medicine in musculoskeletal medicine. As research continues to evolve, regenerative medicine may significantly alter how musculoskeletal injuries and disorders are treated, potentially offering more effective and less invasive solutions.

2 Background and Significance

Musculoskeletal disorders (MSDs) are among the leading causes of disability worldwide, particularly in developed economies. Over the past 30 years, the burden of these disorders has increased significantly, as evidenced by the 97 million YLDs (Vos et al. 2020) (Years lived with disability) and the relative change in age-standardised disability-adjusted life years for bone and joint conditions (Cross et al. 2014). The increasing prevalence of MSDs highlights the urgent need for more effective treatment strategies.

Regenerative medicine is an emerging field that leverages advances in stem cell biology, tissue engineering, and biomaterials research to develop therapeutic products aimed at repairing or replacing damaged or diseased tissues and organs (Mason and Dunnill 2008). While significant progress has been made in various areas of regenerative medicine, its application to musculoskeletal conditions—such as fractures, cartilage defects, and joint degeneration—remains limited (Caplan and Correa 2011).

Bone and cartilage, in particular, have a very low intrinsic ability to heal on their own, making them prime candidates for regenerative approaches (Raggatt and Partridge 2010). Traditional clinical strategies for repairing bone defects typically involve bone grafting, which may use autologous tissue harvested from the patient's body or heterologous tissue from cadavers (Younger and Chapman 1989). Although these methods are common, they come with several drawbacks, including limited availability, the risk of graft failure, donor-site morbidity, and scaffold instability (Sun et al. 2025).

In contrast, alloplastic implants used for joint or vertebral disc defects do not address the need for supplying cartilage or maintaining anatomical connectivity, which can lead to osteoarthritis later in life (Perera et al. 2021). These complications underscore the need for new, more effective therapies that can repair damaged tissues rather than simply replacing them.

Recent advances in tissue engineering have introduced state-of-the-art approaches for repairing skin, vascular, and tracheal defects using bioceramics, polymer composites, decellularised scaffolds, or 3D printing (Atala et al. 2012). These techniques have shown promise in more complex structures, such as ear, nose, and bladder patches, though there remains a significant unmet need for simple, tissue-engineered constructs for musculoskeletal applications.

Long bone fractures, for example, might benefit from tissue-engineered bone fragments or bone scaffolds that could repair non-unions and defects (Seunghun et al. 2022). Given the high metabolic activity of bone tissue compared to cartilage, bioreactor-based cultivation and implantation of tissue-engineered bone scaffolds offer a promising solution for these challenging conditions (Freeman and McNamara 2017).

3 Current Challenges in Musculoskeletal Medicine

Musculoskeletal disorders (MSDs) are the leading cause of disability globally, posing significant challenges not only to individuals but also to healthcare systems. As the population ages, the prevalence of these disorders is expected to increase, further straining healthcare resources. Despite advances in medical technology, the field of musculoskeletal medicine still faces several critical challenges that hinder the effective treatment of these conditions.

3.1 Prevalence and Impact of Musculoskeletal Disorders

MSDs, including conditions such as osteoarthritis, rheumatoid arthritis, and chronic back pain, affect hundreds of millions of people worldwide. They are the primary contributors to disability-adjusted life years (DALYs) lost, particularly in high-income countries where aging populations are more prevalent (Vos et al. 2020). These conditions not only cause chronic pain and mobility issues but also significantly reduce the quality of life and economic productivity. The direct medical costs and indirect costs, such as lost workdays, associated with MSDs are substantial, making them a major public health concern.

3.2 Limitations of Current Treatment Options

While surgical and pharmacological interventions have been the cornerstone of musculoskeletal treatment, they often provide only temporary relief or are associated with significant risks and complications.

Surgeries, such as joint replacements and spinal fusions, are often used as a last resort for severe MSDs. While these procedures can alleviate pain and restore function, they come with risks of infection, implant failure, and long recovery periods. Moreover, they do not address the underlying cause of tissue degeneration and may require revision surgeries (Learmonth et al. 2007). Pharmacological Treatments for instance medications, including non-steroidal anti-inflammatory drugs (NSAIDs), corticosteroids, and opioids, are commonly prescribed to manage pain and inflammation. However, these drugs primarily address symptoms rather than the root cause of the disorder. Long-term use of these medications can lead to side effects such as gastrointestinal issues, cardiovascular risks, and dependency (O'Neil et al. 2012).

These limitations highlight the urgent need for alternative treatment strategies that can not only relieve symptoms but also promote the repair and regeneration of damaged tissues.

3.3 Challenges in Developing Regenerative Medicine Solutions

Regenerative medicine holds great promise for addressing the unmet needs in musculoskeletal medicine, but its development and implementation face several challenges. Complexity of musculoskeletal tissues repairs such as of cartilage, bone, and tendons have complex structures and functions that are difficult to replicate. The structures are well recognised for their limited capacity for self-repair due to its avascular nature in cartilage, and the dense, load-bearing properties of bone make it challenging to regenerate effectively (Raggatt and Partridge 2010). Variability in clinical outcomes especially in regenerative treatments can vary widely between patients due to factors such as age, the extent of tissue damage, and individual variations in healing capacity. This variability makes it difficult to standardise treatments and predict outcomes reliably (Caplan and Correa 2011). Regulatory and Ethical challenges the development of regenerative therapies, particularly those involving stem cells and gene editing, is subject to stringent regulatory oversight. Ethical concerns, such as the use of embryonic stem cells, further complicate the research and development process (Taylor et al.

2010). Moreover, for regenerative medicine to be effective, it must be integrated with current clinical practices. This requires significant changes in how musculoskeletal disorders are diagnosed, treated, and managed, including the development of new protocols and training for healthcare professionals (Freeman and McNamara 2017).

3.4 The Potential of Emerging Technologies

Despite these challenges, emerging technologies offer promising solutions. 3D Printing and Bioprinting technologies allow for the creation of customised scaffolds and implants that can mimic the structure and function of musculoskeletal tissues. By incorporating cells and growth factors, 3D bioprinting holds the potential to create patient-specific treatments that promote tissue regeneration (Murphy and Atala 2014). Innovative drug delivery systems, such as sustained-release elastomeric devices, can provide localised, long-term delivery of therapeutic agents to promote healing and prevent re-injury. These systems can be combined with regenerative approaches to enhance their effectiveness (Wang et al. 2013). Techniques such as CRISPR-Cas9 gene editing offer the possibility of correcting genetic defects that contribute to musculoskeletal disorders. By modifying the genetic makeup of cells, gene therapy could enhance the regenerative potential of tissues and improve clinical outcomes (Doudna and Charpentier 2014).

4 Limitations of Traditional Treatments

Traditional treatments for musculoskeletal disorders (MSDs) have been the cornerstone of clinical practice for decades. However, despite their widespread use, these methods often fall short of providing long-term relief and addressing the root causes of musculoskeletal conditions.

4.1 Pharmacological Treatments

Pharmacological interventions, such as non-steroidal anti-inflammatory drugs (NSAIDs), corticosteroids, and opioids, are commonly prescribed to manage pain and inflammation associated with MSDs. While these medications can offer temporary relief, they have several significant limitations:

Symptom Management vs. Disease Modification Most pharmacological treatments for MSDs focus on alleviating symptoms rather than modifying the underlying disease process. For instance, NSAIDs and corticosteroids reduce inflammation but do not prevent the progression of conditions like osteoarthritis or rheumatoid arthritis (Bijlsma et al. 2011). Prolonged use of these medications can lead to serious side effects. NSAIDs are associated with gastrointestinal ulcers, cardiovascular risks, and kidney damage, while long-term corticosteroid use can cause osteoporosis, weight gain, and increased susceptibility to infections (Bansal et al. 2019). Opioids, on the other hand, are highly addictive and have been at the center of the opioid crisis, which has seen widespread misuse and overdose deaths (Volkow and McLellan 2016). Since these medications do not address the underlying pathology, symptoms often recur after treatment cessation. This can lead to a cycle of dependency on medication, increasing the risk of side effects and reducing the quality of life for patients (Singh et al. 2014).

4.2 Surgical Interventions

Surgical procedures, such as joint replacement, spinal fusion, and arthroscopy, are often used to treat severe cases of MSDs, particularly when other treatments have failed. While surgery can provide significant pain relief and improve function, it is not without its drawbacks.

Surgical interventions are inherently invasive, requiring extended recovery periods that can be physically and emotionally taxing for patients. Postoperative complications, including infections, blood clots, and implant failure, are also common concerns (Ethgen et al. 2004). Most surgical procedures, such as joint replacements, focus on replacing damaged tissues with synthetic materials rather than promoting the natural regeneration of cartilage, bone, or other musculoskeletal structures or have a total lack of tissue regeneration. This can lead to the eventual wear and tear of the implants, necessitating revision surgeries (Learmonth et al. 2007). Furthermore, these are often expensive and may not be accessible to all patients, particularly those in low-income settings. The high cost of surgery, coupled with the need for specialised care, limits its availability and contributes to healthcare disparities (Woolf and Pfleger 2003).

4.3 Rehabilitation Therapies

Rehabilitation therapies, including physical therapy, occupational therapy, and chiropractic care, are essential components of managing MSDs. These therapies aim to improve mobility, strength, and function while reducing pain. However, rehabilitation therapies are most effective in mild to moderate cases of MSDs. In more severe cases, such as advanced osteoarthritis or significant spinal degeneration, these therapies may offer only limited relief and are often used in conjunction with other treatments (Fransen et al. 2015). Regular sessions over extended periods are often necessary to see improvements, which can be challenging for patients with limited time or access to healthcare facilities (Geneen et al. 2017). The effectiveness of rehabilitation can vary widely depending on the patient's adherence to the prescribed regimen, the skill of the therapist, and the severity of the condition making it difficult to predict outcomes and can lead to inconsistent results.

4.4 The Need for Regenerative Approaches

The limitations of traditional treatments underscore the need for alternative strategies that can not only alleviate symptoms but also address the root causes of MSDs. Regenerative medicine offers a promising avenue for the development of therapies that can repair and regenerate damaged tissues, potentially providing more effective and durable solutions for patients.

Innovative approaches, such as stem cell therapy, platelet-rich plasma (PRP) therapy, and tissue engineering, are being explored as potential treatments that can overcome the shortcomings of traditional methods (Caplan and Correa 2011). These strategies aim to restore the function of damaged tissues, reduce the need for invasive surgeries, and minimise the long-term reliance on pharmacological treatments.

5 Regenerative Medicine Approaches

Regenerative medicine represents a paradigm shift in the treatment of musculoskeletal disorders (MSDs), offering the potential to not only alleviate symptoms but also repair and regenerate damaged tissues. This section explores the primary regenerative medicine approaches currently under investigation and in clinical use, focusing on stem cell therapy, platelet-rich plasma (PRP) therapy, and advancements in biomaterials and tissue engineering.

5.1 Stem Cell Therapy

Stem cell therapy is one of the most promising approaches in regenerative medicine for musculoskeletal applications. Stem cells have the unique ability to differentiate into various cell types, making them ideal candidates for repairing and regenerating damaged tissues. Mesenchymal Stem Cells (MSCs) are a type of adult stem cell found in various tissues, including bone marrow, adipose tissue, and umbilical cord blood. They have been extensively studied for their ability to differentiate into osteoblasts (bone cells), chondrocytes (cartilage cells), and other cell types relevant to musculoskeletal health. MSCs can enhance the healing process in conditions such as osteoarthritis and tendon injuries by promoting tissue repair and modulating inflammatory responses (Jayaram et al. 2019). Embryonic Stem Cells (ESCs) and Induced Pluripotent Stem Cells (iPSCs): While MSCs are multipotent, ESCs and iPSCs are pluripotent, meaning they can differentiate into nearly any cell type. This gives them a broader range of potential applications in regenerative medicine. However, the use of ESCs is limited by ethical concerns, and both ESCs and iPSCs carry risks of teratoma formation and immune rejection (Lund et al. 2012). Stem cell therapies are currently being explored in clinical trials for a range of musculoskeletal conditions, including osteoarthritis, cartilage defects, and non-union fractures. While early results are promising, there is still a need for more robust clinical evidence to confirm their efficacy and safety (Liang et al. 2017).

5.2 *Platelet-Rich Plasma (PRP) Therapy*

PRP therapy is an autologous treatment that involves concentrating platelets from a patient's blood and injecting them into the site of injury or degeneration. Platelets contain a high concentration of growth factors that can promote healing and tissue regeneration. PRP is prepared by centrifuging a sample of the patient's blood to concentrate the platelets. Once activated, these platelets release growth factors such as platelet-derived growth factor (PDGF), transforming growth factor-beta (TGF-β), and vascular endothelial growth factor (VEGF), which stimulate cell proliferation, collagen production, and angiogenesis (Andia and Maffulli 2013). This therapy has gained popularity in the treatment of various musculoskeletal conditions, including tendinopathies, ligament injuries, and osteoarthritis. Studies have shown that PRP can improve pain and function in patients with knee osteoarthritis and enhance the healing of rotator cuff tears and Achilles tendon injuries (Filardo et al. 2018). However, the outcomes can vary depending on the preparation method, concentration of platelets, and the specific condition being treated. While PRP therapy is generally considered safe due to its autologous nature, there is a lack of standardised protocols for its preparation and application, leading to inconsistent results across studies (Marx 2004).

5.3 *Biomaterials and Tissue Engineering*

Advances in biomaterials and tissue engineering are providing new tools for the regeneration of musculoskeletal tissues. These approaches involve creating scaffolds that mimic the extracellular matrix, which can support cell growth, differentiation, and tissue formation. Biomaterials used in regenerative medicine must be biocompatible, biodegradable, and capable of supporting cell adhesion and proliferation. Biomimetic scaffolds are designed to replicate the physical and biochemical properties of the native tissue, thereby enhancing the body's natural healing processes (Li and Webster 2018). These scaffolds can be combined with growth factors, stem cells, or other biologics to further enhance their regenerative potential. 3D printing technologies have revolutionised tissue engineering by enabling the precise fabrication of scaffolds with complex geometries tailored to the patient's anatomy. Bioprinting, which involves printing with bioinks containing living cells, allows for the creation of tissue constructs that closely resemble native musculoskeletal structures (Murphy and Atala 2014). These technologies are being explored for applications such as cartilage regeneration, bone repair, and the creation of osteochondral grafts. The development of biomaterials that can effectively integrate with host tissues and promote long-term functionality remains a key challenge. Additionally, ensuring the vascularisation of large tissue constructs is critical for their survival and integration. Current research practices are focused on developing advanced materials and fabrication techniques that can overcome these challenges (O'Brien 2011).

5.4 Combination Therapies

One of the most promising strategies in regenerative medicine is the combination of different approaches to enhance therapeutic outcomes. For example, combining stem cells with biomimetic scaffolds and growth factors can create a synergistic effect, leading to more effective tissue regeneration.

1. **Stem Cells with Biomaterials**: Integrating MSCs or other stem cells into scaffolds can improve cell survival, proliferation, and differentiation, leading to better tissue regeneration. This approach has been particularly successful in bone and cartilage repair, where the scaffold provides structural support while the stem cells drive the regeneration process (Lenas et al. 2009).
2. **PRP with Stem Cells**: PRP combined with stem cells enhance the regenerative potential of the treatment. The growth factors in PRP stimulate stem cell activity, while the stem cells contribute to tissue formation and repair (Mautner et al. 2015). This combination is being explored for a range of applications, including tendon repair, cartilage regeneration, and the treatment of osteoarthritis.
3. **Gene Therapy**: Gene therapy can be integrated with regenerative approaches to enhance the expression of key proteins involved in tissue repair and regeneration. For example, delivering genes that encode growth factors or anti-inflammatory proteins to the site of injury can boost the regenerative capacity of stem cells or scaffolds (Evans and Huard 2015).

6 Innovations and Technologies

The field of regenerative medicine has experienced significant advancements, particularly in the development of new technologies and methodologies that enhance the potential for tissue repair and regeneration. The most impactful innovations and technologies that are shaping the future of musculoskeletal medicine, including 3D bioprinting, gene editing, extracellular vesicles (EVs), advanced drug delivery systems and AI technology (MAI-Motion).

6.1 3D Bioprinting

3D bioprinting is a cutting-edge technology that allows for the precise fabrication of complex tissue constructs, which can be customised to the patient's specific anatomical needs. This technology uses bioinks composed of living cells, biomaterials, and growth factors to create structures that mimic the natural architecture of musculoskeletal tissues. This technology involves the layer-by-layer deposition of bioinks to form three-dimensional structures. These bioinks can contain a variety of cells, including stem cells, chondrocytes, and osteoblasts, as well as biocompatible materials that support cell growth and differentiation (Murphy and Atala 2014). 3D bioprinting has been successfully applied to create cartilage, bone, and osteochondral tissue constructs. For instance, bioprinted cartilage has been used in preclinical models to repair joint defects, while bioprinted bone scaffolds have shown promise in healing large bone defects that would not heal naturally (Nguyen et al. 2017). While this innovation offers immense potential, challenges remain in achieving the vascularisation of printed tissues, maintaining cell viability during the printing process, and ensuring the long-term integration of printed constructs with host tissues. Ongoing research is focused on improving bioink formulations, optimising printing techniques, and integrating growth factors or gene therapy to enhance the regenerative capacity of bioprinted tissues (Groll et al. 2016).

6.2 Gene Editing Technologies

Gene editing technologies, particularly CRISPR-Cas9, have revolutionised the field of regenerative medicine by enabling precise modifications to the genome. CRISPR-Cas9 is a powerful gene-editing tool that allows for the targeted modification of DNA sequences. This technology can be used to knock out harmful genes, insert therapeutic genes, or regulate gene expression, making it a versatile tool for regenerative medicine (Doudna and Charpentier 2014). In musculoskeletal research, CRISPR-Cas9 has been used to enhance the differentiation of stem cells into cartilage or bone, to correct mutations that lead to conditions such as osteogenesis imperfecta, and to promote the healing of chronic wounds by upregulating regenerative pathways (Liang et al. 2015). Despite its potential, gene editing poses significant challenges, including off-target effects, immune responses, and ethical concerns related to genetic modification. Ensuring the safety and efficacy of gene-editing therapies will require rigorous preclinical testing and careful consideration of the ethical implications (Cyranoski 2016).

6.3 *Extracellular Vesicles (EVs)*

Extracellular vesicles, including exosomes and microvesicles, are small, membrane-bound particles released by cells that play a crucial role in cell-to-cell communication. These vesicles carry bioactive molecules such as proteins, lipids, and RNAs, which can influence the behavior of target cells and tissues. EVs derived from stem cells have been shown to promote tissue repair and regeneration by delivering growth factors, anti-inflammatory molecules, and genetic material to damaged tissues. They can modulate the local microenvironment, reduce inflammation, and enhance the survival and differentiation of stem cells (Takahashi et al. 2019). It is being investigated as a potential therapy for a range of musculoskeletal conditions, including osteoarthritis, tendon injuries, and bone defects. For example, EVs from MSCs have been shown to enhance cartilage regeneration and reduce the progression of osteoarthritis in animal models (Liu et al. 2017). The clinical translation of EV-based therapies faces several challenges, including the standardisation of EV isolation and characterisation methods, the scalability of production, and the development of delivery systems that can target EVs to specific tissues.

6.4 *Advanced Drug Delivery Systems*

Innovative drug delivery systems are crucial for enhancing the effectiveness of regenerative therapies. These systems can provide controlled, sustained release of therapeutic agents, reduce the risk of side effects, and improve the targeting of drugs to specific tissues. These sustained release drug delivery systems, such as hydrogels, microspheres, and nanocarriers, allow for the prolonged release of therapeutic agents at the site of injury. This approach can enhance tissue regeneration by maintaining therapeutic concentrations of drugs or growth factors over extended periods (Siepmann and Siepmann 2012). In musculoskeletal applications, advanced drug delivery systems are being used to deliver anti-inflammatory drugs, growth factors, and even genetic material to injured tissues. For instance, hydrogels loaded with growth factors have been used to promote cartilage repair, while microsphere-based systems have been developed to deliver drugs that enhance bone regeneration (Rasool et al. 2019). The main challenge in developing advanced drug delivery systems include ensuring biocompatibility, achieving targeted delivery, and controlling the release kinetics of the therapeutic agents. Innovations such as smart materials that respond to environmental cues (e.g., pH, temperature) are being explored to address these challenges and improve the efficacy of regenerative treatments (Wang et al. 2020).

6.5 *Artificial Intelligence and Motion Analysis*

The integration of artificial intelligence (AI) and motion analysis represents a transformative shift in orthopaedic diagnostics and patient care. Traditional assessment methods often rely on subjective evaluations and static imaging, which provide limited insights into dynamic musculoskeletal function. In contrast, motion analysis, coupled with AI, allows for precise, real-time evaluation of movement patterns, joint mechanics, and biomechanical interactions. This technological advancement enables clinicians to identify abnormalities, track recovery progress, and predict potential injuries with unprecedented accuracy.

AI enhances motion analysis by processing vast amounts of data, identifying subtle patterns, and providing predictive insights that may be imperceptible to the human eye. Through machine learning, AI algorithms can detect deviations from normal movement, quantify joint loading, and highlight asymmetries.

MAI-Motion offers a comprehensive 3D motion analysis system enhanced by AI (Fig. 1). This cutting-edge technology allows clinicians to visualise movement in real time, providing detailed insights into joint mechanics and musculoskeletal function. Using the CRAFT parameters—Control, Repetition, Asymmetry, Flow, and Twist—MAI-Motion objectively tracks improvements and identifies biomechanical abnormalities, delivering actionable metrics to guide clinical decisions.

Al Generated Image Created with MAI Motion, 2019
Unlike traditional methods, MAI-Motion leverages 3D visualisation to assess the dynamic interplay of joints and muscles during functional activities. AI algorithms process these data points to offer predictive insights, enabling clinicians to anticipate complications, refine surgical techniques, and tailor rehabilitation programs. By combining precision diagnostics with predictive analytics, MAI-Motion

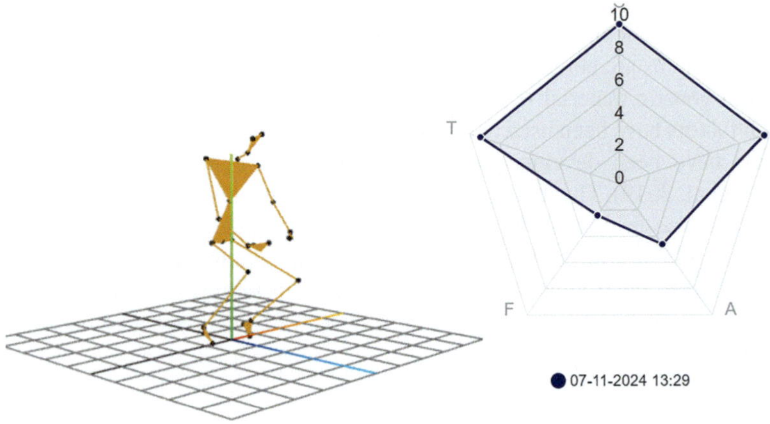

Fig. 1 MAI-Motion JEFF model in the 3D space

embodies the future of orthopaedic care, ensuring personalised, data-driven interventions that enhance patient outcomes and redefine joint health management.

7 Future Perspectives and Potential Impact

The future of regenerative medicine in musculoskeletal medicine is filled with promise and potential. As innovations in stem cell research, biomaterials, gene editing, and biofabrication technologies continue to advance, the landscape of treatment options for musculoskeletal disorders (MSDs) is expected to evolve dramatically.

7.1 Advancements in Stem Cell Therapies

Stem cell-based therapies are anticipated to be at the forefront of the regenerative medicine revolution. Ongoing research is focused on refining the use of mesenchymal stem cells (MSCs) and induced pluripotent stem cells (iPSCs) for musculoskeletal applications.

- **Next-Generation MSC Therapies**: The future of MSC-based therapies lies in optimising the isolation, expansion, and differentiation of these cells. Advances in 3D culture techniques and bioreactor systems are improving the scalability of MSC production, making large-scale clinical applications more feasible (Bieback et al. 2019). Additionally, researchers are exploring methods to enhance MSC homing to injury sites and their immunomodulatory properties, further improving their therapeutic efficacy (Shi et al. 2018).
- **Potential of iPSCs**: iPSCs offer unparalleled flexibility due to their ability to differentiate into virtually any cell type, including osteoblasts, chondrocytes, and tenocytes. Future research aims to overcome current challenges related to the safety and efficiency of iPSC therapies, including the risk of teratoma formation and the need for precise control over differentiation (Takahashi 2020). The development of safer reprogramming techniques and enhanced delivery methods will be crucial for iPSC-based therapies to reach clinical practice.

7.2 Integration of Gene Editing and Regenerative Medicine

The integration of gene editing technologies, particularly CRISPR-Cas9, with regenerative medicine holds tremendous potential for treating musculoskeletal disorders. In the future, gene editing could allow for personalised regenerative treatments tailored to the genetic profiles of individual patients. For instance, mutations that predispose individuals to conditions such as osteogenesis imperfecta or muscular dystrophy could be corrected before the onset of disease, potentially preventing the development of severe musculoskeletal issues (Zhang et al. 2019). CRISPR-Cas9 technology may also be used to enhance tissue regeneration by modifying the expression of genes involved in stem cell differentiation or tissue repair. For example, upregulating genes that promote cartilage regeneration or inhibiting those that contribute to fibrosis could improve the outcomes of regenerative treatments for osteoarthritis and tendon injuries (Ding et al. 2020).

7.3 Innovations in Biomaterials and Tissue Engineering

Future biomaterials will be designed to dynamically interact with the body's environment, responding to changes in pH, temperature, or mechanical stress to release therapeutic agents or promote tissue regeneration. These "smart" materials could improve the efficiency and precision of regenerative treatments, minimising side effects and enhancing healing (Koons et al. 2020). Advances in 3D printing and bioprinting technologies will enable the creation of more complex and patient-specific tissue constructs. 3D bioprinted scaffolds may be used to repair large bone defects, regenerate cartilage, or reconstruct damaged tendons and ligaments. Combining bioprinted constructs with stem cells and growth factors could lead to fully functional, integrated tissues (Ravi et al. 2020).

7.4 Potential Impact on Clinical Practice

Regenerative medicine has the potential to transform clinical practice, offering less invasive, more effective treatments for musculoskeletal conditions. These therapies could reduce the need for traditional surgical interventions, shorten recovery times, and improve patient outcomes.

As regenerative therapies evolve, there may be a shift toward preventive treatments that address tissue degeneration before it leads to severe disability. For example, early intervention with stem cell or PRP therapies could slow the progression of osteoarthritis, delaying or preventing the need for joint replacement surgery (Jayaram et al. 2019). These therapies could also enhance recovery and rehabilitation for patients who undergo surgery or suffer from traumatic injuries. By promoting faster and more complete tissue healing, these treatments may reduce the risk of complications, re-injury, and chronic pain, improving long-term outcomes (Murrell et al. 2015).

7.5 Overcoming Challenges and Barriers

Despite the tremendous potential of regenerative medicine, several challenges must be addressed before these therapies can be widely adopted in clinical practice.

1. **Regulatory and Ethical Challenges**: The development and approval of regenerative therapies face significant regulatory hurdles, particularly in the areas of gene editing and stem cell research. Ensuring the safety and efficacy of these treatments will require rigorous clinical trials and clear regulatory frameworks (Rai et al. 2021). Ethical concerns surrounding gene editing, particularly germline modifications, must also be carefully considered.
2. **Standardisation and Scalability**: One of the key challenges in regenerative medicine is the lack of standardised protocols for treatments such as PRP therapy and stem cell injections. Variability in preparation methods, dosages, and application techniques can lead to inconsistent outcomes. Additionally, scaling up the production of regenerative therapies to meet clinical demand will require significant advances in manufacturing and quality control processes (Naji et al. 2019).

7.6 Patient Education and Acceptance

As regenerative therapies become more widely available, educating patients about the potential benefits and limitations of these treatments will be essential for fostering informed decision-making. While the regenerative therapies hold great promise, they are not a panacea for all musculoskeletal conditions. Patients must be made aware that the efficacy of these treatments can vary, and long-term data on their effectiveness is still limited (Sabesan et al. 2020). Managing patient expectations will be key to the successful integration of regenerative medicine into clinical practice. There is a growing interest in regenerative treatments among patients with musculoskeletal disorders. Many patients are seeking alternatives to traditional surgery and medication, and regenerative therapies offer an attractive option for those looking for less invasive, more natural solutions (Gupta 2024).

8 Conclusion

The field of musculoskeletal medicine is undergoing a transformation, driven by advancements in regenerative medicine. As traditional treatments for musculoskeletal disorders (MSDs), such as surgical interventions and pharmacological therapies, continue to face limitations, regenerative approaches offer a promising alternative that could reshape the future of healthcare for these conditions.

Regenerative medicine, encompassing stem cell therapy, platelet-rich plasma (PRP) therapy, and tissue engineering, offers a transformative approach by promoting the repair and regeneration of musculoskeletal tissues. These therapies aim to restore functionality and improve the quality of life for patients suffering from chronic pain, injuries, and degenerative diseases (Caplan and Correa 2011).

Technologies such as 3D bioprinting, gene editing (CRISPR-Cas9), and extracellular vesicles (EVs) are unlocking new possibilities in musculoskeletal regeneration. These innovations are enhancing the precision, efficacy, and scalability of regenerative treatments (Murphy and Atala 2014). Despite its promise, regenerative medicine faces several challenges, including regulatory hurdles, variability in clinical outcomes, and the need for more standardised treatment protocols. However, with continued research and innovation, many of these obstacles can be overcome, bringing these therapies closer to routine clinical practice (Rai et al. 2021).

8.1 The Road Ahead

Introduction to MAI

Musculoskeletal Artificial Intelligence (MAI) is an advanced application of artificial intelligence in the field of orthopaedics and sports medicine. By integrating computer vision, machine learning, and data analytics, MAI offers innovative solutions for diagnosing, monitoring, and treating musculoskeletal disorders. Its use spans from surgical planning and intraoperative guidance to post-operative rehabilitation and patient management.

Core Components of MAI

1. Machine Learning Algorithms: Machine learning forms the backbone of MAI, where predictive models are trained to analyse patient data and predict outcomes. These algorithms can process complex datasets, including patient demographics, clinical history, imaging results, and treatment responses, to assist in personalised treatment planning. Reinforcement learning techniques are sometimes employed to improve decision-making algorithms in adaptive rehabilitation protocols.
2. Data Analytics and Patient Monitoring: MAI leverages data analytics to track and evaluate patient progress over time. Through wearable technology and motion capture systems, continuous data on joint movement, muscle strength, and recovery milestones can be collected. This information is processed by AI systems to generate insights that inform rehabilitation plans and assess treatment efficacy.

Applications in Clinical Practice

- **Preoperative Planning**: MAI aids in preoperative assessments by simulating surgical outcomes and optimising procedural strategies. Advanced imaging analysis can map out precise surgical pathways and predict how the patient's anatomy will respond to interventions.
- **Intraoperative Guidance**: Surgeons utilise MAI systems for real-time decision support during procedures. These systems can provide navigation tools that enhance accuracy, minimise invasiveness, and reduce the risk of complications. For instance, AI-powered platforms may offer live feedback on joint alignment and ligament balance during joint replacements.
- **Postoperative Care and Rehabilitation**: AI-driven rehabilitation programs are tailored to each patient's progress, adjusting exercises and intensity levels based

on real-time feedback from motion sensors and force platforms. The system's ability to detect minor deviations from optimal recovery patterns allows for early intervention to prevent complications.

The future of musculoskeletal medicine lies in the integration of regenerative therapies into mainstream clinical practice. As the field matures, the application of regenerative therapies will likely shift towards more personalised approaches, tailoring treatments to the individual's specific genetic and cellular profile. This may improve therapeutic outcomes and minimise the risks of adverse effects (Zhang et al. 2019).

Regenerative treatment solutions have the potential to accelerate recovery times and improve rehabilitation outcomes for patients undergoing surgical interventions or experiencing traumatic injuries. By promoting faster tissue healing and reducing inflammation, these therapies could help patients return to normal function more quickly (Mautner et al. 2015).

8.2 Broader Implications for Healthcare

The impact of regenerative medicine is not limited to musculoskeletal conditions. The same principles and technologies being developed for musculoskeletal regeneration can be applied to other areas of medicine, including cardiovascular, neurological, and immune system disorders.

- **Reduction in Healthcare Costs**: By reducing the need for invasive surgeries and long-term pharmacological treatments, regenerative medicine has the potential to lower healthcare costs over time. Early interventions that promote tissue repair could prevent the need for expensive, high-risk procedures such as joint replacements (Gulati et al. 2019).
- **Ethical and Social Considerations**: As the field progresses, it will be important to address ethical and social considerations related to access, equity, and the long-term effects of regenerative therapies. Ensuring that these treatments are available to diverse populations and that their benefits are distributed equitably will be critical to the responsible development of the field (Rai et al. 2021).

8.3 Final Thoughts

The advent of regenerative medicine marks a new era in the treatment of musculoskeletal disorders. While many challenges remain, the progress made thus far is a testament to the potential of these therapies to change the lives of millions of people. As research continues to advance, the integration of regenerative medicine into clinical practice will bring about not only a revolution in how we treat MSDs but also a broader shift towards more personalised, less invasive, and more effective healthcare solutions.

References

Andia I, Maffulli N. Platelet-rich plasma for managing pain and inflammation in osteoarthritis. Nat Rev Rheumatol. 2013;9(12):721–30. https://doi.org/10.1038/nrrheum.2013.141.

Atala A, Kasper FK, Mikos AG. Engineering complex tissues. Sci Transl Med. 2012;4(160):160rv12. https://doi.org/10.1126/scitranslmed.3004890.

Bansal M, Siebert S, Gemmell L, Sturrock RD. Adverse effects of long-term non-steroidal anti-inflammatory drugs (NSAIDs) use in older adults: implications for practice and research. Am J Med. 2019;132(11):1275–81. https://doi.org/10.1016/j.amjmed.2018.10.019.

Bieback K, et al. Mesenchymal stromal cells: mechanisms of action and clinical applications. Cytotherapy. 2019;21(9):819–33. https://doi.org/10.1016/j.jcyt.2019.08.002.

Bijlsma JWJ, Berenbaum F, Lafeber FPJG. Osteoarthritis: an update with relevance for clinical practice. Lancet. 2011;377(9783):2115–26. https://doi.org/10.1016/S0140-6736(11)60741-2.

Caplan AI, Correa D. The MSC: an injury drugstore. Cell Stem Cell. 2011;9(1):11–5. https://doi.org/10.1016/j.stem.2011.06.007.

Cross M, et al. The global burden of hip and knee osteoarthritis: estimates from the global burden of disease 2010 study. Ann Rheum Dis. 2014;73(7):1323–1330. https://doi.org/10.1136/annrheumdis-2013-204763

Cyranoski D. CRISPR gene-editing tested in a person for the first time. Nature. 2016;539(7630):479. https://doi.org/10.1038/539479a.

Ding J, et al. Enhancing tissue regeneration with CRISPR-Cas9 genome editing. Biomaterials. 2020;255:119832. https://doi.org/10.1016/j.biomaterials.2020.119832.

Doudna JA, Charpentier E. The new frontier of genome engineering with CRISPR-Cas9. Science. 2014;346(6213):1258096. https://doi.org/10.1126/science.1258096.

Ethgen O, Bruyère O, Richy F, Dardennes C, Reginster JY. Health-related quality of life in total hip and total knee arthroplasty: a qualitative and systematic review of the literature. J Bone Joint Surg. 2004;86(5):963–74. https://doi.org/10.1016/j.joca.2004.06.009.

Evans CH, Huard J. Gene therapy approaches to regenerating the musculoskeletal system. Nat Rev Rheumatol. 2015;11(4):234–42. https://doi.org/10.1038/nrrheum.2015.19.

Filardo G, Kon E, Di Martino A, Marcacci M. Platelet-rich plasma therapy in the treatment of knee osteoarthritis: a double-blind, randomized, placebo-controlled trial. J Shoulder Elb Surg. 2018;27(1):107–15. https://doi.org/10.1016/j.jse.2017.11.031.

Fransen M, McConnell S, Harmer AR, Van Der Esch M, Simic M, Bennell KL. Exercise for osteoarthritis of the knee: a Cochrane systematic review. Br J Sports Med. 2015;49(24):1554–7. https://doi.org/10.1136/bjsports-2015-095424.

Freeman FE, McNamara LM. Endochondral Priming: A developmental engineering strategy for bone tissue regeneration. Tissue Eng Part B Rev. 2017;23(2):128–141. https://doi.org/10.1089/ten.TEB.2016.0197

Geneen LJ, Moore RA, Clarke C, Martin D, Colvin LA, Smith BH. Physical activity and exercise for chronic pain in adults: an overview of Cochrane reviews. Cochrane Database Syst Rev. 2017;4:CD011279. https://doi.org/10.1002/146518.

Groll J, et al. A definition of bioinks and their distinction from biomaterial inks. Angew Chem Int Ed. 2016;55(38):11376–8. https://doi.org/10.1039/C6CS00133F.

Gulati A, et al. Osteoarthritis: potential for clinical and economic impact. BMJ. 2019;364:l1751. https://doi.org/10.1136/bmj.l1751.

Gupta A. Management of rotator cuff injuries using allogenic platelet-rich plasma. J Orthop Surg Res. 2024;19(1):1–12. https://doi.org/10.1186/s13018-024-04657-4.

Jayaram P, Ikpeama U, Rothenberg J, Malanga G. Bone marrow–derived and adipose-derived mesenchymal stem cell therapy in primary knee osteoarthritis: a narrative review. PM R. 2019;11(2):177–91. https://doi.org/10.1016/j.pmrj.2018.06.019.

Koons GL, Diba M, Mikos AG. Materials design for bone-tissue engineering. Nat Rev Mater. 2020;5(8):584–603. https://doi.org/10.1038/s41578-020-00248-5.

Learmonth ID, Young C, Rorabeck C. The operation of the century: total hip replacement. Lancet. 2007;370(9597):1508–19. https://doi.org/10.1016/S0140-6736(07)60457-7.

Lenas P, Moos M, Luyten FP. Developmental engineering: a new paradigm for the design and manufacturing of cell-based products. Part II. From genes to networks: tissue engineering from the standpoint of systems biology and network science. Biotechnol Adv. 2009;27(6):606–23. https://doi.org/10.1016/j.biotechadv.2008.11.004.

Li J, Webster TJ. 3D printed biomaterials for tissue engineering and medical applications. Acta Biomater. 2018;86:1–7. https://doi.org/10.1016/j.actbio.2017.12.027.

Liang P, et al. CRISPR/Cas9-mediated gene editing in human tripronuclear zygotes. Protein Cell. 2015;6(5):363–72. https://doi.org/10.1007/s13238-015-0153-5.

Liang C, Xu H, Li G, Liu T. Stem cell therapy for treating osteoarthritis: a review of clinical trials. Front Biosci. 2017;22:519–26. https://doi.org/10.2741/4500.

Liu X, et al. Extracellular vesicles derived from mesenchymal stem cells ameliorate osteoarthritis in a preclinical model. J Orthop Res. 2017;35(12):2807–13. https://doi.org/10.1002/jor.23568.

Lund RJ, Närvä E, Lahesmaa R. Genetic and epigenetic stability of human pluripotent stem cells. Nat Rev Mol Cell Biol. 2012;13(12):732–44. https://doi.org/10.1038/nrm3311.

Marx RE. Platelet-rich plasma: evidence to support its use. J Oral Maxillofac Surg. 2004;62(4):489–96. https://doi.org/10.1016/S0030-4220(04)00061-9.

Mason C, Dunnill P. A brief definition of regenerative medicine. Regen Med 2008 Jan;3(1):1–5. https://doi.org/10.2217/17460751.3.1.1. PMID: 18154457.

Mautner K, Malanga G, Smith J, Shiple B, Ibrahim V, Sampson S. A call for a standard classification system for future biologic research: the rationale for new PRP nomenclature. PM R. 2015;7(4S):S53–9. https://doi.org/10.1177/2325967115592611.

Murphy SV, Atala A. 3D bioprinting of tissues and organs. Nat Biotechnol. 2014;32(8):773–85. https://doi.org/10.1038/nbt.2958.

Murrell WD, et al. Regenerative treatments to enhance orthopedic surgical outcome. PM R. 2015;7(4S):S41–52. https://doi.org/10.1016/j.pmrj.2015.01.015.

Naji A, et al. Mesenchymal stromal cells: from basic research to clinical applications. Int J Biochem Cell Biol. 2019;110:1–21. https://doi.org/10.1016/j.biocel.2019.01.006.

Nguyen AH, et al. Bioprinting of osteochondral tissues: a frontier in regenerative medicine. Tissue Eng A. 2017;23(17–18):855–71. https://doi.org/10.1089/ten.tea.2016.0280.

O'Brien FJ. Biomaterials & scaffolds for tissue engineering. Mater Today. 2011;14(3):88–95. https://doi.org/10.1016/j.mattod.2011.04.002.

O'Neil CK, Hanlon JT, Marcum ZA. Adverse effects of analgesics commonly used by older adults with osteoarthritis: focus on non-opioid and opioid analgesics. Am J Geriatr Pharmacother. 2012;10(6):331–42. https://doi.org/10.1016/j.amjopharm.2012.10.008.

Perera K, Ivone R, Natekin E, Wilga CA, Shen J, Menon JU. 3D bioprinted implants for cartilage repair in intervertebral discs and knee menisci. Front Bioeng Biotechnol. 2021;22(9)754113. https://doi.org/10.3389/fbioe.2021.754113. PMID: 34746106; PMCID: PMC8570130.

Raggatt LJ, Partridge NC. Cellular and molecular mechanisms of bone remodeling. J Biol Chem. 2010;285(33):25103–8. https://doi.org/10.1074/jbc.R109.041087.

Rai R, et al. Regulatory and ethical challenges in regenerative medicine. Trends Biotechnol. 2021;39(8):836–43. https://doi.org/10.1016/j.tibtech.2021.05.007.

Rasool M, et al. Nanocarriers-based drug delivery systems for treatment of infectious diseases: current status and future prospects. Mater Today. 2019;27:38–59. https://doi.org/10.1016/j.mattod.2019.04.025.

Ravi M, et al. Advances in bioprinting for cartilage regeneration. Biomaterials. 2020;258:120306. https://doi.org/10.1016/j.biomaterials.2020.120306.

Sabesan V, et al. Regenerative medicine: what can it do for me? Int J Complement Altern Med. 2020;13(6):244–50. https://doi.org/10.15406/ijcam.2020.13.00523.

Seunghun S. Lee, Xiaoyu Du, Inseon Kim, Stephen J. Ferguson, Scaffolds for bone-tissue engineering, Matter, 2022;5(9):2722–59. ISSN 2590–2385, https://doi.org/10.1016/j.matt.2022.06.003.

Shi Y, et al. Mesenchymal stem cells: a new strategy for immunosuppression and tissue repair. Nat Commun. 2018;9(1):217. https://doi.org/10.1038/s41467-018-04719-4.

Siepmann J, Siepmann F. Microparticle-based drug delivery systems: manufacturing, characterization, and application. J Control Release. 2012;161(2):351–62. https://doi.org/10.1007/s11095-012-0732-3.

Singh JA, Noorbaloochi S, MacDonald R, Maxwell LJ. Chondroitin for osteoarthritis. The. Cochrane Database Syst Rev. 2014;1:CD005614. https://doi.org/10.1002/acr.22381.

Sun J, et al. Advances in 3D-printed scaffold technologies for bone defect repair: materials, biomechanics, and clinical prospects. BioMed Eng OnLine. 2025;24:51. https://doi.org/10.1186/s12938-025-01381-w

Takahashi K. Induced pluripotent stem cells: past, present, and future. Cell Stem Cell. 2020;26(6):850–64. https://doi.org/10.1016/j.stem.2020.02.008.

Takahashi Y, et al. The role of extracellular vesicle microRNAs in mesenchymal stem cell senescence and rejuvenation. J Cell Mol Med. 2019;23(9):6462–71. https://doi.org/10.1002/jcmm.3011.

Taylor P, Warnock M, Raff M, Wilmut I. The regulation and ethical guidelines of stem cell research. National Academies Press; 2010. https://doi.org/10.17226/12923.

Volkow ND, McLellan AT. Opioid abuse in chronic pain—misconceptions and mitigation strategies. N Engl J Med. 2016;374(13):1253–63. https://doi.org/10.1056/NEJMra1604339.

Vos T, et al. Global burden of 369 diseases and injuries in 204 countries and territories, 1990–2019: a systematic analysis for the global burden of disease study 2019. Lancet. 2020;396(10258):1204–22. https://doi.org/10.1016/S0140-6736(20)30925-9.

Younger EM, Chapman MW. Morbidity at bone graft donor sites. J Orthop Trauma. 1989;3(3):192–5. https://doi.org/10.1097/00005131-198909000-00002.

Wang H, et al. Controlled release of anti-inflammatory cytokines for targeting inflammatory bone loss in experimental periodontitis. J Control Release. 2013;172(3):666–72. https://doi.org/10.1016/j.jconrel.2013.08.018.

Wang H, et al. Smart materials for bone tissue engineering. Adv Mater. 2020;32(1):1902783. https://doi.org/10.1002/adma.202003377.

Woolf AD, Pfleger B. Burden of major musculoskeletal conditions. Bull World Health Organ. 2003;81(9):646–56. https://doi.org/10.1590/S0042-96862003000900007.

Zhang H, et al. Advances in CRISPR technology and its potential for treating genetic disorders. Cell. 2019;179(3):615–32. https://doi.org/10.1016/j.cell.2019.08.011.

MIX
Papier aus verantwortungsvollen Quellen
Paper from responsible sources
FSC® C105338

If you have any concerns about our products,
you can contact us on
ProductSafety@springernature.com

In case Publisher is established outside the EU,
the EU authorized representative is:
Springer Nature Customer Service Center GmbH
Europaplatz 3, 69115 Heidelberg, Germany

Printed by Libri Plureos GmbH
in Hamburg, Germany